Exploring and Mapping Alaska

The Russian America Era, 1741–1867

ALEXEY POSTNIKOV
AND MARVIN FALK

Translation by Lydia Black

UNIVERSITY OF ALASKA PRESS
Fairbanks

University of Alaska Press
P.O. Box 756240
Fairbanks, AK 99775-6240

ISBN 978-1-60223-251-8 (cloth); ISBN 978-1-60223-252-5 (electronic)

Library of Congress Cataloging-in-Publication Data

Postnikov, A. V. (Aleksei Vladimirovich)
[Russkaia Amerika v geograficheskikh opisaniiakh i na kartakh. English]
Exploring and mapping Alaska : the Russian America era, 1741-1867 / Marvin Falk & Alexey Postnikov ; translated by Lydia Black.
pages cm. — (Rasmuson library historic translation)
ISBN 978-1-60223-251-8 (hardback) — ISBN 978-1-60223-252-5 (electronic)
1. Discoveries in geography—Russian. 2. Discoveries in geography—American. 3. Cartography—Alaska—History. 4. Russians—North America—History. 5. North America—Discovery and exploration—Russian. 6. Alaska—Discovery and exploration—American. I. Falk, Marvin W., 1943-, editor. II. Title.
GA401.P6813 2015
526.09798'09033—dc23

Cover Design by Paula Elmes
Interior design by Mark Bergeron, Publishers' Design and Production Services, Inc.
Cover illustrations: Blossom's barge on Arctic coast, oil painting, UAF Rare Books.

This publication was printed on acid-free paper that meets the minimum requirements for ANSI / NISO Z39.48,Ä̀1992 (R2002) (Permanence of Paper for Printed Library Materials).

Contents

List of Figures

Preface

This book is an account of the discovery and mapping of Alaska before its transfer to the United States in 1867. It is a history full of competing commercial interests, state secrets, and lost manuscripts leaving a record full of riddles. The existing records are widely scattered, not only in Russia but also across Western Europe and North America. Alexey Postnikov labored diligently to find, thoroughly document, and interpret these rich resources.

Professor Postnikov worked with Richard A. Pierce, a leading scholar of Russian America and a publisher of translations. Together, the two professors planned to publish this work simultaneously in both a Russian and a translated English edition. They began their collaboration in 1997, but along the way Pierce decided to publish his English translation following the 2000 publication of the Russian. He turned to our UAF colleague, Professor Lydia T. Black, for help. With the passing of Dr. Pierce in 2004, the full project fell to Dr. Black. She was ideally suited to complete the project with her skills as a leading anthropologist, translator of her native Russian, and historian.

The translation of Postnikov's text was one of Lydia's last projects before she, too, passed away in 2007. Realizing that she would be unable to complete the project, she asked me to "English" the text and to see the project through to publication. One of Dr. Black's challenges was the inaccessibility of Richard Pierce's papers. Among them were almost all of Postnikov's illustrations.

Since Professor Postnikov had sent his original illustrations to Dr. Pierce and was not in a position to re-gather them, I have substituted maps and illustrations from the collections of the Rasmuson Library as appropriate. I have updated portions of the text to reflect scholarship published since Dr. Postnikov completed his manuscript for publication.

For those with an interest in continuing the search for historical maps, a wealth of material has become available online. There is not yet a common search engine across collections that can find historical maps efficiently, but how maps are described and displayed is constantly improving. The most innovative is the David Rumsey site (davidrumsey.com), which combines historical maps with Google

Earth and other modern mapping systems. At this time there are over 48,000 digitized maps on the site, but relatively few are of Alaska.

The most comprehensive site for Russian America is the Library of Congress's "Meeting of Frontiers" (memory.loc.gov/intldl/mtfhtml/mfdigcol/mfdigcol.html), which has contributions from major Russian repositories as well as the Library of Congress and the Rasmuson Library. However, their search and display technologies are out of date and somewhat difficult to use.

Map collections around the world are rapidly being digitized and can be searched, sometimes institution by institution and sometimes as consortia. An example of such a consortium is the Alaska Digital Archives (vilda.alaska.edu), which includes a small selection of maps from collections around the state of Alaska. The best place to track developments in online map access is an international historical mapping site (maphistory.info).

Acknowledgments

Elmer Rasmuson's bequest to the University of Alaska provided financial backing for this project. The Rasmuson Library and the University of Alaska Press have provided support and encouragement to the translation program since 1983. Jim Ketz and Katherine Arndt helped with fact-checking for Lydia Black while she prepared her draft. Sarah Hurst made sense out of the welter of text files that Lydia provided, meshed text and footnotes, and translated missing bits. Sylvie Savage aided in turning academic prose into a style better suited to a modern reader.

The occasional comments incorporated into the text are indicated by [AP]—Alexey Postnikov, [LB]—Lydia Black, and [MF]—Marvin Falk.

Marvin Falk, Ph.D.
Coauthor

CHAPTER I

The Russian Advance Toward the Pacific Ocean

On his 1507 world map,[1] the cartographer Martin Waldseemueller (1470–1518) showed the American continents separated from Asia by an expanse of ocean, thus predicting the existence of the still unknown Pacific Ocean with an opening to the north. Just five years later the Spanish explorer Vasco Núñez de Balboa, crossing the Isthmus of Panama, discovered the Pacific Ocean. It took an additional two centuries before a strait opening to the Arctic Ocean was proven through the discovery of northwestern America by Russian pioneers and seamen who cut the difficult trail through Siberia to the Far East.

In North America the Asian Pacific side differed substantially from the European Atlantic side. The Atlantic Ocean separates the Old and the New Worlds by an immense expanse of water, whereas the northern Pacific Ocean separates Eurasia from North America by a mere fifty-three miles at the Bering Strait. Although the eastern shores of America may have been reached by Irish monks and Vikings, there were no sustained contacts between the inhabitants of Europe and the Americas before Columbus.

In contrast, Eurasia and North America were joined together during the ice ages by a landmass now called the Bering Land Bridge. The first people to penetrate America did so from northeastern Asia and gradually populated both American continents.[2] What became the Bering Strait formed as a result of postglacial melting, flooding the land bridge. In the course of millennia, close trade and cultural relations continued to be maintained across Bering Strait.

Thus, in this special ethnohistoric region—Beringia—contact between the Old and New Worlds was continuous, though the rest of the world had no knowledge of it. Starting in the mid-eighteenth century, Russian pioneers, fur hunters (*promyshlenniki*),[3] and traveling scientists also came to North America, securing the expansion of the Russian state eastward. There are substantial differences between these developments and the first contacts between the Spanish and the Indians of the Caribbean islands.

Members of the Christopher Columbus expedition suddenly found themselves in a totally foreign ethno-cultural environment. They interacted with the Natives who, according to the then-prevalent Christian teaching, had no rightful place on this earth. This was held to be evidence that they were not human and thus justified the genocide of local Native populations by the "civilized" Europeans.[4]

Russians, too, shed a great deal of Native blood in their conquest of Siberia and later movement a farther east, but the nature of this movement and their relationship with local populations was different. The process of colonizing Siberia, Kamchatka, the Aleutian Islands, and coastal Alaska by Russia lasted almost three centuries. Due to the length of this period, Russian pioneers, *promyshlenniki*, and Cossacks integrated with the natural and ethnic environment of the regions they moved into. The several generations of these pioneers who replaced each other over time not only lived next to and with the Natives, but actively exchanged information with them.

Geographic information was needed for the continuing eastward advance. Local place names were particularly important, as they provided a means of orientation in an unknown territory. They were diligently collected by the pioneers and in many instances gradually became accepted as their own. Because of this, Native geographic names, though often in distorted form, can be found on modern maps of Siberia, the Aleutian Islands, and Alaska.

As early as the sixteenth and seventeenth centuries, Russians moving toward the east occupied territories beyond the Urals, which separate Europe from Asia. Siberia was joined to Russia as a result of Ermak's incursions of 1581–1585. The settling of the immense expanses of Siberia was accomplished by people from many different social strata of the Russian population. Cossacks, fugitive serfs, small-time entrepreneurs, and noblemen moved to the east, into the virgin, sparsely populated localities where they were free to be their own bosses. They overcame harsh natural conditions and actively interacted with local tribes and nations who helped them adapt to the environment in their new home and showed them how to use the natural resources.

Resettlement of ordinary Russian people generally took place in a peaceful manner. Governmental initiatives established a Russian administration, imposed *iasak* (head tax) upon the local population, and recovered precious metals for the benefit of the state. Sometimes the local nomadic and hunting populations offered armed resistance, but there were also instances where Natives voluntarily joined Russians to defend against hostile attacks by their neighbors or to gain support in feudal and kin group conflicts. These processes were not accompanied by the extermination and dislocation of local populations as was the case with the Indians of

North America in the course of European colonization moving inland from the coasts of the Atlantic Ocean.[5]

The history of inhabitation of Siberia is well reflected in modern historical and geographical literature.[6] Russian pioneers reached the Pacific Ocean coast early. In 1639 a detachment led by Ivan Moskvitin traveled down the Ulya River to the Sea of Okhotsk. The following year Cossacks from his detachment sailed north on the ocean to the mouth of the Okhota River and south to the Shantar Islands. While sailing to the south, the Cossacks learned about the Amur River and the Natives inhabiting the area, who were then called Gilyak (Nivkh in modern nomenclature). The Okhotsk *ostrog* (fortified outpost) was founded in 1647. This *ostrog* would play an important role in the future.

In 1648, Fedot Alekseev and Semyon Dezhnev, advancing east along the coast of the Arctic Ocean from the Kolyma, passed through Bering Strait. It was then that the Russians first discovered the northeastern extremity of Asia, the closest point to America. Dezhnev named it *Bolshoi kammnoi nos* (Great Rocky Promontory), where "a goodly number of Chukchi live."[7] He reported that across from this promontory were two islands inhabited by "the toothed ones" (labret-wearing people, Eskimos).[8] By "*Bolshoi kamennoi nos*" did Dezhnev mean the entire Chukotka Peninsula?[9] From Dezhnev's report it is evident that he received some information about the Chukotka Peninsula from the local inhabitants, the Yukagir and the Chukchi. He also learned from them that in Chukotka "the shore ice is not carried out to sea every year."[10]

Following the official incorporation of the Amur region into Russia, 1649–1652, the attention of the Russian pioneers was focused mainly on the southern areas, which were more promising with respect to the development of agriculture. However, interest in the Okhotsk shore and Kamchatka did not wane.

The existence of Kamchatka was already known from verbal accounts, but more precise information contained on maps and other documentary sources dates from the 1690s. Ivan Golygin brought back information about Kamchatka, and, according to some data, the peninsula was drawn on maps as early as 1672.[11] Exploration truly began with the journeys of Luka Morozko in 1695–1696 and of Vladimir Atlasov in 1696–1699.[12]

EARLY MAPS OF SIBERIA

The Russians came to Siberia with the necessary skills to compile geographic descriptions and elementary cartography, representing localities in original Native tradition.[13] There is extensive documentary evidence that testifies to various charts and graphic representations compiled in Siberia.[14] In particular, a number of instructions to the pioneers have been preserved, and it is possible to reconstruct the methodologies used in field reconnaissance. This served as the basis for compilation of the charts the pioneers were required to submit along with their reports.

Dezhnev mentions compiling a chart when submitting information about his voyage of 1648. The commanders sent to Siberia to establish fortifications and to hold new lands received orders demanding that "charts" were to be submitted. For example, a 1594 order given to Prince Pytor Gorchakov, who was being sent to "Siberian cities with various supplies and for organization of local affairs," instructed him, after consulting with other commanders, "to seek out a location suitable for establishing the new city . . . occupy the city and make a chart representing it as well as describing the various fortifications."[15]

The initial cartographic investigations were conducted by the Russian pioneers along rivers and trails that served as basic "survey routes," to which all other geographic information was linked. In most instances, members of the local populations (Siberian Tartars, Evenk, Yakut, Yukagir, Chukchi, Eskimos, and others) served not only as guides, but were also providers of information about localities outside the survey routes. The results of such exploratory investigations were incorporated into the description of the charts, which then served as the basis for the compilation of maps. The Siberian surveys reflected the rich tradition of detailed geographic descriptions used by the Pomory people of the Russian North from whom many famous pioneers originated. They were skilled in compiling manuscript guides for coastal navigation in the northern seas. Elementary manuscript charts, transmitted from one generation to the next and constantly being upgraded and corrected, developed into the methodology used for the itinerant cartography employed in the explorations of Siberia, the Far East, the Aleutian Islands, and Alaska.

As early as the fourteenth to fifteenth centuries the Pomory sailors of northern European Russia had acquired the magnetic compass from the Novgorodians.[16] There are testimonies by foreign authors of the sixteenth century that the Pomory provided not only descriptions, but also charts of considerable stretches of the coasts of the northern seas. For example, the Dutch who encountered Russians near Kolguev Island in 1594 acquired a chart of the coast from the White Sea to the Pechora River from a Pomor skipper.[17] The Dutch cartographer Gerhard Mercator in a letter to the English geographer Richard Hakluyt reported that when compiling his map of Russia he received data about the north from a Russian.[18]

The Russian archives contain information about the compilation of hundreds of charts dating from the 1640s to the 1670s made in the process of studying and exploring Siberian lands.

The combined cartographic resource would include charts compiled by the pioneers, the cartographic materials created in the course of diplomatic missions, drawings related to the construction of fortifications, exploration for useful minerals, the imposition of taxes (*iasak*), exploring river and land communication routes, and also land maps created to establish property boundaries.[19] Even areas that were difficult to access in the north or in the mountains were explored and represented on the maps.

Every step in the settling of Siberia was accompanied by cartographic activity, and the descriptions of the charts and the charts themselves were widely accessible. However, as with early Russian cartographic representations in general, most have been lost.

Starting in the 1590s, Tobolsk, then the main administrative center in Siberia, and the Kazan department (from 1637, the Siberia department) in Moscow, received local and regional geographic descriptions and charts. These were compiled on orders from central as well as local officials.[20] All these tasks were executed in the original pre-Petrine Russian cartographic style.[21]

The numerous regional descriptions and geographic charts of Siberia formed a basis for compiling composite maps of Siberia. Thus, in 1626, in connection with the creation of a new Great Chart of the Muscovite State (1626–1627), the Tobolsk commander was entrusted with delivery to Moscow of the materials pertaining to Siberia. This assignment from Moscow was executed in 1633 when, along with the *Rospis sibirskim gorodam i ostrogam* (Description of the Siberian Cities and Fortified Outposts),[22] the first general chart of what was then known of Siberia was created, which, unfortunately, has not survived.[23]

The *rospis* (explanatory text accompanying a map) contains a detailed description of Western Siberia and the southern defensive line of the Russian State. Eastern Siberia is represented very fragmentarily, and the description ends with a mention of the routes along the Lower Tunguska and of the dispatch of noblemen from Mangazeya to the Lena for *iasak* collection.[24]

According to V. N. Tatishchev: "in the time of Tsar Alexei Mikhailovich [reign 1645–1676] a general land map of Russia and several particular maps were compiled; and from the general map it is evident that the compiler understood the Latin tongue, as he used many Latin words and he also divided the map by degrees."[25] The chart, compiled in the fall of 1667 by the volunteer effort of Tobolsk commander P. I. Godunov, is in itself a direct continuation of the geographic and cartographic study of Siberia that began in the 1620s and 1630s.

This is the earliest map of Siberia that has survived, and became world renowned following Adolf Nordenskjold's 1887 discovery of Swedish copies of it. This 1667 chart has been published repeatedly. In 1962 the original of the *rospis* that accompanied the chart[26] was published. Prior to that, only copies had been available.[27] Even now, the original of this general schematic chart is known only from manuscript copies. It presents a number of unsolved puzzles.[28] A comparison of all extant copies shows the similarities of the cartographic representation of Siberia (the gross reworking by Schleising excepted), except for insignificant differences and distortions in the transmission of geographic place names.

The chart shows an immense territory to the east of the Volga and Pechora that encompasses the whole of Siberia and the Far East. The chart of 1667, like most other early Russian maps, is oriented toward the south, and the meridian and the latitude net are absent. The cartographic image at first glance appears rather naive, but nevertheless it realistically represents the position and basic outlines of the branching river systems. Besides the major rivers, such as the Ob, Yenisey, Lena, Olenek, Kolyma, and Amur, many tributaries and small rivers and streams flowing into the ocean are shown. The Ural Mountains are represented relatively correctly. Specific lands and population distributions (Kalmyks, Bukhars, Mungals, Saians, Kyrgyz, Bashkirs, and others) are indicated by ethnonyms in the corresponding places. Characteristically, the outline of the Asian continent, bounded by ocean in the north and east, is represented in a rectangular form.

Here, for the first time (since Waldseemueller), we encounter a categorical assertion, expressed cartographically, that a direct sea passage from the Arctic Ocean to the Pacific is feasible. This idea was advanced by the chart's author on the basis of Dezhnev's voyage in 1648.[29] Evidence for the limitation of the geographic data available to the Russians is the absence of the large peninsulas of Taimyr, Chukotka, Kamchatka, and the Anadyr and Kolyma Rivers, which debouch into the Pacific and Arctic Oceans, respectively.

The chart was compiled on "the testimony of men of various ranks who visited Siberian towns and fortified outposts and know the local environments, the routes, and the lands truly . . . and also on the testimony of the visiting Bukhara men and Tartar noblemen."[30] The main part of the description contains an enumeration of the cities, fortresses, winter habitations, and rivers with their tributaries; states the distances between population points, in versts and days of travel; and mentions bordering countries and their peoples, particularly the kingdom of China and the "Indian (Native) land" that lies "beyond the rock," that is, beyond the mountains.

When the chart of Siberia was being compiled in 1667, the first effort was made to establish standard symbols in Russian cartography. In the chart's description, the original *rospis*, are listed "signs for recognition on the chart of cities, fortified outposts (*ostrogi*); suburbs (*slobody*); rivers, lakes, neighborhoods (*volosti*); winter camps (*zimovia*); and nomadic camps (*kochevia*); specifically: V—*volost*, G—*gorod*, Z—*zimovie*, K—*kochevie*, M—monastery, O—*ostrog*, R—river, S—*sloboda*, and W—*ozero* (lake).

B. P. Polevoi, analyzing the chart of Siberia of 1667 and its corresponding description, advanced the hypothesis that a number of regional charts were appended to the surviving composite chart, which together formed a chart book or an atlas of Siberia and that the number of such charts corresponds to numbers in the chart's description; that is, there were 22 of them.[31]

Though the chart of Siberia of 1667 is known only in manuscript form, it is interesting to note the testimonies of Semyon Remezov, Nicolas Witsen, and Georg Schleising that this cartographic work had indeed been printed. In the absence of surviving printed examples, this is still unresolved and researchers do not hold a unified position on this matter.

The chart of 1667 was of significance in the history of world cartography because for the first time a true picture of Siberia as a whole was reliably represented. Though this chart represents a cartographic declaration of Russia's incorporation and assimilation of lands to the east, to the Pacific Ocean itself, it also testifies to the intensive accumulation of data about the nature and population of the region. The chart of 1667, composed by Siberians, was received in the capital as an important geographic document. This is attested to specifically by the compilation on its basis of a number of later composite maps in the 1770s to 1790s, both in Moscow and in Tobolsk.

From the creation of the map of Tartaria in 1570 by Abraham Ortelius to the time when Nicolas Witsen published his map of 1690, Western European science had no reliable sources of information about the Asiatic part of the Russian State. For this reason, foreign atlases and maps by Mercator, Ortelius, and Hondius reflected, as far as Siberian territory is concerned, primarily legendary data, based on the materials of Pliny, Strabo, Plano Carpini, Marco Polo, and even the Holy Bible.[32] For example, shown on one of the better maps of the time, that of Tartaria of 1570 by Ortelius, are, near the mouth of the Ob River, the ten tribes of Israel, taken by the Salmanassar into Assirian captivity, etc.

A new version of the 1667 chart of Siberia was compiled in 1673. It has much in common with the earlier version but is somewhat more detailed in content. The compilation of this map was evidently connected with a cartographic and geographic project conducted in the summer of 1673 in Tobolsk with the support and participation of the Metropolitan Kornely.[33] This Chart of Siberia of 1673 is preserved in three copies—a Russian one with inscriptions in the Russian and Latin languages, executed apparently by L. Klishin; and two Swedish copies, by E. Palmquist dated 1673 and by Johan Sparwenfeld dated 1687.

In spite of some variations, all three copies confirm the general impression that this cartographic work was a composite one, incorporating new and expanded data about Eastern Siberia with an unknown ancient map of Western Siberia.[34] Several rivers emptying into the Sea of Okhotsk are shown in the northeast and a rounded outline of a not very large peninsula (apparently the Kamchatka Peninsula). The unknown author of this chart once again confirms the idea of a free northern passage by sea from Europe to the Pacific.[35]

This text appears to be a working-over of the *rospis* for the chart of 1667. It is divided into eight parts. It contains valuable historical and geographical data that, specifically, confirm that Semyon Dezhnev's voyage around the Chukotka Peninsula (or by portage in the northeasternmost extremity of it): ". . . From Kolyma River and around the land . . . to the rock [that is, the Chukotka Peninsula—AP, they run under sails, and having passed beyond the rock, they arrive at the River

Anadyr and here they take the fish bone [walrus tusks—AP] and that rock they pass with the utmost effort."[36]

Another general chart of Siberia surfaced 270 years after its creation. Its compilation is linked to the name of the famous diplomat, one of the best educated men of his age, a native of Moldavia who adopted the Russian state as his second homeland, Nikolai Milescu-Spafary. Spafary's chart, brought to the attention of the scholarly world by Leo Bagrow,[37] in its content is closely linked with Spafary's geographic works and his journeys of 1675–1678 as the head of the Russian embassy to China. Spafary's map is compiled in the ancient tradition of Russian geographic charts. It encompasses all of Siberia, parts of bordering states in the south and southeast, and the European part of Russia beginning at the Black Sea. A dotted line indicates the embassy's route.

Spafary's chart was a notable step forward in comparison with the charts of 1667 and 1673 in respect to its representations of some details of Siberian geography and a more precise relationship between a series of geographic features. In particular, the coastline of the Arctic Ocean (especially eastward from the mouth of the Lena) is much closer to the truth, and the Anadyr River is shown for the first time. Here Spafary demonstrates his intimate acquaintance with the materials of the Diplomatic Desk (the equivalent of a Russian Department of State of the day) by his evident use of the descriptions and routes of Petelin, Baikov, Khabarov, Poliakov, and other travelers and pioneers of the seventeenth century. A unique feature of the chart of 1678 is its representation of a vast mountain massif stretching from Lake Baikal in a northeasterly direction and protruding far into the ocean.

The opinion of modern scholars about this protrusion or rocky obstacle is divided. Some treat this promontory as the Chukotka Peninsula; others see in it a representation of Kamchatka. Spafary himself, however, recorded in his travel journal (notation of September 22, 1675) the geographic concepts of Siberians about a mountainous promontory that allegedly extended even to America.[38] We believe it possible that this representation of a mountain range from Baikal to the sea reflects the distorted knowledge on the part of the (local) Russians about the existence of the massif system of the latitudinal ranges (Yablonovoy, Stanovoy) and the longitudinal one (Verkhoyansky) that in their entirety form the main watershed between the Pacific and Arctic oceans.

On June 18, 1687, a new composite chart of all of Siberia was completed. It was compiled in Tobolsk as a supplement to the Godunov chart of 1667 but was based on fresh and detailed cartographic materials and the results of measured surveys of Siberian lands conducted in 1683–1687.[39] A surviving small copy, 27 × 17 cm (10.5 × 6.6 in.), shows a branching net of Siberian rivers, a multitude of population points, and peoples and their "lands." In spite of its small scale, this chart differs from all the previous ones in its richness of detail and considerably greater precision in its outline of the main contours of Northeastern Asia, particularly in the region from the Lena River to the Amur.

The relationships of rivers and the courses of them are shown. The Tobol, Iset, and Ob rivers, compared with the Irtysh, Yenisey, Lena, Omoloy, Yana, Indigirka, Alazaya, Kolyma, and Anabara rivers, are represented more reliably. An attempt is made to show the uneven, cut-up coastline of the Arctic Ocean to the Eastern Sea (the Pacific Ocean), and promontories, bays, and peninsulas have acquired recognizable outlines. In particular, to the east of Lena are shown Capes Bykovsky, Svyatoy, Shelagsky, and, apparently, the modern Cape Dezhnev, as are shown also Omoloev and Chaun inlets. A ridge with the inscription "the length of this rock is not known to anybody" extends into the ocean, as on Spafary's chart, but without doubt it is treated as the southern part of the Chukotka Peninsula, because beyond it are shown the Anadyr River, the bay of the same name, Oliutorsky Island, and Kamchatka River. A four-day portage from Anadyr to the Kolyma tributary, the Bludnaya River, is indicated.[40] On this chart of 1687, for the first time, the Kamchatka Peninsula and the Sea of Okhotsk with eight rivers that empty into it are clearly delineated and so precisely that there is no room for any other interpretation.[41]

The map of Siberia and the European part of the Moscow State, compiled by Andrei Vinius, occupies a special place in the cartography of Siberia and the extreme northeastern reaches of Asia. The date of its compilation has not been established precisely and various modern researchers date it either to 1680–1683 or 1689. Vinius served in Russia as an interpreter of the Dutch language and later as a clerk in the Apothecary and Diplomatic departments. His map survives in a copy made by Semyon Remezov.[42] In its technique, Vinius's map is far above any known early Russian charts, and it is through this map, thanks to Nicolaas Witsen, that Western European cartography acquired reliable information about Siberia.

NICOLAAS WITSEN

Though Witsen's map is oriented toward the south, like the vast majority of early Russian geographic charts, it is the first one to employ the geographic net, though it has decorative rather than mathematical significance. The map evidences the influence of Spafary's geographic concepts and of his chart of Siberia. In the northeast of Asia, cut off by the map's frame, two peninsulas are shown thrusting into the sea. The first is located between the rivers Lena and Kolyma, the second between the Kolyma and Anabara. On the famous 1690 map by Witsen, the unbounded mountainous peninsulas to the east of Kolyma—Cape Tabin and Icy Cape—are very similar.[43]

The Witsen map created a tradition of uncertain representation of the northeastern corner of Eurasia, suggesting the possibility that in this region Asia and America are joined. The influence of Russian geographic charts on the formation of Western European cartographic representations of Siberia has been researched in detail by many scholars, in particular by studying maps compiled by Sparwenfeld

(1688–1689), Witsen (1690), and European Jesuit missionaries in China, Ferdinand Verbiest (1673–1676) and Antoine Thomas (1690).[44]

I (Postnikov) had the good fortune to discover in the Newberry Library in Chicago a previously unknown map of Siberia that differs not only from Witsen's map but also from the other Russian and foreign maps studied earlier. This is the *Carte générale de la Siberie et de la Grande Tartaria . . .* (Fig. 1) contained in the spectacular manuscript atlas *Cartes Marines* in the Edward E. Ayer Collection. Comparison of this work with the known maps of the period suggests that it is possibly a unique French copy of an unknown Russian map dating to the end of the 1670s or the beginning of the 1680s. That a Russian source was used when the map was compiled is evidenced by the following:

1. All the toponyms on this map are French transliterations of Russian geographic place names current in Siberia in the seventeenth century.
2. Besides the seventeenth-century place names, the map shows such names as the Golden Horde, designating the Mongol-Tartar State that extended its power to a part of Southern Siberia. The presence of such names is typical of composite Russian geographic charts of the seventeenth century and specifically of the works of Semyon Remezov.
3. The detailed representation of the hydrographic net, and especially of the Amur River system (this name was used only in Russian sources), is similar in detail to the way these elements are represented on the Russian charts at the end of the seventeenth century.

Shown in the northeastern corner of this map are the River Lena with its tributaries, the Kolyma, and other river basins linked to the Arctic Ocean and separated by a promontory from the rivers of the Pacific Ocean basin, which is in part under the cartouche that gives the map's title. On this promontory are the Kamchatka River and even a city called Kamchatka.

This map cannot be later than the year 1689, when the Treaty of Nerchinsk was concluded between the Moscow State and the Qing Empire (China). The results of this treaty were obligatorily shown on all Russian maps compiled at a later date and are missing on the map under discussion. A much earlier date is improbable because of the relative complexity and reliability of the geographic content, which appears to fall between the 1670s and 1680s.

The map, however, has several peculiarities that set it apart from the Russian charts of the seventeenth century. In accordance with the strict canon of Western European cartography, the map is oriented toward the north. It has a geographic net in some arbitrary projection, which appears to be the closest to the pseudo-conical one; and the cartouche indicates a scale: one degree of longitude equals 104½ Russian versts (69.3 miles).

COSSACK CONTRIBUTIONS

By 1800, the central government began paying more attention to the northeastern extremity of Asia, probably encouraged by the foray into Kamchatka in 1697 by Vladimir Atlasov. In 1701 he testified in Moscow as follows: ". . . across from that promontory that cannot be rounded is an island, and when the sea freezes in winter the foreigners from that island come; they speak their own language and bring poor quality sables resembling weasels, and he, Vladimir himself, saw three of these sables. The tails of these sables are about a quarter of an arshin [seven inches] long with black and red stripes."[45]

In the following years the Anadyr cossacks continued to supply similar information to their superiors. In their submissions we now find direct references to Native informants. For example, Pyotr Popov, in the report on his foray into the Chukotka Peninsula in the year 1711, talked about the existence of a large island directly across from Anadyrsky promontory [he means the American mainland—LB] where:

> people are toothed [wear labrets—LB] and are of totally different faith and customs as well as language. The language is distinct from the Chukchi one. Since ancient times and to this day there is no peace between the Cape Chukchi and those island people. They attack each other with arms. They [the island people] are armed with bows the same as the Chukchi. And he, Pyotr and his company, saw these island people taken prisoner by the Chukchi, about ten in number. Besides their natural teeth, these people insert small pieces of walrus tusks into their cheeks, next to their natural teeth. From the promontory to the island in summer it takes about a day rowing a baidara. In winter, by reindeer, unloaded it also takes one day. On that island there are various animals, sables, martens, foxes in variety, Arctic foxes, wolves, wolverines, polar bears and sea otters. They keep large herds of reindeer. They subsist on sea mammals, berries, and roots. On that island there is timber, cedar, pine, spruce, fir trees, and broad-leaved trees. And he, Pyotr and his comrades, saw branches of these trees in the Chukchi baidaras and yurts. These people live in the same way as the Chukchi and have no authorities whatsoever. Pyotr Popov's main informant, the Chukchi Makachkin, maintained that there are three times as many "island" inhabitants as there are Chukchi inhabiting the promontory. He added that . . . he, Makachkin, was on that island in many years on raids and that the Chukchi call that island the Great Land.[46]

Statements by local inhabitants recorded somewhat later then spoke about two islands.

Captain Pyotr Tatarinov wrote in 1718:

> across from the Chukchi promontory there is an apparently small island to which they [the Chukchi] run in the Chukchi skin baidaras in the time span from morning to noon. There is no timber whatsoever on that island and the number of people is small. Though they look somewhat like the Chukchi, they speak a different language. Beyond the sea from that island is the Great Land, which may be reached in one day by rowing when the weather is calm. The people there are very numerous and also speak a distinct language. They live in earthen dwellings, and subsist on caribou, fish and various sea animals, such as whales, walrus, seals, belugas and others. They wear sable, fox and caribou clothing. In that lands there are great rivers which fall into the sea; various timber grows there, specifically larch, pine, fir trees and large cedars, which are not found near the Chukchi habitations.[47]

Such vivid descriptions of the lands and peoples to the east of the Chukotka Peninsula are interpreted by modern Russian historians and geographers with assurance to be the earliest reliable information about the islands in Bering Strait, Alaska, and the aboriginal inhabitants of these lands that were obtained by Europeans. In this they follow Leo Berg, who was the first to offer the interpretation.[48] This information began to be reflected in Russian geographic charts, one of the earliest being the *Karta mest ot r. Yeniseya do Kamchatki lezhashchikh* [Map of the Localities Situated from the Yenisey River to Kamchatka], which was compiled by Selenga District Commissar Fyodor Beyton (Beton) in 1710–1711. Though the outline of Kamchatka, and of the territories adjacent to it, is rather erroneous, the map shows portions of land to the east of Russia's Asian territories, in the Pacific Ocean.

The inscription on the northern land here permits us to consider it part of Alaska: "Land inhabited by people called in the Chukchi tongue *kykymei*, who resemble the Yukagir. Their clothing consists of fox parkas. They are very *izlikny* [may be "stout," may be "stealthy," I do not know—LB]. They fight with bows. There are sables and foxes, and their timber is pine and birch."[49] The northeastern extremity of Asia is shown with considerably more authority on the *karta iakutskago dvorianina Lvova* [The Chart of the Yakut Nobleman Lvov] (Fig. 2), dating to about 1710. Here the Chukotka Peninsula, the Diomede Islands, and Alaska are clearly discernible, shown as a long promontory with the inscription: "Great Land, where live the people called *Kigin Eliat* by the Chukchi, who speak a distinct tongue, and wear sable, fox, and caribou parkas; animals are here of all sorts, sables, foxes, and caribou. Their yurts are in the earth, and their fighting weapons are bows. Their timber consists of pine, larch, spruce, and birch, and they have a fort."[50]

This chart proved to be one of the very few whose content was reflected on a foreign map. In 1725 the regions to the east of the Chukotka Peninsula were included in the atlas by the Nuremberg cartographer J. B. Homann as *Geographica nova ex Oriente gratiossima, duabus tabulis specialissimis contenta, quarum une mare Caspium, altera Kamtzadaliam seu terram Jedso curiose exhibet, Editore Jo.*

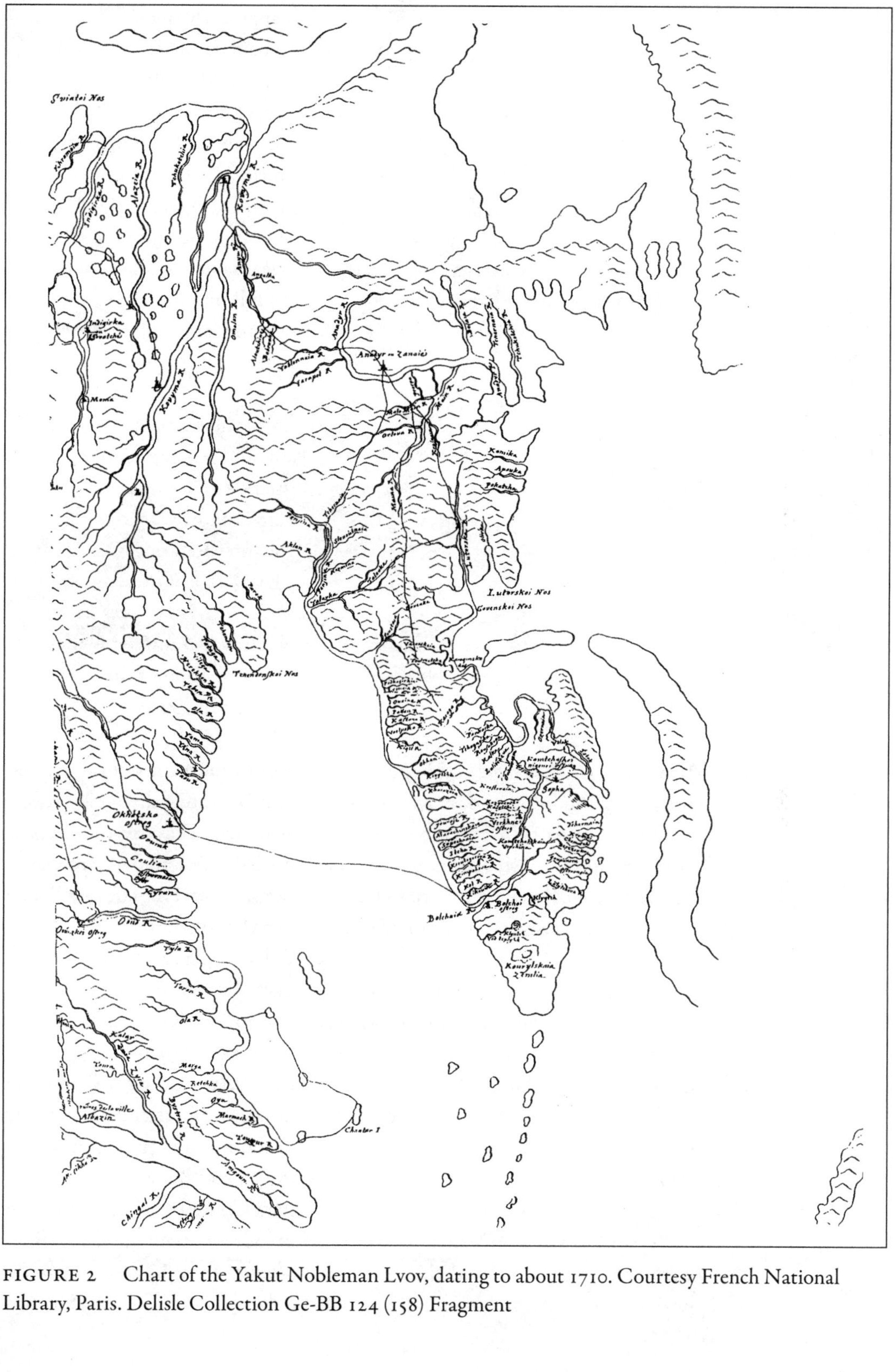

FIGURE 2 Chart of the Yakut Nobleman Lvov, dating to about 1710. Courtesy French National Library, Paris. Delisle Collection Ge-BB 124 (158) Fragment

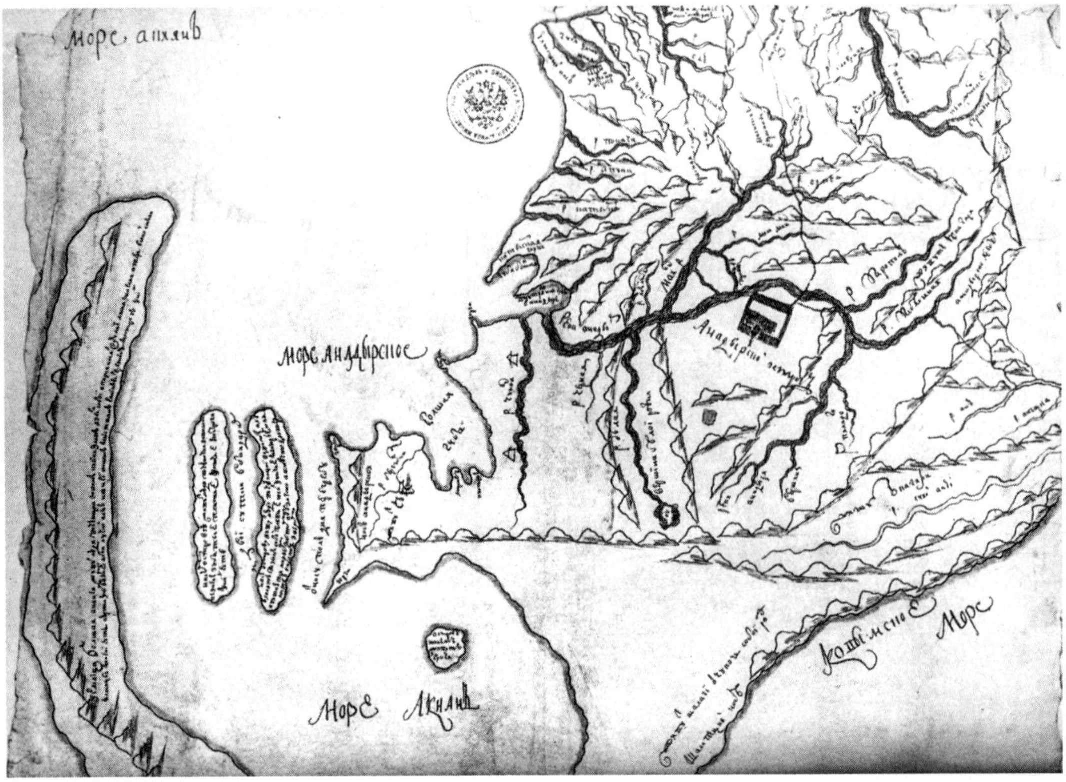

FIGURE 3 Map by Ivan Lvov about 1710. RGADA. F. 192

Bapt. Homann S.C.M. Geogr. Norimbergae [A new and most important geography of the East, containing two images of special content, one of the Caspian Sea, the other of Kamchadalia. that is the Land of Iedzo, finely drawn. Editor Jo[hann] Bapt[ist] Homann, S. C. M. geogr[apher] of Nuremberg]. This map was prepared on the basis of Lvov's data by order of Peter I and delivered to Homann by Yakov Vilimovich Bruce, a close collaborator of the emperor (Fig. 3).[51]

SEMYON REMEZOV

Information about the islands, the strait, and the Great Land east of the Chukotka Peninsula was also reflected in the work of the famous Siberian cartographer Semyon Ulyanovich Remezov, especially his chart of Kamchatka (not earlier than 1712–1714) in the *Sluzhebnaya chertezhnaya kniga* [Official Book of Charts, i.e., atlas—LB]. Here, to the east of Chukotka, an elongated stretch of land is shown with an inscription stating that "this land is but newly explored" and there are "many fortified settlements" and the inhabitants are *kynyntsy*. Similar terminology appears in other Russian sources, even in the second half of the eighteenth century—*kykhmaltsy, kynyntsy, zemlitsa Gigmalskaya* or *Kygmalskaya*, land of Kygmyn, Cape Kyng-Myn, Kiginelyat people, and so on [some of these refer to Point Hope, Alaska, Kygmyn—LB].

As was proposed by the Russian historian of geography L. S. Berg, all of these variations in terminology derive from and have their origin in the Chukchi name for America—Land of the Kyymlyt—that is, the land of Eskimos. The American ethnographer Dorothy Jean Ray, in her 1975 work, *The Eskimos of Bering Strait, 1650–1898*, narrows down the possible locality to which these terms may apply. In particular, she suggests that on the chart of Kamchatka in Remezov's *Atlas* [*Sluzhebnaya chertezhnaya kniga*] of Siberia the term *kynyntsy* refers to the Eskimo settlement Kingigan (contemporary name for the settlement of Wales).[52] Already at the beginning of the second decade of the eighteenth century Russian cartographers represented the western part of Alaska on the basis of information supplied by pioneers, noblemen, and Native peoples.[53]

Most copies, as well as the original Russian charts of Siberia, including those of its northeastern extremity, were preserved by the efforts of the talented Siberian scholar, author of works on geography, ethnography, and the history of Siberia, gifted cartographer, architect, and artist, Semyon Remezov (1642—after 1720). His activity was a kind of summary of the development of pre-Petrine cartography, the Russian people's movement eastward, and their active collecting of information from local tribes on the newly encountered territories that were adjacent to them.[54]

Remezov's main cartographic works, as well as copies and versions by him of works by other authors dedicated to the Siberian lands of the Russian State, are found in three remarkable manuscript compilations of maps, charts, and drawings: *Chertezhnaya kniga Sibiri* (1699–1701), preserved in the Rumiantsev Collection in the Russian State Library in Moscow; *Khorograficheskaya chertezhnaya kniga* (1697–1730), preserved in the L. S. Bagrow Collection, Harvard University, USA; and *Sluzhebnaya chertezhnaya kniga*, finished by the Siberian cartographer's sons, preserved in the Russian National Library in St. Petersburg.

The compilation of the first and most outstanding of these atlases—*Chertezhnaya kniga Sibiri*—was entrusted to Remezov in compliance with a legislative act of the Russian government—the Boyar Decision of January 10, 1696. This document is of great interest as an early example of Russian "governmental instruction" about cartography, and therefore it is fitting to cite its text here in full.

> January 10. The Boyar Decision.
> About making a chart of Siberia, on linen, indicating therein cities, settlements, peoples, and distances between natural areas.
>
> On the order of the Great Lords [*Pyotr and Ivan—AP*], the Boyar Prince Ivan Borisovich Repnin and colleagues has ordered that the Great Lords' royal edicts be sent to all Siberian cities to the effect that all counties [uezdy] of the

> Russian villages and all townships [volosti], including iasak-paying townships, be mapped on a linen chart, and distances between cities in versts or by days of travel be indicated, and the same applies to Russian villages and townships and iasak-paying townships and the nearest city. It is to be shown on which rivers these same cities and counties and iasak-paying townships are located, and this is to be shown specifically on the chart.
>
> Tobolsk is ordered that a good and experienced master there is to compile a [composite] chart of all of Siberia which is to have below it an inscription stating the distance between cities, in versts or by days of travel; the counties linked to each city are to be delineated and a description made stating which peoples are nomads and live in what localities; also indicate where, from which side, and what kind of people approach the border regions; such charts must be made of the [Siberian] cities because the Siberia department has no charts of Siberian cities and therefore all information is lacking. Once the charts are completed, they are to be sent to the Siberia department without delay. Charts are to be made in the size of three arshins [*84 inches*] in length and two arshins [*56 inches*] in width, drawn with the best possible skill. The great chart of all of Siberia is to be made in the size of three arshins in height and four arshins [*112 inches*] across.[55]

It should be noted that though the law cited above was formally approved by the young Peter together with his brother Ivan, the content of the cartographic principles therein belongs purely to the traditional Russian past. And Peter I would dedicate his entire life to changing, and in some instances to fighting, this past, as well as to changing the governing of the state, which in 1721 was declared the Russian Empire. The Boyar Decision of 1696 was proclaimed on the brink of the eighteenth century, on the eve of the drastic Petrine reforms. It was in some measure their precursor and served as evidence that the central government sought to become better acquainted with the territory of the immense state and determine its place in relation to its neighbors, and thus, on the world map.

The latter especially interested Peter, who had been fascinated with geography since childhood. At the very outset of the eighteenth century he entertained the possibility of a voyage from the Arctic Ocean to China and India. His collaborator, master shipwright Fyodor Saltykov, proposed in 1713 to dispatch ships out of Arkhangelsk (European Russia's port on the White Sea) to explore the northern sea route and to continue the voyage "to the mouth of the Amur and along [the coasts] of Japan and China."[56] Peter, well-educated for his times and standing in awe before European science, widely perused foreign, especially Dutch, maps and atlases, even though these, with a few exceptions when based on Russian data, presented in practice only legendary notions about the northern part of the Pacific Ocean. The question of the existence of a strait between Asia and America engaged many European scholars, but in practice they did not have at their disposal any real data needed to answer it.

It has been demonstrated above that the Siberians, who had already traversed the strait in 1648, showed on most of their charts not only a water expanse linking the Arctic and the Pacific oceans, but even showed the islands in the Bering Strait and the Great Land now identified as Alaska. It is necessary to emphasize, however, that the authors of these charts had no way of knowing that their Great Land was, in fact, a part of the immense American continent, and the majority of them probably did not even know that such a continent existed. Information about Dezhnev's voyage was brought to the attention of the central government only after 1736, when Gerhard Müller, a participant in the Vitus Bering expedition, found in the Yakutsk archive a later report by Dezhnev containing various petitions where he mentioned that he had rounded the eastern tip of Asia by sea and found himself on the Anadyr River.

PETER THE GREAT

Peter I had at his disposal Russian charts of Siberia, including the giant (213 × 277 cm) [83 × 108 in.] Map of Siberia from the Chinese Border, the so-called Ekateringof map, prepared by Remezov in response to the decree of 1696 (repeated in 1697) in Moscow in 1698, now located in the Petrine Gallery of the State Hermitage in St. Petersburg.[57] These were available to Peter for the solution of the question about the linkage of Asia and America. The chart, along with the other Russian materials, should have been evidence for the existence of the sea passage at least from the mouth of the Kolyma to the mouth of the Amur. However, as we saw earlier, foreign maps, even those based on Russian materials, left open the question of the Asia and America linkage.

Peter I, impressed during his journeys abroad of the achievements of European progress, was not inclined to trust information from "the people of the past," whose charts were supplied mostly by uneducated, bearded Siberians, especially if they contradicted foreign maps. This mistrust may have been aggravated by the fact that even Russian geographic charts sometimes fostered doubts that passage by the northern sea route into the Pacific Ocean was possible. Various areas on the northern and northeastern coasts of Siberia had "impassable capes, promontories and rocks." One such cape appears immediately to the east of the mouth of the Lena on the chart by Remezov, now in the Petrine Gallery.

To conclude, the overview of the early charts of the territories of Siberia and the northeastern regions of Asia produced cartography outside the Western European scientific framework of the time. Even in their appearance the Siberian charts of the seventeenth to early eighteenth century sharply differed from contemporary foreign maps created within Ptolemy's cartographic paradigm that was then dominant in European geography. Russian charts lacked the geographic net and consistent scale and projection for all parts of the cartographic image.

The absence of these geographic map parameters, taken for granted today, is due to the fact that no instruments were used to determine the geographic coordinates of various localities, nor was the topographic survey employed. It is known that only during Spafary's mission to China (1675–1678) the coordinates (latitude) of several points in Siberia were determined and a compass route survey was carried out. For this purpose, the Russian embassy was issued with an astrolabe and compass. However, Spafary's surviving map has no astronomical or geodesic data, and it differs but little from other Russian charts of the time. The spatial organization of the charts within the framework of a single cartographic canvas was, instead of geographic coordinates and projections, main rivers, land routes, and the seas which served as the routes of travel. But their orientation and placement relative to other features were probably established through the use of the compass, with which the Russian pioneers were well acquainted.

The Siberian cartographer assigned great importance to geographic place names and to various explanatory texts. The absence of scale was compensated for in some measure by indicating the length of this or that river or land route, expressed in days of travel or distance in versts. Prominent on these geographic works were the names of localities, natural features, population points, tribes, and nations. Rather lengthy texts were incorporated, apparently excerpts from the official reports by the pioneers.

Once in a while these texts and place names on the charts provided historical information (sometimes a legend) which was linked to the territory being mapped. The charts were evidently a representation of textual descriptions in graphic form. On the other hand, traditionally, detailed textual descriptions—*rospisi* and *knigi*—were appended to the charts. These included, besides detailed accounts about all elements of the charts' contents, sources, symbols used, and methods of compilation. This practice is most fully expressed in the work of Remezov, whose explanatory texts appended to his maps and atlases of Siberia are not only samples of historical and geographic descriptions in depth, but also a treatise on the traditional Russian methodology of chart compilation.

Russian cartography and geography had yet another characteristic that outlived all changes—those linked to Peter's reforms and to later changes and revolutions in Russia—and has carried forward even in very recent times. We have in mind the constant drive by the official powers of Russia to limit, by all possible means, the free spread of geographic materials and maps. In many instances, documents of this type were considered secret and their acquisition or copying, especially by foreigners, was made very difficult. In this connection, Adolf Nordenskiöld's discovery in the Swedish archives of information about his countrymen's attempts to obtain and copy Russian charts of Siberia is of great interest. For example, two copies of the Godunov chart of Siberia of 1667 were made in Moscow, one by the head of the Swedish embassy in Russia, Lt. Col. Fritz Cronman, and one by a noncommissioned officer, Claes Johansson Prytz.

In his introduction to a report about the embassy, Prytz writes: ". . . the appended land map of Siberia and adjacent lands I copied 8 January 1669 in Moscow from a poorly preserved original which was loaned to me for a few hours by Prince Ivan Alekseevich Vorotynsky on the condition that I may examine it but under no circumstances copy it."

Ambassador Conman in a letter to the Swedish king Charles XI, dated February 10, 1667, reported as follows: "The map of all these Siberian lands up to China, which recently was sent here on His Majesty's orders by Tobolsk commander Godunov, was shown me, and having received permission to keep it overnight, I copied it."[58]

Eric Palmquist, a military agent in the Swedish embassy of 1673 in Moscow, managed to obtain many materials about Russia, among them sixteen geographic maps and plans of cities, including the general Siberian charts of 1667 and 1673. In the dedication to his album Palmquist mentions that he collected Russian materials "not without effort and difficulties" and that he ". . . himself in various places made secret observations and sketches, putting himself at risk, and also obtained some information for money from Russian subjects."[59]

The Dutchman Nicholas Witsen and Swedish linguist Johann Gabriel Sparwenfeld obtained Russian sources in a different manner. Having visited Russia in 1664–1665 as a member of the Dutch mission of Jacob Boreel, Witsen established correspondence of many years duration, with Moscow, Astrakhan, Georgia, Poland, Isfahan, Constantinople, and Peking. In his own words, he "assembled volumes of diaries and notes in which are the names of mountains, rivers, cities and towns, together with a magnitude of drawings executed to my order."

These collections of materials, in which Russian data occupy a most important place, provided the basis for the creation of his book and the compilation of maps of Asia.[60] One of Witsen's Russian correspondents, an artist at the court of Tsar Alexei Mikhailovich, Stanislav Loputsky, transmitted to him a copy of the chart of Siberia of 1667 and a chart of Novaya Zemlya.[61] Sparwenfeld, while in Moscow, established friendships with many influential people, who provided him with geographic and cartographic materials.[62]

So, foreign geographers and cartographers actively collected Russian material and, in spite of their deficiencies from the point of view of the scholarship of the times, used them when compiling composite maps of Siberia and Asia. This use, however, encountered one problem that could not be solved with precision on the basis of the traditional Russian geographic charts: the placement of information derived from these charts on the world map within the system of geographic coordinates (latitudes and longitudes). Lacking data about the coordinates of various points in Siberia, especially in its eastern part, foreign geographers were forced to adopt an approximate solution to this fundamental problem. They used the approximate data contained in Russian charts and descriptions about distances (or rather the time needed to overcome such distances) between certain population points and translated these into the geographic language of latitudes and longitudes in a specific cartographic projection. Naturally, such translation could not be accomplished with geodesic precision. As a result, even on the best foreign map of Siberia, that of Witsen in 1690, the errors in establishing of longitudes ranged from –23°39′ (Yakutsk) to +4°27′ (Tobolsk) and of latitudes from +7°20′ (Irkutsk) to –2°30′ (Cape Chelyuskin).[63]

In this, as in all other areas of Russian science, technology, economics, politics, and the way of life in general, Peter I decidedly directed Russia into the mainstream of Western European civilization. Geography and cartography played an important role in the reforms undertaken by the first Russian emperor, permitting not only a declaration, on the world map, of the appearance of the vast Eurasian empire, but also its inclusion in ambitious geographic projects, the most obvious of which, for Russia, was the question of the existence of the Strait of Anian, and in connection with it, the study and cartography of the extreme regions of northeastern Asia and the northwestern extremity of North America.

NOTES

1. Martin Waldseemueller, *Universalis Cosmographia*. Preserved at Schloss Wolfegg, Wuertemberg.
2. Current research indicates that this migration occurred sometime after 16,500 years ago. On this topic, see David M. Hopkins, ed., *The Bering Land Bridge*. 1967.
3. The *promyshlenniki* hunted furs directly or procured them through barter or the supervision of hunting parties. Some operated independently but more often worked on a contract basis. This term will be employed throughout since there is no precise English equivalent.
4. But not completely. In fact, Pope Paul III promulgated a Papal Bull in 1537 declaring that the Natives of the Americas were human beings and entitled to their liberty. Also, Spanish-born Dominican priest Bartolomé de las Casas successfully campaigned for Native rights in the early sixteenth century. [LB]
5. *Russkie ekspeditsii po izucheniyu severnoy chasti Tikhogo okeana v pervoi polovine XVIII v.* (hereafter cited as *Russkie ekspeditsii . . . XVIII v.*) *Sbornik dokumentov* [*Russian Expeditions for the Study of the Northern Part of the Pacific Ocean in the First Half of the Eighteenth century. Collected Documents*]. Moscow: Nauka, 1984:8.
6. A. V. Efimov, ed., *Atlas geograficheskikh otkrytii v Sibiri i v Severo-zapadnoy Amerike v XVII–XVIII vv.* (*Atlas of Geographic Discoveries in Siberia and in Northwest America, Seventeenth to Eighteenth Centuries*). 1964 (hereafter cited as *Atlas . . .*). *Istoriya Sibiri* (*History of Siberia*). Moscow: 1968, vol. 2, A. V. Efimov, *Iz istorii velikikh russkikh geograficheskikh otkrytii* (*From the History of the Great Russian Geographical Discoveries*). 1971, and many others.
7. Ibid.,:138.
8. Original documents of the voyage of Semyon Dezhnev published by T. D. Lavrentsova. *Russkie arkticheskie ekspeditsii XVII–XX vv.* p. 131.

9. B. P. Polevoi, *250-letie otkrytiya Alyaski, Izvestiya Vsesoyuznogo Geograficheskogo obshchestva* [250th anniversary of the discovery of Alaska, *News of the All-Union Geographical Society*] (hereafter cited as Izv. VGO), vol. 114, no. 5, Leningrad, 1982:409. [Polevoi's interpretation is not universally accepted. The majority of experts believe that he meant only the promontory striking out into the Bering sea—he at that time had no notion of the entire Chukchi Peninsula—LB]
10. *Russkie arkticheskie ekspeditsii . . . ,* 137. Polevoi, ibid.
11. Grekov, 1960:334.
12. V. I. Grekov, *Ocherki iz istorii russkikh geograficheskikh issledovanii v 1725–1765 gg.* (*Essays from the History of Russian Geographical Research in 1725–1765*), Moscow, 1960:8.
13. For analysis of the specifics of the Russian cartographic tradition, see A. V. Postnikov, *Razvitie krupnomasshtabnoi kartografii v Rossii* (*The Development of Large-Scale Cartography in Russia*), Moscow: Nauka, 1989.
14. E. E. Zamyslovsky, "Chertezhy Sibirskikh Zemel' XVI–XVII vv ("Charts of the Siberian Lands Sixteenth–Seventeenth Centuries"). In *Zhurnal Ministerstva narodnago prosveshcheniia* [*Journal of the Ministry of People's Enlightenment*], vol. 257 June (1891), pp. 334–347. A. I. Andreev, "Chertezhy i karty Rossii XVII veka, naidennye v poslevoennye gody ["Charts and Maps of Russia of the Seventeenth Century, Discovered in the Post-War Years"]. In *Voprosy ekonomiki i klassovykh otnoshenii v Russkom gosudarstve XII–XVII vekov* [*Issues of Economics and Class Relationships in the Russian State of the Twelfth to Seventeenth Centuries*]. Number 2 (Trudy Leningradskogo otdeleniya Instituta istorii Akademii Nauk SSSR). Moscow/Leningrad, 1960:80–90.
15. *Russkaya istoricheskaya biblioteka, izdavaemaya Arkheograficheskoi komissieyu* (*Russian Historical Bibliography Published by the Archeographic Commission*), vol. 2, St. Petersburg, 1875:106. D. M. Lebedev, *Ocherki po istorii geografii v Rossii XV i XVI vv.* (*Essays on the History of Geography in Russia in the Fifteenth and Sixteenth Centuries*), Moscow, 1956:217.
16. B. P. Ivanov, "K istorii kartografirovaniya russkogo Severa" ("On the History of the Cartography of the Russian North"). In *Vestnik Kharkovskogo universiteta. Seriya geologo-geograficheskaya* (*Kharkov University Bulletin. Geological and Geographical Series*). Issue 2, no. 25, Kharkov 1967:97. Postnikov, ibid.,:16.
17. B. A. Rybakov, "Geograficheskie znaniya" ("Geographical Knowledge") In *Ocherki russkoi kultury XVI v.* (*Essays on Russian Culture of the Sixteenth Century*), Part 2, Moscow, 1977:213.
18. Ivanov, ibid.
19. L. A. Goldenberg, *Izograf Zemli Sibirskoi. Zhizn i trudy Semyona Remezova* (*Isograph of the Siberian Land. The Life and Work of Semyon Remezov*), Magadan, 1990:24.
20. Zamyslovsky, ibid.
21. An impression of these works may be gained from the series of preserved charts, particularly those of the River Upsa of 1655, by R. Starkov, a portion of the course of the Yenisey River of 1684 by A. Korenev, portions of Yenisey and Ilim counties (*uezdy*) adjacent to the Tunguska River of 1685 by A. Galkin and F. Rosputin, of the Lena townships (*volosti*) of 1694, and of Dauria of 1689 by A. I. Beiton and of 1699 by I. Peteli. L. S. Bagrow, *Karty Aziatskoi Rossii. Istoricheskie zametki* (*Maps of Asiatic*

Russia. Historical Notes). Petrograd, 1914:10. Leo Bagrow, "A Few Remarks on Maps of the Amur, the Tartar Strait and Sakhalin." In *Imago Mundi*, vol. XII (1955):127–136. *Atlas*... nos. 35–37.

22. A. Titov, *Sibir v XVII veke. Sbornik strannykh russkikh statei o Sibiri i prilezhashchikh k nei zemliakh* (*Siberia in the Seventeenth Century. Anthology of Odd Russian Articles on Siberia and the Lands Adjacent to it*). Moscow, 1890.
23. Y. A. Limonov, "Rospis pervogo obshchego chertezha Sibiri. (Opyt datirovki)" (Description of the First Complete Chart of Siberia, in *Problemy istochnikovedeniya* [*Problems of Sourcing*]), vol. 8, 1959:360.
24. Information about the Lena was already on hand in Yeniseisk and Mangazeia in 1619, while more extensive data were brought by the detachments of S. Navatsky, A. Dobrynsky, and M. Vasiliev, who were on the Viliui, Chona, and Lena in 1628–1631. Goldenberg, ibid.,:25.
25. V. N. Tatishchev, *Izbrannye trudy po geografii Rossii* (*Collected Works on the Geography of Russia*), Moscow, 1950:215.
26. L. A. Goldenberg, "Podlinnaya rospis chertezha Sibiri 1667 g." ("The Original Description of the Chart of Siberia of 1667."). *Trudy instituta istorii estestvozniya i tekhniki AN SSSR* (*Works of the Institute of the History of Natural Science and Technology of the Academy of Sciences of the USSR*). Vol. 42, no.3, Moscow, 1962:252–271.
27. Titov, ibid., and others.
28. These copies are: Swedish copies by Claes Johansson Prytz of 1669, Fritz Cronman of 1669, and Eric Palmquist of 1673; a German copy by Georg Schleising of 1690, and the Russian ones by Semyon Remezov of 1697 and 1702. Bagrow, ibid., 1914:11. Leo Bagrow, "The First Russian Maps of Siberia and their Influence on the West-European Cartography of N.E. Asia." In *Imago Mundi*, vol. 9 (1952):83–93. A. V. Efimov, *Iz istorii velikikh russkikh geograficheskikh otkrytii v Severnom Ledovitom i Tikhom okeanakh, XVII—pervaya polovina XVIII v.* [*From the History of the Great Russian Geographical Discoveries in the Arctic and Pacific Oceans, Seventeenth–First Half of the Eighteenth Centuries*], Moscow, 1949:75. Goldenberg, ibid., 1990:28.
29. M. I. Belov, *Semyon Dezhnev*. Moscow, 1955. R. Fisher, *The Voyage of Semen Dezhnev in 1648: Bering's Precursor, with Selected Documents*. London, 1981. Goldenberg, ibid., 1990:28.
30. Goldenberg, ibid., 1962.
31. B. P. Polevoi, "Gipoteza o *godunovskom* atlase Sibiri 1667 g" ("Hypothesis on the Godunov Map of Siberia of 1667."). In *Izvestiya AN SSSR. Seriya geograficheskaya* (*News of the Academy of Sciences of the USSR. Geographical Series*), 4 (1966):123–132. B. P. Polevoi, "Commemorating the 300th Anniversary of the Godunov Map of Siberia." Translated by James R. Gibson. *The Canadian Cartographer*, vol. 8, no. 1 (June 1971):19–26. B. P. Polevoi, "Sibirskaya kartografiya XVII v. i problema Bolshogo chertezha" (Siberian Cartography of the Seventeenth Century and the Problem of the Great Chart), in *Strany i narody Vostoka* (*Countries and Peoples of the East*), no. 18 (1976):213–227. This position is not generally accepted and was especially harshly criticized by L. A. Goldenberg. See, for example, Goldenberg, ibid., 1990:28.
32. Bagrow, ibid., 1952:88.
33. Goldenberg, ibid., 1990:32–33.

34. V. I. Grekov, "O chertezhe vsei Rossii do Kitaiskogo tsarstva i do Nikaskogo" ("On the Chart of All of Russia to the Chinese Empire and to Nikasky.") In *Izvestiya AN SSSR, Seriya geograficheskaya*, 2, 1959:82. M. Mikami, "Russkaya karta Sibiri 1673 g." (The Russian Map of Siberia of 1673.") In *Dzimbun tiri*, vol. 16, no. 1, 1964:19–39.
35. Goldenberg, ibid., 1990:33.
36. Cited after O. M. Medushevskaya, *Kartograficheskie istochniki XVII–XVIII vv. Uchebnoe posobie po istochnikovedeniyu SSSR* (*Cartographic Sources of the Seventeenth–Eighteenth Centuries, Instructional Textbook on Sourcing for the USSR*), Moscow, 1957 10–11.
37. Leo Bagrow, "Sparwenfeld's Map of Siberia." *Imago Mundi*, vol. 4 (1947):69.
38. Goldenberg, ibid., 1990:34–35.
39. This chart was first published as end papers of Y. N. Semyonov's book about the conquest of Siberia. See Yuri Semyonov, *Siberia: Its Conquest and Development*, London, 1963. Published also Bagrow, ibid., 1947:70–71, Belov, ibid., 1955:116–117 and 118–119, Andreev, ibid., 1960:56–57, *Atlas* . . . 1964, no.34. The date of the compilation of this chart was established by L. A. Goldenberg. At present, the Chart of Siberia of 1687 is in the L. S. Bagrow Collection, Harvard University, USA.
40. M. Mikami, "Russkaya karta Sibiri 1687 g." In *Dzimbun tirigaku-no semondai*, Tokyo, 1968:425–436. Goldenberg, ibid., 1990:35.
41. Belov, ibid., 1955:118.
42. A. I. Andreev, "Pervye karty Sibiri XVII veka" ["The First Maps of Siberia of the Seventeenth Century," in *Globus. Geografichesky ezhegodnik dlya detei* [*The Globe. Geographical Annual for Children*]. Moscow-Leningrad, 1939:309–313.
43. Goldenberg, ibid., 1990:36.
44. J. F. Baddeley, *Russia, Mongolia, China. Being Some Record of the Relations Between Them from the Beginning of XVIIth Century to the Death of the Tsar Alexei Mikhailovich, A.D. 1602–1676* . . . 2 vols. And maps, Macmillan: London and New York, 1919. Leo Bagrow, "Sparwenfeld's Map of Siberia." *Imago Mundi*, vol. 4 (1947):65–70. Leo Bagrow, "The First Russian Maps of Siberia and Their Influence on the West European Cartography of N. E. Asia." *Imago Mundi*, vol. 9 (1952):83–93. Leo Bagrow, "A Few Remarks on the Amur, Tartar Strait, and Sakhalin." *Imago Mundi*, vol. 12 (1955):127–136." Akio Funakoshi, "O karte Sibiri perioda Kansi, khranyashcheisya v lichnom arkhive Lo Chzhen'yuya" ["On the Map of Siberia of the Kansi Period, Preserved in the Personal Archive of Lo Chzhenyuya"]. In *Takho takukho*, no. 33 (Tokyo 1963):199–218. L. A. Goldenberg, "Izuchenie kart Sibiri XVII–XVIII vv. v Iaponii ("The Study of Maps of Siberia of the Seventeenth–Eighteenth Centuries in Japan"). In *Voprosy istorii* (*Historical Issues*) 8 (1981):163–168.
45. N. N. Ogloblin, "Dve 'skazki' V. Atlasova ob otkrytii Kamchatki" ("Vladimir Atlasov's Two 'Fairytales' About the Discovery of Kamchatka"). *Chteniia v Imperatorskom obshchestva istorii i drevnostei rossiskikh pri Moskovskom universitete* (*Lectures at the Imperial Society of the History and Ancient Times of the Russians at Moscow University*). Book 3, Moscow, 1891:12.
46. *Pamiatniki Sibirskoi istorii XVIII v.* (*Relics of Eighteenth-Century Siberian History*), Book 1, 1700–1713, St. Petersburg, 1882:458–459.
47. *Pamiatniki* . . . 1882:224.

48. L. S. Berg, *Otkrytie Kamchatki i ekspeditsii Beringa, 1725–41* (*The Discovery of Kamchatka and Bering's Expeditions*) Moscow-Leningrad, 1946:49. M. I. Belov, *Istoriia otkrytiya i osvoeniya Severnogo morskogo puti* (*The History of the Discovery and Mastery of the Northern Sea Route*). Vol. 1. Moscow, 1956 :186. M. I. Belov, "Novye dannye o sluzhbakh V. Atlasova i pervykh pokhodakh russkikh na Kamchatku" ("New Data About the Services of Vladimir Atlasov and the First Russian Journeys to Kamchatka."). In *Letopis Severa* (*Chronicle of the North*), vol. 2:79–107. *Atlas*. . . 1964:viii–ix. B. P. Polevoi, 250–letie otkrytiya Alyaski. In *Izvestiya VGO*, vol. 114, no. 5, 1982:410.
49. This map was published by A. V. Efimov, *Iz istorii russkikh ekspeditsii na Tikhom okeane (pervaya polovina XVIII v.)* (*From the History of the Russian Expeditions in the Pacific Ocean [First Half of the Eighteenth Century]*). Moscow, 1948. *Atlas*. . . 1964:54, 38–39.
50. Efimov, ibid., 1948:98–99. *Atlas*. . . 1964:55, 39.
51. E. F. Varep, "O kartakh, sostavlennykh russkimi, v atlase I. B. Gomana 1725 g." ("On the Maps Created by Russians in I. B. Goman's Atlas."). *Izvestiya VGO*, vol. 91, no. 3, 1959:290–291. *Atlas*. . . 1964:58, 40–41.
52. D. J. Ray, *The Eskimos of Bering Strait, 1650–1898*. Seattle and London: University of Washington Press, 1975:13.
53. Goldenberg, ibid., 1990:363–364.
54. Semyon Remezov's legacy is examined in detail in the works of L. A. Goldenberg: *Semyon Ulyanovich Remezov: Sibirsky kartograf i geograf, 1642-posle 1720 gg* [*Semyon Ulyanovich Remezov: Siberian Cartographer and Geographer, 1642-after 1720*]. Moscow: Nauka, 1965; Russian Maps and Atlases as Historical Sources. Cartografica, 3, Toronto, 1971 ; "The Atlases of Siberia by S. U. Remezov as a Source for Old Russian Urban History." In *Imago Mundi*, vol. 25 (1971):39–46.; Goldenberg, ibid., 1990, and other works.
55. *Polnyi svod zakonov Rossiskoi imperii* (*Complete Code of Laws of the Russian Empire*), Issue 1, vol.3, Law no. 1532.
56. *Russkie ekspeditsii . . . XVIII v.*, 1984:9.
57. A. V. Grigoriev, "Podlinnaya karta Sibiri XVII veka (raboty Semyona Remezova)" (The Original Map of Siberia of the Seventeenth Century (the Work of Semyon Remezov)], in *Zhurnal Ministerstva narodnogo prosveshcheniya*. Part II (1907):374, 381. Baddeley, ibid., 1919, Map 2.
58. A. E. Nordenskiöld, "Pervaya karta Severnoi Asii, osnovannaya na deistvitelnykh nablyudeniyakh" ("The First Map of Northern Asia Based on Actual Observations."). In *Zapiski Voenno-topograficheskago otdela Glavnogo shtaba* (*Notes of the Military Topographical Department of the Chief Headquarters*), vol. XILIV, Part 7, 1889:9.
59. Cited after Goldenberg, ibid., 1990:38. Eric Palmquist's album was published in Stockholm in 1898.
60. Johannes Keuning, "Nicolaas Witsen as a Cartographer." *Imago Mundi*, vol.11, 1954:97.
61. P. Alekseev, *Sibir v izvestiyakh zapadno-evropeiskikh puteshestvennikov i pisatelei, XIII–XVII vv.* (*Siberia in the Reports of Western European Travelers and Writers, Thirteenth–Seventeenth Centuries*), vol. 1, part 1, Irkutsk:1936:63, 64.
62. Leo Bagrow, "Sparwenfeld's Map of Siberia." *Imago Mundi*, vol. 4, 1947:66.
63. Keuning, ibid., 1954:101.

CHAPTER II

Are America and Asia Joined?

Grandiose geographic studies and cartographic works were planned for the vast state that was proclaimed the Russian Empire in 1721. These were undertaken to guarantee the success of Peter's reforms and his foreign policy. They were carried out almost exclusively by state agencies and were financed, almost totally, by the state. They were strictly directed and controlled by the highest Imperial authorities (for the most part by the Ruling Senate and the Admiralty College). The importance of the cartographic work was addressed in a special chapter of Peter's General Regulations of 1720, entitled "On land maps and royal charts."

TRAINING NEW SURVEYORS AND CARTOGRAPHERS

In order to create geographic maps of the same quality as those in Europe, it became necessary to train qualified geodesists (surveyors), hydrographers, and cartographers. With this in mind, as early as 1698 a school of mathematics and surveying opened in Moscow at artillery headquarters. Little information about this school survives today. The school burned down in 1699, and it is not mentioned again in the documents of the period.[1]

The systematic training of geodesists substantially increased under Peter I from 1701, first at the new Moscow School of Mathematics and Navigation, and later, from 1715, at the Naval Academy in St. Petersburg, where in 1716 a special course in geodesy was created. These educational institutions served as scholarly centers in marine, geographic, geodesic, and cartographic sciences until the establishment in 1724 of the Petersburg Academy of Sciences. Here the first Russian specialized textbooks and teaching aids were developed. Various mathematical and astronomical tables were prepared for publication.[2]

New textbooks were prepared at the Moscow School of Mathematics and Navigation. These included: *Arithmetic, the Science of Calculation*, by the Russian mathematician Leonty Magnitsky (1669–1739), published in 1703, which also contained information on geodesics and astronomy; *Geometry in Slavic Survey Science*, translated by Yakov Bruce in 1708; *Textbook of Sea Navigation*, by Avraam Degraf, in translation by Kopievsky, 1701; and *Practical Geometry with Illustrations*, translation by Kvart, 1709. The last two presented the methodology of constructing maps in reduced scale with the aid of an overlay grid or proportional graph ("by squares"). This method was widely employed until recent times in small-scale cartography.

The Moscow school was divided into the "mathematical" and "navigational" divisions. The following subjects were offered in 1705–1715: Russian grammar, arithmetic, geometry, trigonometry (both linear and spherical), navigation ("plane"—short-distance navigation; "Mercator"—navigation with the aid of charts; and "spherical"—navigation along the arc of a great circle), keeping a ship's log, applied astronomy, elementary geodesy, geography, and fencing. As one may discern from this list, the school provided professional as well as general education.

At first the Moscow School of Mathematics and Navigation was under the authority of the Armory, which earlier, in the seventeenth century, had been the main source for Russian draftsmen—the main authors of the traditional pre-Petrine cartographic works. In 1706 the school was transferred to the jurisdiction of the Department of the Navy, and in 1712 to the Admiralty office headed by the commander of the Russian Navy, Admiral-General Fyodor Apraksin. In 1702 the school was located in the Sukhareva (Sretenskaya) Tower, and it remained there until its dissolution by decree on December 15, 1752.[3] In 1701 the Sukhareva Tower also housed the second astronomical observatory established in Russia (the first one was created by Bishop Afanasy [A. A. Lyubimov] in Kholmogory in 1692). The Sukhareva Tower observatory had at its disposal astronomical telescopes, quadrants, sectors, nocturnals, and cross-staffs. Instructor James David Bruce (Iakov Vilimovich Bruce), Peter's collaborator and the founder of the observatory, made sure that it was a base where the students of the Moscow School of Mathematics and Navigation could practice applied astronomy.[4]

The prominent faculty who taught geodesics, cartography, and navigation were Professor of Mathematics of Aberdeen University Andrew D. Farquharson and navigators Stephen Gwyn and Richard Gries, who had been invited to come from England back in 1698. Magnitsky taught arithmetic, geometry, and trigonometry. Magnitsky also assisted Bruce in the instruction of students in astronomy and geodesy.[5] An outstanding cartographer, engraver, and publisher of the early eighteenth century, Vasily Kipriyanov, joined the teaching staff in 1701 and was Magnitsky's assistant and in charge of the school library. By a decree on May 30, 1705, the Moscow Civil Press was created and Kipriyanov was appointed its manager.

The press, under the general supervision of Bruce, was entrusted with publication of "books, maps, and various broadsides." Its main product turned out to be maps and engravings. In essence this press was not only a printing house but also the first Russian scientific, research, and cartographic production enterprise. In 1715 Peter I established the second civil press and ordered that "its manager will be . . . Vasily Kipriyanov and he is to be designated librarian."[6] Kipriyanov left the Moscow Navigation School and from that time on devoted himself solely to the work of compiling, engraving, and publishing maps.[7]

In 1715 all the students of the Navigation School (a total of 293), together with Farquharson and Gwyn, were transferred to the Naval Academy in St. Petersburg. From that time on most geodesists were trained at the latter institution. The Moscow school turned to general education, but with an emphasis on mathematics, and Magnitsky continued to direct it for the next 25 years. The Moscow Mathematics School was closed in 1752 when the Moscow Cadet Corps was established.

Since Peter's geodesists were trained by Farquharson and Gwyn, it was the English topographic and geodesic school that influenced the formation of new methods of cartography in eighteenth-century Russia. In England at that time the astrolabe survey was dominant,[8] in contrast to France, where the topographic drawing board was preferred. The astrolabe methods found easy acceptance in Russian field cartography because of the similarity of land measurement by means of the astrolabe to the methods employed in compass route surveys along the rivers and trails that were traditional in Russian practice.

The instruments studied were mostly angle-measuring devices used at that time in the progressive countries: quadrants, sectors, astrolabes, surveying compasses, distance-measuring chains, and so on. While in the realm of applied field cartography the new professional habits of Russian geodesists were mainly influenced by English angle-measuring (in modern terminology, tachometric surveys), the French school of mathematical geography was dominant in general cartography. The latter considered the determination of latitude and longitude of various points on the globe's surface and geometrically precise representation of lands and regions of the globe on a map to be the main goal of geographic science, concepts that were almost completely absent in traditional Russian cartography.

The courses in geography were limited to elementary concepts of mathematical geography, which included the understanding of cartographic projections and, to a very limited extent, a survey of all parts of the world. Specialist training at both the Moscow School of Mathematics and Navigation and the Naval Academy took place on a grand scale. In 1712 the navigational school had 517 students, and by 1715 the number had increased to 719.

The Moscow School provided specialist training in practically every branch of the economy of the rapidly developing state, but above all it trained navigators for the young Russian navy. The education of officers, as far as hydrographic surveys were concerned, was on a par with the training of geodesists. Many navigators, graduates of the navigation school as well as of the Naval Academy, actively participated not only in sea and coastal surveys but also in land surveys.[9] They took the leading role in exploration in eastern Siberia, the northern Pacific Basin, and along the coasts of northwestern America.

However, the immediate thrust into unknown lands was led by nonspecialists: cossacks, *promyshlenniki*, peasants, exiles, and other ordinary Russians. They provided reconnaissance surveys and cartography in the pre-Petrine tradition throughout the eighteenth century in spite of the constant growth in the ranks of professionally prepared experts in surveying and cartography.

New notions and revolutionary changes always penetrated the Russian culture with great difficulty, while the Russian reformers, due to the maximalism that was characteristic of the national character in all periods, even in very recent ones, were inclined to reject the experiences of their predecessors or, in the best cases, treat it with great skepticism. This nihilistic approach cost the Russian people a great deal and continues to do so. This attitude led the professionally trained navy officers, geodesists, and scholars of the St. Petersburg Academy of Sciences to reject charts compiled by the *promyshlenniki*.

Thus, in this period, two parallel, and in practice, separate, directions were developing in the geographic study and cartography of the new eastern lands and waters of the Russian Empire. The first, traditional approach transmitted rather precisely the topology (placement of features in relation to each other) and local place names, but lacked the new astronomical and geodesic data. It was therefore "unscientific" from the point of view of the educated professionals and central authorities.

The second approach rested on the scientific expeditions and efforts of the naval officers and geodesists. The overwhelming majority of them were not Siberian old-timers but newcomers. Therefore they "discovered" new lands in the same way as was done by their Western European colleagues, not bothering much about ascertaining from the local "savages" who inhabited these lands the place names that were current among them.[10] The history of the geographic exploration and cartography of the Bering Strait, Aleutian Islands, and coastal America in the eighteenth century is to some extent the history of the rapprochement, and, in the end, merging of the two approaches.

VITUS BERING'S FIRST KAMCHATKA EXPEDITION

For a long time Peter I was engaged in formulating plans to solve, once and for all, the question of whether a strait existed between Asia and America. Of interest is the testimony of John Perry, who worked in Russia from 1698 to 1715 and who participated in the design and construction of the first canals, the basis for the formation of a single internal water communication system in European Russia. He "heard more than once the tsar expressing his intent to send men to provide a reliable map of the land . . . to ascertain if it is possible for ships to pass around Novaya

Zemlya to the Tartar Sea east of Ob, where ships may be built for dispatch to the shores of China and Japan."[11] In 1719–1722, geodesists Evreinov and Luzhin were ordered by Peter to find, in addition to various other tasks, if "America and Asia are joined." They were not able to carry out this assignment. However, their voyage to the shores of Kamchatka and to the Kuriles, in the course of which they described the northern islands and, on the basis of information obtained, compiled a chart of the entire Kurile Islands chain, provided a great deal of new geographic data.[12]

On December 23, 1724, the Admiralty College received Peter's decree to organize the First Kamchatka Expedition. The text of this decree and the commentaries on it by the Admiralty College clearly show the importance Peter assigned to the expedition's geographic and cartographic work. In the first paragraph an order is given "to search out those geodesists who had been to Siberia and have returned." There is a notation in Peter's hand: "there is great need for a navigator and sub-navigators who have been to North America." This notation points out exactly the main goal of the expedition: to reach the American coast.[13] Three weeks before his death, on January 6, 1725, the Russian emperor, in his own hand, wrote an instruction to the expedition's commander, Captain Vitus Jonassen Bering (1681–1741), and entrusted its implementation to Admiral Apraksin. Although this instruction has been carefully studied and published numerous times by many researchers, we deem it necessary to reproduce this historical document here:

1. At Kamchatka, or in another locality, one or two decked boats are to be built.
2. These boats are to sail close to the land which extends to the north, as it is expected that apparently this land is part of America (because it is unknown where it ends).
3. And search (to the point) where it joins (alternate reading: approaches) America, and then go toward any city in European possession; should a European ship be encountered, inquire how this coast is called, record this in writing, and obtaining true knowledge, chart it, and then return hither.[14]

Developing Peter's instruction further, the Admiralty College issued its own admonition to Bering, stating that the expedition commander would receive a chart compiled in 1721 in order to acquaint himself with the local regions and that he should engage the help of one or two men from among the geodesists now there [in Siberia—AP].[15] The chart mentioned by the Admiralty College was apparently one compiled from the expedition to Kamchatka and the Kurile Islands by Evreinov and Luzhin. Besides this map, Bering had at his disposal the map of "Kamchdalia" from Homann's atlas, discussed earlier, which was based on Lvov's data. On this map a large mythical land is depicted near the eastern coast of Kamchatka. A study of Homann's map permitted B. P. Polevoi to advance a daring hypothesis, supported by Raymond H. Fisher in *Bering's Voyages: Whither and Why*, on where exactly Peter proposed searching for America.[16]

The questions in the title of Fisher's book, "whither" and "why" did the emperor of Russia dispatch Bering's expeditions, have long engaged scholars' attention. Many are inclined to the view that the content of Peter's instruction does not provide a clear picture of the problems he had set before the First Kamchatka Expedition. All participants and contemporaries of the expedition (particularly Gerard Müller and Ivan Kirilov) believed that the expedition was supposed to study the Asiatic shore to the north of Kamchatka, find the area where Asia "joins" America, and, should it be able to do so, ascertain if it was, indeed, America. For this reason it was necessary either to reach any European possession or to meet with a European vessel.

Many historians shared this view (and many still do), among them Polonsky, Berg, Kushnarev, and Pasetsky.[17] Other researchers see its priorities differently, taking as their point of departure the more general political, economic, military, and strategic aims of Peter's administration, and tend to view the expedition's goal in a wider perspective. Thus, in Belov's opinion, Bering and his companions were expected not only to clarify if the two continents were joined, but also to find a route from Arkhangelsk to the Pacific and to America.[18] Pokrovsky believed that the expedition "had to reach the Spanish colonies in America" and establish trade relations with them. Andreev and Lebedev supported this conclusion.[19]

Grekov believed that the expedition was not to search for a strait between Asia and America but to "investigate the route to America adjacent to Asia and to ascertain who was Russia's nearest neighbor on the American continent."[20] Barratt inclines to the view that through Bering's expedition, Peter took a decisive step in establishing a Russian naval presence in the Pacific Ocean.[21] Polevoi reinterpreted Peter's intentions, basing his argument on a new reading of the imperial instruction. Having compared it with Homann's map of "Kamchadalia," he began to argue that Bering should have sailed not to the north, but to the east along the southern coast of the land trending north.[22]

Polevoi and Fisher after him had in mind the unnamed land on Homann's map to the east of Kamchatka that trends to the north. They believe that Peter had precisely this land in mind when he wrote that the expedition should sail "along the land stretching to the north." Since in this view the main goal of the expedition was to reach America, then, in the opinion of Polevoi and Fisher, this land had to be rounded by taking a course to southeast and east. According to this hypothesis, however, members of the Bering expedition failed to understand the real meaning of Peter's order to follow the land that trends to the north.[23]

The Polevoi-Fisher hypothesis has been examined favorably and in detail in a monograph by Bolkhovitinov, who states, contrary to the opinion of the vast majority of historians of geography, that the authors of this concept "provided the most convincing and well-argued interpretation of the meaning of Peter I's instruction." However, he stresses that "nevertheless, this version remains a hypothesis as no matter what Peter I's possible and suggested intent was," the expedition sailed north.[24]

The instructions issued to Bering by Peter and the Admiralty College are silent in regard to a detailed methodology for the expedition's activities and do not put forward any demand in respect to the content of summary charts and descriptions. It is abundantly clear that members of the First Kamchatka Expedition had to proceed on the basis of theoretical projects then current in Russia, the by-then developed navigational practice, and their own professional skills. As noted above, in 1714 Saltykov had already presented to Peter a proposal on activities beneficial to the state. In his Chapter 25, an idea is developed "On the exploration of a free sea passage from the Dvina River to the mouth of the Amur and even into China."

One item contained in this chapter states general guidelines for expeditionary route descriptions in new territories as follows:

> When these vessels succeed and reach the sea, the sailors are to be instructed to describe the coasts and river mouths which they might find from the seaside and establish in these rivers the depth and currents, what kind of bottom there is, latitude and longitude and along which compass direction there is a distinguishing feature: a tree, a mountain, a building or similar things; distances in which they are situated, climate, places which are suitable for landing and which are not; what kind of timber is found in these areas, and also along the shores. They are to note bays formed by the sea, peninsulas or sea bars, sand banks, rock formations, headlands, where anchorages are to be found, in what latitudes are islands in the sea, their length and width, what people occupy them, what are their riches, forests, animals, birds; on such islands, rivers are to be described, landing places, and where fresh water is to be found, rocks in the sea, those submerged and those above water.[25]

Members of the expedition were trained and experienced, fully satisfying the requirements of the time. Bering was a graduate of the Naval Cadet Corps in Amsterdam, considered then the best institution of its kind. He had sailed to the East Indies aboard a Dutch vessel, and since 1703 was an active participant in many operations conducted by the Russian Navy.[26] The other navigators were, in the main, experienced seamen, graduates of the Moscow School of Mathematics and Navigation and of the Naval Academy. They were trained in latitude determination, on the basis of astronomical observations (primarily the meridian altitude of the sun) with the aid of sextants and quadrants, and longitude, on the basis of dead reckoning and also by the difference in latitude between observation points.[27] Without entering into the details of the much-studied story of the First Kamchatka Expedition, we will examine it in a general outline along with aspects of the methodology involved.

The expedition left St. Petersburg on January 24, 1725, and reached Okhotsk two years later in January 1727. A vessel named *Fortuna* had been constructed there beforehand. The expedition sailed on the *Fortuna* to Bolsheretsk, located at the mouth of the Bolshaya River on the western shore of Kamchatka. The sea route from Okhotsk to Kamchatka was a familiar one since the voyage of Sokolov and Treska had taken place in 1717.[28] However, the passage from the Sea of Okhotsk into the Pacific was as yet unknown to the Russian sailors. For this reason, the expedition's goods and supplies were transported overland by dog teams from Bolsheretsk to Nizhne-Kamchatsk *ostrog.* There, under Bering's supervision, a double-masted boat was constructed between April and June 1728 and named *Sviatyi Arkhangel Gavriil* (St. Archangel Gabriel).

On July 13, 1728, the expedition members sailed aboard the *Gabriel* to execute Peter's instruction. As is evident from Sopotsko's analysis of the original expedition's log, the hydrographic work carried out by Bering and his navigators was outstanding in comparison to the usual sea voyage descriptions of the time. For example, according to the then-existing rules for the keeping of the ship's log, astronomical observations of latitude were to be carried out once every twenty-four hours, along with calculations of longitude, which were recorded to one minute as was standard. Aboard the *Gabriel*, astronomical observations were taken twice (and sometimes, weather permitting, thrice) in the course of every twenty-four hours. The coordinates were entered in the log with precision to one-hundredth of a minute. Bearings (directional azimuths) in respect to shore contours and orientation points were recorded not in compass directions (that is, with precision up to eight degrees, as was the custom in the eighteenth century) but in degrees, and their values were recorded with precision to the minute. The time of observation was usually recorded with precision to the hour, but Alexei Chirikov and Pyotr Chaplin recorded it with precision to the minute.[29]

What is striking is the attention to detail and thoroughness of recording in the log of all the observations made. A hydrographic survey of the coast was carried out systematically and diligently. On some days the seafarers took bearings on up to eight points of orientation, as these shores were practically unknown to the Europeans. The log's records of bearings taken on coastal features are so detailed that Sopotsko was able to reconstruct precisely on a modern chart the route of the First Kamchatka Expedition in the Pacific and Arctic Oceans and to duplicate it in the course of an experimental voyage of the yachts *Motherland* and *Russia* in 1976–1977.[30]

Bering's voyage began the tradition of Russian scientific hydrographic research. It was also one of the first to introduce to the map of the northern part of the Pacific Ocean place names foreign to the local areas, such as St. Lawrence Island and Cross Bay, so named by members of the expedition. This, however, does not mean that Bering and his navigators were contemptuous of geographic information received from the Natives. It is known, for example, that on August 8, 1728, not far from the entry into the Bering Strait (evidently somewhere between Capes Yakun and Ching-An, about latitude 64°32′ north), Bering, Martin Spanberg, and Chirikov had a conversation with Chukchi Peninsula inhabitants who came out to the vessel in their skin baidara. From them they received precise information that "our land" (that is, the Chukotka Peninsula) "turns almost immediately from here to the left (in other words westward) and stretches far; it is inhabited by our Chukchi. There is no promontory extending into the sea from our land, our entire country is even."[31] Following this meeting, the *Gabriel* continued along the coast of the Chukchi Peninsula and entered the strait that today bears Bering's name.

Having reached latitude 65°31′ north, on August 13, 1728, Bering called an officers' council to work out a plan for further expedition activities. The information obtained from the Natives played a significant role in the discussion. Chirikov, in his opinion presented to Bering, stated that even though

> according to testimonies of Chukchi inhabitants the coastline trends from the Chukotka Peninsula between N and NW, but . . . inasmuch as we have no information on which degree of latitude was visited by any of the known European nations in the Northern Sea at the eastern coast of Asia, we have no way of knowing precisely if the sea separates Asia from America unless we reach the mouth of the Kolyma or the pack ice. It is known that ice is always present in the Northern Sea. It behooves us, therefore, to sail (unless hindered by the ice or the coast does not lead westward to the Kolyma River mouth), in accordance with the order of His Imperial Majesty, ever respected and of blessed memory, to the areas indicated in the said order from H.I.M. Should the land prove to trend again to N, we ought, up to the 25th day (of this month), to seek out a location suitable for wintering, best of all on the land across from the Chukotka Peninsula where, according to testimony by the Chukchi, received through Pyotr Tatarinov, timber is found.[32]

Apparently Chirikov is using information from the Natives of the Chukotka Peninsula about the Great Land to the east that is reflected in various eighteenth-century reports by the Anadyr cossacks and on the charts that were discussed earlier. In particular, such data were delivered to the Anadyr *ostrog* in 1711 by the Yakutsk nobleman Pyotr Popov and were transmitted to Chirikov by the trade manager of this *ostrog*, Tatarinov. In 1736 Müller found the testimonies of Popov and of the Chukchi in the Yakutsk archive.[33]

Taking into account the lateness of the season, Bering did not find it feasible to follow Chirikov's advice to attempt to sail along Dezhnev's route in reverse to the mouth of the Kolyma. Instead, the expedition commander, following Spanberg's opinion, sailed northward until August 16. Having reached latitude 67°18′ N while in the Chukchi Sea, he reversed course. The Russian voyagers were not able to sight the American coast because of the fog and they placed on the map only an island in the Bering Strait (August 16) named in honor of St. Diomede (Big Diomede). On September 1, 1728, the expedition returned to the Nizhne-Kamchatsk *ostrog*.

Having wintered in Kamchatka, in 1729 Bering made an attempt to sail eastward in search of the land he had heard about from the local inhabitants. Strong winds and fogs forced the expedition to return without achieving any significant results. On March 1, 1730, Bering and his officers returned to St. Petersburg. The Russian administration rated the expedition's efforts highly. As soon as August 14, 1730, Bering was granted by a royal edict the rank of captain-commodore, outside of the regular promotion order. In addition, in June 1732 he received a monetary award of 1,000 rubles.[34] On August 23, 1730, Chirikov was promoted to the rank of captain-lieutenant, also outside of the regular promotion order, and on October 16, 1732, to the rank of captain third rank. Spanberg received the rank of captain-lieutenant while the expedition was in the field on December 15, 1727. Chaplin, "for his work with the Siberian expedition," was promoted, out of order, to midshipman and, on October 23, 1730, to under-officer.[35]

Expedition accounts were delivered to the Admiralty College by Bering, and the summary map to the Senate. After preliminary examination of these documents, the conclusion was reached that Bering's expedition proved the existence of a sea passage from the Atlantic into the Pacific Ocean via the Arctic Ocean along the Siberian coast. On the basis of this conclusion, on March 16, 1730, the newspaper *Sankt-Peterburgskie vedomosti* published a brief item that stated that Bering reached 67°19′ N latitude, "and found that, indeed, the northeastern passage exists. If, in this northern land, the ice did not present an obstacle, it would be possible, out of Lena, to reach by water Kamchatka, Japan, China, and the East Indies. In addition, he learned from the local inhabitants that 50 or 60 sixty years earlier a vessel arrived at Kamchatka from the Lena."[36]

We see that this news item is not wholly, or not predominantly, based on direct hydrographic data obtained by the expedition, which did not penetrate appreciably to the west, but relied heavily on the ancient concepts of the Siberians, recorded, as discussed above, on many seventeenth-century charts, including those of Dezhnev's 1648 voyage. A reference to the latter is clearly evident in the excerpt from the news article cited here. In that same year, news about Bering's expedition was published in the Copenhagen newspaper *Nye Tidender*.[37] The article apparently was derived from the published one in the *Sankt-Peterburgskie vedomosti*. In 1740, information about Bering's voyage appeared in a book by Weber.[38]

In spite of the initial high evaluation of the results of the First Kamchatka Expedition, the leadership of the Admiralty College, as well as the majority of scholars, in time ceased to acknowledge as fact that Bering's voyage confirmed the existence of the Northeastern Passage. It seems to me that besides the many reasons widely discussed in scholarly literature in connection with changes in the leadership of the Russian Navy, a certain role may have been played by the staunch refusal of educated fleet commanders and scholars to accept geographic information provided by "uncivilized" Siberians as reliable. Bering himself, however, apparently held that the question of the separation of Asia and America had been settled. Immediately upon his return to St. Petersburg, he actively engaged in preparing the next expedition that had as its main goal the item in Peter's instruction which was, indeed, not carried out: a thrust to the coast of the American continent. The main results of the First Kamchatka Expedition are reflected in the summary chart (Fig. 4) compiled by Chaplin in late 1728 to early 1729, apparently under Chirikov's guidance. It is preserved to this day in three manuscript variants dating to the 1730s.[39]

Besides this map, a general map from Tobolsk to the Chukotka Peninsula was produced, illustrated by drawings of Siberian peoples with indications by inscriptions of their areas of settlement. This early thematic map is treated in modern literature, and quite rightly so, as an ethnographic work.

The cartographic results of the expedition appeared in printed form for the first time on the general map of the Russian empire compiled, as accurately as possible, by Ivan Kirilov, high secretary to the ruling Senate in St. Petersburg. It was published in 1733–1734 as part of his *Atlas of Russia*.[40] Outside of Russia, the summary map

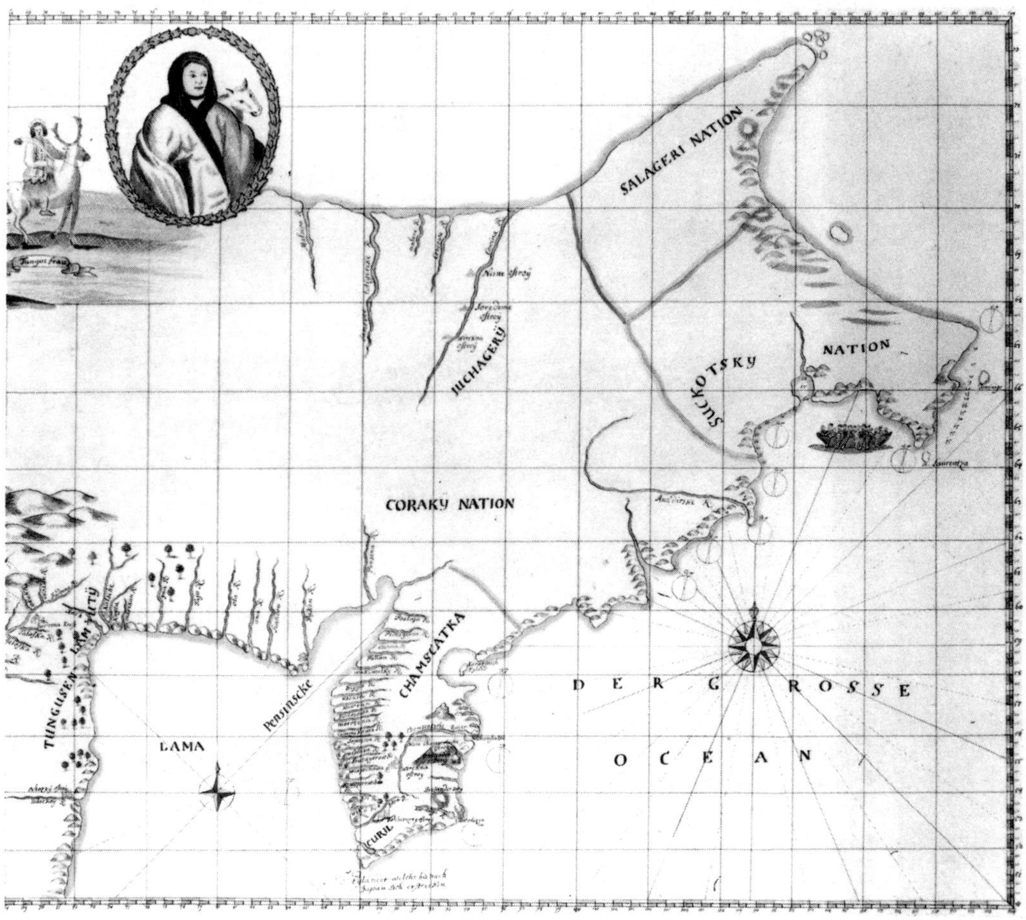

FIGURE 4 Eastern portion of a manuscript map of Bering's First Kamchatka Expedition. Courtesy of the James Ford Bell Library, University of Minnesota

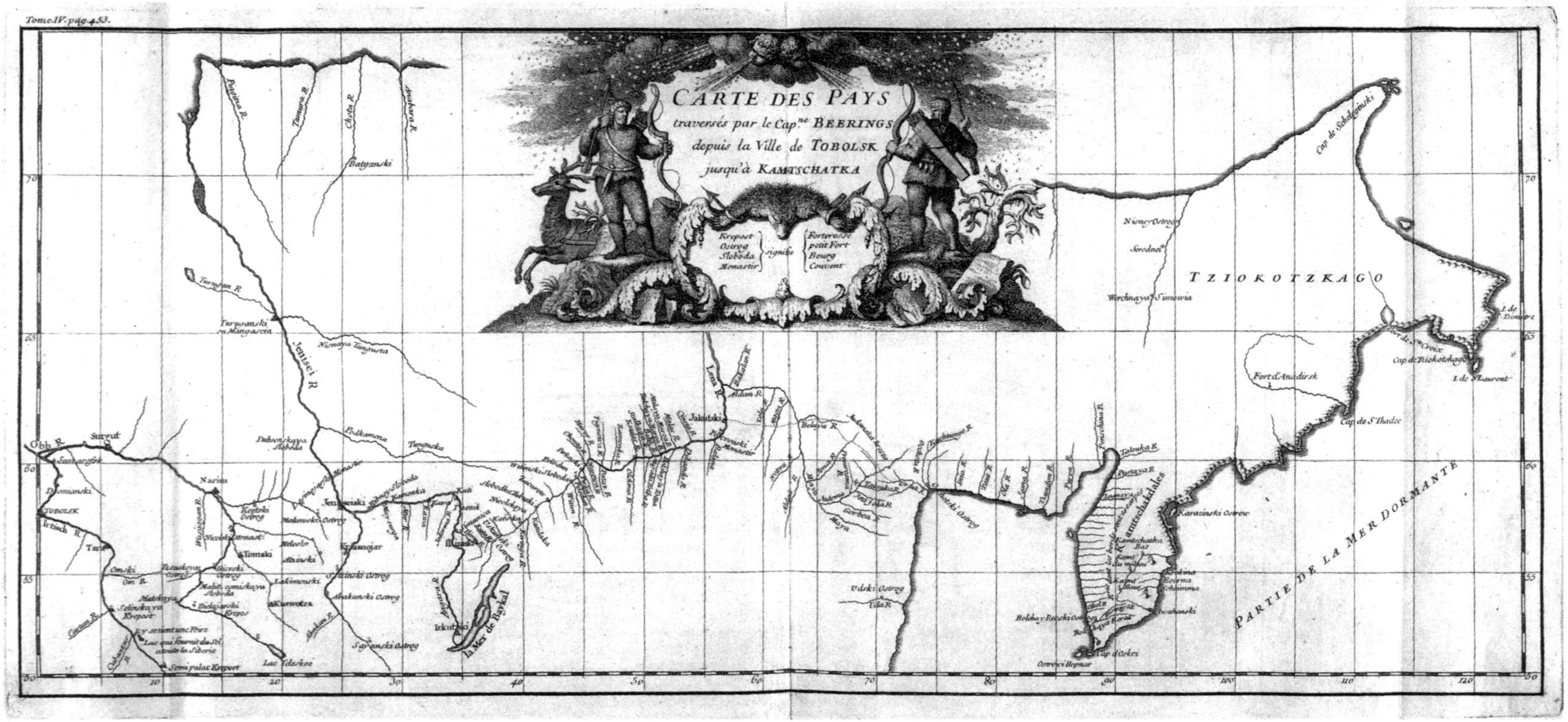

FIGURE 5 "Carte des Pays traversè par le Cap.ne Beerings depuis la Ville de Tobolsk jusqu'a Kamschatka." In Pere Jean Baptiste Du Halde, *Description . . . de l'Empire de la Chine*, vol. 4, p. 453. Paris, 1735. UAF Rare Book B0025

(Fig. 5) of the First Kamchatka Expedition was published in the multi volume work on the geography, history, and culture of China by Jean Baptiste Du Halde (1735) and in the atlas of China by d'Anville.[41] In this way, the northeastern extremity of Asia was represented on maps for the first time, drawn on the basis of geographic coordinates, to scale, and in projection.

SHESTAKOV EXPEDITION

The Great Land—that is, America—continued to engage the attention of the Siberian and Petersburg authorities. Significant progress was made by the government expedition led by the cossack commander Afanasy Fyodorovich Shestakov. This enterprise was conceived as an exploratory action, clearly of colonial character. Empress Anna Ioannovna issued a personal decree signed on March 23, 1727: "A call to the hostile foreigners who neighbor the Siberian land to become subjects of Russian sovereignty, and also to traitors who formerly were subjects. The Yakutsk cossack commander Afanasy Shestakov is to be dispatched to deliver this call, as well as to find new lands."[42] This is clear evidence of the royal drive to expand the empire's territorial possessions, to speed up the drive to the northeast, and to obtain new sources of revenue for the treasury.

Fifteen "admiralty servicemen" were assigned to Shestakov's detachment, specifically navigator Yakov Gens, sub-navigator Ivan Fyodorov, geodesist Mikhail Gvozdev, ten sailors, apprentice Speshnev, and a copyist. On the Siberian governor's initiative, a military cossack troop, commanded by Captain of the Dragoons Dmitry Pavlutsky, was ordered to join Shestakov. When, on November 28, 1728, the Shestakov-Pavlutsky detachment left Tobolsk, it mustered 591 men.[43] The Admiralty College issued to Gvozdev, for his cartographic and geodesic work, "a measuring chain, a drafting kit with instruments, a quadrant, an azimuth-compass, dividers, one for map-making and a simple one as well as a tripod, a plane scale, tables, maps, plane and Mercator ones, and one theodolite."[44] On the way, before reaching the Yudomsky Cross, on August 12, 1729, Shestakov's command met the returning First Kamchatka Expedition.[45]

Okhotsk became the main base of operations of the Shestakov-Pavlutsky detachment, both at sea and on the coasts of the Pacific Ocean, while they used Bering's *Gabriel* for the sea voyages. After Shestakov's death on March 14, 1730, from a wound received in an encounter with the Chukchi on the River Egacha, Pavlutsky assumed command. That same year, he received interesting geographic information from the Anadyr cossack Melnikov, who had marched into the Chukchi land to "call" them to become *iasak*-paying men and for "investigation of the sea islands."

While on the Chukotka Peninsula in April 1730, Melnikov had managed to "call in" two men who paid their *iasak* in red fox furs. As this Anadyr cossack reports, at that very same time, "while he, Melnikov, was present, there came to visit this very same Cape Chukchi two men from the island, those who wear teeth (labrets) of walrus ivory, and these toothed men told him, Melnikov, verbally that their settlement on the island where they live could be reached from the Great Chukchi Cape in one day's travel and, from that island to the next one, called the Great Land, the distance is also one day's travel. These islanders said further that on the Great Land there are all kinds of animals, sables, foxes, river beavers, wolverines, lynx, and wild reindeer, also every kind of tree, and there is a considerable number of reindeer-keeping and foot (riderless) people." The Eskimos gave the cossack fourteen "ore-yellow" pebbles and the Cape Chukchi five pieces of rock, blue in color.[46] The "toothed people" refused to take Melnikov to their island because of the scarcity of food and lack of dogs.[47] On the basis of this information, which had aroused Pavlutsky's interest, he reported to the Siberian governor Pleshcheev on November 26, 1730, that "presently we were dispatched out of Anadyr *ostrog* to call in and pacify . . . hostile non-*iasak*-paying Chukchi and to reconnoiter the Sea Islands and the Great Land."[48]

Because of a number of other pending matters, linked to international conflicts and participation in these by the Russian cossacks, this enterprise stretched into the year 1732, when it was entrusted to the *Gabriel.*

GVOZDEV'S VOYAGE

This historic voyage to the American shores has been the subject of research by Polonsky, Sokolov, Divin, Goldenberg,[49] and others. The most detailed treatment is by Goldenberg, who carefully researched the entire documentary corpus concerning this expedition. In particular, Goldenberg established that the official head of the expedition was Gvozdev, by an order of Pavlutsky dated February 11, 1732, because the former leader of the party, Gens, was removed from command. Pavlutsky, explaining to Gens why he was forced to take such action, wrote: "Thou, a navigator, while living in Kamchatka, does mix in matters which are of no concern to you, something which is outside of your brief, you drink and engage in other unseemly behavior, while the assignments given you are not carried out; thou livest in Kamchadal settlements in idleness, following only your vain lusts, but bringing no benefit whatsoever to the dispatched party."

Pavlutsky ordered Gens to transfer to Gvozdev's authority all noblemen, soldiers, and supplies. Sub-navigator Fyodorov was to hold command only over sailors and "was not to mix in any other matters as we know that you are at present almost blind and your legs are causing you pain, and you are unable to carry out the assignment entrusted to you." He should proceed "to sea together" with Gvozdev aboard the *Gabriel* to the "previously ordered destination."[50] Subsequently, the grievously ill Fyodorov played no significant role in the preparation and execution of the voyage.[51]

On July 23, 1732, the expedition members, aboard the *Gabriel*, sailed out of the mouth of the Bolshaya Kamchatka River (later Bolshaya) and, having four days later rounded the Kamchatka Cape, set out toward the Chukotka Peninsula. The voyage north was along the Asiatic coast, and on August 5 they "reached the Chukotka Peninsula, on the southern side." Here, about three versts off shore, they were becalmed and forced to drop anchor. The search for the islands in the Bering Strait was based on the information of one of the expedition members, an experienced skipper called Moshkov, who had sailed in this region with Bering. It is assumed that the expedition's first landfall took place somewhere between Cape Dezhnev and Cape Litke.

During the three-day stay, Gvozdev, accompanied by sailors, twice went ashore, and saw a stream, two local inhabitants, and a herd of reindeer. Two baidaras approached the vessel, but the Chukchi did not respond to efforts to enter into a conversation: "having looked at the vessel, they returned to shore." On the fourth day near the Chukotka Peninsula, with "*fair wind*," the expedition continued the search for the islands. First they searched "off the south side of the Chukotka Peninsula to the Anadyr side," but then, on August 9, following Fyodorov's advice, the search to the south was abandoned and the vessel returned to their earlier anchorage where a boat was sent ashore for fresh water. On August 11 another attempt to stand off the Chukchi shore was made, but the threat of calm forced the expedition to return to the Asian coast. Here, until August 15, Gvozdev observed the environment and local population and attempted to call in the Chukchi to pay *iasak*.

The latter effort was fruitless, as the Natives remembered the encounter with Pavlutsky's party out of Anadyr. In response to a request made by the interpreter Buslaev, one of the Natives raised his spear and said: "Who are you to ask *iasak* from us? I am big myself among my kinsmen and command them. What *iasak* do you want? We had a fight with the captain, and you meet us at the wrong time."

The expedition sailed on August 15 and two days later investigated the island that today is called Big Diomede. That same day, because of a shift in the wind, the *Gabriel* once again returned to the Asiatic shore and, apparently, stood off Cape Dezhnev near the Eskimo settlement of Nuukan (modern Naukan).[52]

Here, Gvozdev once again tried to obtain from the Natives, but without discernible success, information about the islands, population, and so on, prescribed in the expedition's program. Then the expedition sailed to Big Diomede for the second time. When Gvozdev, accompanied by sailor Petrov, and ten servicemen landed, a fire fight ensued with the local inhabitants, who were Chukchi, according to Gvozdev. The skirmish ended without damage, and it did not impede conversation between members of the expedition and the locals. The latter provided information that "our kinsmen went with the Reindeer Chukchi to fight the captain (Pavlutsky) and slaughtered all of you there."

It is clear that in this instance, members of the expedition encountered not the Chukchi but Eskimos, but Gvozdev summarily refers to all whom he encountered first on the Asian cape, on the islands, and then the American inhabitants, as Chukchi. Thus, when asked about the Great Land, the islanders "did not say about that land if it is an island or not" but, with absolute certainty, said that "our own Chukchi" live there. During the parley, they told the interpreter: "We know not any *iasak* and never paid such." From the first island visited, members of the expedition saw the Great Land. The island (Big Diomede) itself, according to their description, "is not big, and there is no timber whatsoever."

On August 20 the vessel dropped anchor at the second island, now called Little Diomede. No parley took place, as the two parties sent to investigate, Buslaev in a baidara and Permyakov in a sloop, accompanied by ten cossacks, were met by bow and arrow fire. It is interesting to note that here we find one of the earliest mentions of the use by the Russians, in pursuit of an expedition's goals, of a Native northern craft—the baidara (umiak). It, together with the baidarka, played a significant role in the exploration of the Aleutian Islands and Alaska.

According to Gvozdev's report, submitted in 1743, on August 21, 1732, "in the 3d hour after midnight, the wind rose, and we hoisted anchor, set the sails, and sailed for the Great Land, approached that land and dropped anchor, but there were no dwellings across there on land." Thus, members of Gvozdev's expedition became the first Europeans to sight the Alaska coast. To judge by the area they described, which also appears on Spanberg's map of 1743, this took place near the shore of the modern Seward Peninsula, to the northeast of the westernmost promontory of the American continent (Cape Prince of Wales), approximately 66°45′ N latitude. Afterwards, the vessel sailed along the coast southward and approached Cape Prince of Wales, where the travelers saw "at the south end toward the western side, inhabited dwellings about a verst and a half distant." According to Berg, this was the Eskimo settlement which, on modern maps, bears the name Kingegan. (Wales).

The vessel then proceeded "along the shore, on the southern side, where it became shallow, and having reached a depth of 7 and then of 6 sazhen, we turned, sailing by the wind in order to stay close to that land." To judge by this description and Spanberg's map, this occurred at the entry into Kaviak Bay (Port Clarence) at 65°15′ N latitude. Farther along, with the course southwest and a strong north wind, on August 22 the expedition approached a fourth island. "Due to strong weather, it was impossible to stand at anchor off this island; when we approached the island, we struck sails and were carried away without sails from that fourth island." It has long been established that this was King Island, or Ukivok, not far off the Alaska coast.[53] Some researchers hold that the expedition moved southward along the coast to Cape Nome and even to Norton Sound.[54]

To this day, it has not been established what Gvozdev meant by "the third island," as, in his documents, there is direct mention only of the two in the Bering Strait and the island identified as Ukivok. Polonsky, Golder, and Efimov held that members of Gvozdev's expedition considered the Great Land to be this "third island." On the other hand, Grekov, on the basis of analysis of Spanberg's map, and the report by the cossack Ivan Skurikhin (1741), advanced persuasive arguments that "the third island" meant the offshore rock Feruei, shown on Spanberg's map. Goldenberg, who wholly supports Grekov's position, also thinks it possible that the concept of the Great Land as an island, widely current at the time, may have been shared by members of Gvozdev's expedition in 1732.[55]

The expedition vessel was approached by an Eskimo in a single-hatch baidarka—a *kukhta*—which Gvozdev described: "(it is made) entirely of skin, the top is also skin, only one man can sit in it, he has a shirt sewn from whale gut which is tied to his *kukhta* and also tied at his head and hands, to prevent water from entering. When the sea wave swamps him, the water cannot enter the *kukhta*." The islander said that on the Great Land "live our Chukchi," while the trees there are "also of spruce" and there are "caribou, marten, foxes and river beavers." Gvozdev's descriptions complement the reminiscences of Skurikhin, although the locations differ. Thus, he maintained that the Native approached the vessel not from the island but from the Great Land.

Skurikhin also talks about barter with the "foreigner" conducted by the geodesist (Gvozdev), something the latter chose to be silent about in his report. In exchange for needles, thimbles, and large beads, the Eskimos' "presents were two marten parkas, three cross foxes and about ten red fox." According to Skurikhin, Gvozdev shared the gifts equally with the sub-navigator (Fyodorov), as the presents they gave the Native were from their personal stores, not from the government's.[56]

Here we have evidence of barter in furs between the Russians and an Alaska inhabitant, an exchange that was destined to play the defining role in the economics and development of the geographic investigations of Russian America. Skurikhin's testimony also gives us a glimpse of the extent of geographic knowledge of the Eskimos through Gvozdev's expedition's informant. He "showed them the direction where they ought to seek Kamchatka."

After the Eskimo returned to the island, Gvozdev and Fyodorov, "with general consent," based on a petition submitted by members of the expedition, agreed to return to Kamchatka as the season was late and supplies were scarce. On the way to Kamchatka, according to Skurikhin, the expedition halted at two islands that "were sighted on the outward bound voyage." Gvozdev and servicemen hunted there, got several caribou and 100 pud (3,395 lbs.) of walrus ivory: "fish bone, called walrus tooth." The return voyage was not easy, and during a spell of foul weather, one mast was broken. With great difficulty, on September 27 or 28, 1732, the *Gabriel* entered the mouth of the Kamchatka River and concluded the first European voyage to the shores of America from the Asian side.

Soon after the expedition's return, Fyodorov died. Five months later, on June 22, 1733, Gvozdev submitted a report and the log to the Port of Okhotsk office. These documents were delivered to Skornyakov-Pisarev (1695–1737), then Okhotsk commandant, on December 23, 1733. No chart was submitted because, according to Gvozdev, "due to the poor quality of the vessel's log, it was not possible to construct any kind of a chart." Although neither Pavlutsky nor the Okhotsk office informed the Admiralty College or the Tobolsk or Irkutsk Province offices about the results of a sea voyage to the Great Land, rumors about the expedition spread rapidly. One Kazantsev, who visited Kamchatka in 1732–1733, delivered to Tobolsk a chart and a note about the voyage of the *Gabriel* to America.

So far, the original chart has not been found.[57] However, the possibility cannot be ruled out that Kazantsev's data were the basis for information that appeared in France. The informant was Joseph Nicolas Delisle, who served at the Petersburg Academy of Science by invitation of Peter I. Golder found among the Delisle papers a note: "Navigation and discoveries by the Russians in the Eastern (Pacific) Ocean undertaken between the two voyages of captain Beering [*sic*] during the years 1731 and 1732" and also "A sketch illustrating Gwosdef's [*sic*] Voyage."[58] The note has a subheading: "New information regarding the Eastern lands furnished by Mr. Feodor Ivanitch Soimonof, March 1, 1738."[59]

It is probable that the interview given by Soimonov to a French newspaper took place after the meetings of the Admiralty College on February 12 and 14, when the accuracy of L. Petrov's information about the Gilyak land was discussed. Ober-Shterkrigskomissar Soimonov was present at these meetings and had the opportunity to become acquainted with the official report by Petrov through Admiral Gordon and also to interrogate the author of the report face to face.

Delisle presented Soimonov's account as follows:

> Captain-Commander Bering, on his return from his first voyage, met ten sailors who were sent to the Eastern Ocean. They went on board the boat left by Captain Beerings [*sic*] and, following his route, they went to Kamchatka and even beyond that, where they discovered two gulfs A, B. From there they steered east and found the island C and a large body of land D, a half day's distance from the land F. While they were near this land, there came to them a man in a small boat similar to those of Greenland. He was asked what country that was and whether there were any fur-bearing animals, but he could not give them any satisfactory information. For two days they sailed along the shore of this land D, E of this country going in a southerly direction. They attempted to make a landing, but a storm came up, forcing them back to Kamchatka.

The note continues with a description of a voyage toward the Amur by the same persons, an island across from the Amur mouth, confused information about a navigator who died and about a German who replaced him and conflicts between members of the crew that led to an investigation in St. Petersburg.[60] The data cited above clearly refer to Gvozdev's voyage, and the schematic drawing is an attempt to interpret cartographically his arrival at the coast of Alaska during the voyage of 1732, although it also reflects, to all appearances, the routes of other Shestakov-Pavlutsky 1730–1732 expeditionary detachments.

An investigation took place in 1738 in St. Petersburg regarding the accusation of crimes against the state made by the sailor Petrov against Gens, Gvozdev, and Speshnev. The accused were arrested and placed in the Tobolsk jail. The accusation proved to be false. Petrov was imprisoned in the Kronstadt fortress, where he submitted to the Kronstadt Port commander on February 9, 1738, a report about his voyage with Gvozdev and Fyodorov to the American coast and about the riches in fur-bearing animals in the lands that they passed.[61]

This report aroused the interest of the Admiralty College. On February 14, 1738, the college decided to free Gvozdev and Gens from imprisonment and demanded that the latter immediately be sent to St. Petersburg with journals, descriptions, and charts of the voyage. However, Gens died in jail on October 23, 1737, long before this decision was made. Reporting on June 15, 1738, Buturlin stated that no descriptions of the voyage, logs, or charts were found among the papers left after his

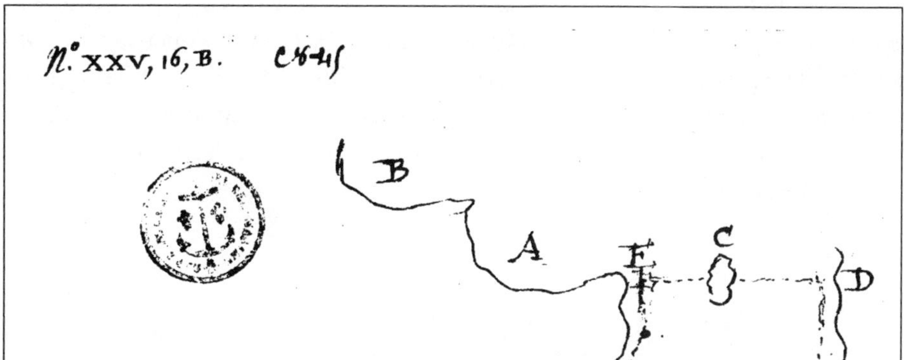

FIGURE 6 Gvozdev's Voyage. Courtesy of the Archives Nationales, Paris. Delisle Manuscripts, xxv, 16 A

death. The Admiralty College demanded that Gvozdev be sent to the capital. On December 23, 1738, an executive order was issued to the Siberian guberniya office to institute a search in Tobolsk and Yakutsk for any descriptions, journals, or charts left by Fyodorov and to dispatch them to St. Petersburg. Buturlin responded that no documents belonging to Fyodorov, who died in 1733, were in the Yakutsk office and their whereabouts were not known.[62]

Not until October 1743 did Gvozdev, together with navigators Khariton Yushin and Rodichev, compile, on Spanberg's orders, a chart of the voyage to the Great Land of 1732 on the basis of Fyodorov's "private" journal, which was located by Spanberg. The original of this chart has not been preserved, but its general outline is reflected in another chart signed by Spanberg (Fig. 6). This 1743 chart reflects the results of Gvozdev's expedition in 1732 and also, in part, of the voyages of Bering-Chirikov. Four copies of this chart have been preserved. They differ among themselves only in insignificant details.[63]

In printed form, testimony about Gvozdev's voyage to the shores of northwestern America appeared for the first time in an atlas by Homann's heirs, where, on the world map, there appears a notation "detecta a Gwosedew 1730."[64] It should be noted, however, that the interest in Gvozdev's voyage was linked, without any doubt whatsoever, with the conclusion of the Second Kamchatka Expedition, which we shall now examine.

BERING'S SECOND KAMCHATKA EXPEDITION

Preparing the Expedition

The results of the First Kamchatka Expedition did not fully satisfy the Russian government. Nor was Bering, its leader, satisfied. As early as December of 1730 he had expressed "considerations" about the desirability of dispatching a new expedition that would sail specifically "from Komchatka [*sic*] east." His proposal was based on circumstantial evidence obtained in Kamchatka "(large pine timbers of wood that does not grow in Komchatka have been washed ashore on the island called Karaginskoi," and some other similar data). Allegedly, this was evidence "that America or any land lying in between is not too distant from Komcharka [*sic*]." Bering supposed that the distance between these lands would not exceed 150 or 200 miles.[65] Already in this proposal he outlined a rather large program of geographic exploration that encompassed, in addition to the proposal to reach the American shores, "a sea voyage to the mouth of the River Amur" and a study of the rivers and northern coasts of Siberia, "as these lands are under the sovereign scepter of the Russian Empire."

The discussion of this new project assumed a scale never seen before. The Admiralty College, the Ruling Senate, and the Petersburg Academy of Sciences engaged in the in-depth planning of the expedition's goals and the preparation of instructions. The organization of the expedition was discussed in the broad context of the expansion of the borders of the empire, the growth of industry and trade, and the improvement of the navy. In this connection, Count Nikolai Golovin, vice admiral and chief inspector of the fleet, proposed, in a "Submission" to Empress Anna Ioannovna of October 12, 1732, a change in the strategy of geographic exploration by organizing circumnavigating voyages by Russian navy vessels. It was proposed to dispatch from Petersburg two frigates and a cargo vessel "which are to sail across the Great Ocean-Sea around Cape Horn into the South Sea and between the Japanese islands even to Komchatka [*sic*]."

On the basis of the experience of the Dutch sailors, who covered the distance to Japan and back in sixteen to eighteen months, it was assumed that Russian ships could complete the course to Kamchatka "in 11 months or less." In addition, it was stated that "this route is well known to any fine navigator or naval officer, and there is nothing to fear on such a voyage and in those seas except the privateers. They are not to be feared as each frigate will be equipped with at least forty cannons and a double contingent of upper- and under-officers, and will have plenty of ammunition of all kinds; thus, these frigates will be well able to defend themselves and offer proper resistance."

The Admiralty College, quite reasonably, considered that such circumnavigating voyages would offer excellent schooling for regular officers and sailors manning the vessels and provide training for those Natives who might be recruited to replaces losses among the crews. For this reason, "in Kamchatka or in Siberia several young men who are natives are to be taken onto the two frigates to replace those who might have died; they will learn, within 6 months, the sailor's craft, and when the frigates return, these people may even be converted to the Christian faith." Naturally, the greatest benefit to the state from circumnavigating voyages to America was seen in the natural resources of that continent, "as there are mines, exceedingly rich, of silver and gold, known to exist but it is not known how much gain they bring to the kingdoms of Spain, England and Portugal nor how important for those kingdoms are this commerce and sea-voyaging to these lands."[66]

This idea had to wait until much later. Instead, the alternative offered by the Ruling Senate, to initially proceed by land, was accepted. Its proponent was Ivan Kirilov, the Russian geographer, cartographer, and statesman, mentioned several times above, who was then high secretary of the senate.

In my opinion, this was not predicated on Kirilov's preference for Russia's continental development, as stated by Barratt,[67] but, much more likely, on Kirilov's urge to plan all geographic and cartographic work from proven Russian experience, existing methodology, and available specialist personnel. In this respect, his theoretical debates with Delisle about the general methodology of the creation of the astronomical and geodesic basis for Russian geographic maps are very interesting. Delisle insisted on the need to create a triangulation net. Kirilov was concerned with the need to map the immense territory of Russia within a realistic historical time frame. With knowledge of pre-Petrine traditions of Russian cartography, the backbone of which were reconnaissance compass surveys along the rivers and trails, Kirilov considered it possible to create a geodetic basis for maps by constructing a skeleton survey net of baseline moves from points whose geographic coordinates were determined astronomically.[68]

In the case of preparation for the Second Kamchatka Expedition, Kirilov sought to ensure that the rich experience acquired by the Siberians in their pioneering geographic explorations would not be lost. This required that this experience be channeled into Western European geographic science. Those who were actually to carry out the policy of expansion of Russian possessions needed to be equipped with contemporary scientific knowledge. The Ruling Senate's executive order to the Admiralty College, dated May 15, 1732, and signed by Kirilov, prescribed the following: "the training of the cossack children in navigation are at his, Bering's, discretion, as already 3 Admiralty navigators and 6 seamen have been assigned to the voyage. To these are to be assigned young cossack children to be trained in navigation in order to create (a body of) local navigators and seamen; besides, a special school is to be opened in Okhotsk to teach not only literacy, but also mathematics and navigation. Small stipends for the students' subsistence are to be given out."[68]

In the preparation of the expedition, considerable attention was paid to planning its scientific program for the collection of meteorological, astronomical, geographical, natural science, historical, and ethnographic materials. Leading scholars of the St. Petersburg Academy of Sciences participated in formulating a special instruction to the Second Kamchatka Expedition. The scholars were naturalist Johann Georg Gmelin, historian and geographer Müller, astronomer and geographer Delisle, and mechanic Daniel Bernoulli. This instruction of November 28, 1732, the first of its kind in Russia, became the basis for all subsequent instructions and guidelines for scientific expeditionary field researches undertaken in the study of regions in the northern part of the Pacific and Alaska. For this reason, we shall consider this instruction in greater detail.[70]

Geographic problems were of primary importance in the expedition's work. Mapping of land and the seas was stressed. It was emphasized that "astronomical observations require that the geodesists who will take observations for the map on land and of the coasts are to be under the supervision of the professor of astronomy." Louis Delisle de la Croyere, a new "extraordinary" member of the academy, was assigned to the expedition as an astronomer.

Joseph-Nicolas Delisle, the author of this particular section of the instruction, proposed that the best method of forming the astronomical and geodesic base for the cartography of new lands was triangulation, as he had stated earlier in his project for the creation of the general map of Russia. He instructed the geodesist to develop the survey "by establishing triangles over the entire area he wishes to represent, actually measuring only some sides of the triangles and calculating the rest by the angles and so on." But even this maximalist critic, who negatively evaluated the simplified methods proposed by Kirilov, understood that under field conditions, "accepted as permissible," establishing a line by compass or even actually measuring the distance or course by rope or chain, or even estimating them by one's own guess or according to the testimony of the inhabitants."

Delisle included one more point in the academy's, as well as the general government's, instruction to the expedition, based on his blind faith in the existence of the mythological lands of Gama, Iesso, and Company Land. The search for these lands was declared to be almost the foremost among Bering's goals. Delisle expressed his conception of these lands on a map (Fig. 7) he specially constructed for the expedition: *Carte dressee en 1731 Pour Servir a la recherche des Terres et des Mers situees au Nord de la Mer Sud* that in mid-1732 was examined and discussed in the senate.[71]

Delisle's map and his note, entitled "Information about a new land map which represents the eastern Ocean and the shortest route between Asia and America," had an important role in the history of the Second Kamchatka Expedition. These materials were a compilation of information then current in Russia and in Western Europe regarding the eastern and northeastern coasts of Asia, southwestern coasts and northern part of the North American continent, and also, located between them, the hypothetical lands. Delisle, in compiling this map, used materials from the expedition of Evreinov and Luzhin, the First Kamchatka Expedition, and early Dutch and Portuguese voyages.

The earliest information about Iesso, located to the north of Japan, was brought to Europe by the Jesuits not later than 1566. It seems that the Japanese called all lands to the north "Iesso," designating islands by this term as well as the mainland. This place name, sometimes found in differing forms such as Ezo, Esso, and Company Land, was confirmed by the Dutch voyage of H. C. Schaep and Maarten Gerritsz Fries (alternatively Vries and de Vries) aboard the vessels *Brekens* and *Castricum* in 1643. This expedition was dispatched by the Dutch East India Company, which, since 1619, had had a base on Java. The vessels visited the islands of Honshu, Hokkaido, Iturup, Urup, Kunashir, Shikotan, and Sakhalin, which Fries assumed to be part of Japan. They formed the opinion that Iturup was located near America, or was part of it, and it was given the name "Company Land."

The chart (Fig. 8) compiled by the Dutch in the course of this voyage was the cause of the false geographic concepts about this region. The mythological Da Gama Land was represented by Delisle after the Portuguese seafarer João da Gama, who allegedly discovered it in 1590. In 1630 and1649 this land was drawn on Portuguese maps by the well-known cartographer João Teixeira. However, by the early eighteenth century, cartographic opinions about these lands began to change. Thus, when the first information about Kamchatka reached Europe, some geographers concluded that the Japanese called it Iesso. This was the point of view held by Philipp Johann Strahlenberg, and is reflected on the 1725 map of Kamchatka in Homann's *Atlas* discussed above.[72]

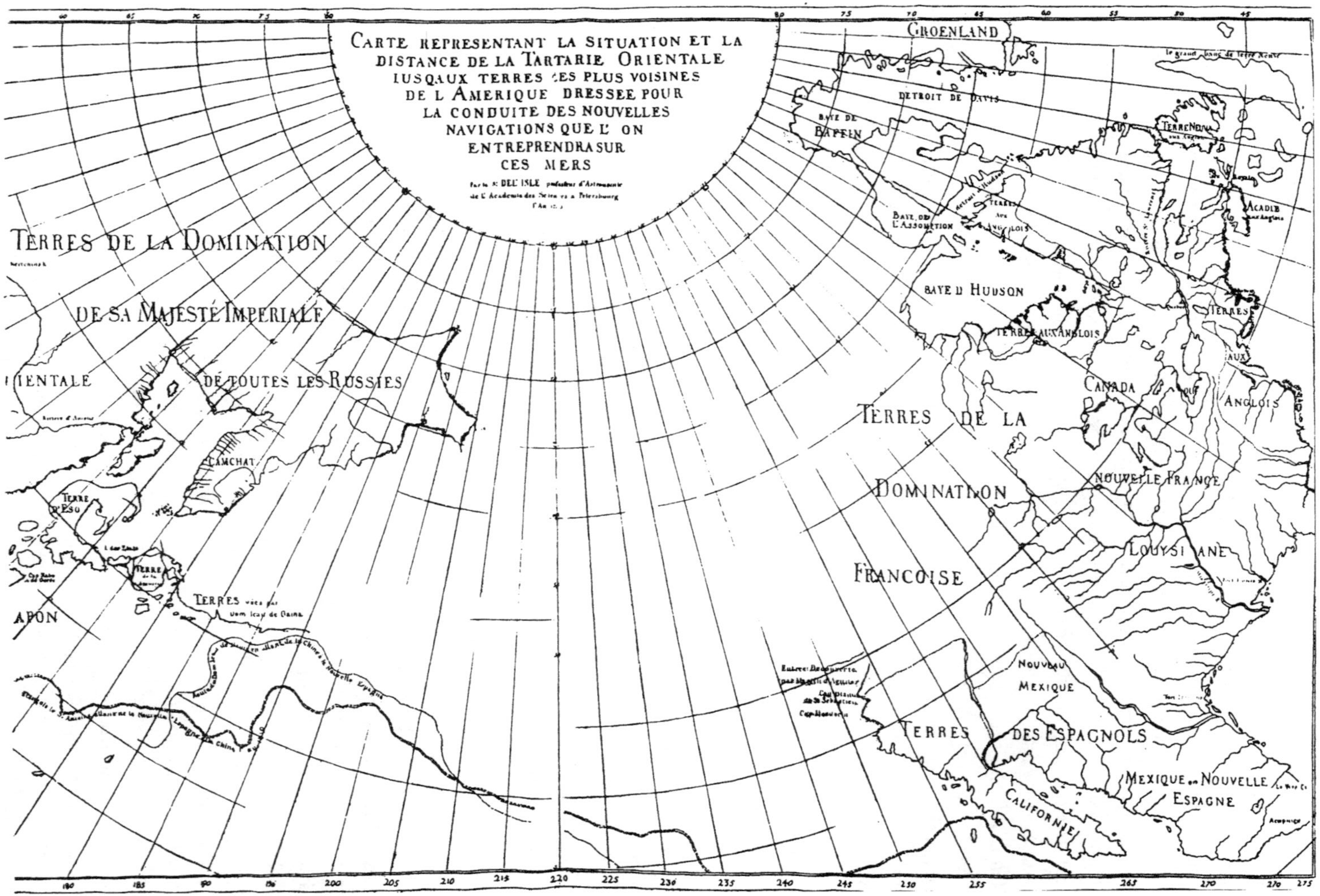

FIGURE 7 Delsile's map. BAN Kartograficheskii Fond, reproduced in Efimov, map 78

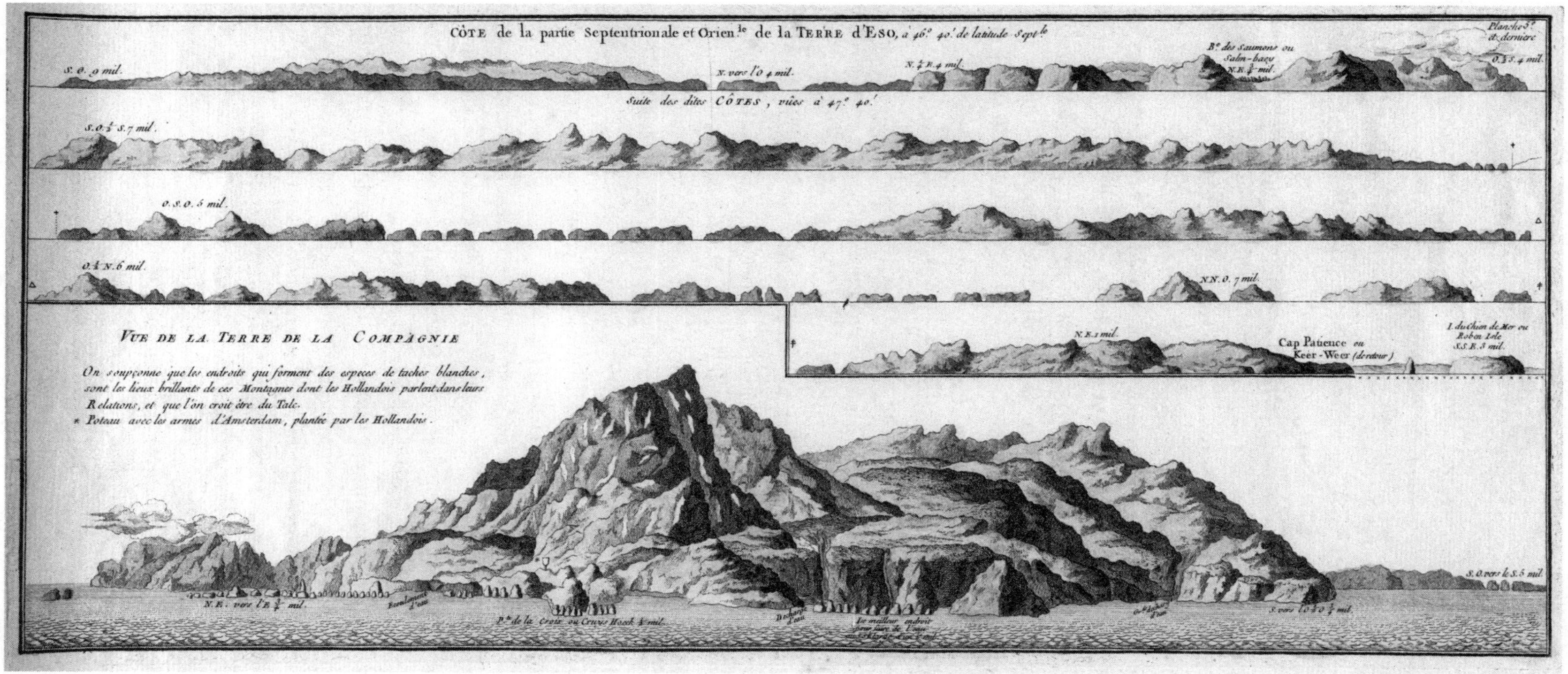

FIGURE 8 The Dutch discovery of Company Land in 1643 as rendered in Bauche, *Exposé des Découvertes au nord de la Grande Mer.* 1753 UAF Rare Book C0028

In spite of the presence of the mythological lands, in search of which Bering's expedition wasted precious time, the map was useful because, for the first time, distances approached realistic scale, and possible sea routes between Asia and America were suggested. It also indicated the relationship of the then-known coastal sections of northern America and northeastern Asia.[73]

Returning to the discussion of the academy's instruction, it should be noted that considerable attention was devoted to the observation of natural phenomena and of the peoples inhabiting the territories Bering's expedition was to visit. Specifically, a set of meteorological observations was to be conducted: "in every locality, temperature was to be carefully measured at the preset hours of the day and recorded, indicating day and month, in order to establish the mean temperature for each month. Also, the wind, direction and strength are to be noted with reference to the changes in the terrain. When all these data are compiled I think that the reason may be found why in these areas the eastern winds are cold, and the western ones warm."[74] An interesting measure to judge the "benefit of climate and air" was proposed—to establish population numbers resident in a locality by age categories: under ten years of age, between ten and twenty, and so on.

As far as oceanographic observations were concerned, it was deemed "necessary everywhere to note the color of water, its salinity, depth, transparency or brightness . . . the height of waves and their duration, but most of all the flood and ebb tides corresponding to 10 lunar phases or faces and all the particulars of the shore localities are to be recorded. As well, the elevations of the barometer . . . are to be accurately recorded in order to establish the mean height and to see if it is the same as in other seas. It is not in violation of the principles of hydrostatics for different seas to have different pressure even when they are connected if one assumes that the waters flow from the highest to the lowest (barometric pressure) constantly. The currents are also to be diligently observed."[75]

What is striking is the depth and detail set forth in the instruction with regard to descriptions of relief, soils, geology, minerals, vegetation, and fauna. Thus, speaking of the mountain relief, it was prescribed that the distances of ranges and separate peaks from the coast be indicated, the elevations or relative heights, and the steepness, breakup, and orientation of the main ranges and their spurs were to be described: "are they ascending gradually and gently, are the mountains high or low, are they distant from each other, or there is no mountain, do the large mountains trend from west to the east, or from east to south, are both sides protruding, which protrudes to the north, which to the south, and does the smaller one face east or west?" The river valleys were to be characterized by their form, and proclivities between the mountains and their orientation and relationship to surrounding elevations were to be noted.

As far as Kamchatka was concerned, specific attention was to be given to volcanic phenomena and their periodicity was to be described: "are there any signs or distinctive phenomena which occur beforehand . . . what kind of materials are extruded from the burning liquid, to what distances . . . when water is extruded sometimes from these mountains, is it salty or fresh, or sour, foul, or bitter, what are its advantages and disadvantages? Is there an increase in the depth of the water, or a decrease, at sea and in the rivers at such times, are there great storms at such times at sea, and for how long is the fire emitted?" Earthquakes were to be described, their periodicity, the linkage if any to volcanic phenomena, the presence of thermal waters and their temperature, and of mineral water and its characteristics and density.

In geology, first of all, it was prescribed that the extent of the strata be noted as well as any presence of fossils. In Siberia it was prescribed to "search for mammoth bones" and append a description of "in what strata the earth is arranged where the bones are," their orientation, and the characteristics of the orientation of these strata were to be noted. In order to establish the peculiarities and distribution of sedimentary rocks, test pits were to be drilled either with an auger or by mining gimlet. Caves were to be located and carefully described.

The description of the animal world and vegetation was to be specifically focused on hitherto unknown species. The most interesting samples of flora, fauna, "copper, gold, silver, iron ores," and so on were to be collected and delivered to the academy upon the expedition's return, with a precise indication of the locations where they were collected.

It should be noted that this instruction is one of the earliest examples of instructions for collecting native information about the area: "so that traditions or rules current among the local peoples about this or that natural occurrence or this or that phenomenon are noted, as this item is part of our study, first of all local inhabitants' knowledge or information is always to be elicited in detail, about their fiery mountains, the shaking of the earth, their understanding about the mountains and of the Kamchatka iron water (springs) and its uses, about diseases that are frequent there and the medicines used against these sicknesses, as well as other matters spontaneously brought up, their causes, are diligently and accurately to be noted."

The section that today would be called ethnography consisted of ten points. It asked: "what is the beginning or origin of the people, their traditions about ancient peoples, their settlements and migrations . . . Their faith, customs and habitual behavior, unchristian or Islamic acts, domestic, marriage and wedding (customs) and the rest." The political organization of the local peoples and their military and trade alliances were to be studied, and the "fruits of the earth and income, worldly sciences, war units and battles" were to be described. Population numbers by male and female genders were to be noted, and their clothing (the latter illustrated by drawings), languages, writing systems, buildings, and "old and new vessels, earthenware or clay, idols or figures," the character and history of settlements, "all and every kind of rocks, any stone ruins or manors, ancient burials or cemeteries."

Following the established practice of the Siberian pioneers, "the names of each country or province, of rivers, cities, and others, natural ones to each dialect, are to be accurately written down, including the etymology, that is, the right one as appears in the people's own language, should it be available."[76]

Concluding the examination of the academy's instruction to the Second Kamchatka Expedition, it ought to be noted that it was not practicable to carry it out in regard to the American coasts because of the extremely limited time at the disposal of Bering and Chirikov, though the expedition's naturalist, Georg Wilhelm Steller (1709–1746) did collect materials that enriched European science of the time. The expedition's Siberian detachments were able to implement the tasks set out in the instruction. Probably the most important scientific materials in the history and ethnography of the peoples of Siberia were collected by Müller, who personally carried out those sections of the instruction that he himself had written.

In my [AP] opinion the academy's instruction occupies a significant place in the history of Russian science, a place still not fully appreciated. It is the first in a series of eighteenth-century instructions for studies of the environment and society in various regions of Russia by means of a standard questionnaire. The best known are the 1738 questionnaire by Tatishchev containing 192 points, the "geographical queries" by Mikhail Lomonosov containing thirty questions (1761), and the "Questions demanded for the St. Petersburg Army Cadet Corps, regarding geographic information from all cities and provinces of the Russian State needed for preparation of the geographic description of the same." The latter was composed by Müller who, upon his return from Siberia, became the academic dean of the St. Petersburg Army Cadet Corps.[77]

Let us now return to Bering's expedition. On December 28, 1732, the Ruling Senate issued a general decree pertaining to the organization and tasks of the Second Kamchatka Expedition. This document summed up and confirmed the recommendations of all the departments that participated in the development of the expedition's program. This decree prescribed the study and cartography of the coasts of the Arctic and Pacific oceans, a final determination of the absence of an isthmus to America, a search for islands, the thrust to the American coasts to the east and southeast from Kamchatka, a "search for the route to Japan," and other tasks set out in special instructions. The need for strict confidentiality was stressed, both in carrying out the expedition tasks and in the utilization of the obtained materials: "until it is ordered to go to press, nothing is to be published, either secretly or openly, nor announced or written up, to prevent the knowledge being learned in foreign parts before it is known here."[78]

On the basis of the Senate order, the Admiralty College issued to Bering, on February 28, 1733, a general instruction regarding the outfitting of the Second Kamchatka Expedition, construction of vessels in Okhotsk and Kamchatka, an investigation of the American coast, and the search for the route to Japan. The instruction specified that two vessels were to be used:

> on one, Captain-Commodore Bering, on the second, Captain Chirikov, are to proceed together in the search for American coasts and islands with utmost effort; you, Mr. Captain-Commodore, are to act in general agreement with Captain Chirikov in navigational sciences and to include in the council the professor dispatched from the Academy of Sciences. When you find, as is expected, the American coasts, you are to visit them and to establish exactly which people inhabit them, what that locality is called, and to ascertain that, indeed, this is the American coast. Having done so and having ascertained the facts, enter them all on the chart and then proceed for the same reconnaissance along the coast as long as, in your judgment, time and the situation permit so as to return under local climatic conditions to Kamchatka . . . Also, during the course of the entire voyage a journal is to be kept in accordance with the navy regulations, bearings are to be taken whenever possible, and the coasts, islands, and fairways are to be recorded in accordance with navigational sciences so that, on that basis, a detailed and true chart may be compiled.

The necessity for confidentiality in carrying out the expedition's work was reaffirmed and even "this instruction is to be kept secret, and for public use another special instruction is being issued to you."[79]

Bering and Chirikov Sail

The Pacific Ocean squadron of the Second Kamchatka Expedition consisted of two detachments. The first, under the command of Bering,[80] was to proceed to the American coast; the second, commanded by Spanberg, was to sail along the Kurile Islands to the coasts of Japan and also to describe the shores of the Okhotsk Sea up to the mouth of the Amur River. For the American voyage, two vessels were constructed in Okhotsk, the *St. Peter* and the *St. Paul*. On September 8, 1740, the two vessels sailed for Kamchatka, for Avacha Bay, where depth soundings had been taken beforehand and quarters for the expedition had been constructed. The port that was established in Avacha Bay was named, in honor of the expedition vessels, Petropavlovsk. Plans of Avacha Bay and the Petropavlovsk Harbor, compiled by members of Bering's detachment, have been preserved.[81]

On May 4, 1741, a council of officers of the *St. Peter* and the *St. Paul* was held where, in accordance with the Senate decree and the Admiralty College instruction, it was decided

> in accordance with the sections pertaining to the search for the American coast, the course out of this harbor will be set first south by east to east, on the right compass; this compass heading is to be adhered to until 46° North latitude, then from that latitude on east by north, to be held until land is sighted. As soon as the land is sighted, on the course either southeast by east or east

> by north, the coast is to be followed if it should trend from east to north or from north to west. If the land should trend between south and east, it is to be left behind and the course set for east until land is reached. Then we are to sail along this land northward, as stated above, up to the 65° latitude. If, with God's aid, and time permitting, we reach safely the 65°, we are to sail westward until the Chukotka land is sighted in order to establish the distance between America and the Chukotka land and thence sail to this harbor.[82]

In accordance with this decision, the vessels left the harbor on June 4, 1741, and sailed toward the João da Gama Land shown on Delisle's map. The expedition members aboard the *St. Peter* under Bering numbered seventy-seven, and those aboard the *St. Paul* under Chirikov numbered seventy-five.

On June 20, in thick fog, the vessels lost sight of each other and, from then on, proceeded on the voyage separately. Chirikov, on the *St. Paul*, held his course to east by northeast, and on July 15, 1741, sighted land. In his journals Chirikov recorded the event thus: "At 2 o'clock after midnight we saw ahead land, with high mountains, but it was not very light yet, and we hove to; by 3 o'clock the land became more clearly visible."[83] The vessel was in the vicinity of modern-day Bartolome Cape on Baker Island, to the west of Prince of Wales island, at 55°20′ latitude. On July 17 Chirikov stood off Yakoby Island (near Chichagof Island in the Alexander Archipelago) and sent his longboat ashore under navigator Avram Dementiev with ten crew (three sailors, five soldiers, and two interpreters). They waited for the party to return until July 23, when the yawl, under boatswain Saveliev, carrying one sailor and two craftsmen, was dispatched to shore, but they too vanished without a trace. It is supposed that the Russian sailors perished as they approached the shore, unable to cope with the unfamiliar strong tidal currents, so common on these coasts, or were killed or imprisoned by the local Indians. Today the prevalent view is that the men vanished at the entry into the Lisyansky Strait.

Having lost both of his boats and thus the ability to come close to shore, Chirikov called an officers' council on July 26. At it "all agreed that due to the misfortune which has befallen us, to whit the loss of the yawl and the small boat under Fleet Master Dementiev with accompanying servicemen, the voyage is not to be continued, and on this day the return to Avacha commenced."[84] For a long time, Chirikov could not bring himself to leave the spot where the members of his crew disappeared, and the *St. Paul* cruised along the American coast until the provisions were almost exhausted. On August 6 Chirikov was forced finally to set a course for Kamchatka. Homeward bound, the vessel sailed along the Aleutian Islands and took bearings on Unimak, Unalaska, and a few other islands.

On October 12, 1741, the *St. Paul* entered Petropavlovsk Harbor. Out of seventy-five crew, only fifty-one returned, of whom only Chirikov and Ivan Yelagin, navigator (d. 1766), were officers.[85]

Bering, aboard the *St. Peter*, having parted with the *St. Paul* in the fog, continued the search for the João da Gama Land until June 25, but having become convinced that this land did not exist, set a course to the northeast and on July 16, in the latitude of 58°14′ north, the Russian seafarers "sighted land, with high ridges and over them a tall peak . . . and these ridges and the peak are covered with snow." This was the American coast in the area of Kayak Island (Fig. 9). The expedition conducted a hydrographic survey of the island, which Bering named St. Elias, and of the adjacent mainland coast.[86] Steller, accompanied by the cossack Fyodor Lepikhin, went on a yawl to acquire drinking water and to investigate the new land.

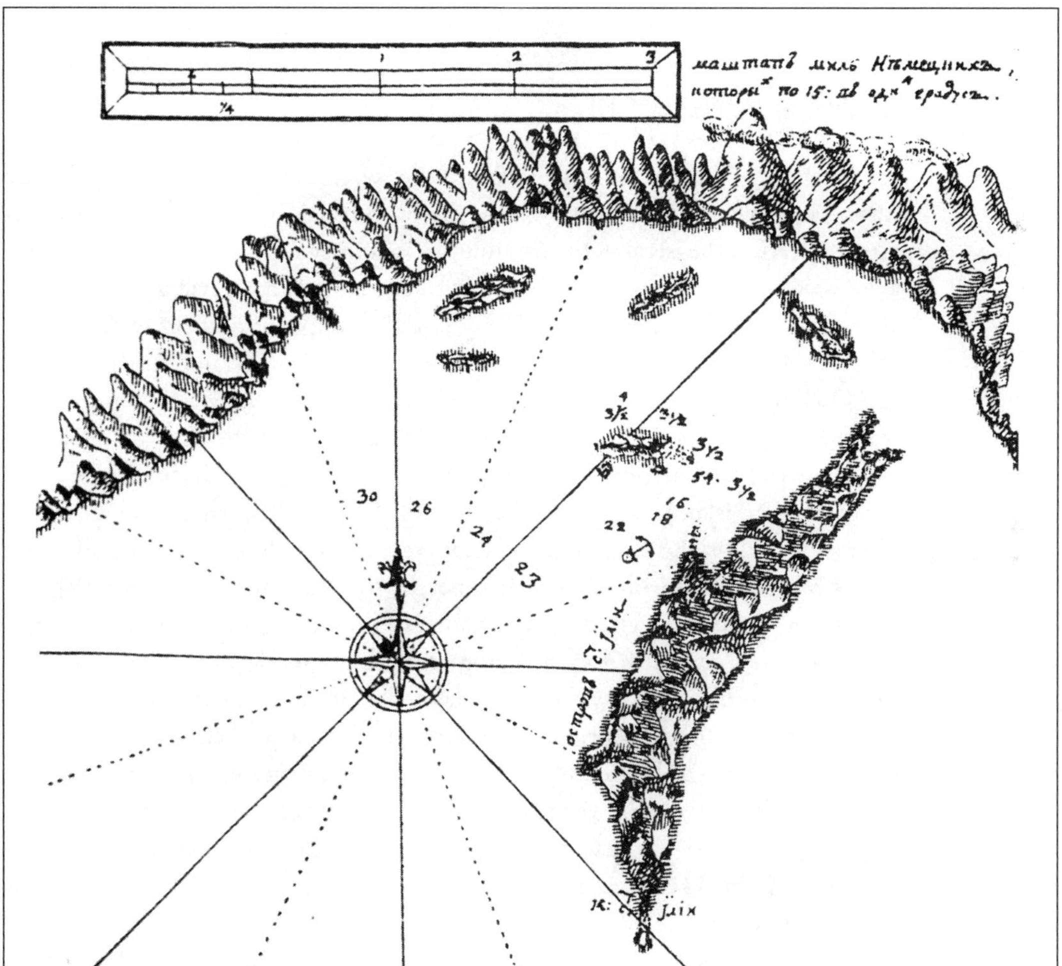

FIGURE 9 Manuscript map of Kayak Island, 1741. Reproduced in *Russkie ekspeditsii . . . XVIII v.* 1984:220

Navigator Sofron Khitrovo, with fifteen crew, was sent beyond the island "to measure the fairway, as it is evident that there lies a bay which ought to be investigated as, in an emergency, it will provide shelter from high winds."

In the log of the *St. Peter* the results of this first scientific field excursion on the northwest coast of North America is noted in a rather humdrum and laconic style:

> At 11 o'clock after midnight, the yawl returned with water and reported that they saw fire places and human traces and live foxes running around on shore, and Adjunct Steller brought various grasses. Moreover, they found an earth dwelling resembling an underground storage, but saw no people; inside there was smoked fish, a wooden fire drill, and broken arrows. On the order of Captain-Commodore Bering, gifts from the government stores were sent to that dwelling, 16 1/2 arshin of green dyed cloth, two knives, 20 large beads, and two Chinese iron pipes.

When by evening Khitrovo returned, he stated that beyond the island there was a navigable channel, "its bottom is sandy, in places silt and sheltered from all directions." On the island Khitrovo found a "summer shelter made of planks with a floor also of wooden planks, covered with earth; from that shelter they brought as evidence a basket, a wooden spade, and a rock on which copper is pulverized, and another rock with a slit."[87]

On July 21, 1741, the expedition started on its homebound voyage along the coast of the Alaska Peninsula and the Aleutian Islands. The *St. Peter* closely approached the islands of Kodiak and Ukamok (modern-day Chirikof). In September they approached an island, apparently the modern-day Nagai Island, to take on water. They called this island Shumagin in memory of the sailor Shumagin, who died and was buried there. This island and those nearby that are now called collectively the Shumagin Islands were charted by members of the Bering Expedition.[88] On these islands the first contact between Bering's crew and American Natives took place. The latter approached the vessels in kayaks. On Bering's orders, they were given "4 1/2 arshin of red Chinese cloth, 2 mirrors, 3 Chinese pocket books, beads, and small bells . . . an iron kettle and several needles." In return the Natives passed from their baidarkas "2 shaved (whittled) sticks, one of which had feathers tied to one end, these feathers were determined to have been falcon ones, while to the other end was tied a bird's claw, and two hats, made of wooden bark, to one of which was tied a small (figurine of a) man made of bone."[89]

The final stage of the return voyage of the *St. Peter* took place under extraordinarily difficult conditions. Severe storms gave way to cold, thick fogs, fresh water and provisions were exhausted, and scurvy was rampant aboard the vessel, carrying off many lives. In the end, on November 5, 1741, the vessel was cast ashore on an island that subsequently, in memory of the captain-commodore, was called Bering Island.

As Sven Waxell (1701–1762), master of the *St. Peter*, reported to the Admiralty College, "Toward December 8 of the past year of 1741, Captain-Commodore Bering died, by God's will, from scurvy, from which he suffered about four months without relief; he was buried on that island where we and the crew wintered."[90] Those of the *St. Peter*'s crew who had wintered there and remained alive dismantled the vessel and constructed a new one out of the wreckage, aboard which the surviving forty-six members of the Bering Expedition returned to Petropavlovsk on August 26, 1742.

One of the main tasks of the Second Kamchatka Expedition was the one entrusted to Spanberg to reach the coasts of Japan and China. This detachment sailed in 1738 aboard the brigantine *Archangel St. Michael* out of Bolsheretsk, along the Kurile Islands to Urup and back. In 1739 Spanberg and William Walton, on the *Archangel St. Michael* and the *St. Gabriel*, completed the voyage out of Bolsheretsk to the shores of Japan. In the course of this voyage the Russians visited Hokkaido Island for the first time. Between 1738 and 1742, Spanberg's detachments charted the Kurile Islands, collected the first data on Japan, and investigated part of the eastern shore of Sakhalin Island and the Okhotsk Sea littoral. Though the results of this work had no direct relation to the approach to the American coasts from Asia, they were instrumental in establishing true outlines of the land and sea expanse between these two continents. Specifically, Spanberg's expedition confirmed Chirikov's conviction "that the Ian de Gamma [*sic*] land does not exist as the area where it is supposed to be we crisscrossed in its entirety." It also established that the "Company Land" or "Uruppu" was just one of the Kurile Islands, Iturup.[91]

The Heritage of Bering's Voyage

Besides the widely known and thoroughly studied geographic and cartographic results, the Second Kamchatka Expedition laid both the methodological and the practical foundation for further scientific generalizations of materials on the waters and lands lying between Asia and America in the northern part of the Pacific Ocean. As rightly noted by such leading students of the problem as Belov, S. G. Fyodorova, and Goldenberg, there emerged within Russian cartography several traditions of representation of these areas.[92]

There are two basic trends. The first is represented by the descriptions and cartographic works that to a considerable degree continued the pre-Petrine traditions of Russian geographic materials (charts/sketches and various descriptions—*rospisi*). Usually, these had a nonmathematical foundation or a very weak one (scale, projection, and geographic net), but at the same time, they fixed geographic objects by names derived from local toponyms—Chukchi, Eskimo, and, later, Aleut ones.

The second trend, having its principal beginnings in the First Kamchatka Expedition, acquired significant force after the Second Kamchatka Expedition. We noted that the instructions issued to the Second Kamchatka Expedition reflected the desire to utilize Native data in greater depth, particularly in recording Native toponyms. On the other hand, the administrative departments that supervised the geographic investigation and cartography of Russia made the first step toward the establishment of centers whose task was to train qualified cadres of seamen and geodesists stationed in remote borderlands. This permitted the Russian pioneers and mariners to obtain at least the minimum of knowledge necessary to compile charts according to European geographic tradition. Thus, in accordance with the Senate's decree signed by Kirilov, Bering opened a school of navigation in Okhotsk. This school later played an important role in the training of Russian seamen for the Pacific. Devier, the successor to Skornyakov-Pisarev as commandant of the Okhotsk Port, opened a navigational school at Bolsheretsk in Kamchatka. This school existed until Devier's return to St. Petersburg in 1742.[93]

The basic work of assembling the main expedition materials, their analysis, and summation, as well as compiling the general and summary charts, was performed by the Admiralty College and its associated scientific and educational center, the Naval Academy. This kind of work focused initially on the Second Kamchatka Expedition, and the subsequent government undertakings of this kind were carried on in secret. The results long remained unknown even to the Academy of Sciences and, of course, to the ordinary *promyshlenniki*, who, in the wake of Bering's crew's return with valuable furs obtained during the winter on Bering Island, commenced the procurement of "soft goods" in the Komandorskie (Commander) and Aleutian Islands, and by the 1770s they reached the Alaska Peninsula.

The Imperial Admiralty College entrusted the task of compiling summary charts based on materials of the Second Kamchatka Expedition to the Russian hydrographer Alexei Nagaev (d. 1781). Having received the expedition's journals, logs, descriptions and charts, the Admiralty College, in the words of the historian of Russian geography Vasily Berkh, "directed A. I. Nagaev and Lieutenant Afrosimov to compile detailed charts of the Kamchatka Sea and the American coast and also to compile a chart on the basis of notes by Captain Spanberg and Midshipman Shelga [*sic*, Shelting]. Nagaev was engaged in this task for the entire year of 1745."[94] However, the initial cartographic interpretation of the results of the Second Kamchatka Expedition was performed, quite naturally, by its main participants, foremost among them Chirikov and Waxell. We shall now examine these materials.

After the *St. Paul*'s return in 1741 from the voyage to the American coast, Yelagin, the expedition's navigator, compiled, with Chirikov's guidance, a summary chart of the voyage under the title "*New Chart compiled on the basis of the log kept aboard the vessel Apostle St. Paul under the command of Captain Alexei Chirikov during the voyage in 1741 from the Harbor of the Holy Apostles Peter and Paul in the land of Kamchatka to search out America's western coasts; outward bound routes are marked in black ink, homeward bound in red color. The location at the end of each day is marked by the letters O and O**, *O signifies the location as calculated, O* the location corrected to the observed latitude. The mouth of the River Moozemmlek, which* *flows out of America, and the land in its vicinity have been placed on this chart from the published map by the Nuremberg geographer Ioann Baptista Homman* [*sic*] *and at the edge of these locations marked by letter <H> were added the northern region of California, part of the interior eastern shore, the Hudson Bay Company Fort and the upper reaches of several rivers of North America from the map compiled in years past by the professor of astronomy Ludvig* [*sic*] *Delisle de la Croyere; the borders of these areas are marked by the letter L; the rest of the coasts shown on this chart are those located in the course of the above voyage; as the distance between Kamchatka and the land previously found by us (which we sighted and described on the 15th day) calculated on the outward bound voyage differed from the one calculated homeward bound, it is shown on the chart twice to explain where it is located by both reckonings. This chart was composed, on the order of Captain of the Fleet Chirikov, (commander) of the above-mentioned vessel, by Navigator Ivan Yelagin.*" The original is signed by Captain of the Fleet Alexei Chirikov.[95]

The Chirikov-Yelagin chart is outstanding in its strict objectivity in representation of the voyage of the *St. Paul* and the hydrographic work by the expedition. Only those elements are shown that have been directly observed by expedition members. Absent are elements guessed at (coastlines, islands), not to speak of mythological Da Gama Land, Iesso, and others. The Second Kamchatka Expedition filled in the places that these imaginary places had occupied so that no place was left for them on the maps of the Pacific Ocean. On the other hand, existing features are shown, such as Hudson Bay, the Mississippi River, and to the south, "Newly found part of American coast lying toward California."

The authors of this chart represented these features based on observed latitudes and longitudes of the stretches of the New World coast approached by the *St. Paul*. The mainland character of this coast was evident from many signs indicating a large landmass as reported by Chirikov in his account to the Admiralty College:

> Near the land along which we sailed and observed for about 400 versts, we saw whales, sea lions, walruses, dolphins, birds, white-bellied ducks and a multitude of another kind which have crooked red beaks, and many seagulls of different species. On land everywhere are high mountains, the seaward shores are steep, and the offshore depth is great; on the mountains in the vicinity where we approached land, as stated above, tall timber is plentiful; snow was seen occasionally on the mountains. The farther north we sailed, the more plentiful became the snow on the mountains. At the point where we turned homeward we saw to N E 3/40 on the coast very high mountains, greater in height than those on Kamchatka, all covered with snow.[96]

The Chirikov-Yelagin chart shows only four islands between America and Kamchatka, and even a hint that the American continent may stretch in that direction is absent. The authors approached the presentation of the results of the hydrographic description composed aboard the *St. Paul* from the point of view of marine cartography quite correctly. This is evident from the use of different signs to indicate points of orientation whose coordinates were calculated (very approximately) and those coordinates established through the determination of latitude by sextant and the observation of the midday altitude of the sun. The discrepancy between indications of the American coast as established when outward and homeward bound was preserved. This offered the possibility of a later analysis of the geographic results in conjunction with the use of other materials (including from Bering's voyage), to which the authors had no access when compiling the chart.

Chirikov and Yelagin were also apparently involved in the compilation in 1742 of a chart that is the first attempt to generalize the results of the Second Kamchatka Expedition and to compare the latter with other data, including the mythological ones. This chart is preserved in the archive of the navy but has been little utilized by researchers until the present time.[97] The inscription within the cartouche is clear evidence of the analytic character of this cartographic work based on materials by Chirikov, Spanberg, and the Siberian detachments of the Second Kamchatka Expedition. The inscription indicates specifically that the coasts of the Arctic Ocean and the mouths of the rivers that enter it, "beginning from Arkhangelsk city eastward to the River Kovyma were entered from new descriptions by naval officers, and from the River Kovyma, Chukotskoi and Shelaginskoi lands to the Land of Kamchatka from previous maps."

With respect to the voyage to the American shores, information clearly is derived from the Chirikov-Yelagin chart, as is evident from the inscription: "From Kamchatka land Avacha Bay to the western coast of North America, discovered by him, the routes are laid down from the description by Captain Chirikov, the outward bound marked in black ink, homeward bound in red color. The coast is indicated on this chart in two locations, as there is a discrepancy between outward and homebound calculations. Some locations of this America are shown marked with the letter L, and the rest of the coast was newly found along these routes."

On this chart, in contrast to other Chirikov-Yelagin charts, the lands of Esso and Da Gamma are indicated by a faint dotted line against a pale pink background with the following explanation:

> To the southeast of Kamchatka is shown the land which was sighted en route from Khina (China) to New Spain by Don Ioann de Gamma and was so named, but the captain mentioned here [Chirikov—AP] sailing to America did not find any land where it was supposed to be . . . The routes from Okhotsk along the Penzinskoe Sea to the Bolsheretskoe mouth are shown after the description by Captain Spanberg, who sailed from that (river) mouth to Japan and back by this route where, according to old maps, the land of Iesso ought to have been, but he sailed in the open sea between islands he had discovered.

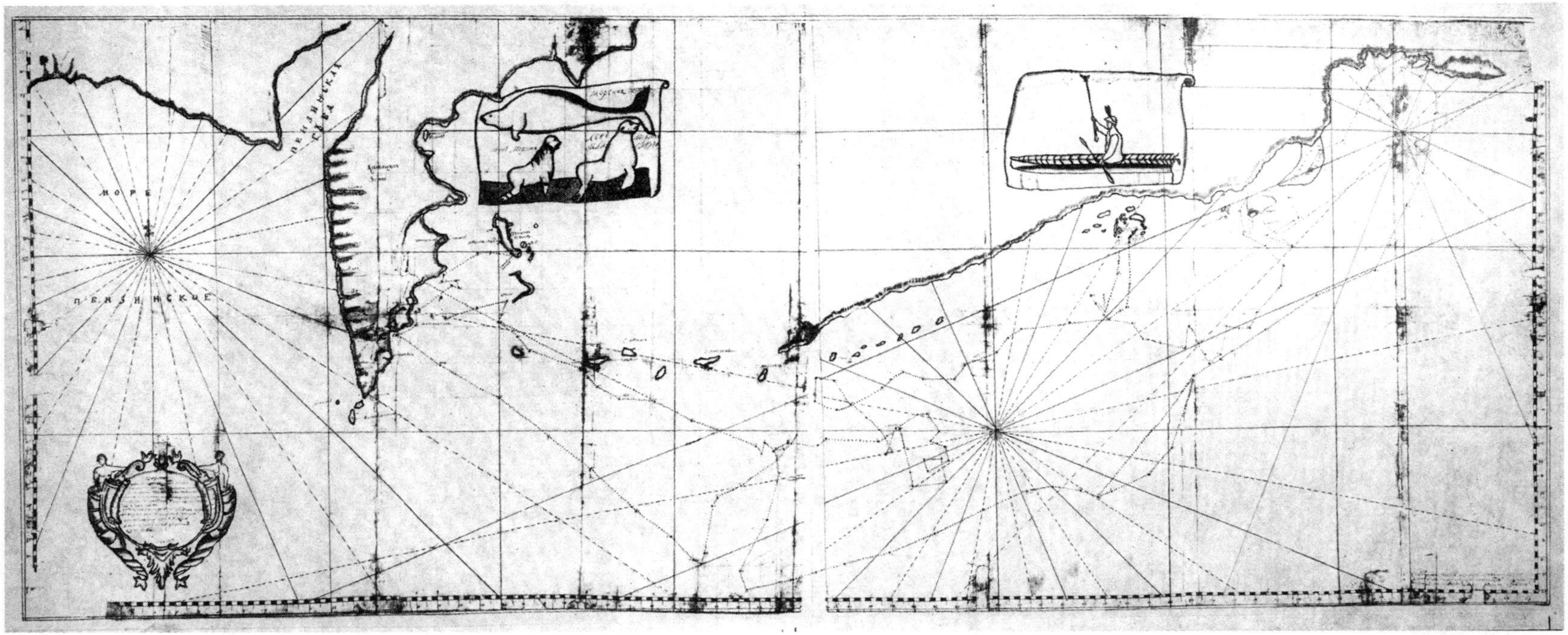

FIGURE 10 Bering Expedition map drawn by Sofron Khitrovo who was on both Bering expeditions. RGAVMF. F. 133. Op. 4. No. 99

It is obvious that one of the aims of this chart, which was compiled according to the inscription "in the year 1742 in November, 19th day, from various charts," was to demonstrate the nonexistence of the mythological islands in the Pacific.

An essentially different interpretation from the one on the original Chirikov-Yelagin charts of the land and seas between Kamchatka and America appears on the chart (Fig. 10) by Waxell and Khitrov of 1744. This is the "Chart of sighted American Land and islands newly found under the command of the late Mr. Captain-Commodore Bering compiled from the journal of Lieutenant Waxell during the campaign of the years 741 and 742 [*sic*]. Original, by the hands of Lieutenant Sven Waxell and Fleet Master Sofron Khitrov. In the year 1744."[98]

All versions of Waxell's chart, in contrast to those of Chirikov-Yelagin, show a section of the American coast stretching far to the west from "Cape St. Elias" to the island "Markyan" toward Kamchatka. This interpretation reflects the opinions of members of Bering's detachment who, because of poor visibility, assumed the chain of the Aleutian Islands to be parts of the American shore. It is possible that the genesis of this opinion was influenced by information acquired in Kamchatka about the closeness of the Great Land. However, the influence of the mythological Da Gama Land must be considered.

Some versions of the chart representing the summary results of Bering's detachment's voyage show, besides the routes of the *St. Peter* and geographic contours, illustrations of sea cows, fur seals, sea lions, and an "American in a skin sea boat." Also, in contrast to the Chirikov-Yelagin charts, the chart compiled by Bering's detachment does not juxtapose the new geographic information with the contours of the then-known North America.

As early as November of 1742 Chirikov began work on a chart would summarize the results of the Second Kamchatka Expedition, but at that time he did not have at his disposal any materials from the Bering detachment or Waxell's charts. As soon as Chirikov received these materials he continued this work,[99] and while he was still in Siberia he created "A chart compiled on the order of Fleet Captain Alexei Chirikov, in Mercator projection . . ." In the summer of 1745, prior to his departure from Yeniseisk for St. Petersburg, Chirikov transmitted this chart to Waxell. The original chart signed by Chirikov is preserved.[100]

This chart utilized not only data from all detachments of the Second Kamchatka Expedition (including the Arctic ones), but also information obtained on Gvozdev's voyage, as an inscription states: "the land shown across from the eastern corner of Chukotka to the east lying between 65 and 68 degrees N latitude is drawn from the chart of geodesist Gvozdev." The coastlines directly observed on the voyages of both the *St. Peter* and the *St. Paul* are represented by clearly delineated contours, while a hypothetical interpretation of the continuous American shore, taken from Waxell's chart, is shown in a weak wash as unconfirmed coastline joining the American coast in the north after Gvozdev's chart.

As noted above, by 1744 the Admiralty Department had created, attached to the Naval Academy, an authoritative commission whose task was to summarize the materials of the Second Kamchatka Expedition and to compile a general chart. Initially Nagaev supervised this work. All leading officers were part of the commission. Its work was concluded under the guidance of Chirikov in 1746. The commission's most basic contribution was a "General map of the Russian Empire, the northern and eastern shores of Siberia (lying) adjacent to northern ice and eastern oceans, with western American islands and the island of Japan newly found in the course of a sea voyage."[101]

The chart shows the signatures of the compilers, Vasily Krasilnikov and Pyotr Rastorguev, whose role apparently were that of draftsmen. The main authors were members of the Second Kamchatka Expedition: "Fleet Captain Alexei Chirikov, Lieutenant Khariton Laptev, Fleet Captains Stepan Malygin and Dmitry Laptev, Lieutenant Dmitry Ovtsyn, Lieutenant Sofron Khitrovoi [*sic*], Fleet Lieutenant Ivan Yelagin." Comparison of this chart with that compiled by the order of Fleet Captain Chirikov in 1745 shows that the two charts are without doubt related. The only essential difference, from the cartographic point of view, is the fact that the Naval Academy map of 1746 is in conical projection, while the 1745 chart is in the basic projection for navigational charts, the conforming cylindrical Mercator projection. The sources and their interpretation are identical. Even the errors are repeated. Thus, in both cases, the Diomede Islands are represented twice, once according to Bering and once according to Gvozdev materials.

In 1728 Bering located only one island, the most westerly in the strait. It is represented on both summary charts of 1745 and 1746, but there are also shown, at the Prince of Wales Cape latitude, three islands from the Spanberg-Gvozdev chart (modern-day Big Diomede, Little Diomede, and the cliff Fereuey).[102]

As far as the hypothetical coastline of America stretching toward Kamchatka is concerned, the Naval Academy map shows it in a graphically more definitive manner than that on the chart of 1745, a dotted line instead of wash. However, this did not signify unquestioning acceptance by the authors of the single point of view expressed on Waxell's chart. It expressed the need for further exploration in future of the shoreline and of the ocean between the regions visited by the vessels of the Second Kamchatka and Gvozdev expeditions.

In their entirety, the charts discussed here that summarized the work of all the Russian expeditions in the northeast from 1725 to 1743 are evidence, to cite Chirikov, that "not in a small part of the globe's land and water expanse, many lands and islands which until the present time were unknown are now apparent . . ." In June 1746, based on a summary of the results, Chirikov completed a detailed project for study of the Far East in which he particularly focused on the discovery in the Bering Island region "and in other areas" of various animals. He stressed the possibility that:

> near other islands, sea otters, seals, fur seals, and whales are present, and the land where geodesist Gvozdev saw inhabitants is in direct line distant from the mouth of Kamchatka River not less than 1,700 verst . . . from Anadyrsky *ostrog* by way of the Anadyr River and then by sea more than 1,100 verst, and from the mouth of the Anadyr River, only by sea, more than 600 verst . . . Near the Chukchi land in the sea are walruses, whales, seals and other sea products. This may be expected as well at the land sighted by geodesist Gvozdev, because of the short distance (to it).[103]

To continue the explorations and to clarify geographical questions left unsolved, Chirikov considered a new voyage to the American coast necessary:

> When time permits, even if not soon, even to the land sighted by Gvozdev, situated above the 64 degrees northern latitude, and though it is distant from Kamchatka, and its climate is cold so that the earth cannot bear fruit, moreover rumors are circulated that there are many sables and other animals, one ought to sail there to obtain true information.[104]

This statement clearly shows that the results of the Second Kamchatka Expedition were the object of very close attention by the Russian authorities and particularly by the leadership of the naval forces of the empire. The results obtained permitted the compilation of objective geographical maps of the explored regions and projected directions for future investigations. However, as usual, these results were treated as the most important state secret, as is stated, for example, by Golovin's message addressed to the Admiralty College on November 6, 1742: "As it is imperative to keep various events on that voyage in utmost secrecy, I am forwarding the report received from Chirikov, together with an original chart and journal and other equally confidential documents, and I suggest that the State Admiralty College keep it most secret, under seal, until such time as H(er) I(mperial) M(ajesty)'s executive order (to release the documents) will follow. I have thus reported to H.I.M."[105]

This government effort to preserve in secrecy the results of the Second Kamchatka Expedition left even the Academy of Sciences in ignorance, at least as far as cartography is concerned.

The publication in 1745 "through the efforts and labor of the Imperial Academy of Sciences" of a "Russian Atlas consisting of nineteen special maps which represent the Russian Empire with the adjacent border lands, compiled according to the principles of geography and according to the newest observations, with an appended General map of this Great Empire" did not, in practical terms, reflect any results of the Second Kamchatka Expedition. In this respect, the "Russian Atlas" is typical of a series of Russian printed geographic creations, which, out of considerations by the regime, until the most recent times, did not reflect the true state of geographic knowledge and basic cartography in the country. Such an unfortunate tradition brought keenly felt harm, most of all to Russian prestige, as it did not reflect Russia's priority in many geographic explorations and in the cartography of regions not studied by European scholars.

DELISLE-BAUCHE MAPS

These government measures, sufficiently effective inside the country, could not prevent a leakage of information to Western Europe regarding the geographic results of the Second Kamchatka Expedition. The data appeared there in the form of newspaper notes, articles, and maps. Specifically, relatively accurate information about the organization and history of the Kamchatka expeditions was published in 1747 in Copenhagen, by von Haven. The author, who visited Russia twice, met, in 1744, with Spanberg and "a Kurland nobleman" who had returned from Siberian exile. It is apparent that he obtained his information from them. Von Haven's publication also contains excerpts from Waxell's report to the Admiralty College dated November 15, 1742.

Very distorted information about the results of the Second Kamchatka Expedition was brought to Europe by Joseph Nicolas Delisle, who returned to Paris in 1747. The French astronomer, deeply interested in Russian geographic materials, had exported from Russia a huge collection of manuscript maps compiled on the basis of surveys by Peter's geodesists, which is now kept in the National Library and in the Archive of the Hydrographic Department in Paris. He was, however, only very superficially acquainted with the original materials of the Second Kamchatka Expedition.

On April 8, 1750, Delisle read a paper at the Royal Academy of Sciences in Paris in which he gave a rather brief and imprecise overview of the history of Russian voyages to American coasts, and focused his main attention on the imaginary discoveries of an Admiral Bartolomeo de Fonte. The latter allegedly sailed to discover the Northwest Passage and found a bay on the coast of North America at 53° N latitude, which cut deeply into the continent's interior. Subsequently this bay came to be identified with a no-less-mythological inland sea, allegedly discovered in 1592 by the navigator Juan de Fuca.

This propaganda effort for the entirely novel geography of the northwest coast of America and the unknown territories to the east was joined by Delisle's brother-in-law, cartographer Philippe Buache. They compiled and published the *Carte Generale des decouvertes de l'Amiral de Fonte et autres Navigateurs Espagnols, Anglois et Russes pour la recherche du Passage a la Mer du Sud* in 1752 (Fig. 11). This map shows Bering's expedition's routes, and, in the region of Alaska, the inscription: "Land sighted by Spanberg in the year 1728 [?!—AP] and now frequently visited by the Russians." Below is the inscription: "Land sighted by Russians under Chirikov's command in 1741." To the south of these lands are shown the hypothetical "Strait of Anian," "de Fonte Lake," and so on. Not too far from Kamchatka is shown, as a large island, "Bering's Land, discovered by him in the year 1728," evidently representing Da Gama Land, of which Delisle was so enamored. The Fonte Strait on this map begins in the river "R de Los Reyes," continues past Cataracte into Hauter des Terres, and the big Lac de Fonte located west of Hudson Bay and linked with the latter by a waterway.

This fantastical Delisle-Buache map immediately aroused savage criticism by their contemporaries. In 1753 English geographer John Green published a work that exposed the stories of de Fuca and de Fonte as being totally fantastical and demonstrated that not only did the Delisle map fail to represent the real results of the Kamchatka expeditions, but it even transmitted the mythological opinions of de Fonte incorrectly.[106] John Green's publication was accompanied by the *New Chart of North and South America*, compiled by the geographer Thomas Jefferys, which, although it did not fully reflect the results of Bering's expeditions, at least represented the geography of America on the basis of the then-available real information and not on any legendary data. In 1757, the *Noticia de la California* by the Jesuit Father Andreas Burriel was published in Madrid. This work was also critical of Delisle's fantasy.

In spite of the fact that the above-mentioned critical works were not alone and that the opinions of Green and Burriel were shared by many scholars, the mythological geography of de Fonte and de Fuca in Delisle/Buache's interpretation found its reflection on many well-known maps of the eighteenth century, and the name de Fuca became enshrined in the name of the strait that only has one thing in common with de Fuca's history: its latitude.[107]

In Russia, Delisle's geographic opus immediately provoked a totally negative reaction. Entrusted with the task by Razumovsky, the president of the Academy of Sciences, Müller, himself a member of the Second Kamchatka Expedition, wrote a rebuttal of Delisle's paper in which he pointed out the intentional distortions perpetrated by the French scholar and gave a general outline of the real history of the expedition. This rebuttal was published in French in 1752 and 1753 as a separate pamphlet.[108]

Müller 's critique of Delisle's concepts notwithstanding, the opinions of the French geographer did influence, to a certain extent, the first map of Russian explorations in the northern part of the Pacific, compiled by Müller in 1754 and published, in French, by the Russian Academy of Sciences[109] (Fig. 12).

A Russian version of this published map, prepared by the academy's cartographer, Ivan Truskot, also exists. It is entitled "A chart representing discoveries by Russian seamen in the northern part of America with adjacent locations, made by the Titular Councillor of the Imperial Academy of Sciences Truskot." It is obvious that on these maps the distribution of land in the Pacific, and especially the outline of the American coast stretching toward Kamchatka, is derived from the charts of the Bering-Waxell detachment. Routes of the First as well as of the Second Kamchatka Expedition are indicated. An incorrect date is given for "The coast discovered by geodesist Gvozdev in 1730." Evidently the academy's cartographers, when compiling these maps, did not have at their disposal original materials from Bering's expeditions. As a result, they were forced to utilize foreign published materials, in particular the map by Delisle-Buache. This is attested to by the fact that on Müller's map, the mythological information about de Fuca and de Fonte is present ("Entree decouv. par Iean de Fuco en 1592" and "Pretendue R. de las Reyes de l'Amiral de Fonte en 1640, Suivant Mr. Dehse" [*sic*, Delisle?]). Müller's map was published in the English translation of his survey of Russian voyages to America, and this only served to support the mythological geography of Delisle/Buache.[110]

THE LEGACY OF THE SECOND KAMCHATKA EXPEDITION

Thus, the publicly available cartographic representations of the Bering expeditions did not adequately reflect the results that had been achieved. Nevertheless, their significance for natural sciences was already appreciated in the eighteenth century due to the activities of the academy's personnel attached to the Second Kamchatka Expedition. Among them were Müller, Gmelin, Delisle, de la Croyere, Steller, Fisher, four geodesists, two artists, an interpreter, and five students. Among the latter were Stepan P. Krasheninnikov and geodesist A. D. Krasilnikov, who independently completed several scientific works. The government provided the academy's detachment with exceptionally favorable working conditions. Siberian governors and commanders were ordered to render all possible assistance.

Between 1732 and 1743 Müller and Gmelin investigated enormous areas of Siberia, from Tobolsk to Yakutsk. They conducted numerous observations and a great deal of research in history, geography, ethnography, archaeology, paleontology, and meteorology; they collected a huge scientific store of information about the cities, factories, industries, agriculture, useful minerals, and natural resources of Siberia, and the lifestyles of the peoples who inhabited it. Müller established the foundation studies of the history of the journeys by Russian pioneers and seafarers in the seventeenth and early eighteenth centuries. It was he who located, in the Siberian archives, documents pertaining to Dezhnev's voyage of 1648, and brought to the capital a mass of materials on Siberian history, which eventually formed the collection known as "Müller's Portfolios," now for the most part in the Russian State Archive of Ancient Documents (RGADA). This collection remains an inexhaustible source for historians.

Most fruitful were the efforts of Steller, the academy adjunct in natural sciences. He kept a journal that remains a valuable source for the history of the Second Kamchatka Expedition.[111] The journal was first published in St. Petersburg in 1793, 50 years after its author's death. Steller's observations were varied and interesting.

He provided Europe's first descriptions of Alaska and the Aleutian Islands and a considerable amount of valuable data about the Natives who, up to that time, had practically no contact with the Europeans. In other works, Steller gave a vivid and scientifically precise description of the appearance and ecology of animals he had observed during the voyage and the forced wintering on Bering Island, including a new species—the sea cow, exterminated by the *promyshlenniki* toward the end of the eighteenth century. This extinct species has been named after him. His name has also been given to a peak on Bering Island. Upon the expedition's return, Steller, this time on his own, undertook several journeys in Kamchatka, mostly solitary ones. Steller died in 1746, traveling from Solikamsk to Irkutsk.[112]

Materials collected by these scholars resulted in the appearance of scientific publications on a scale unprecedented in the eighteenth century. Krasheninnikov, who, upon his return from the expedition, became the first Russian academician-geographer, published a summary of his research in Kamchatka: *Description of the Land of Kamchatka*, in which he also included some data by Steller. Steller's own treatise, *Steller's History of Kamchatka*, first appeared in English in 2003.[113] Other publications included Müller's numerous works, such as *History of Siberia*; *Description of Sea Voyages in the Icy and Eastern Seas Conducted from the Russian Side*; *Information about the Siberian Trade*; and other works on the history of geographic research, economics, and geography. Gmelin's *Flora of Siberia* and *Journey through Siberia* are among many works produced by other members of the academy's detachment.

The overall result of the Second Kamchatka Expedition was the development of geographic knowledge and the establishment of geographic science in Russia and the beginning of the planned and systematic study of the history and natural environment of Siberia.[114] In respect to the Pacific Ocean and northwest coast of America, the Second Kamchatka Expedition brought the "meeting of the Old and the New Worlds" out of Asia. For the Natives of the Aleutian Islands and Alaska, the period of development without contact with European civilization came to an end. The Natives themselves, and the natural resources of their lands, especially the animal world, became the objects of cruel exploitation, and even their very existence came under threat.

NOTES

1. V. K. Sergeev, "Moskovskaia matematiko-navigatskaia shkola (1701–1752)." In *Voprosy geografii*, no. 34, 1954:150–151.
2. S. E. Fel', "Kartografiia Rossii XVIII v. (kak sintez astronomo-geodezicheskikh i graficheskikh protsessov)." In *Trudy Instituta istorii estestvoznaniia i tekhniki AN SSSR*, vol. 37, no. 2, Moscow 1961:244.

3. *Polnyi svod zakonov Rossiiskoi imperii. 1 vypusk* (*Complete Code of Laws of the Russian Empire. 1st Series*), vol. 13, no. 10062.
4. Sergeev, ibid., 1954. A. V. Postnikov. *Razvitie krupnomasshtabnoi kartografii v Rossii* (*The Development of Large-Scale Cartography in Russia*), Moscow: Nauka, 1989:36.
5. P. G. Kulikovsky, *M.V. Lomonosov—astronom i astrofizik* (*M.V. Lomonosov, Astronomist and Astrophysicist*). Moscow/Leningrad [*sic*, no date].
6. M. D. Khmyrov, "Svedeniya o Vasily Kipriyanove, bibliotekare Moskovskoi grazhdansoi tipografii pri Pyotre I" ("Information about Vasily Kipriyanov, Librarian of the Moscow Civil Printing House under Peter I"). In *Russkii arkhiv* (*Russian Archive*), no. 8/9 (1866).
7. A. V. Borodin, "Moskovskaya grazhdanskaya tipografiya i bibliotekari Kipriyanovy" ("The Moscow Civil Printing House and the Kipriyanov Librarians"). In *Stati i materialy po istorii knigi v Rossii* (*Articles and Materials on the History of the Book in Russia*). Moscow/Leningrad, 1936:59–87.
8. A. W. Richeson, *English Land Measuring to 1800: Instruments and Practices*. Cambridge, Massachusetts, 1966.
9. Postnikov, Ibid., 37.
10. The formation of different traditions in representing the Chukotka Peninsula, the Bering Strait and islands therein, and the Great Lands, is discussed by M.I. Belov and S. G. Fedorova. See M. I. Belov, "O sostavlenii Generalnoi karty Vtoroi Kamchatskoi ekspeditsii 1746 g." ("On the Creation of the General Map of the Second Kamchatka Expedition of 1746"). In *Geograficheshy sbornik* (*Geographical Collection*), vol. 3, 1954:142; and S. G. Fedorova, "K voprosu rannikh russkikh poselenii na Aliaske" (On the Issue of the Early Russian Settlement in Alaska"). In *Letopis Severa* (*Chronicle of the North*), vol. 4. Moscow, 1964:111.
11. J. Perry, *Sostoyanie Rossii pri nyneshnem tsare* (first published in English as *The State of Russia Under the Present Czar*). Moscow, 1871:40.
12. O. A. Evteev, *Pervye russkie geodesisty na Tikhom okeane* (*The First Russian Geodesists on the Pacific Ocean*). Moscow, 1950.
13. Cited after *Russkie ekspeditsii . . . XVIII v.* (*Russian Expeditions . . . of the Eighteenth Century*), 1984:33.
14. *Polnyi svod zakonov Rossiiskoi imperii*. Series 1, vol. 7, no. 4679.
15. *Russkie ekspeditsii . . . XVIII v.*, 1984:37–40.
16. B. P. Polevoi, "Pyotr I, Nikolai Vitsen i problema soshlas li Amerika s Asiei"("Peter I, Nicolaas Witsen and the Problem of Whether America was Joined with Asia"). In *Strany i narody Vostoka* (*Countries and Peoples of the East*), no. 17, book 3, 1975:29. R. H. Fisher, *Bering's Voyages: Whither and Why*. Seattle and London, 1977.
17. A. S. Polonsky, "Pervaya kamchatkskaya ekspeditsiya Beringa 1725–1729 g." (Bering's First Kamchatka Expedition"). In *Zapiski Gidrograficheskago Departamenta Morskago ministerstva* (*Notes of the Hydrographic Department of the Marine Ministry*). Part 8, 1850:535. L.S. Berg, *Otkrytie Kamchatki i ekspeditsii Beringa (1725–1742)* (*The Discovery of Kamchatka and Bering's Expeditions*). Moscow/Leningrad: Akademii nauk SSR, 1946:88–89.
18. M. I. Belov, *Istoriya otkrytiya i osvoeniya severnogo morskogo puti* (*The History of the Discovery and Mastery of the Northern Sea Route*). Moscow, 1956:256.

19. A. A. Pokrovsky, "Ekspeditskya Beringa" ("Bering's Expedition"). Moscow, 1941:25. A. I. Andreev, "Ekspeditsiya Beringa." In *Izvestiya VGO* (*News of the All-Union Geographical Society*), vol. 75, no. 2, 1943:31. D. M. Lebedev, *Geografiya v Rossii Petrovskogo vremeni* (*Geography in Russia in the Era of Peter I*). Moscow/Leningrad, 1950:94.
20. V. I. Grekov, *Ocherki iz istorii russkikh geograficheskikh issledovanii v 1725–1765 gg.* (*Essays from the History of Russian Geographical Research in 1725–1765*). Moscow, 1960:20–21.
21. G. Barratt, *Russia in Pacific Waters, 1715–1825. A Survey of the Origins of Russia's Naval Presence in the North and South Pacific.* Vancouver and London, 1981:10.
22. B. P. Polevoi, "Iz istorii otkrytiia severo-zapadnoi chasti Ameriki" ("From the History of the Discovery of the Northwest Part of America"). In *Ot Aliaski do Ognennoi zemli* (*From* B. P. *Alaska to the Land of Fire*). Moscow, 1967:113.
23. B. P. Polevoi, "Pyotr Pervy, Nikolai Vitsen i problema soshlas li Amerika s Aziei." In *Strany i narody Vostoka*, no. 17, book 3, Moscow, 1975:29.
24. N. N. Bolkhovitinov, *Rossiya otkryvaet Ameriku. 1732–1799* (*Russia Discovers America. 1732–1799*). Moscow: Mezhdunarodnyie Otnosheniia, 1991:12–15.
25. *Russkie ekspeditsii . . . XVII v.*, 1984:22.
26. A. A. Sopotsko, *Istoriya plavaniya Beringa na bote Sv. Gavriil v Severnyi ledovityi okean* (*The History of Bering's Voyage on the Gabriel to the Northern Icy Ocean*). Moscow, 1983:6. N. Nielsen, "Vitus Berings Stordoad." *Danmarkposten* 6 (1942).
27. F. A. Shibanov, "Russkaya polevaya astronomiya v 18 veke" ("Russian Field Astronomy in the Eighteenth Century"). In *Uchenye zapiski Leningradskogo gosudarstvennogo universiteta. Seriya geograficheskikh nauk* (*Scholarly Notes from Leningrad State University. Geographical Sciences Series*). no. 226, issue 12, kartografiya, 1958:5–6.
28. A. I. Alekseev, *Okhotsk—kolybel russkogo Tikhookeanskogo flota* (*Okhotsk—the Cradle of the Russian Pacific Fleet*). Khabarovsk, 1958:26–34.
29. For a precise record of this expedition, see *The Journal of Midshipman Chaplin: A Record of Bering's First Kamchatka Expedition*. Edited and translated from the Russian by Tatiana S. Fedorova, Peter Ulf Møller, Viktor G. Sedov, and Carol L. Urness. Aarhus, 2010. [MF]
30. Sopotsko, ibid., 1983:18.
31. *Russkie ekspeditsii . . . XVIII v.* 1984:84.
32. Ibid., 85.
33. G. F. Müller, *Sochineniya i perevody* (*Works and Translations*). Part 1, St. Petersburg 1758:104.
34. RGADA, fond 248, book 543, folios 275–276.
35. *Russkie ekspeditsii . . . v XVIII v.* 1984:290.
36. V. I. Grekov, *Ocherki iz istorii russkikh geograficheskikh issledovanii v 1725–1765 gg.* Moscow, 1968:38. Sopotsko, ibid., 1983:31.
37. P. Lauridsen, *Vitus Bering og be russiske Opdagelsesrejser fra 1725–1743*. Kobenhavn, Gylenda Iske boghanels forlay, 1885:35.
38. F. Weber, *Das verandertern Russlands dritter Theil.* Vol. 1, Hannover, 1738, 247 pp; vol. 2 Hannover, 1740, 240 p. Sopotsko, ibid., 1983:32, 225.
39. On publication of this chart, see Grekov, ibid., 1960:41; Efimov, *Atlas . . .* 1964:45–45, maps 63, 63; A. V. Efimov, *Iz istorii velikikh russkikh geograficheskikh otkrytii* (*From

the History of the Great Russian Geographical Discoveries). Moscow, 1971:244; Sopotsko, ibid., 1983:22–23.

40. On this map and its reproductions, see Lebedev, *Geografiya v Rossii Petrovskogo vremeni*. Moscow/Leningrad, 1950:247, 249]); M. G. Novlianskaya, *I. K. Kirilov i ego Atlas vserossiiskoi imperii* (*I. K. Kirilov and his Atlas of the Russian Empire*). Moscow/Leningrad, 1958; *Atlas*. . . 1964:48–49, maps 71, 72, and others.
41. J. B. Du Halde. *Nouvel atlas de la China, etc.*, Paris, 1737.
42. RGADA, fond 248, book 690, folio 287 and 287 verso. L. A. Goldenberg, *Mezhdu dvumya ekspeditsiyami Beringa* (*Between Bering's Two Expeditions*). Magadan, 1984:33.
43. Goldenberg, ibid.
44. *Russkie ekspeditsii . . . XVIII v.* 1984:61.
45. V. N. Berkh, *Pervoe morskoe puteshestvie rossiyan, predprinyatoe dlya resheniya geograficheskoi zadachi soedinyaetsiya li Aziya s Amerikoi? I sovershennoe v 1727, 28, i 29 godakh pod nachalstvom flota kapitana 1-go ranga Vitusa Beringa* (*The First Sea Voyages by Russians Embarked on to Solve the Geographical Problem of Whether Asia was Joined with America. And Completed in the Years 1727, 1728, and 1729 under the Leadership of Fleet Captain First Class Vitus Bering*). St. Petersburg, 1823:84.
46. RGADA, fond 248, opis 113, book 666, folio 30 and 30 verso, 438 and 438 verso. Goldenberg, ibid., 1984:82.
47. Grekov, ibid., 1960:54.
48. RGADA, fond 248, book 666, folio 31.
49. A. Polonsky, "Pokhod geodezista Mikhaila Gvozdeva v Beringov proliv 1732 goda" ("Geodesist Mikhail Gvozdev's Voyage to the Bering Strait in 1732"). In *Morskoi sbornik*, vol. 4, no. 11, 1850. A. P. Sokolov, "Pervy pokhod russkikh k Amerike 1732 goda" ("The First Voyage by Russians to America in 1732"). In *Zapiski Gidrograficheskago departamenta Morskago ministerstva*, part 9, St. Peterburg, 1851. V. A. Divin, *K beregam Ameriki. Plavaniya i issledovaniya M.S. Gvozdeva, pervoot-kryvateliya Severo-Zapadnoi Ameriki* (*To the Shores of America. The Voyages and Research of M.S. Gvozdev, the First Discoverer of Northwestern America*). Moscow, 1956. Published in English as *To the American Coast: the Voyages and Explorations of M.S. Gvozdev, the Discoverer of Northwestern America*. Edited by J. L. Smith. Anchorage, 1997. L. A. Goldenberg, *Mikhail Gvozdev*, Moscow, 1982. L. A. Goldenberg, *Mezhdu dvumya ekspeditsiyami Beringa*. Magadan, 1984.
50. RGAVMF, fond 216, *delo* 3, folios 91 and 91 verso. Goldenberg, ibid., 1984:116.
51. Ibid., 1984:128–129. L. S. Berg, *Otkrytie Kamchatki i ekspeditsii Beringa. 1725–1742*. Moscow, 1946:100.
52. Goldenberg, ibid., 1984:131–132.
53. Goldenberg, ibid., 1984:120–121.
54. Y. M. Svet and S. Fedorova, "Captain Cook and the Russians."In *Pacific Studies*, vol. 11, no. 1, 1978:5.
55. Goldenberg, ibid., 1984:133–134.
56. Goldenberg, ibid., 1984:135.
57. Goldenberg, ibid., 1984:137.
58. Frank A. Golder, *Russian Expansion on the Pacific 1641–1850* . . . , 1914:154, 299–301.
59. Ibid.

60. Golder, ibid., 1914:299.
61. RGAVMF, fond 216, opis 1, no. 24. Leningrad, 596–598.
62. RGAVMF, ibid., Leningrad, 602–604; *Russkie ekspeditsii . . . XVIII v.* 1984:291–292.
63. RGADA, fond 192. Irkutskaya guberniya, no. 36 ; Rossiiskii gosudarstvennyi voenno-istoricheskii arkhiv (Russian State Military History Archive) [further—RGVIA], fond 846: Voenno-ucheny arkhiva (Military Educational Archive) [further—VUA]. no. 23431; RGAVMF, opis 16. Arkhiv tsentralnogo kartograficheskogo proizvodstva (Archive of the Central Cartographic Production Company), no. 69/70. A copy of the map from RGVIA and an enlarged section of it are published in: *Atlas . . .* no. 69, 70:48.
64. *Atlas mapparum geographicarum generalium & specialium centum folii compositum et quotidianis usibus accomodatum accedit introductio in gheographiam mathematicam, naturalem & historicam.* Norimbergae, impensis Homannianorum heredum (1728–1793), no. 3. *Planiglobii terrestis mappa universalis . . . designata a G. M. Lowizio . . . 1746.*
65. Rossiisky gosudarstvenny istorichesky arkhiv [further—RGIA], fond 1329, opis 1, no. 35, folios 289–291. *Russkie ekspeditsii . . . XVIII v.* 1984:97. First published in *Zapiski Gidrograficheskago departamenta Morskago ministerstva*, part 9, 1851:435–436.
66. *Russkie ekspeditsii . . . XVIII v.* 1984:113–115.
67. Barratt, ibid., 1981:28.
68. For details, see A. V. Postnikov, *Razvitie krupnomasshtabnoi kartografii v Rossii.* Moscow, 1989:43–45.
69. *Russkie ekspeditsii . . . XVIII v.* 1984:112.
70. The original instructions formulated by the professors of the Academy of Sciences and presented to the Senate are in RGADA, fond 248, book 666, folios 193–228. In the current work copies received by Bering from the Admiralty College were used, published in *Russkie ekspeditsii . . . XVIII v.* 1984:117–119,121–122; RGAVMF, fond 216, opis 1, no. 27, folios 46, 53–61, 66–71.
71. This map was first published by F. A. Golder in *Bering's Voyages. An Account of the Efforts of the Russians to Determine the Relation of Asia and America*, vol. 2. 1925:72–73. Another version of this map is in RGADA, fond 192, opis 1, *Karty generalnye Rossii* (General Maps of Russia), no. 29. *Russkie ekspeditsii . . . XVIII v.* 1984:293.
72. F. A. Golder, *Russian Expansion on the Pacific. 1641–1850 . . .* Cleveland, 1914:118, 127–128. Grekov, ibid., 1960:58. *Russkie ekspeditsii . . . XVIII v.* 1984:293–294.
73. Grekov, ibid., 1960:57–60.
74. *Russkie ekspeditsii . . . XVIII v.* 1984:122.
75. Ibid.
76. Ibid., 122.
77. This is discussed in greater detail in Postnikov, ibid., 1989:46–48, 52–53, 80.
78. *Russkie ekspeditsii . . . XVIII v.* 1984:124–132.
79. *Russkie ekspeditsii . . . XVIII v.* 1984:138–144.
80. For the personal side of Bering's expedition, see *Until Death Do Us Part: the Letters and Travels of Anna and Vitus Bering.* Fairbanks: University of Alaska Press, 2008. Bering's wife and two of his children accompanied him as far as Okhotsk. [MF]

81. *Russkie ekspeditsii . . . XVIII v.* 1984:213, 215.
82. Ibid., 1984:214, 216.
83. D. M. Lebedev, *Plavanie A. I. Chirikova na paketbote Sv. Pavel k poberezhiyam Ameriki* (*A. I. Chirikov's Voyage on the Small Boat St. Paul to the Shores of America*). Moscow, 1951:209.
84. *Russkie ekspeditsii . . . XVIII v.* 1984:217, 297–298.
85. For a biography of Chirikov, see Vasilii A. Divin, The Great Russian Navigator, A. I. Chirikov. Translated and annotated by Raymond H. Fisher. Fairbanks, 1993. [MF]
86. The map remained in manuscript. A copy was reproduced in *Russkie ekspeditsii . . . XVIII v.* 1984:220.
87. From the log of the vessel St. Peter, *Russkie ekspeditsii . . . XVIII v.* 1984:236–237.
88. A copy of this manuscript chart has been published (see *Russkie ekspeditsii . . . XVIII v.* 1984:221).
89. *Russkie ekspeditsii . . . XVIII v.* 1984:244.
90. *Russkie ekspeditsii . . . XVIII v.* 1984:267.
91. *Russkie ekspeditsii . . . XVIII v.* 1984:223, Barratt, ibid., 1981:37.
92. M. I. Belov, "O sostavlenii Generalnoi karty Vtoroi Kamchatskoi ekspeditsii 1746 g." In *Geografichesky sbornik*, vol. 3, Moscow/Leningrad, 1954. S. G. Fedorova, "K voprosu o rannikh russkikh poseleniyakh na Aliaske." In *Letopis Severa*, vol. 4, Moscow, 1964. L. A. Goldenberg, "Zdes byl geodesist Gvozdev 1732 goda" ("Geodesist Gvozdev was here in 1732"). ("Kartograficheskie traditsii XVIII v. v izobrazhenii pervootkrytii v Beringovom prolive"; "The Cartographic Traditions of the Eighteenth Century in the Depiction of the First Discoveries in the Bering Strait"). In *Izvestiya VGO*, vol. 114, issue 6, 1982:526–532.
93. Barratt, ibid., 1981:47.
94. V. Berkh, *Zhizneopisanie admirala Alekseia Ivanovicha Nagaeva*. St. Petersburg, 1831: 19.
95. Arkhiv Vneshnei Politiki Rossii skoi Imperii [further—AVPRI], fond 339 [Rossiskaya Amerikanskaya Kompaniya—further RAK], opis 888, no. 921/1. RGAVMF, fond 1331, opis 4, no. 72, folio 3. Efimov *Atlas . . .* (1964), maps 97 and 98 unfortunately are later copies with the addition of the Aleutian Islands, the names given apparently on *promyshlenniki* testimony. The two charts which we reference in this work are, to judge by all criteria, author's copies by Elagin, compiled not later than 1742.
96. *Russkie ekspeditsii . . . XVIII v.* 1984:226–227.
97. RGAVMF, fond 1331, opis 4, no. 16.
98. RGAVMF, fond 1331, opis 4, no. 99. This chart is known also in authors' copies by Sven Waxell and Sofron Khitrov and also in a copy by Pyotr von Dezin, who compiled it from the watch logs. Like Chirikov's chart, this manuscript chart was reproduced numerous times in the course of various researches. See, for example, *Atlas . . .* 1964, maps nos. 99–102, p. 68–69; V. Andreev, "Dokumenty po ekspeditsii kapitan-komandora Beringa v Ameriku v 1741 g. " ("Documents on the Expedition of Captain-Commander Bering to America in 1741"). In *Morskoi sbornik*, vol. CCLV, no. 5, 1893; Sven Waxell, *Vtoraya Kamchatskaya ekspeditsiya Vitusa Beringa* (*The Second Kamchatka Expedition of Vitus Bering*). Moscow/Leningrad, 1940; Grekov, ibid., 1960;

Efimov, ibid., 1950; Berg, ibid., 1946; M. I. Belov, "Daniya i Vitus Bering" ("Denmark and Vitus Bering"). In *Puteshestviya i geograficheskie otkrytiya v XV–XIX v v.* (*Journeys and Geographical Discoveries in the Fifteenth–Nineteenth Centuries*), Moscow-Leningrad, 1956; G. W. Steller, *Reise von Kamtschatka nach Amerika mit dem Commandeur-Capitan Bering*, St. Petersburg, 1793; and others.

99. The first to draw historians' attention to this fact was L. A. Goldenberg; see Goldenberg, ibid., 1984:213.

100. RGVIA, fond VUA, no. 23466. Published by Efimov, ibid., 1950; *Atlas*... 1964, map 110, p. 74.

101. RGVIA, fond VUA, no. 20227. *Opisanie starinnykh atlasov, kart i planov XVI, XVII, XVIII i (pervoi) poloviny XIX veka, khranyashchikhsya v Arkhive Tsentralnogo kartograficheskogo proizvodstva VMF* (*Description of the Antique Atlases, Maps and Plans of the Sixteenth, Seventeenth, Eighteenth and (first) half of the Nineteenth Centuries Kept in the Archive of the Central Cartographic Production Company of the Naval Flee*t). Leningrad, 1958, no. 82. Published in Atlas... 1964, map 111, p. 75; Divin, ibid.

102. Goldenberg, ibid., 1984:463–464.

103. RGAVMF, fond 230, no. 1, folio 56. See also A. P. Sokolov, "Severnaya ekspeditsiya 1733–43 goda" ("The Northern Expedition of 1733–43"). *Zapiski Gidrograficheskago departamenta Morskago ministerstva*. Part 9. St. Petersburg, 1851:463–464.

104. RGAVMF, fond 230, no. 1, 61v. fond 216, no. 48, folio 742 and 742v. Sokolov, ibid., 1851:465. Goldenberg, ibid., 1984:464–465.

105. *Russkie ekspeditsii... XVIII v.* 1984:262.

106. John Green, *Remarks in the Support of the New Chart of North and South America*. London, 1753.

107. Henry R. Wagner, *The Cartography of the Northwest Coast of America to the Year 1800*, vol. 1. Berkeley, 1937:159–160. Carl Wheat. *Mapping the Transmississippi West 1540–1861*, vol. 1, *The Spanish Entranda to the Louisiana Purchase, 1540–1804*. San Francisco, 1957:131–142.

108. G. F. Müller, "Lettre d'un officier de la marine Russienne... Nouvelle bibliotheque germanique, ou histoire litteraire de l'Allemagne, de la Suisse et de Pays Nord, par Mr. Samuel Formay. Amsterdam, 1753." This work was published in 1752 in Berlin, without the author's name—see J. R. Masterson and H. Brower, *Bering's Successors, 1745–1780: Contributions of Peter Simon Pallas to the History of Russian Exploration toward Alaska*... Seattle, 1948:2. Grekov, ibid., 1960:166–167. For a modern translation of Müller with scholarly analysis, see *Bering's Voyages: the Reports from Russia*. Translated with commenatary by Carol Urness. Fairbanks, 1986.

109. Nouvelle Carte des decouvertes faites par des Vaisseaux Russiens... A St. Petersbourg a l'Academie imperiale des sciences 1758 in AVPRI, fond 339, RAK, opis 888, no. 922/2. Formerly in RGADA, fond 192, Irkutskaya guberniya, no. 50.

110. G. F. Müller, *Voyages from Asia to America for Completing the Discoveries of the North West Coast of America. To Which is Prefixed a Summary of the Voyages Made by the Russians on the Frozen Sea, in Search of a North East Passage, Serving as an Explanation of a Map of the Russian Discoveries, Published by the Academy of Sciences at Petersburg*... London, 1761.

111. G. W. Steller, *Iz Kamchatki v Ameriku* (*From Kamchatka to America*). Leningrad, 1927. Full translation of the preserved original journal (from German into English): *Georg Wilhelm Steller: Journal of a Voyage with Bering, 1741–1742*. Edited with an introduction by O. W. Frost, translated by Margritt A. Engel and O. W. Frost. Stanford, 1988.
112. *Ocherki istorii geograficheskoi nauki SSSR* (*Essays on the History of Geographical Science in the USSR*). Moscow, 1976:39. *Russkie ekspeditsii . . . XVIII v.* 1964:299.
113. Georg Steller, *Steller's History of Kamchatka: Collected Information Concerning the History of Kamchatka, Its Peoples, Their Manners, Names, Lifestyles, and Various Customary Practices*. Translated by Margritt Engel and Karen Willmore. Fairbanks: University of Alaska Press, 2003.
114. Grekov, ibid., 1960:130–160. *Ocherki istorii geograficheskoi nauki v SSSR*. Moscow, 1976:41–47.

CHAPTER III

Mapping the Distribution of Water and Land in the North Pacific (1750–1800)

THE ERA OF THE *PROMYSHLENNIKI*

Despite the considerable scientific legacy of the Kamchatka expeditions, the cartographic results were only available to those at the highest levels of authority in Russia. These charts showed only approximate locations of small segments of the islands and coasts of Alaska. Published maps continued to be dominated by the hypothetical constructs of Delisle-Buache.

The Siberian administration and the Admiralty College remained interested in the geography and cartography of the northern Pacific, but for the three decades following the Second Kamchatka Expedition, most active exploration was done by parties of Russian *promyshlenniki*. They, like their predecessors of the seventeenth century, continued to move eastward in search of valuable furs. Except this time, they had to venture out on the open sea.

In the wake of the Kamchatka expeditions, Siberians wishing to search for fur seals, walruses, and other valuable marine animals had opportunity to acquire elementary skills in open-sea sailing through the new navigational schools in Okhotsk and Kamchatka. However, the role of these schools remained limited, and the Russian *promyshlenniki* sailing in the Eastern Sea (Pacific Ocean) continued to rely on earlier pioneer methods used in exploration and coastal navigation.

Thus, in spite of an essential change in the character of the advance eastward by the *promyshlennik*, their past experience, especially in the utilization of Native geographical knowledge, and the compilation of descriptions and charts, was not wasted. In the decades following the Second Kamchatka Expedition, more than 100 fur-procurement parties were organized. As these moved gradually along the

Aleutian Islands toward America, they delivered to Siberia and Central Russia not only valuable furs, but also geographic descriptions and schematic charts of the islands they visited.

The Admiralty College and St. Petersburg Academy of Sciences were always interested in these materials and, working in secret, attempted to link them to a system of coordinates (latitude and longitude) with the data provided by the Kamchatka expeditions. Eventually, five government expeditions were launched to provide new data. The best geographic and cartographic results were obtained by the expeditions led by Pyotr Krenitzin and Mikhail Levashov and by Joseph Billings and Gavriil Sarychev. In the 1770s to 1790s English, Spanish, and French expeditions also appeared off the northwestern coasts of America, establishing precise locations for the geographic information provided by the *promyshlenniki* and Russian mariners. The most important in this respect were the expeditions of James Cook and George Vancouver.

Emelian Basov

The start of the eastward movement after the Second Kamchatka Expedition was initiated by the sergeant of the Okhotsk Port Command, Emelian Basov. In the winter of 1743–1744 he sailed to the Commander Islands on the *St. Peter* and procured furs on Bering Island. In the course of his second voyage in June 1746 he found and explored Medny (Copper) Island to the southeast of Bering Island.[1] Though Basov's voyage was accomplished at "his own cost," the Okhotsk Port Office issued an official instruction to him dated June 1, 1741, which contained not only various demands, but also official support of his enterprise.

Specifically, he was given the right "to demand from the Kamchatka commander one blacksmith for working iron, two men for carpentry work, whom he is to pay the proper rate, and anyone suited to act as interpreter. Should there be on Kamchatka a compass, he can take one man who is knowledgeable in its use; should there be available old baidaras, he may take them on his voyage having given a receipt for them."[2] This government support is quite simply explained: besides the fur procurement for his own and his companions' benefit, Basov, when visiting "the islands in the sea, was to bring *iasak* by all possible kindness and friendliness under Her Majesty's High and All Powerful Hand." With this goal in mind, he was issued a "gift treasury" from which he was to draw presents "for princelings and the best men." He was strictly admonished: "Not to advance by force of arms against any people whatsoever, except in self-defense; if he must do so he will be under threat of heavy penalty. At all times he must take care that his men are not exposed to a sudden hostile attack, to loss and harm."

During winter on Bering Island, the servicemen under Basov's command discovered "the landing place of the Kamchatka expedition of Captain-Commander Bering, where the vessel was wrecked." At this spot they found the "government material" left by the expedition, but the servicemen, used to discipline, "did not dare to take these things without permission."[3] Basov's expedition was the first to visit Medny Island.[4]

Basov and members of his party returned with a rich cargo of furs. This served as an additional stimulus for other expeditions to the Commander and Aleutian islands. In particular, a member of Spanberg's expedition to the Japanese coasts, navigator Mikhail Nevodchikov, on the *St. Evdokim*, "reached the islands of Attu, Agattu and Semichi." The return voyage was accomplished in two baidaras, as the ship was wrecked. The voyagers brought a Native boy, Temnak, to Kamchatka (later on baptized with the name Pavel). From him it was learned that the inhabitants of the islands called themselves "Aleut."[5]

The official report on this expedition included a question that needed an answer: "at what distance are these islands from the land Kamchatka, at what latitude . . . the above-named Nevodchikov will report the details with a log and chart." This order was repeated on June 28, 1749, at the highest level issued from the Siberian desk to the Irkutsk provincial chancellery on the compilation of a description and a chart of the Near Aleutian Islands.

The decree demanded, in particular, that the Aleut boy brought by Nevodchikov be interviewed specifically about whose subjects or under whose power the inhabitants of these newly located islands were, whether they "pay any tribute or are they self-governed without any overlords, what their rights and laws are, how large and populous each are, what their armament is, what resources they utilize and with whom the peoples of each of their own lands and islands maintain commerce, and whether there are merchants, whether there are important and abundant minerals and animals, and in general whatever he (the Aleut) is able to convey, is to be taken down in writing."[6] Here we once again encounter a short questionnaire serving as the basis for interviewing local inhabitants and describing new lands.

Nevodchikov carried out the order and drew the newly located islands of the Aleutian chain on the chart of Near Islands. Nevodchikov delivered this chart, together with his report on the voyage, to the Bolsheretsk chancellery, and on May 4, 1751, the M. V. Nevodchikov chart was dispatched to the Senate. Where this chart is today has not been determined.[7] The results of the voyage to the Near Aleutian Islands were highly valued in the capital. On June 20, 1752, Nevodchikov, by an order issued by the Siberian desk, was appointed under-officer of the Okhotsk Port command, and a year later, on August 28, 1753, he received the rank of apprentice navigator and was ordered to serve in the Pacific.

Nevodchikov was an active participant in the work of the government expedition directed by Krenitzin.[8] Having experience in compiling charts since the Second Kamchatka Expedition, Nevodchikov participated in summarizing the results of Ivan Sindt's expedition. Between 1769 and 1771 he sailed the galiot *St. Paul* on the Okhotsk-Kamchatka route. Later, in Okhotsk, he instructed students in basic cartography.[9] The life of this Siberian, rich in events, whose entire life was bound up with the Pacific Ocean and the Far East, allows us to see the formation of a local professional cadre of navigators who were responsible for a considerable part of the description and cartography of the Aleutian Islands.

In 1748–1749, Ivan Bakhov, a merchant from Ustiug, and Yakutsk merchant Semyon Novikov visited the Commander Islands and rediscovered the sea route from the mouth of the Anadyr River to Kamchatka. [Merchants had been sailing from Anadyr to Kamchatka since the seventeenth century—LB] The *promyshlenniki* sailed out of the Anadyr River on the vessel they had built, the *St. Perkun and Zanat*. They were wrecked near the place where Bering's *St. Peter* went down. The remnants of the ship, rigging and iron parts, were used by the Bakhov-Novikov party to build a new vessel out of driftwood. They called it the *St. Kapiton*. The *promyshlenniki* reported to Empress Elizaveta Petrovna in February 1750 that they "undertook a second sea trek proceeding off the unlucky Commander Islands, holding to the northeast side to look for an unknown land or island which proved to be but a deception of the eye, then returned to the second island, from where, without prolonging the stay, we sailed for Kamchatka, arriving there on August twentieth, 1749."[10]

The merchants Novikov and Bakhov, with "Timofei Perevalov, acting sub-lieutenant" on this expedition,[11] understood the significance for future explorations in the Pacific Ocean of the route they pioneered and petitioned the empress "to accept this report about a new route as special, serving the benefit of the State, and show us great generosity by the rewarding and releasing to us the state property which we took on that island in the sea called Commander, overlooking our bold act."[12]

The *promyshlenniki* conclude their report to the empress with a plea that, in spite of the hardships they had suffered, they be permitted in the future to initiate sea voyagers, "and to grant full benevolent support as those who sail on their own, and not under (naval) command, (we) wish to avoid the danger of active and worthy simple people engaged in sea voyaging not being treated properly when, as a result, the completion of the tasks assigned to us is doubted, knowledge is lost, and our kind always finds himself under accusations."[13]

The results of the voyage by Novikov, Bakhov, and Perevalov were shown on a chart the sailors submitted to the commander of the Nizhne-Kamchatsk *ostrog*, Lieutenant Timofei Ivanovich Shmalev, who, in turn, having studied the results and possibly augmented the chart, forwarded it to his superiors. The chart reached the Senate (apparently after 1750) and survives in four versions.[14] This work deserves special analysis.

The versions differ in their arrangement and some details of their content, but all are apparently based on a copy in the Russian State Archive of Ancient Documents (RGADA, fond 192, opis 1, Irkutskaya guberniya, no. 10, part 1), specifically the *Chart compiled in Yakutsk county of Kamchatka and the River Anadyr, with adjacent areas of the Chukchi land, also part of North America and a small part of Japan*.

There is a statement that the chart was compiled by Perevalov, who "presented it to Lieutenant Timofei Ivanovich Shmalev."

On other versions Perevalov makes self-deprecating remarks, stating that the chart "was compiled with little sense by the last-among-men cossack subaltern." Apparently he was conscious of the fact that the work was not up to the cartographic standards of the time. Indeed, the chart brings to mind pre-Petrine geographic sketch charts; the presentation has the character of a rough draft. In spite of these drawbacks, the chart shows a reliable outline of all the coastlines in the northern part of the Pacific known at that time and even attempts a summary of information about this region obtained by various expeditions in the 1730s and 1740s.

In his general treatment of the land and sea between Kamchatka and the territory named as "The American Land, investigated and described in the year '741 during the sea voyage by Captain Bering" (one version also shows "the coast observed by Captain Chirikov"), the author quite clearly follows Sven Waxell's version, as the "American Land" is represented as extending almost to Kamchatka. Two versions show the routes of the Bering and Chirikov vessels.

The chart shows the result of Gvozdev's 1732 voyage, though with an incorrect date. Specifically, across from the Chukotka a "land sighted by geodesist Gvozdev in 1730 during his voyage from Kamchatka" is shown. In the strait between this land and Chukotka (the Bering Strait) "Magli" ("Mogli") and "Ngalgi" ("Igalgi"), islands are shown, "inhabited by toothed (labret-wearing) Chukchi."

Across from these islands, on the mainland, there is an inscription: "part of North America, here live people called kykhmatlsy." Two versions show the routes of "the sea voyage of geodesist Gvozdev in the year '730." The chart also shows a crude representation of the Kurile Islands and the northern part of Japan. To the northeast of Bering Island is an inscription: "Land sighted in the year '748, on the 13 day of July during the sea voyage from the mouth of the Anadyr River to Bering Island by the merchants Novikov and Bakhov and from the Anadyr subaltern Perevalov."

This part of the chart clearly contradicts the report submitted by the merchants to the empress stating that they were unable to find "the unknown land or island which appeared" and that having become convinced that this was a "deception of the eye" they turned back. Apparently Perevalov was not convinced of the absence of the land and therefore showed it on his chart. One version of this chart shows Pavlutsky's raid across the Chukotka Peninsula in 1743.

It is difficult to maintain that Perevalov was the sole author of this chart, especially since on one of the versions Lieutenant Shmalev is named as compiler.[15] On another version Perevalov's name has been erased from the chart's title.[16] Regardless of who the participants were in the creation of this chart, it is clear they were well acquainted with the voyages undertaken in the 1730s–1740s. They had access to the Bakhov-Novikov materials, charts by Sven Waxell and Martin Spanberg, descriptions and charts arising out of Gvozdev's voyage in 1732, and information provided by the Chukchi and Eskimos about the islands in the Bering Strait and about the lands where the *kykhmaltsy* lived. This is the first mention of the islands' Native names.

Despite the absence of any mathematical grounding, all versions of the chart allow a hypothesis about the land and sea distribution in the northern part of the Pacific based on real information on hand at the end of the 1740s. The chart does not show any mythical lands of the type of Iesso, Company Land, etc., but it is also missing the Near Aleutian islands of Attu, Agattu, and Semichi, by then known from the voyage of Nevodchikov, 1745–1747. The chart examined here may be considered to be an interesting attempt to summarize the available geographic information. The fact that four versions of it are present in Russia's central archives testifies to the fact that it aroused interest.

The Anadyr sea route to Kamchatka increased the flow of Russian *promyshlenniki* toward the Aleutian Islands. In the beginning, fur-procuring parties visited Bering and Medny islands, where they wintered and put up the meat of sea mammals for continuation of their voyages. The next year they moved to one or another of the Aleutian Islands, where they used Aleuts to obtained valuable furs. To ensure their safety as well as to obtain a guarantee that island inhabitants would accept Russian citizenship, the *promyshlenniki* used the ancient Siberian practice of taking hostages. Some of these were subsequently brought to Kamchatka, taught to read and write Russian, and became trusted guides and interpreters.

The Siberian and central Russian authorities supported this movement eastward. The appointment of Vasily Miatlev as Siberian governor in 1753, a man well educated and knowledgeable in marine science, is clear evidence for this. In June 1753 Miatlev presented a project for renewal of the Kamchatka Expedition to the Senate. The project included the construction of a fortress and a shipyard at the mouth of the Anadyr River. In response, a decree was issued December 28, 1753, "*by H. I. M., autocrat of all Russia, from the Governing Senate of the Admiralty College*" to renew the Kamchatka Expedition. The leadership was entrusted to Lieutenant-General V. A. Miatlev.

The decree underscored the necessity of collecting and analyzing the materials of previous voyages.[17] Copies of all the available charts of the Second Kamchatka Expedition were made at the Naval Noble Cadet Corps under the supervision of Alexei Nagaev (who had worked with this material in the 1740s). Only some of the originals were available in St Petersburg. Copies were made and all the originals were shipped to Tobolsk—in total thirty-seven original journals and sixty-three original charts.[18] These journals and charts were destroyed in a fire in Tobolsk in 1787. Today, scholars have only copies and the very few originals that had remained at the Senate or the Admiralty College.[19]

On Miatlev's recommendation, preparations for the renewed Kamchatka Expedition were led in 1754 by Fyodor Soimonov,[20] who later became governor of

Siberia in 1757. Miatlev's plan to use the mouth of the Amur River as the base for further marine voyages was blocked due to unshakable opposition by the Chinese government, which refused the Russians permission to establish navigation on the Amur.[21] For this reason, in the summer of 1763 when Soimonov was once again in Moscow, he formed a new plan for reviving the Kamchatka Expedition, proposing to dispatch vessels to America from of the mouth of the Anadyr.[22]

His proposal was accepted by the Senate on March 11, 1764, but due to tensions in Russian-Chinese relations, the activities of the renewed Kamchatka Expedition, or the Nerchinsk Expedition, as it was then called, were canceled on the pretext that excessive spending was to be avoided. Although the Miatlev and Soimonov plans did not materialize as actual voyages, the efforts to collect and analyze the materials of all previous expeditions prepared the documentary basis for new exploits in the Pacific Ocean.

The measures undertaken by Soimonov for economic development of the region were also instrumental.[23] The governor of Siberia was concerned with preparing a qualified body of sailors, not only for future government expeditions to America, but also for those undertaken by the "promyshlenniki at their own expense." He continued the practice of organizing local navigational schools that began back with Bering's educational institution in Okhotsk.

In the meantime, an ever-growing number of private sea voyages were launched to sail along the Aleutian chain. In 1755–1756 navigator Pyotr Bashmakov sighted eight Aleutian Islands, sailing on the vessel *St. Jeremiah* [the *Jeremiah* was rebuilt out of a wreck in the Aleutians—LB]. Some of these island groups received the names Rat and Andreanof islands. Though it has not been determined precisely on which islands Bashmakov made his landfalls, he apparently landed on Adak Island.[24] On the basis of Bashmakov's description, Bancroft identified one of these islands as Tanaga.[25]

On the summary chart of the Aleutian Islands, the *promyshlenniki* noted that the *Jeremiah* was wrecked at the island of Adak. Out of the wreckage a new vessel was built and named the *St. Peter and Paul*. On it the sailors returned to Kamchatka in August 1755.[26] Between 1756 and 1758 navigator Bashmakov visited thirteen islands to the east of the Near Aleutians.

On this voyage baidaras were used in investigating the islands, for hunting, and for in-shore navigation by the Russian *promyshlenniki*. The policy of hostage-taking continued, as did the collecting of unusual Native artifacts, in particular they gathered "a (model) baidarka for small children, so called by them, and also from these peoples arrows [spears or harpoons—LB] cast from a board by hand and one hat which they wear when at sea."[27]

Bashmakov's voyage began a period of intensive investigations and cartography of the Aleutian Islands by private fur-procuring parties. Of major significance was the voyage of the Yarensk traders Stepan Glotov and cossack Sergei Ponomarev aboard the *St. Julian*: "a sea voyage to the known and unknown sea islands."

The sailors reached the most distant islands of the Aleutian Chain, Umnak and Unalaska, which were named the Fox Islands.

By order of the Kamchatka office, the members of the fur-procurement party, besides bringing new peoples under the *iasak* payment system and taking from them "hostages from among the best men, as far as is seemly," were obliged "to investigate and search out land and sea curiosities and foreign items, gold and silver ores, pearls, stones, lead, iron, mica and other valuable items."

Having wintered on Medny Island and "after procuring for provisions sea cows, seals and sea lions and put up dried meat," the expedition sailed on August 1, 1759, along the Aleutian Islands "without touching on those that were known" and exactly one month later arrived at an island "lying to the northeastern side, called by the local inhabitants Umnak . . . Not far from this island, about 15 versts distant, is a second island, greater in circumference than the first, called Unalashka."

The travelers learned from the Natives that Unalashka was inhabited by about 300 men and that to the east there were eight more islands "on one of which there is standing timber (but of what kind they were unable to explain in the Russian language) and also there are animals, sea otters, foxes silver, black, cross and red ones, which are found on all islands in considerable quantities. On the island where there is standing timber there are caribou, bears, wolves and ermine." In this information provided by the Aleuts we clearly discern data about the Alaska Peninsula or an island close to it with flora and fauna that is continental in character. Once again the inhabitants of the islands and coasts of the northern Pacific demonstrate the breadth of their geographical knowledge.

In their report submitted to the Bolsheretsk chancellery, Ponomarev and Glotov included a brief description of the environment and animal world of the islands they visited and noted particularly that there were lakes and streams which were entered by the same kind of fish as those on Kamchatka: silver salmon and Dolly Varden. And that halibut and cod were found in the sea. Commercially valuable animals were listed: "sea otters; silver, black, gray, cross and red foxes of various qualities." They also noted that the islands were totally treeless and that "the inhabitants use driftwood in the construction of their dwellings, sod houses forty *sazhen* in length (280 feet) and even longer."

Interesting ornithological information was provided, some of which was clearly derived from Native observations of long duration. In particular, on both islands they reported that there were "pelagic cormorants, tufted puffins, murres, Anas hyemalis, oyster catchers, river ducks, American goldeneye, mallards, drakes, and for the most part white geese which always winter here but in the spring take off to return next winter season, and gray and black geese, migrating ones for sure, seagulls all year round, and also eagles."

The report also contains a large amount of data on human population, social structure (*toyons* [headmen] and their kinsmen), language, everyday life, hunting and fishing methods, and fire-making. According to the *promyshlenniki*, as they were departing from the islands in 1762, "because of the voluntary inclination of these people to citizenship and because of our kindness and civility, they decided to be (Russian) subjects, wishing that Russians would always come with their vessels and stating that they would pay *iasak* in full, and they of their own free will gave as hostage a nephew of the toyon of this island, named Mushkal, aged about 12 or 13 years, whom we named Ivan." The boy was safely brought to Bolsheretsk.

The list of islands to the east ("distant ones") as well as Unalashka and Umnak to the west given to our travelers by the Natives of these two islands is of considerable interest. The list contains local place names and a brief description for practically all the large Aleutian Islands (primarily listing commercially valuable animals). Many of these still retain the place names first given to the Russian sailors by the inhabitants of Unalashka and Umnak. For lands to the east of the islands where members of the Ponomarev and Glotov company wintered, fourteen place names are listed (Akutanak twice), specifically "Alakhshak," characterized as "populous, with plenty of standing timber, foxes, bears, caribou, by all indications martens, wild boar found where the Bechevin vessel wintered [Bechevin Bay, Alaska Peninsula—LB]. Also Kadiak, to the south side [of the Alaska Peninsula—LB] where there are foxes, sea lions, and standing timber up to five hands wide."[28]

On the basis of the data cited above and information provided by other merchants, Pyotr Shishkin, a member of the Glotov-Ponomarev Expedition, compiled a chart of the Aleutian Islands known up to that time. The *Chart of the islands of Umnak and Unalashka compiled by Totma trader Pyotr Shishkin, participant in the voyage of the vessel Julian in 1759–1762* (Fig. 13) was presented to the Admiralty College and reported to the empress, Catherine II.

It is indicated on the chart that it was compiled on September 1, 1762, at the Nizhne-Kamchatsk *ostrog*. In reality it is a rather conditional sketch, without adherence to a scale or strict orientation. The Aleutian Islands themselves are indicated, though their general orientation (to the north-northeast) is imprecise. On the map itself the length of Unalashka and Umnak is given: "about 300 versts" for Unalashka and "about 250 versts" for Umnak. On both islands volcanoes are shown ("the peak is burning strongly"). Kodiak Island is shown at the extreme southeast and along its coast is the inscription: "a man from this island, Kashmak, joined the company."

The Alaska Peninsula is shown as a big island to the north of Kodiak with the elucidation: "Alakshak, rather many people there, the vessel of Bechevin, Alaska." In all, 55 islands are indicated, not counting Commander (Bering) and Medny Islands and Alaska with two islands immediately adjacent to it: "Tugidak" and "Kuguk." Shishkin's work, as far as the Aleutian Islands are concerned, presents us with an original cartographic description, indicating names for some of the islands, and their sequential location proceeding from Kamchatka. The author also attempted to indicate the islands' size and the general outline for the biggest.[29] In the northern part of the chart where the Bering Strait ought to be, three islands are drawn,

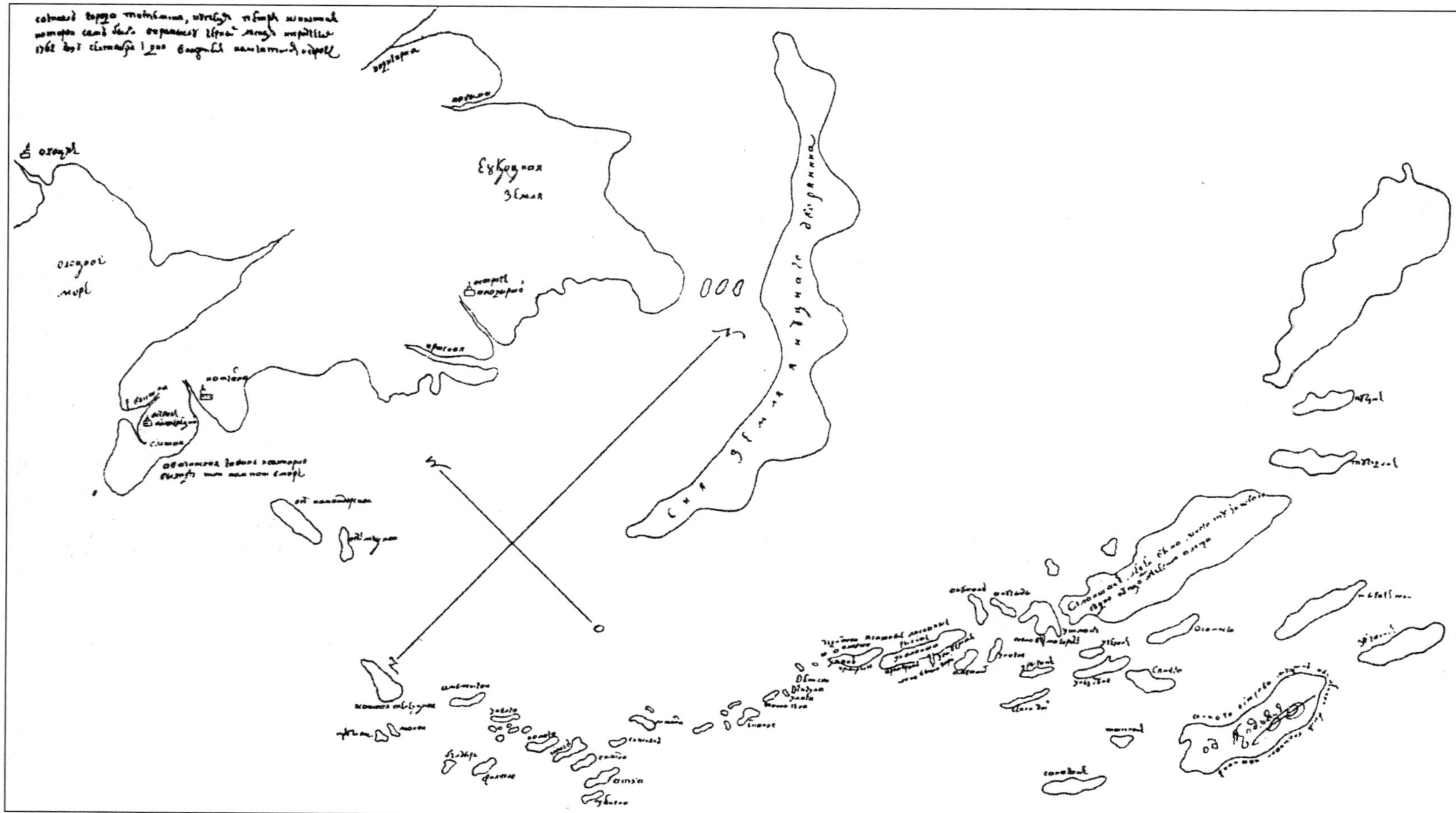

FIGURE 13 Shishkin's map of Umnak and Unalaska Islands. AVPRI F. 339. IAK Op. 888. No. 923/3

and to the east of these an elongated, extensive "land of the Yakut nobleman." This indicates the use of the chart by Ivan Lvov, about 1710, which was discussed earlier.

The voyage of Selenga merchant Andrean Tolstykh on the vessel *St. Andrean and Nataliya* in 1760–1764 was of major significance in the exploration of the northern part of the Pacific Ocean. In 1761 he was active on the island of Adak and five other islands: Kanaga, Tanaga, Chetkhina (Great Sitkin), Tagalak, Atkha (Atka), and Amlia. These islands are still called the Andreanof Islands. Tolstykh prepared a detailed description of these islands and their inhabitants.

Tolstykh and his assistants, Maxim Lazarev and Pyotr Vasiutinsky, as well as other members of the command, at once established friendly relations with the islanders who were brought "under the power of H.I.M, and *iasak* payment." As Tolstykh states in his report, when the Russians were ready to depart, all the "toyons unanimously declared that we did not offer any harm to anyone and showed nothing but friendliness."[30]

In 1763 Glotov continued the explorations he had begun two years earlier, on the vessel *St. Andrean and Nataliya*. Guided by information obtained from the Aleuts about distant islands, he reached Kodiak by October. His party remained in Kodiak until May 24, 1764. Glotov reported on his voyage to the commander of Nizhne-Kamchatsk *ostrog* on August 22, 1766. He described the Kodiak natural environment and its people, stating that "they are called kanagyst."[31] This first description by a European of one of the largest islands off the coasts of Alaska and its Native inhabitants is of significant historical, geographical, and ethnographic interest. A special copy prepared for Gerhard Müller survives in the archives and has many of Müller's annotations in German. The winter Glotov and his party spent on Kodiak marked the beginning of the Russian colonization of the islands off the northwest coast of America.

In the 1760s and 1770s, several summary maps were prepared on the basis of *promyshlenniki* voyages. These attracted considerable attention on the part of the Siberian as well as the central authorities in Russia. One of these is a schematic chart mainly based on information from the 1760s, though apparently at a later date supplemental information from the voyages of the 1770s was added (Ivan Soloviev on the vessel *Paul* 1770–1775 and Pankov on the vessel *Aleksandr Nevsky* 1770–1774).[32]

The compilation of the initial version of this cartographic work is linked to the names of Pyotr Shishkin and Savva Ponomarev.[33] It is most likely, however, that several of the *promyshlenniki* participated in its composition and the chart is the result of a collective effort. In contrast to the chart of the newly discovered islands of Umnak and Unalashka of 1762, discussed earlier, on this chart the Aleutian Islands are oriented more or less correctly (approximately by latitude) and are shown in greater detail. Although coordinates, scale, and projection are absent, this chart is a geographically illustrated register of the islands. The information included shows that Russian sailors were amassing considerable information about the lands between Kamchatka and Alaska.

The shortcomings of the maps are compensated for by the many explanatory notations containing extremely diverse information of value to future seafarers. Among these are the local islands' names, sometimes rendered in several forms, and the appearance of Russian names for some of the islands. Thus, for the island of Amlia, a place name surviving into our times, other versions are given: "also Amlan or Amlag." Tanaga Island is indicated as "Taniakha or Tanakhan"; Chagulak Island—"Chigula, in Russian Shapka"; Kyska—"Kiska or Kysa"; Agattu—"Krugloi also Igattaku"; and so on.

The chart contains historical data about the voyages of the *promyshlenniki* and pilotage information they obtained. Thus at the island of Attaku is the inscription: "In the year '755 Trapeznikov's vessel wintered here on the left side in a bay where up to 10 vessels can stand, the bottom is shingle (pebbles)." Druzhinin reported, "A rock, covered at high tide, but open at ebb, on it sea otters are caught." Between the islands of Amlia and Seguam there is a warning for sailors: "in this strait the current is very swift from the tidal change of high and ebb; the current is N and S." Between the Alaska Peninsula ("Alyaksa or Alyaska, in bay "A" Krasilnikov's vessel wintered, on her were Druzhinin and Pankov, the skipper was quartermaster Sharypov") and Kodiak Island ("Kadiak, alder trees are found here") is the following inscription: "here, as declared by Soloviev and Glotov, there are many small islands and some larger ones, how many they were unable to state."

About Unalashka it is stated that "there is a larger plan of this island on a separate sheet showing 4 bays." This plan apparently did not survive. The contours of Alaska's northeastern coast were left incomplete on the map.

Russian sailors observed natural phenomena. They paid particular attention to volcanos. At Seguam Island we find the following information: "Seugam or Sakhagamun, all ridges are flaming, the fire can be seen always, and thundering noise occurs from time to time, before bad weather it always thunders, but when it rains, there is an increase in flames." The islands of Amukta and Yunaska are marked thus: "Amukhta or Aliautak burns with fire, while Aniuska burns but little."

This chart is a valuable historical and geographical source. One of the aims of its creation was apparently to represent the study of the Aleutian Islands by various fur-procuring parties. The evidence for this is the fact that island groups are divided by the sources used in the compilation of this chart: "According to the testimony of Prokopy Lisenkov," "According to the testimony of Druzhinin, Pankov and Soloviev," and "According to the testimony of Glotov."

Another generalized map of the Aleutian Islands is not as well known as those examined above. This multicolored chart, in watercolors, is in the Russian State Archive of the Navy in St. Petersburg.[34] The cartouches that bore the chart's title are lost, though the signature certifying the chart of D. I. Checherin (1763–1781) survives.

The chart covers Kamchatka to Atkha and Amlia, the Near Islands, the Rat Islands, and the Andreanof Islands, with the rest in an inset that bears an extended title: *The Location of Islands in Forward Direction. Those beyond Atkha and Amlia are Estimated on the Basis of the Testimony of Faithful Subjects who Live in those Islands and also of the Three Men Brought aboard a Merchant Vessel from the Distant (Islands of) Shemalga (and) Umnak who have Learned Russian Speech Somewhat; from the Promyshlennye Men what Kind of Animals are Found on Each Island.*

The chart has a latitude/longitude grid with compass lines and two drawings of Aleuts in single-hatch and two-hatch baidarkas. The inset chart is more schematically executed and lacks the grid. An inscription at the Andreanof Islands states "found and brought into citizenship by the Selenga merchant Andrean Tolstykh." It is indicated that "The chart was copied by the Sergeant Stepan Tentinov." This work may be earlier than the chart of 1767 by Vasily Shilov discussed above. What sets it apart from other generalized charts by the *promyshlenniki* is the direct statement that information was obtained from the inhabitants of the Aleutian Islands.

News brought by the *promyshlenniki* about the formerly unknown lands in the Pacific Ocean and their charts attracted the attention of the Siberian authorities, the central administration of Russia, and also Russian scholars. This whole corpus of new information delivered to the central institutions left open a crucial question: how did it integrate into the cartographic results of the Kamchatka expeditions of Bering and Chirikov and, consequently, what was its real location within the system of the geographic grid? Neither the charts nor the descriptions provided by the *promyshlenniki* provided, and could not provide, direct answers to those questions, as few astronomical observations of latitudes were conducted when these charts were compiled. One exception is Shalaurov's map presently in the manuscript division of the Library of the Academy of Sciences in St. Petersburg. It clearly indicates that astronomical observations were taken and also shows which instruments and methods were used.

In order to answer those questions a new expedition was needed. Reactivating elements of the Kamchatka Expedition in 1753 might be the answer. Soimonov, desiring to clarify the geography of the North Pacific, dispatched an expedition under a member of Bering's Second Kamchatka Expedition, Lt. Ivan Sindt in 1761.

Sindt's Voyage

Sindt's instructions were "to sail along the coast between north and east to the Chukotka Peninsula, lying to east, across from which there is a small island called St. Lawrence. From this region proceed directly east or very slightly north toward the American continental land."[35]

In Yakutsk, the "chief commander of Anadyr," Colonel Friedrich Plenisner, also a member of Bering's Second Kamchatka Expedition, ordered the expedition

> upon arrival at the Chukotka Peninsula, to sail east or only one rhumb to the north, directly towards the American coast, that is east or east-north; arriving

> at the American coast, continue along it northward until a river or a location where fresh water is available is sighted, and stand there. You are to investigate and describe as you have been directed; having spent some time there, you are then to proceed further north along the coast for about two weeks in order to more precisely determine how the American land continues (that is if it is an island or the mainland itself) and again you are to go on shore in order to investigate the local circumstances and to describe everything in detail; from there you are to return as time permits to establish a winter camp.

Sindt's instructions also took into account the possibility of contrary winds interfering, in which case "return and on the American shore, in a convenient location, and for safety, construct a fortress and winter there."[36]

Sindt's expedition continued from September 18, 1764, to September 24, 1768, with the galiot *St. Paul*, which became damaged and was reconstructed en route as the *St. Catherine* in 1765. The first two years were spent along the Okhotsk Sea littoral (coastline) and off Kamchatka. Only in June 1766, did Sindt sail toward the northwestern coast of America. The *St. Catherine* reached 65° N latitude and just past 40° E longitude from Kamchatka.

The main, possibly the single, achievement of this voyage was the discovery of St. Lawrence Island.[37] In thick fog, Sindt assumed the capes and headlands of St. Lawrence Island he sighted to be separate islands.

Sindt's expedition returned to Kamchatka on September 21, 1766. The chart of the voyage to the American coast, compiled upon the return to Kamchatka utilizing the vessels' logs, were sent, in turn, to Pyotr Krenitzin and Plenisner, and then to the governor of Siberia, Denis Chicherin, who then dispatched them to the Admiralty College. The chart had to be redrawn because the original was found to be "in many places smudged and later judged not suitable for presentation."[38] Copies of the chart of Sindt's voyage of 1764–1768 (Fig. 14) can be found in the manuscript department of the Russian Academy of Sciences Library in St. Petersburg and in the Russian State Archive of the Navy.[39]

The results of this expedition were viewed negatively by the majority of scholars, and rightly so. The report and the chart submitted to the Admiralty College are evidence that Sindt exaggerated the number and size of the islands he had sighted. Jacob von Staehlen believed that the islands sighted by Sindt's expedition were between 56° and 57° N latitude, while A. R. Vorontsov in 1788–1789 assumed that these lands were between 60° and 65° N latitude.

One point on which Sindt's contemporaries all agreed was that the expedition changed the geographic concepts about Anadyr Bay and the American coasts across from it.[40] Sindt added to the confusion and because of that he failed, absolutely. The expedition's materials created a tradition of false representation of the northern part of the Pacific Ocean and this caused difficulties for later mariners.

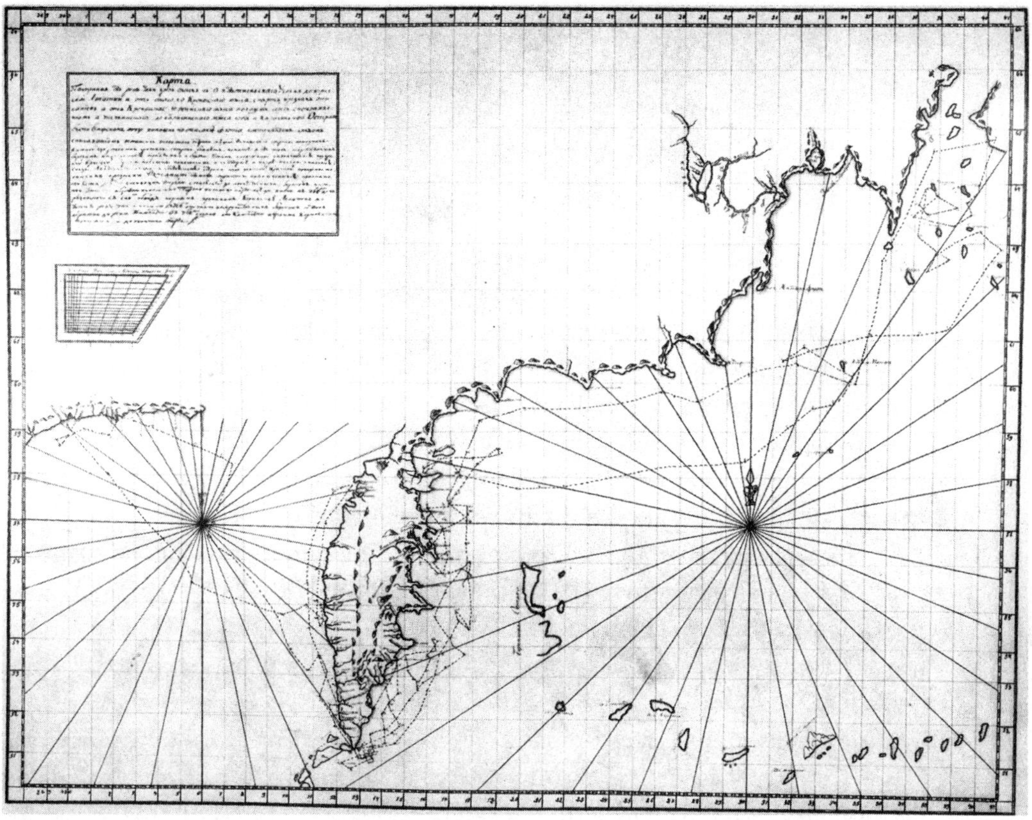

FIGURE 14 Map of Sindt's voyage, 1764–1768. BAN Otel rukopisci. Op. 35. Karta 365

The partisans of this tradition were mainly scholars and cartographers of the St. Petersburg Academy of Sciences who trusted "enlightened" seamen instead of the *promyshlenniki* or the Natives.

THE KRENITZIN/LEVASHOV EXPEDITION

Even as Sindt was sailing toward the American coast, a new secret expedition under the leadership of Petr Krenitzin (1728?–1770) and Mikhail Levashov (1739–1775?) was in active preparation.

On May 4, 1764, Empress Catherine II signed a decree by the Admiralty College authorizing the organization of an expedition to describe new islands found in the Pacific Ocean. The goal of the expedition was the precise determination of the lands "possessed" by the *promyshlenniki*. It stressed that

> as acquisition of these possessions was accomplished by persons lacking navigational knowledge and science, their descriptions and notes are not sufficient

> for the extraction of all possible benefit; we order hereby, with appendices of all information received, our Admiralty College to act in accordance with the proposition by Governor Chicherin by sending immediately, in accordance with their judgment, the needed number of officers and navigators, entrusting the command to the senior one whose knowledge of marine science, and diligent attention to it, is known.[41]

The decree stressed the extreme confidentiality of this expedition and demanded that "this enterprise be conducted in secret manner, not making known this order until the proper time, even to the Senate, entrusting the conduct of the preparations only to the over-secretary and one person able to make copies." In order to observe secrecy, the expedition was officially named the "Commission for the description of forests along the Rivers Kama and Belaya."

Captain-Lieutenant Krenitzin was appointed the expedition's commander and Captain-Lieutenant Levashov his assistant. The Admiralty College instruction dated June 26, 1764, mandated that the expedition check the information supplied by Glotov and Ponomarev about the islands of Unimak and Unalaska and investigate based on data supplied by local inhabitants. The route of the expedition was prescribed. They were to strictly follow the Aleutian Islands "keeping the course in sight of the islands which stretch from the Aleutians [Near Islands—LB] to Unimak and link the islands sighted as you sail by means of bearings, accurate rhumbs and distances, observing en route all the signs and circumstances pertaining to the description of these islands and to your routes, recording everything in the log."

They were to enlist up to five persons for the voyage, Kamchatka *promyshlenniki* or cossacks who "have become acquainted with the customs of the Aleutian inhabitants and know their language somewhat," and it was stressed that it was imperative to engage in the expedition the "Nizhne-Kamchatsk cossack (Ivan) Savin Ponomarev." It was expected that through Ponomarev information would be obtained from the inhabitants of Unimak and Unalaska about the sixteen islands farther to the east or northeast and "which stretch far, as is indicated in the report by this self-same cossack to the Bolsheretsk office and also on the chart compiled by the merchant Pyotr Shishkin."

There was great insistence on the need to question local inhabitants about "the course for and distance to the unknown timbered islands, such as Alakshan, the populous one where the vessel of the Russian merchant Bechevin wintered, to Kadiak and Tygachtany or Shugachtany." In this instruction by the Admiralty College we encounter for the first time recommendations regarding methods of utilizing Native geographic information in the compilation of charts. "When these persons, or those on other islands, will indicate lands where islands not sighted lie, by gesture, these are to be recorded diligently and discreetly through establishing compass bearings and rhumbs and the recording (of) the same in the log. As far as distances are concerned, if it is impossible to measure these, it is to be found out

. . . how many days it takes to reach one (of these islands), so that you might be able to plan your courses."[42]

The necessity of going ashore was stressed and Alakhshak was recommended as the most suitable location for wintering over, if it became necessary to do so, as "on the chart (that of P. Shishkin) an inscription indicates that the vessel of the Russian merchant Bechevin wintered there." Wintering over was deemed desirable, as such a circumstance would allow the members of the expedition to explore the island and adjacent lands, get acquainted with its population, and establish "how distant from it is the American mainland" as well as to conduct surveys and cartographic work.

The return route from the islands lying in the east was to be planned in such a way as to "pass by all of the 16 islands reported to cossack Ponomarev by the Umnak inhabitants, sailing from one to the other, specifically from one wooded island to the next wooded one, calculating the distance by sea most precisely, projecting your course in a manner which would bring all the treeless islands into sight, enabling you to take bearings and thus to tie in these points to your voyage."

The Admiralty instruction paid great attention to the need for the expedition to establish the location of the distant Aleutian Islands in relation to the "American mainland." It also mandated clarification of whether any nearby territories or settlements on the mainland were under the control of one of the European nations and also if there were any "large inlets cutting into the mainland from our side or passages to another sea to the east or north."

This instruction by the Admiralty College constitutes direct evidence that by this time Russian hydrographers were taking seriously the descriptions and charts compiled by the Russian *promyshlenniki* and the geographic information provided by the aboriginal inhabitants of the Aleutian Islands and considered it quite possible to use these data in the creation of sea charts within the framework of a professional exploratory expedition.

Previous documentation was gathered and copied. As a result, the Admiralty had Shishkin's chart of the Aleutian Islands, "extracts from the journals of Captain-Commander Bering, Captain Chirikov, Lieutenant Waxell and Master Khitrov and copies of their charts of the Arctic as well as of charts compiled in Siberia." They also had Gvozdev's report to Spanberg about his voyage to America.[43]

It was expected that Krenitsin's expedition would be coordinated with the expedition of Vasily Chichagov, planned in accordance with the ideas of Mikhail Lomonosov—Russian scientist and a cofounder of Moscow University—to sail to the Pacific Ocean from Arkhangelsk across the North Pole. A "secret appendix" was created that dealt with the meeting of the two expeditions in the Pacific Ocean.[44]

In Tobolsk, Denis Chicherin, governor of Siberia, issued additional directions, which ordered the expedition's personnel be augmented by enlisting students of the Siberian navigational schools. To implement this instruction "the students of navigation of the Irkutsk and formerly Tomsk schools as listed on the attached roster are to be examined and of these, nine whom you should find worthy are to be selected

. . . and when your excellency commences your voyage, you will assign them as you deem fit to the vessels of your squadron."[45]

The governor of Siberia saw as the main goal of the expedition the provision of the "detailed description and cartography of several Aleutian islands that have been located by merchants but especially of the large and populous island called Alakshak and the second one, Kadiak, making every effort to circumnavigate the latter and to describe it. It is very necessary to learn whether it is an island or the mainland, as one cannot rely on the testimony of our people who have visited it."

Prior to sailing, he recommended that several of the *promyshlenniki* be enlisted to serve on the expedition vessels because "if landfalls be made in the islands, they would be able to engage in procurement for themselves, and their knowledge of the peoples' customs would serve the expedition for the best."[46] In accordance with this directive, sixteen *promyshlenniki*, among them Glotov, Pushkarev, and Soloviev, joined the expedition.

Krenitsin's expedition had problems and it was only on July 20, 1768, that the galiot *St. Catherine* and the hooker *St. Paul* left the mouth of the Kamchatka River and commenced the voyage to the American coast.[47] By the end of August the expedition reached the distant Aleutian Islands and Krenitzin spent the winter of 1768–1769 on Unimak Island and Levashov on Unalashka.

Throughout this time separate detachments actively explored the islands and the adjacent part of the Alaska Peninsula. For the hydrographic surveys and depth determination the main vessels used by the expedition were Aleut baidarkas and baidaras made in Kamchatka. Krenitzin compiled a special guide for the hydrographic work, issued initially to navigator A. A. Dudin, Sr., but apparently used by everyone engaged in the expedition's work.

The instruction had practically nothing to say about the methods of surveys and descriptions; the main focus was on relations with the local population and security measures. The instruction contained the following: "While on shore, examine the land, timber, animals and birds, and especially settlements; whatever is observed, is to be described and upon your return to the vessel is to be explained in a report; the servicemen are always to be under arms and never to be permitted to leave the coast for the interior; when taking on water, those not so engaged are to act as sentries with guns at the ready." The men were also instructed to persuade representatives of the local population to visit the expedition's vessel "to declare to the commander the manner of their life and condition and their knowledge about other people dwelling in this sea."

In 1769 the expedition returned to Kamchatka. On July 4, 1769, Krenitzin drowned in a tragic accident. Bringing the expedition to its close and organizing its materials was completed under the leadership of Levashov.

The expedition's geographic and cartographic results were of considerable value. The main geographic goal—finding the position of the Aleutian Islands on the chart of the Pacific Ocean—was achieved. Wintering on the islands of Unimak and Unalashka permitted multiple determinations of geographic coordinates based on lunar and solar observations, while the carefully planned course, both outward and home bound, with the frequent taking of bearings on the coasts of the Aleutian Islands, resulted in the establishment of their locations in relation to each other with a degree of precision fully acceptable by the standards of the time.

The latitudes of the Aleutian chain as a whole were also clarified with precision. However, the extent of the chain's reach toward Kamchatka as shown on the expedition's charts proved to be too short because of the inability at that time to determine precise longitude.[48] Forty-five islands in all were placed on the chart with an indication of their coordinates; 200 km of the Alaska coastline were investigated, Isanotsky Strait was explored, and regions north of Unimak and the islands of the archipelago located between Unimak and Unalashka, later named the Krenitzin Islands, investigated.[49]

The Krenitzin/Levashov Expedition laid the foundation for a systematic hydrographic survey of the Aleutian Islands. The cartographic work performed by members of the expedition during the wintering on the islands of Unimak and Unalashka are outstanding in their thoroughness and detail.

The manuscript copies made in Tobolsk of the large-scale plans of Igunok Bay (modern Unalaska Bay) and St. Paul Harbor in that bay (modern Captain's Bay), on the island of Unalashka, executed by navigator Y. I. Shabanov, and which are preserved in the Russian Archive of Military History, show the quality of the work done (Fig. 15).[50]

The summary charts of the Krenitzin/Levashov Expedition, several versions of which are preserved in various Russian archives, were a substantial contribution to the development of geographic concepts about the northern part of the Pacific Ocean.[51]

All the versions of these charts offer data on the islands' inhabitants and resources in the Aleutian Islands and the Alaska Peninsula as well as stating sources for each chart and methods of its compilation. For example, on one of the versions of a summary chart of the expedition[52] the heading reads:

> Mercator chart, compiled aboard the galiot *St. Catherine* under the command of Fleet Lieutenant Pyotr Krenitzin in the course of the voyage in the years '768 and '769 from the mouth of Okhota (Okhotsk) to the island of Aliaksa (Alaska Peninsula) with an outline of the Kamchatka coast and its interior situation up to the mouth at Nizhne-Kamchatsk. The locations sighted were charted according to bearings taken, and others taken from previous charts of Bering and from the catalogue of the Russian atlas; from the mouth at Nizhne-Kamchatsk eastward, up to Aliaksa Island, the islands were placed on the chart on the basis of bearings (recorded) in the journal; some island bays were indicated on the basis of experienced *promyshlennye*; from the strait between Aliaksa and Unimak the coast of Aliaksa eastward together with the

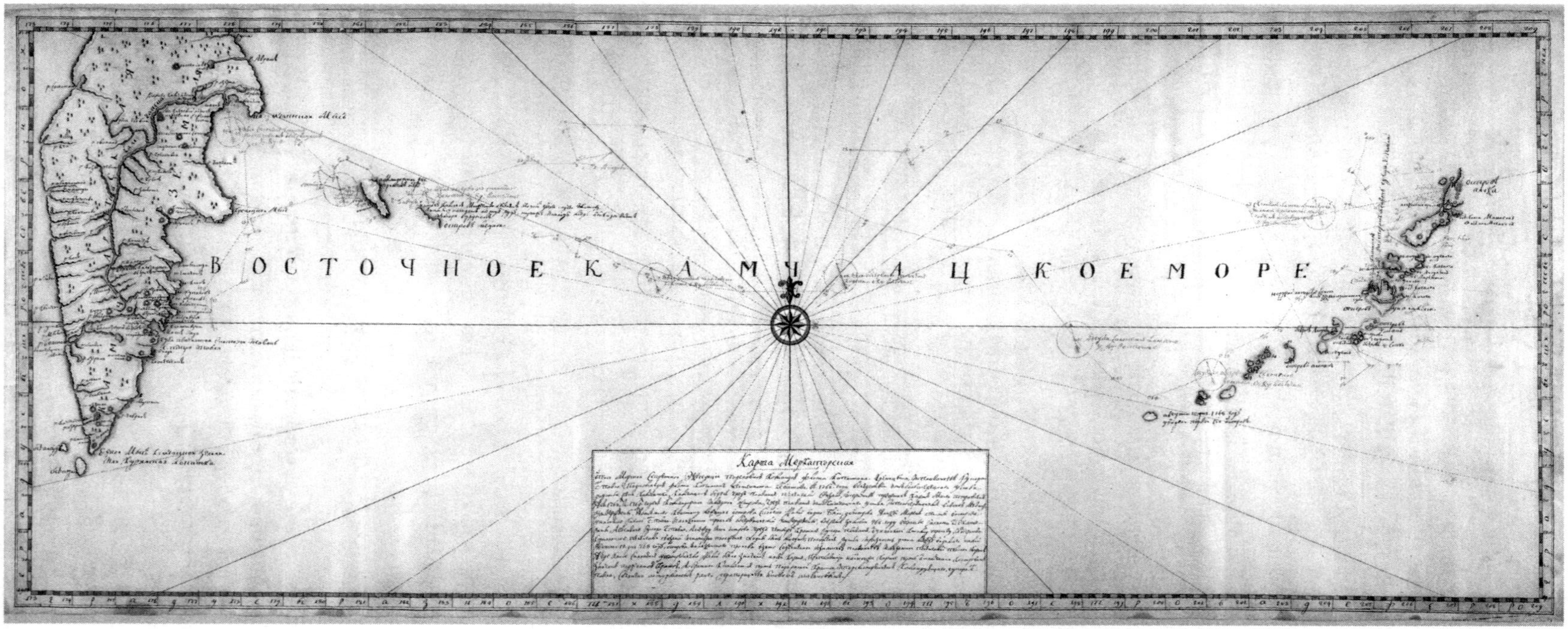

FIGURE 15 Jakoff Shabanov's manuscript map of Krenitzin's and Levashov's Voyage to the Fox Islands in 1768 and 1769. UAF Rare Map G9236 S12 [177–?]

multiple islands sighted were charted by the journal of Navigator Krasheninnikov, who explored the same in baidaras.

From the legend on another version of the summary chart[53] we learn that on Unalashka Island the versions of the chart reflect inconsistencies in the hydrographic recording of the islands' location done while outward and home bound. For example, the island "Attaku" (Attu) has two locations: the western, with the inscription "Island of Attaku charted by computation from the 1st point homeward bound from the strait between Aliaksa and Unimak, by observed (that is determined by astronomical observations) latitude" and the eastern, for which the inscription indicates that the island was "charted by calculation back from the mouth of the Kamchatka River, as by calculation we came ashore, and sailed with favorable winds, so that calculation was not confused, while returning to the (home) shore by calculation from the islands nearby [Near Islands?] in great storms, and the bearings then taken are indicated in the journal."[54]

Besides the cartographic materials, the Krenitzin/Levashov Expedition provided fifty-three watch and shore logs into which were carefully entered the sailors' observations on the geography and population of the Aleutian Islands and part of the Alaska Peninsula as well as all the events that occurred in the course of the voyage, hydro-meteorological observations, and so on. The scientific notes, travel journals, and drawings by Levashov are extremely useful.

The members of the expedition relied on the experience of the *promyshlenniki*. In this they followed the instructions of the Admiralty College and the governor of Siberia. On the expedition's charts all the islands, with very few exceptions, have Aleut names.

Levashov was particularly interested in the way of life, culture, and customs of the peoples in the lands he visited. He attempted to determine the origin of the name "Aleut" and the autonyms of Natives of various islands. He was informed that "the people of these islands, and of Unalaska Island, are called kogolagi, of Akutan and others to the east to Unimak are called kigigus, and those of Unimak and Aliaksa are called katagaeguk. Before the arrival of the *promyshlenniki* the islanders did not know what an Aleut was [that name came from Near islands—LB] but now they call themselves by that name."[55]

Two albums of sketches by members of the expedition (predominantly by Levashov) preserved in the Russian State Archive of the Navy and in the Russian State Archive of Military History are of great value as historical, geographical, and ethnographic sources.[56] These albums contain coastal sketches in India ink of Kamchatka, the Aleutian Islands, and the Alaska Peninsula, which are intended to aid future sailors. Drawings in watercolor of various tools, weapons, household items, hunting and fishing equipment, and other items of Aleut material culture are valued by modern ethnographers and have become standard illustrations of their works.[57]

The Krenitzin/Levashov Expedition remained top secret and its basic materials were never published in Russia. The first foreigner to become acquainted with these materials was, with the Empress Catherine II's permission, the English Historiographer Royal William Robertson. He used part of these materials in his *History of America*, published in London in 1777, and later transferred them to his countryman, William Coxe.[58] Coxe's book, *Account of the Russian Discoveries between Asia and America to which are Added the Conquest of Siberia and the History of the Transactions of Commerce between Russia and China* was published in London in 1780. Information was included about the Krenitzin/Levashov Expedition obtained from Robertson and Gerhard Müller as well as some materials that the author had examined in Petersburg. For a long time Coxe's work was used as a primary source along with archival materials.[59]

Information about the expedition in the Russian language was first published by Peter Simon Pallas as *O rossiyskikh otkrytiyakh mezhdu Aziei i Amerikoi (On the Russian Discoveries Between Asia and America)*, in *Mesiatseslov istorichesky i geografichesky (Historical and Geographical Almanac)* for the year 1781 and reprinted in *Sobranie sochinenii vybrannykh iz mesiatesloivov za raznye gody* (*Collected Essays Chosen from the Almanacs for Various Years*) in 1790.[60] Here Pallas, using Levashov's data, compiled the first geological map of the Pacific Ocean littoral along the coasts of Asia and America.[61]

The original materials, including journals, of the secret government expedition of Krenitzin/Levashov came to be used more widely only from the middle of the nineteenth century when the naval historian A. P. Sokolov published a special paper on the topic.[62] However, even he did not have at his disposal the entire volume of the documentary sources and therefore underestimated the results of the expedition. In later times, researchers who trusted his authority did not pay adequate attention to the study of the materials of this secret expedition to the Aleutian Islands. Only in the twentieth century did Russian scholars begin to introduce these documents into the scientific literature, though clearly not to a sufficient extent. Only the popular science book by I. V. Glushankov focuses solely on the expedition's history.[63]

High Russian naval officers kept the hydrographic and cartographic documents of this expedition secret. Even professional Russian seamen who sailed the northern part of the Pacific in the late eighteenth and early nineteenth centuries did not know of their existence. For example, in 1819 Vasily Golovnin, commander of the sloop *Kamchatka*, in his report to the directors of the Russian-American Company after his voyage to the shores of Russian America, maintained that neither the materials of Krenitzin and Levashov nor of other sailors who had sailed in the northern Pacific had been delivered to the Admiralty College.[64]

Captain Golovnin was in error. The arrival of members of the expedition led by Levashov, and the rich hydrographic materials they were bringing, were impatiently awaited in St. Petersburg, especially by Admiral Alexei Nagaev, who had continued to collect and summarize cartography on the northern Pacific from the Second Kamchatka Expedition forward.

Among the materials of this period, the map compiled in 1765 by the Chukchi Nikolai Daurkin (Fig. 16) occupies a special place. This map recorded his travels in 1763–1764 when he visited St. Lawrence Island and had the opportunity to collect geographic information from the Chukchi who habitually visited islands in Bering Strait and the American mainland across from them. Daurkin had been trained at the Irkutsk Navigational School by Mikhail Tatarinov.[65] However, his map was executed in the traditional pre-Petrine Siberian cartographic style.

In spite of this, the map not only attracted considerable attention in the eighteenth century, which is evident from the number of copies made at the time, but is widely discussed even today by historians, geographers, and ethnographers. Daurkin's[66] map and subsequent verbal information gave birth to a persistent legend about the existence of a Russian settlement somewhere in Alaska's north whose

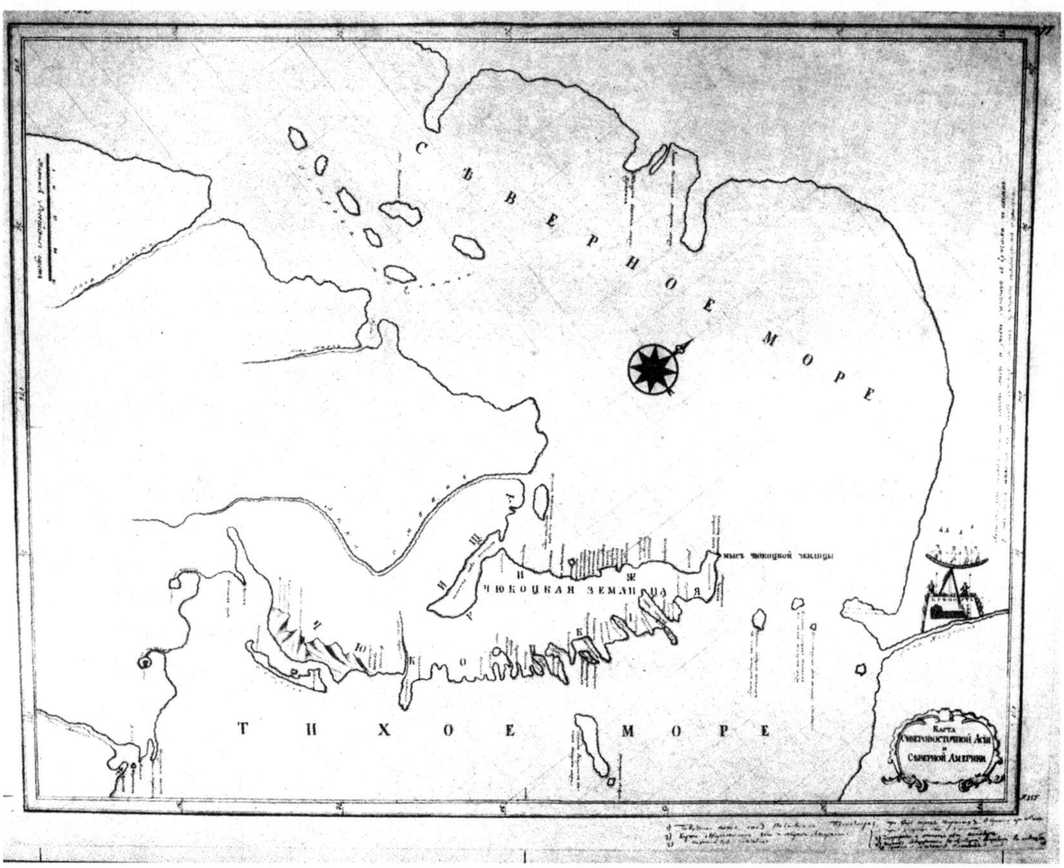

FIGURE 16 N. I. Daurkin's map of Northeast Asia and North America. BAN Otdel rukopisei. Op. 35. Karta 355

inhabitants were the descendants of cossacks who had accompanied Dezhnev or some other Russian travelers, perhaps even Novgorodians, who fled from repression by Ivan the Terrible. [Variant states of Daurkin's chart also attract interest because they are full of valuable data, including Alaska place names—LB]

Later research conducted by the American ethnographer Dorothy Jean Ray, notably field work on the Seward Peninsula, demonstrated that many elements of the drawing on Daurkin's map are in accord with several versions of the folk tale still extant among the Eskimo concerning "the last Siberian invasion." All the tales agree that a battle took place near Kauwerak, which appears on different versions of Daurkin's map as Kheuveren, Kheuv'ren, Khveren', and so on. She explains the ending "–en" as a rendering of the place name in Chukchi.[67]

Even before Ray, Russian researcher M. V. Chernenko[68] identified the River Kheuveren as the River Kuzitrin, but not having the field data later obtained by Ray, he did not associate Kheuveren with the Eskimo settlement of Kauwerak. Fedorova disagrees with Chernenko and identifies Kheuveren with the Koyuk River. No matter how the question of localization of the River Kheuveren is finally resolved, Ray's conclusion that Daurkin's map, in spite of its schematic character, is a milestone, the first after the Gvozdev-Spanberg's chart of 1743, to present the Alaska coast to the north of the Aleutian Islands. This map gives three Native geographic terms, "Tikegan" or "Tikaga" [ancient name Tigara—LB] for modern Point Hope, "Okibian" or "Ukivuk" for King Island, and "Kheuveren" (Kauwerak). For modern Cape Prince of Wales, Daurkin's map presents a new variation of the Eskimo place name, "Kyng-Myn." Thus, no fewer than four local settlements in northwest Alaska were known to the Siberians prior to Captain Cook's voyage.[69]

Although at present the identification of Kheuveren with an ancient Russian settlement in Alaska is considered to be an error, the legend about this settlement played an important role in the geographic explorations within the territory of Russian America in the late eighteenth and early nineteenth centuries and we shall encounter it more than once.

Let us now return to the work of Nagaev, the summary of the materials for the study of the North Pacific provided by the Russian *promyshlenniki*, and the government expeditions. A noticeable push in this direction was provided by the chart submitted to the Admiralty College in 1767 by Ustiug merchant Vasily Shilov.[70] There are no direct data pertaining to Shilov's voyages, his statement that he visited all the islands of the Aleutian Chain, including Amlia, has no documentary confirmation. But the charts he compiled do testify to Shilov's fine knowledge of the then most current cartographic materials on the Aleutian Islands.

It is clearly evident that in the compilation of the *Flat Chart of the Islands Situated in the Eastern Ocean, which were Visited by the Ustiug Merchant Vasily Shilov, with the Addition of the Land of Kamchatka*, its author used the summary charts of the *promyshlenniki* we have discussed earlier [However, Shilov was part owner of many vessels and part owners often sailed—LB]. It is possible that he also used the charts of Bering and Chirikov. On Shilov's chart we encounter for the first time the name "Alyaska," which later became firmly established as the geographic term for the easternmost dry land reached by the *promyshlenniki*.

This chart differs from the earlier *promyshlenniki* charts in the considerably abbreviated inscriptions and by a certain pretension to geographic exactitude in the representation of the Aleutian archipelago. This chart has the geographic grid in the square cylindrical projection with numbered latitudes. One may conclude that Shilov placed the islands and Alaska from two to eight degrees farther north than their real location. However, this became the leading theme of the geographic analysis performed by Nagaev and the Admiralty College.

Having compared the distance from Bering Island to Alaska indicated on Shilov's chart (3,291 versts) to Chirikov's calculated distance from Kamchatka to the place where Dementiev was left (3,984 versts), the Admiralty College simply concluded that "Chirikov in his voyage from Kamchatka to the American islands went further than the location of the Alaska (Peninsula)."

Shilov maintained that he visited the islands from Kamchatka to Amlia himself, and that he hunted sea otters there and charted other islands by calculation; though the bearings from location to location are not completely accurate, the distances between them were measured by log with all possible attention. Shilov himself did not sail to, and did not see, the islands beyond Amlia, to the Island of Unimak and Alyaska, but collected information about them from two Aleuts, Natives of the islands of Umnak and Shelmaga (modern Samalga island). The Aleuts were brought to Kamchatka in 1763 by merchant Bechevin's *promyshlenniki* and, having lived in Kamchatka for three years, accepted the Christian faith and learned the Russian language so well that any information about their people's way of life and the situation of localities around their home islands could easily be obtained from them.

Shilov also provided the information that when sailing along the Aleutian Islands he determined they were

> situated by general bearings almost identically as indicated on Cossack Ponomarev's chart and the distances as measured by Shilov by log; but the place names, except for the island of Umnak and a few others, are different because when Ponomarev sailed from Bering Island to Umnak Island by sea, he did not sight a single Aleutian island and collected their names from Umnak inhabitants among whom at that time not a single one knew his language . . . Shilov, on the other hand, was on those islands in the wake of many other Russian *promyshlenniki* and found there a number of Aleuts who knew the Russian language. From them he obtained the islands' names as written, for the most part different from Ponomarev's.[71]

Obviously researchers studying Native place names in the Aleutian Islands should take Shilov's remarks into account.

The Admiralty College presented Shilov's material to the empress, Catherine II, stressing that his explanation and especially the chart "is considered important, even though it cannot be deemed to be exact nor based on rules pertaining to hydrographic descriptions, but at the initial stage it is a rather major contribution.[72] Shilov and his companion I. Lapin were awarded gold medals for their efforts and diligence."[73]

The Admiralty College entrusted to Vice Admiral Nagaev the task of making a detailed analysis of the descriptions and charts of Shilov and other Russian *promyshlenniki*, and, after comparing these data to those of the Kamchatka expeditions of Bering and Chirikov, to prepare a new chart of the northern part of the Pacific Ocean. Nagaev carried out this assignment with dispatch. As soon as April 26, 1767, a "Mercator chart . . . with explanations to the same"[74] was prepared for discussion by the Admiralty College. Alaska is represented in two alternatives, one by dashes, the second by dotted lines.

The dotted outline carries the explanatory inscription: "Alakshan according to Cossack Ponomarev's chart, stated to be populous, with plentiful timber, animals are bears, caribou, martens and boars, here wintered merchant Bechevin's vessel." The outline indicated by dashes carries the inscription: "A large island, named by Cossack Ponomarev Alakshan . . . and by merchant Shilov Alaska," and a note is added stating that Alaska is shown "according to Shilov's chart." The routes of both the First and Second Kamchatka Expeditions of Bering and Chirikov are indicated and "that part of land found in the year '741 by Captain Chirikov" with "that part of land when on July 18 Fleet Master Avram Dementiev with 14 servicemen, with a ship and a boat remained on shore." Ponomarev's voyage is also indicated.

The chart shows "Kadiak" Island and an inscription next to a small island off the Alaska Peninsula stating that "here, according to the declaration of Cossack Ponomarev, was found grounded a foreign vessel, by all indications a two-masted one."

St. Lawrence Island and the Diomedes are shown in the north, and east of the latter part of the mainland is the following inscription: "island or mainland, undetermined; people named kignyelat [apparently a Chukchi name—LB]. They lead a nomadic life, are not subject to anyone, resemble the Yukagir, wear sable, fox and caribou clothing; animals which are found here are sables, foxes, ermine, lynx, wolverine and caribou in quantity; timber consists of conifers, deciduous trees, birch, spruce; there are mountain ranges and rivers; their houses are underground."

Nagaev's chart clearly indicates that according to the *promyshlenniki* the Aleutian Islands and Alaska were several degrees of latitude to the north of the landfalls made in 1741 by the vessels *St. Peter* and *St. Paul*. This geographic question was debated at a special session of the Admiralty College held on April 27, 1767. The analysis involved not only the charts but also various *promyshlenniki* descriptions presented to the college.

The vice president of the Admiralty College, Lieutenant-General Count I. G. Chernyshev (1729–1797), quite reasonably remarked that "on merchant Shilov's chart are not the newly acquired islands too far to the north and should not they, by his description, be much further to the south because the information on hand states that the ground there is almost never covered by snow?"[75]

Comparison of this information with data from the voyages of Bering and Chirikov which "may be considered accurate because it is based on the art and principles of navigation" permitted the Admiralty College to reach the correct conclusion "that all these newly located islands, those by Bering and Chirikov, cossack Ponomarev, and merchant Shilov, are nothing other than a continuation of the chain, or continuous island links, occupying the space from 51° N to 60° N, the longitude difference being forty-nine degrees. The most distant islands are to the south, at 51° N. According to the return routes of Bering and Chirikov, they cannot be situated further to the south, as these were accomplished by sailing against the winds, tacking, covering a sea space that was from 500 to 600 versts wide."

It was noted that the vessels of the Second Kamchatka Expedition, while tacking on their return voyage, did not cross the outbound routes from Kamchatka to the American shores. Returning to Kamchatka, members of the expedition saw land, "or sure signs of land being nearby, such as floating grass, shore birds, and reached bottom with a lead," only to the north of their course. In the area of tacking, the depth exceeded the maximum length of the lead line (200 sazhen); "for these reasons there cannot be, within the area of their tacking, any but especially frequent islands separated by small straits; on such voyages, (constantly) tacking, they could not avoid sighting such and even less passing between them."

In order to illustrate the geographic content of their conclusions to the empress, Catherine II, the Admiralty College presented her with the chart discussed above, "another chart compiled on assignment by the College by Mr. Vice Admiral and Cavalier Nagaev, in the same scale as that of the chart of Chirikov and Bering, which contains not only the new acquisitions, but also parts of the coasts of Russia, Kamchatka and adjacent to the Russian borders those of China. On the right compass, the chart is fully a Mercator one, drawn by right meridians and latitudes, and appended to it is a detailed annotation."[76]

The contradiction between the information provided by the members of the Second Kamchatka Expedition and the *promyshlenniki* about the climate of the lands they found in the Pacific Ocean remained unresolved for the Admiralty College. Members of the Russian fur-procurement crews maintained that "the climate there is very moderate, so that there is no real winter there, while Bering and Chirikov, in contrast, sailing to the south, experienced cold; thus, the matter remains doubtful, as it is impossible to assume that Bering and Chirikov made incorrect entries in their journals. Also, there is no doubt that they were not only in the same localities but sailed even farther to the south."

At the same time, Russian hydrographers were not inclined to doubt the truth of information supplied by the *promyshlenniki*, who had accomplished numerous voyages along the Aleutian Islands and who had wintered there many times. The Admiralty College, therefore, considered it permissible to "maintain in the last note" that the island climate may be moderated by "sufficient physical conditions, as is the case with other northern lands which are surrounded by the sea and free of cold, as for example England situated approximately in the same latitudes; to the sea surrounding (the islands) it is possible to ascribe with probability underground warmth close to the surface of the sea bottom; this is evidenced by the fire-breathing mountains, of which four have been located in the new-found islands and more probably exist."[77]

From the above we see that the Admiralty College arrived at the unanimous conclusion that the lands sighted by members of the crews of the *St. Peter* and the *St. Paul* were identical with the Aleutian Islands shown on the charts of the *promyshlenniki*. This hypothesis was fully confirmed by the Krenitzin/Levashov Expedition whose materials were delivered to St. Petersburg in 1771.[78]

Nagaev was once again assigned the task of compiling a chart of the discoveries in the Pacific Ocean. Levashov and apprentice navigators from the Tomsk and Irkutsk navigational schools who had participated in the expedition worked on this chart under his guidance.[79] According to Nagaev's report, this chart suffered great damage during a fire in his house.[80] This chart has not been located to date, but there are on hand several charts which possibly served as preliminary ones. In our opinion, two copies of the summary charts of Chirikov's 1741 voyage, published in A. V. Efimov's *Atlas* and mentioned above, belong to this class of materials.[81]

On these charts we see for the first time that part of the distant Aleutian Islands and of Alaska are represented in the mainly correct latitude corresponding to the coordinates determined in the course of the Second Kamchatka Expedition and the Krenitzin/Levashov Expedition, with place names elicited from the Native inhabitants by the *promyshlenniki*. It is significant that the extent of the Alaska Peninsula here is much greater than on the *promyshlenniki* charts, showing the route of Bering's vessel along that section of the coast. Even though only the Aleutian Islands whose situation was determined by the scientific expeditions are shown, that treatment of the entire chain seems above reproach and fully corresponds with the conclusion by the Admiralty College. The charts reflect the geography of the area quite adequately.

The latitudes of the Aleutian Islands on charts of the northern part of the Pacific Ocean were compiled by Russian sailors of the 1770s. One of the most remarkable products of this period is the chart signed "1775, 6th day of September in Okhotsk" by the commandant of the Okhotsk Port, Second-Major Timofei Shmalev, the *Chart of Newly Discovered American Islands, Comp. 1775*.[82] This multicolored manuscript chart is adorned with beautiful watercolor illustrations that are a wonderful source for the ethnography of the Aleuts. Kadiak and the adjacent coast of the Alaska Peninsula appear on this chart, as well as the Aleutian Islands then known ["Aleutian Islands" is applied only to the Near Islands and the inhabitants of each island to the east are called "Americans."—LB].[83]

The islands and the Alaska Peninsula are shown located farther to the south by one to two degrees than is the case. Umnak, Unalashka, and Kadiak are represented with the greatest detail, and Native settlements and those of the Russian *promyshlenniki* are shown.

There is another version of this chart showing a somewhat greater geographic expanse of territory—it extends to Bering Strait in the north and includes a single island within the channel.[84] This chart is in monochrome and more schematic than the one discussed previously. Though the numeration of meridians and latitudes is absent, it is completely obvious that the author of this chart on the whole accepted the correct interpretation of the disposition of land and waters in the northern part of the Pacific Ocean. Alaska is shown as a peninsula with the inscription: "Alyaksa. On this land are the animals: caribou, bears and wolves; black foxes are taken here as well as silvers; timber is birch, aspen, to a thickness of eighteen vershok and more."

Timofei Shmalev, commandant of the Nizhne-Kamchatsk Port and later of the Okhotsk Port, was continuously in touch with sea voyages into the Pacific Ocean. Practically all the reports and charts of the *promyshlenniki* and of the government expeditions passed through his hands. Therefore it is not surprising that his own charts reflected the most objective picture of the new lands being explored by the Russians. In particular, these charts demonstrated how out of step the results of Sindt's expedition were. It cannot be ruled out that one of the goals for the compilation of these charts was to illustrate the errors of Sindt's geographic constructs.

This is even more likely since, upon a request of the Admiralty College, Shmalev compiled a special *addendum* to Lt. Sindt's report in 1775. In his comment, Shmalev expressed misgivings about the number of islands found on Sindt's chart in the northern part of the Bering Sea. Shmalev wrote:

> About the islands sighted by Lieutenant Sindt which stretch to 65° latitude north, from Okhotsk to the east along the longitude of more than 40° as written down in the log and put on the chart without relation to the routes along the same course(s) of the first expedition by Mr. Captain-Commander Bering, I wisely leave it to scholars to ascertain, also if the said islands are steep cliffs and the depth at a distance from them is shallow; that, it seems to me, is indicated incorrectly, as I myself voyaged in the Eastern and Penzhina seas more than once and observed, and heard from other most capable sailors, that wherever there are cliffs, the depth one verst distant from them may be up to one hundred sazhen and more, and also that rocky reefs stretching out

> to sea may be a verst or more in length, but are not very continuous and the shallows are quite distinct.[85]

Concluding his statement, Shmalev writes, "in the end one can only regret that this expedition so necessary and which lasted almost eight years and for which large sums were expended, did not achieve anything useful and, except for the fact that it was dispatched, it does not deserve any consideration."[86]

In addition to Shmalev's charts, several other cartographic works were compiled in the 1770s at the Admiralty College and by the *promyshlenniki*. These objectively reflected the Aleutian Islands and the Alaska Peninsula in Mercator projection from the data of trade and scientific expeditions, especially the *Chart of islands lying beyond Kamchatka compiled by the navigator Afanasy Ocheredin in the year 1775, September 6*[87] and the charts compiled at the Admiralty College by Second-Major Vasily Krasilnikov and Captain-Lieutenant Ivan Izvekov in the years 1777–1778. These works testify to the exceptional attention paid by the high command of the Russian Navy to the northern part of the Pacific Ocean, attention which was probably intensified by the news of the beginning of Captain James Cook's third voyage.

The charts from Izvekov-Krasilnikov capped a stage in summarizing the voyages of the *promyshlenniki*, combining them with the materials of the scientific expeditions (the Kamchatka expeditions and that of Krenitzin/Levashov). This set was compiled utilizing the method, and possibly even the materials, of Admiral Nagaev. Like Nagaev, Krasilnikov began with a *Chart of Portions of the Pacific Ocean and Western Coast of North America which were Approached in 1741 by Captain Chirikov with the Vessel St. Paul.* On this chart the results of the Second Kamchatka Expedition, based on Chirikov's and Bering's materials, were represented in Mercator projection.[88]

He then compiled the *Chart of Newly Found by the Russian Promyshlenniki and cossack Ponomarev in the Pacific Ocean Near Aleutian Islands, Reduced from the Chart Sent Here, Forwarded by Governor of Siberia Mr. Chicherin; the Said Islands Have Been Located on that Chart in order by Captain Lieutenant Ivan Izvekov but at the Admiralty College's the Drafting Office Were Placed Without Exceptions in 1777 According to Observations and Linkages by Bearing and to the Same Scale as the Preceding Voyage of Captain Krenitzin in August of 1768 to the Aleutian Islands, Longitude from Ferro Island.*[89]

The heading of this chart contains a brief description of the methods employed in its creation, something rare for its time. Additional information stating the use of Second Kamchatka Expedition materials appears on the chart itself, where Chirikov's anchorage is shown "when he was approached by 21 wild persons in skin boats."

The westernmost part of the Pacific Ocean waters, adjacent to Kamchatka, is shown on the next chart entitled *Chart of the Part of the Coast of Kamchatka Land, with Part of the Pacific Eastern Ocean and on it Komandorsky (Bering) and Medny*

(Copper) Islands, and the First Near Aleutian Ones, the Iasak-Paying Atta (Attu), Shemia and Anattu [*sic,* Agattu], *and the Voyage from the Mouth of the Kamchatka River Eastward of Captain Krenitzin and Captain-Lieutenant Levashov in the Year 1768 to the Islands of Unimak and Alyaska; Compiled in the State Admiralty College Draft Room in the Year 1777, Longitude is Calculated from the Island of Ferro.*[90]

The easternmost areas reached by the Krenitzin/Levashov Expedition are shown on the *Chart of Newly-Located Distant Aleutian Islands Where in the Year 1768 Fleet Captain Krenitzin, Having Been Separated from Captain-Lieutenant Levashov by a Severe Storm, Neither Knowing Where the Other was, Wintered with the Galiet* [*sic*] *St. Catherine in the Alyaksa Strait at Unimak Island, While the Other with the Hooker St. Paul Wintered at Unalashka Island on the North Side in Igunok Inlet; Then in the Year 1769 through the Unalashka Inhabitants the Galiet* [*sic*] *St. Catherine was Located and was Joined by the Hooker St. Paul in the Alyaksa Strait. Compiled in the State Admiralty College Draft Room in the Year 1777. Longitudinal Difference Indicated from Ferro Island. Second-Major Vasily Krasilnikov.*[91]

To this set of charts also belongs the large-scale *Chart of Unimak Island, the Adjacent Alaska Peninsula and the Strait that Separates Them (now Isanotsky Strait) with Depth Measurements.*[92] The set of charts analyzed here, in its entirety, reflected most fully the cartographic information acquired up to that time about the waters and the lands in the Pacific Ocean between Kamchatka and the Alaska Peninsula.

BEFORE COOK: MAKING SENSE OF THE DATA

By the 1770s the Russian Admiralty College had a concept of the Pacific Ocean between Kamchatka and America which fully corresponded to reality. Unfortunately, this reality was not on the charts being published in Russia for general use. The official authorities prohibited the use of data provided by the secret government expeditions and the self-supporting voyages to America. The *promyshlenniki* also had an interest in secrecy, which guaranteed the monopoly of fur resources and the rapid acquisition of wealth by persons possessing geographic information about the Aleutian Islands and Alaska.

Even the St. Petersburg Imperial Academy of Sciences was in many instances deprived of the use of exploration materials from the northern part of the Pacific Ocean, and therefore the maps it published did not incorporate new information about these regions or, much worse, represented these data with gross errors.

Staehlen

The most glaring example of such a distorted representation is the *Chart of the Newly-Discovered Northern Archipelago by the Russian Sailors in the Kamchatka and Anadyr Seas* (Fig. 17), published by Jacob von Staehlen in French by the Academy of Sciences in St. Petersburg. An English edition was published in 1774.[93] It is based on the academic map discussed here earlier, that of Müller-Truskott of 1754–1758, to which Staehlen added fragmentary information about later voyages that had reached him; in the main these additions were based on the voyages of Lieutenant Sindt in the years 1764–1768.[94]

Staehlen pictured the Aleutian Islands as extending to the north-northeast in such a way that the eastern shore of the Bering Strait became on his map "Alashka Island," while many islands shown in this region, apparently on the basis of Sindt's chart, were arbitrarily assigned the names of some of the distant Aleutian islands (Umnak, Unalashka, and others). Of interest is the fact that this map gives information, though with an erroneous date (1748 instead of 1648) about the voyage of Semyon Dezhnev. The routes are indicated and there is an inscription: "Route followed frequently in ancient voyages, 1748, by three Russian ships, one of which even reached Kamchatka."

On the whole, Staehlen's map once more (following the academic map of Müller-Truskott) created a totally false image of the results of Russian investigations in the northern part of the Pacific Ocean. Its negative impact cannot be exaggerated, especially since the West lacked any other maps. Thus, Staehlen's map (which was also in the possession of Captain Cook, as we shall see below) created a very negative opinion about Russian geographic explorations.

The quality of marine hydrographic and cartographic work by Russian professional seamen by this time was up to international quality standards. The navigational skills of the *promyshlenniki* and their technical equipment had significantly improved. An ever-greater number of their vessels sailed along the Aleutian Islands with licensed navigators on board—graduates of the Siberian schools of navigation.

The Admiralty College's conclusion in the matter of Shilov's chart describes changes in the construction and outfitting of *promyshlenniki* vessels. They had begun with *shitiki* (sewn vessels). Instead of being secured with iron fastenings, tree twigs were used. Later shipbuilding improved to include decked ships with rigging and cables of hemp replacing those made of nettles.[95]

Besides the vessels of their own construction, the *promyshlenniki* quickly adopted the Aleut and Eskimo baidaras and baidarkas which, starting with the Krenitzin/Levashov Expedition, also were widely used as expeditionary vessels for professional hydrographic surveys. Factory-manufactured instruments became standard navigational equipment for ships, even on baidaras. Thus, for example, merchant Kholodilov's vessel's equipment in 1772 included "one compass with rhumb bearings indicated, three regular compasses in copper housings, and two baidara compasses."[96]

We have noted above that the scholars of the St. Petersburg Imperial Academy of Sciences had little access to the original materials of the explorations in the northern part of the Pacific Ocean. There is reason to suppose, however, that Gerhard

Müller was the sole exception and that he was apparently trusted more than other academicians. Müller was a member of the Siberian detachment of Bering's Second Kamchatka Expedition and its official historiographer. He taught at the main Petersburg military establishment of the time, the Noble Ground Corps of Cadets, and was, to judge by all appearances, one of the few foreigners whom the Russians trusted almost as one of their own countrymen, as he was privy to state secrets. He fully justified this trust and, as a result, the overwhelming majority of the materials he assembled and partly reworked pertaining to the history of the Russian geographic explorations in the northern part of the Pacific Ocean remained unpublished or were published, but in much abridged form, in later publications of William Coxe and Peter Simon Pallas.

It is assumed, however, that as early as the 1770s Müller put some of his materials at the disposal of one J. L. S., who stepped forward with a critique of Staehlen's work.[97] J. L. S. also provided information on the history of Russian explorations taken from ships' logs and official reports. In this work, twenty-four voyages by Russian hunters and merchants to Bering and Medny islands, to the Aleutian Islands, and to Kodiak in the period 1745–1770, are described, with data on the geographic location of islands, their environment, and their population.[98] This work found wide distribution only when a greater part of it was published in English translation as part of the work of William Coxe in 1780.[99]

Naturally, the work by J. L. S. could not fully encompass Müller's materials. Coxe, who visited the aged academician in Moscow in 1778, wrote the following:

> His collection of government papers and manuscripts is priceless and is organized in exemplary order. Every bibliophile should regret that Müller, so eminently well qualified to execute the task and who had already marshaled material for it, did not make the public happy with a fundamental history of Russia because due to advanced age he is obliged to entrust to other persons the use of his papers, so painstakingly assembled by him."[100]

Perhaps Coxe would not have grieved so much had he foreseen how widely and fruitfully Müller's portfolios have been and still are, used by Russian historians and geographers.[100]

One must acknowledge, however, that during Müller's lifetime many of the materials he collected, or the sources to which he had access as the official historiographer of Russia, could not be published because of the official policy of secrecy pertaining to investigations in the Pacific Ocean. Even Swedish scholar Samuel Engel, Müller's contemporary, who believed in the possibility of passage via the Arctic Ocean from the Atlantic to the Pacific, understood that the Russian historian was conscientiously following the rules which existed in the country at that time. He wrote that the Russian nation loves "panegyric, not the truth" and that

Müller's history of Russia would have been considerably improved if there were not censorship by the Senate, which was striving to conceal "the lack of effectiveness of the Russian Navy and navigation."[102]

We have seen that it was not the Senate, but the Admiralty College, which was striving to conceal. Engel's conclusion is justified, however, since only academy maps were available to the reading public. Müller reacted with indignation to Engel's accusation and answered it rather sharply, not even refraining from personal attacks.[103] Engel, in his answer published in 1777, cited an unnamed correspondent from Russia who had written that when Müller edited Georg Steller's manuscript "each page was signed by Mr. M. and brackets indicated what had to be omitted with a note on the margin: 'foreigners ought not to know this.'"

It is possible that this foreign correspondent cited by Engel may have been Jean Benoît Scherer, who in his preface to the German edition of Krasheninnikov's book calls it "a castrated and distorted work."[104] We note that Scherer himself did not find it dishonorable to publish Krasheninnikov's work and name Steller as its author. Further, Engel maintained that Müller had been delaying the publication of the history of Siberia by Johann Eberhard Fischer[105] until the president of the St. Petersburg Academy of Sciences ordered it published, declaring that "such works are not written for storage in the archives." Moreover, (according to Engel) Müller refused Fischer access to the archival documents from Yakutsk and Irkutsk until he was ordered to give it to him. Engel concluded his article with a rebuke that "whatever happens in Russia, it is a state secret."[106]

Peter Simon Pallas

Though most of the materials from the voyages by the *promyshlenniki* and state-sponsored expeditions were not available to their contemporaries, in the 1780s the remarkable natural scientist and member of the St. Petersburg Academy of Sciences, Peter Simon Pallas (1741–1811) published a series of articles in the first four volumes of the journal *Neue Nordische Beiträge* (1781–1783). The *Beiträge* dealt with the geography, ethnography, and environment of northern countries. Pallas based his articles on earlier publications and original documents about the Russian investigations in the northern part of the Pacific Ocean. Here, for the first time, wide geographic generalizations were made about this region of the globe.

Pallas wrote using the rich materials derived from his own field observations when he was the head of the academy expedition to Siberia, beginning in 1768 and continuing for six years, traveling to Baikal and even to Kiachta. Hydrographic materials pertaining to the northern part of the Pacific Ocean were open to him to a greater degree than to any other member of the St. Petersburg Academy.[107]

Let us look at the geographic constructions created by Pallas as presented in his work *Erlaeuterungern ueber die im oestlichen Ocean zwischen Siberian und America geschehenen Entdeckungen*.[108] Even though this work was published after Captain

Cook's voyage, Pallas left its contents, dated 1772, unchanged, thus stressing the significance of the progress made by the Russians in the study of the northern part of the Pacific Ocean long before the voyage by the great English sailor.

Pallas speculates for the first time about geological configurations which form the main coastal and island mountains in the northern part of the Pacific Ocean. He proposes that a spur of the mountain ridge which is Kamchatka is linked via the islands of Bering and Medny to the entire island chain (the Aleutians) that lies between Kamchatka and America. At the same time, another spur of this chain extends southward via the Kurile Islands in the direction of Japan and the mountainous coast of the Far East.

Pallas came to the following conclusion:

> Therefore it appears that the Asian mountains which are across from the American mainland have two extensions, one via the point of land on the Chukotka Peninsula which extends at least to 205° of longitude, the other much to the south, the island chain that begins at Kamchatka. On both counts, there could have been a much easier and closer common land connection which in part due to the continuing currents southward from the Arctic Sea, in part due to earthquakes (which even now rage in the island chain extending from Kamchatka which has many volcanoes), but also in part due to global changes and floods, were broken up more and more, and that is the reason the mainland, as well as the islands whose coasts appear broken up and precipitous, were torn asunder by force.[109]

Pallas saw the main arc in the Pacific Ocean as a direct continuation of the Kamchatka chain; to be more specific, of the Stolbovoi and Kronotskoi capes. This mountain-island chain has a general direction first to the islands of Bering and Medny, then in the Aleutian Islands, finally changing its orientation via the Andreanof Islands to the Fox Islands, which, for their part, are structurally linked to the American continent. In Pallas's opinion, this chain closes off part of the Eastern Ocean, which he proposes to distinguish as the Kamchatka Sea.

The question of the general character of land distribution in the northern part of the Pacific Ocean continued to be a lively debate. For example, Count Buffon in his *Histoire Naturelle* expressed the thought that America was joined with Asia somewhere to the north of the Andreanof Islands, but that the intercontinental land bridge was concealed by permanent polar ice, similar to those of Spitsbergen and Greenland.[110] Pallas rejects this construct completely, citing Bering's voyage when the latter reached 67° N as well as "the earlier examples of the Siberian cossack commanders, who, shortly after having explored Eastern Siberia, traveled the route from Kolyma to the Anadyr in small vessels, sailing along the coast and rounding the northeastern extremity of Asia. Of these, the voyage of Dezhnev in the year 1648 is the most trustworthy."[111]

Rejecting also the notion (for example, Daurkin), on some Russian charts of the time, that the shore of the American continent makes an arc to the northeast and reaches the meridian of the Kolyma River mouth, Pallas makes a purely geographic argument:

> Against the notion of a possible landmass which lies parallel to the Siberian coast, one can cite various arguments, for example the absence of trees and the very severe climate of the entire northeastern corner of Asia on that side of the Anadyr which could not be in accord, had there been a protective land; there are also the strong north winds, the currents flowing from the north into the Kamchatka Sea, the sea ice which is being driven through the narrows from the north southward, the walrus which are found along the entire Chukotka coast to the Bay of Anadyr and which, as is well known, like the open sea.[112]

Pallas proposed that the continental character of the eastern shore of that strait was proven. He cites the voyages of Gvozdev and Sindt and their conviction that their expeditions touched upon the American continent.

Pallas summarized all the available information on the geography and geology of the islands visited by the Russians in the northern part of the Pacific Ocean. He noted in particular that Bering and Medny islands had been investigated thoroughly and underscored the presence of high sand dunes in which sea mammal bones and driftwood were found. He linked their origin to the tidal waves caused by earthquakes and volcanic eruptions.

In Pallas's opinion, however, these two islands were not of volcanic origin, and he notes, quoting Steller, "The steep shores of these islands are diminished from year to year through weathering and frost (which splits the coastal cliffs), and the might of the sea, so that there cannot be any doubt that these islands, as well as the others, which without exception show signs of destruction, in the past were much greater and formed a much more cohesive chain. Perhaps some have completely disappeared and plunged into the depths of the sea."[113]

Pallas paid attention to the question of the existence of land between the Aleutian Islands and the Chukotka Peninsula. First of all, he decisively rejected Staehlen's representation of the location of the Aleutian Islands on his chart, and on some of the others, which permitted the notion that the chain trended to north-northeast and reached the Chukotka Peninsula.

In particular, Pallas, quite reasonably, remarks:

> Such maps would have been much more accurate had they taken Bering's chart as their guiding light and taken as their basis the ever more securely-expressed opinion that all of the newly-discovered islands were the coast and the islands that Bering had sighted. Indeed, the more easterly islands on Bering's chart,

which recently received correct navigational coordinates, correspond to the Fox islands, appearing on our newest and best charts; also it was demonstrated long ago by new sea voyages that the proposition, defended foremost by Steller, that all the coasts sighted by Bering when homeward bound constitute a single extension of land from America westward, as has been represented on a number of maps, lacks any basis whatsoever.[114]

Thus, Pallas rejects the Müller-Truskott geography of the northern part of the Pacific Ocean.[115] Discussing the geography of the Pacific Ocean to the north of the Aleutian Islands, Pallas maintained that there were no significant landmasses in the entire expanse up to the Bering Strait. His conclusion was based on the observation that the Aleutian Islands, lacking shelter in the north, were treeless, just like the eastern coast of Kamchatka, which is unprotected against the oceanic storms. Only on Kodiak, sheltered in the north and west by the Alaska Peninsula, were there forests. By Pallas's time it was known that the peninsula had high mountains. Pallas thought the Alaska Peninsula was part of the American mainland, not an island.

Supporting his position, Pallas wrote:

The high timber growth which was found on Kodiak, the frequently observed varieties of land animals which do not occur in the western islands, the islanders' reports from the remote Fox islands such as Unimga (Unimak) and Kodiak, that there is to the north not far away a great land, mountainous and rich in forests, where black foxes are supposed to be plentiful and which the Russians had, therefore, named the 'Land of Black Foxes'—all were taken by me for a long time to be evidence that mainland America is meant here.[116]

Pallas represented his concept of the geography of the northern part of the Pacific Ocean on the map (Fig. 18) he compiled, citing his sources and describing their use. For placement of the Aleutian Islands, Pallas used their list constructed by Müller, where Native place names are given and the islands are listed in sequence from Kamchatka to America.

Compiling his map with the aid of this list, Pallas attempted to place the islands as accurately as possible, though he had to admit that he could not vouch for absolute precision: "Astronomical observations will correct any errors. My wish was to present, insofar as is possible to do at this time, a more reliable picture of this part of the world, as yet so little known, than the one available formerly."[117]

Pallas devoted attention to the description of the environment and population of the Aleutian Islands and Alaska. In particular, he stressed the volcanic character of the islands and the presence of hot sulfur springs. Notable among the vegetation were berry bushes and shrubs, which were especially abundant in the river valleys of Unalaska Island. Trees were lacking on all islands with the exception of

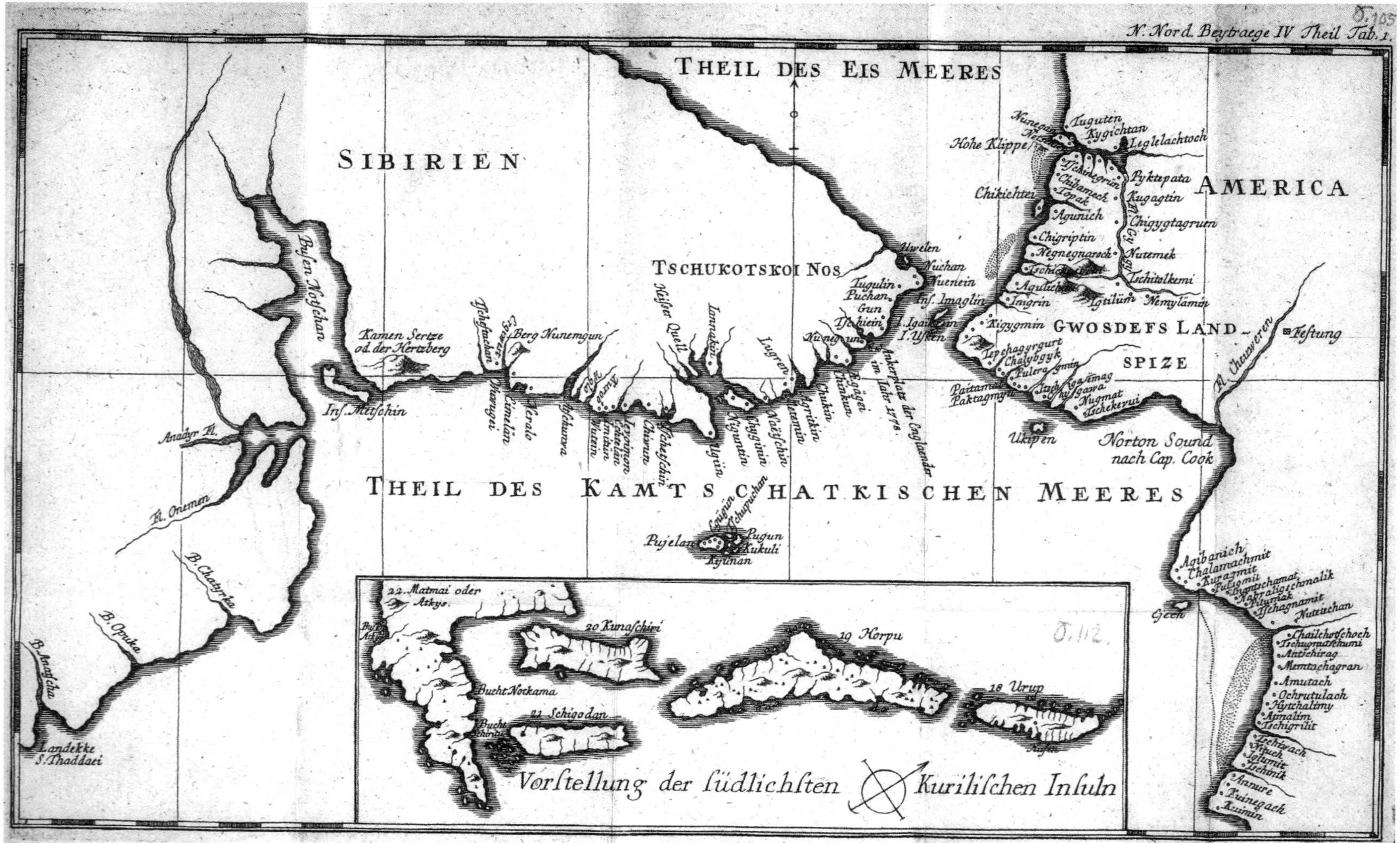

FIGURE 18 Pallas, Map of Bering Strait *Beitraege zur Kenntniss des Russ. Reiches.* IV. Map following p. 330. UAF Rare Book A3468

Kodiak. Foreign trees were found among driftwood. Russians brought from the Aleutian Islands, for example, Japanese camphor wood, which preserved its characteristic aroma.

In the description of the animal world it was stressed that on the Fox Islands a variety of foxes were found: black, silver, red, and cross. Pallas notes that as a result of predatory exploitation of fur-bearing resources the most valuable animal population in the Aleutians—the sea otters—had decreased considerably.[118]

Pallas considered all the human inhabitants of the Aleutian Islands to be of the same origin, but singled out from other Aleuts the inhabitants of the easternmost and most populous ones, believing them to be rather close to Greenlanders and Eskimos in morals, customs, way of life, dwellings, vessel construction, and outward appearance, even though in their language they had but a few words in common with Eskimos and Greenlanders. Moreover, the name "Kanagist" (Kaniag), as they called themselves according to some reports, was similar to the "Karalit" of the Greenlanders and their North American relatives and was nothing but a corruption of this autonym transmitted by Russian sailors.[119]

Pallas's work clearly demonstrated the considerable achievements, by the 1770s, of the Russian *promyshlenniki* and the government expeditions in the study and cartography of the northern part of the Pacific Ocean. In conjunction with the charts discussed above, this work demonstrates the formation of a cartographic representation of the Aleutian Chain, the Alaska Peninsula, and the northernmost part of the Bering Sea, including territories in the Bering Strait region adjacent to the Seward Peninsula and St. Lawrence Island. However, the coast of America between the southern part of the Seward Peninsula and the Alaska Peninsula remained totally unknown to the Russians. To the east and south of Kodiak, the American coast was known to the Russians only fragmentarily and only from the voyages of Bering and Chirikov in 1741.

The characteristics of the waters and landmasses in these areas, as well as the precise longitudinal extent of the Pacific Ocean between the Alaska Peninsula and Kamchatka, were documented by Spanish and English sailors and, above all, by the British scientist, explorer, and hydrographer James Cook.

Spain

We have seen that in spite of many efforts by the Russian authorities to keep the results of the commercial and government expeditions in the northern part of the Pacific Ocean secret, information about them did leak abroad. This information alarmed Spain most of all, which had the pretension that "the right of discovery" to all American lands north of California belonged to her. As early as 1759 the Franciscan monk José Torrubia, who had visited Mexico and the Philippines, warned his countrymen about the growing danger from the "Muscovites," who had "discovered a new way to North America, the way by which the Asiatics had settled the continent," and thus the Russians were a possible danger to the Spanish possessions in America.[120] Similar concern was expressed by Miguel Venegas in the third volume of *Noticia de la California*, prepared for publication by Father Andrés Burriel in 1757.[121] In March 1761 the Spanish ambassador to Russia, Pedro de Gongora y Lujan, Marqués de Almodóvar, received orders to directly approach the Admiralty College and the St. Petersburg Academy of Sciences and, above all, the court of the Empress Elizabeth to clarify the matter.[122]

On October 7, 1761, the Spanish ambassador sent the first reports based on official and unofficial sources from St. Petersburg to the Spanish Minister of Foreign Affairs, Recardo Wall. He reported that the Russians had not yet taken possession of lands on the American continent but might do so in the future, as in 1741 Chirikov and Bering went as far as 45° [!?—AP] N latitude, and the men left by Chirikov on the American coast could have reached California.[123]

The new Spanish ambassador, Alvaro de Nava, Viscount de Herrería, who replaced the Marqués de Almodóvar in St. Petersburg, scared his government with information about a grandiose Russian expedition allegedly being dispatched to the shores of northwestern America. Nava garbled the information which had reached the Spanish embassy about the expeditions of Chichagov and Krenitzin and assumed that the Russian plans constituted a danger to the Spanish possessions in southern California. Herrería's reports moved the Madrid administration to institute urgent measures to settle the lands situated to the north of the Mexican possessions. Thus, in California in the years 1769–1776 there arose a dense network of new settlements and missions. In particular, San Diego, Monterey, and San Francisco were founded in this period. The northern boundary of these newly colonized territories was in the region of 39° N latitude.[124]

Spanish northward expansion was most intense in the years 1774–1775, provoked by reports by Herrería's successor in St. Petersburg, Francisco Antonio de Lacy. Lacy managed to plant a secret agent in an important office in Russia, probably in the Foreign Affairs College. Through this agent he received copies of secret documents relating to Russian discoveries. On March 19, 1773, the Spanish ambassador reported to Minister of Marine Affairs Julián de Arriaga about the Krenitzin/Levashov Expedition and about the discoveries by these sailors in the Aleutian Islands at 235° E longitude. In May of 1773, Lacy transmitted to Madrid the map by Müller-Truscott of 1758 and Arriaga immediately sent it to the viceroy of New Spain, Antonio María Bucareli y Ursúa. In Madrid and Mexico, Lacy, Arriaga, and Bucareli worked out a plan for naval expeditions into the northern part of the Pacific Ocean. In order to guarantee a high professional level for these expeditions, six naval officers were dispatched to Mexico.[125]

On January 25, 1774, the Spanish naval ship *Santiago* (or *Nuevo Galicia*) left the Mexican port of San Blas under the command of Juan José Pérez Hernándes with Estéban José Martínez the first officer. Bucareli instructed the expedition to proceed along the northwestern coast of America to 60° N latitude and take possession of the entire territory for the Spanish Crown. The voyage north commenced out of San Diego on July 11. On July 18 the Spanish vessel approached the coast of Alaska for the first time.

The next day members of the expedition sighted a cape which they named Santa Margarita. Eight leagues to the north, at latitude 54°30′, near the island they named Santa Christina (now Forrester Island) in the Alexander Archipelago, the *Santiago* turned homeward. Apparently the expedition had been in the vicinity of Dixon Entrance but was unable to enter it due to strong wind. Homeward bound, the Spanish stood in some bay which had the form of the letter "C" where they traded with the Indians.

The location of this bay caused much discussion. If one were to trust the astronomical observations by the expedition's staff, then, according to the latitude, the vessel was in the vicinity of the modern Nootka Bay on the west coast of Vancouver Island. But the observations conducted by Pérez were very imprecise and it is therefore believed that the *Santiago* dropped anchor much farther to the south. Martínez, however, always maintained that the vessel stood off Nootka Bay. If the observations were conducted by him, and not by Pérez, it is possible that this is true. Martínez subsequently demonstrated his ability to determine geographic coordinates accurately.

The expedition returned to San Blas on November 3, 1774. It had not achieved anything of practical significance except to confirm that in the region near 55° N latitude, the coastline did not extend farther west than Mendocino. Members of the expedition did not go ashore even once. Apparently, no charts were compiled in the course of this voyage.[126]

A new expedition sailed out of San Blas on March 16, 1775, under the command of Bruno de Hezeta aboard the ship *Santiago* with Juan Francisco de la Bodega y Quadra aboard the *Sonora*. On July 10 both vessels entered a bay which they named Trinidad. The coordinates established by the members of the expedition were 41°07′ N latitude and 19°04′ W longitude. The true latitude of this bay is 41°03′. This error of three to four minutes appeared in all this expedition's subsequent observations. The flagship *Santiago*, commanded by Hezeta, only reached 49° N latitude. Separated from the *Sonora*, she arrived at Monterey on August 29.

Bodega y Quadra continued sailing north and reached the localities north of 58° N. In Alaskan waters, he approached the coasts of Baranof and Chichagov islands; that is, approximately in that area of the American continent's coast visited by Chirikov. Twice he went ashore and proclaimed the land he visited a possession of the Spanish Crown. At a place he named Entrada de Bucareli (57°17′ N latitude), on August 25, 1775, Bodega y Quadra erected a cross to mark the taking into possession of new lands. The expedition returned to Monterey in October 1775.

This Spanish expedition, in contrast to the previous one, produced significant cartographic results. However, like the majority of the Russian materials of the time, they were considered secret by the Spanish authorities and are preserved only in the original manuscript in the National Archive of Mexico and, in copies of most of them, in the Archive of the City of Seville.[127] In addition to the plans of Bucareli (Fig. 19) and Bodega Bays that were appended to Bodega's ship's log, three general charts reflecting the expeditions' results were compiled. Bruno de Hezeta compiled the plans of Puerto de Trinidad and the Entrada de Hezeta.

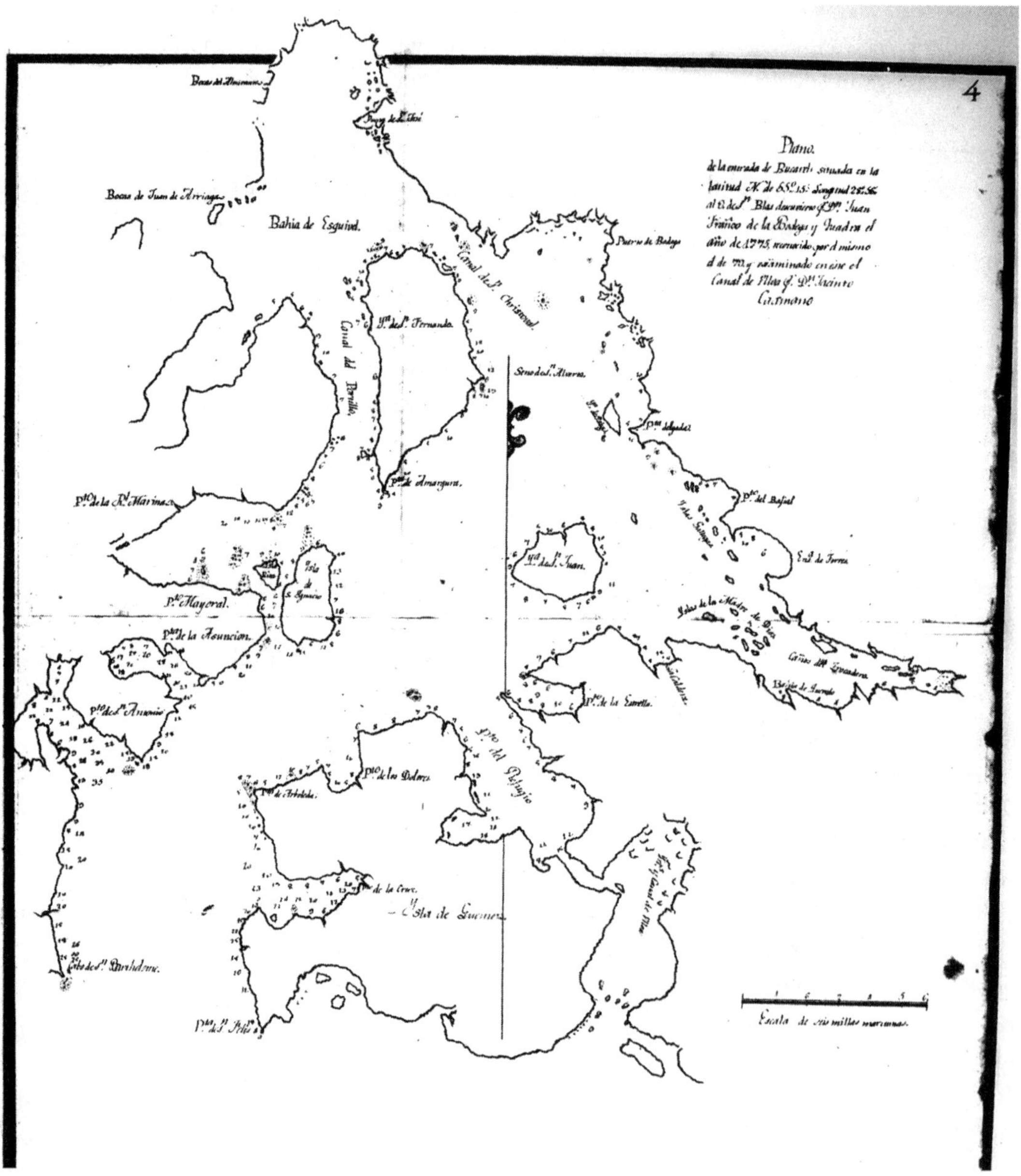

FIGURE 19 A facsimile reproduction of Juan Francisco de la Bodega y Quadra's 1775 plan of Bucareli Entrance. UAF Map G3351.S12 1775 P36

Among the general charts, the one compiled by Bodega y Quadra with participation by Mourelle is of most interest. This chart represents the entire coast from Monterey to 58°20′ N latitude. The charts compiled by Hezeta show the territory up to approximately 50°, but to the north of 48° practically nothing is shown with the exception of a few small islands and separate coastal points. The main cartographic source used by Bodega y Quadra for the northern part of the Pacific Ocean was the map published in 1766 by Bellin, *Carte Reduite*. This map was based on the information contained in the Russian Academy of Sciences map of 1754–1758 by Müller and Truskot. Bodega and Mourelle considered this map to be full of errors.

Like most expeditions of the period, Bodega determined the longitudinal points by dead reckoning. The base meridian was the San Blas meridian. It is not known, however, if the original longitude of this port was corrected in accordance with the astronomical observations conducted in San Jose in 1769, nor is it known what the originating longitude was on Bellin's map of 1766. His 1769 map does not show San Blas at all, as it lies to the south of the lower frame. The expedition's charts were not published at the time, but the English editors of the *London Magazine* published in December 1780 a map they compiled on the basis of Mourelle's description. In 1781 Bodega's materials were used by Daines Barrington for the map he published in his *Miscellanies*. Both of these maps differed little from the chart by Bodega and Mourelle.[128]

One of the results of the Spanish expeditions of 1774 and 1775 was the general outline of the coasts of North America between 38° and 58° N latitude. Regions farther to the north were studied and charted in their general outline during the third circumnavigating voyage of Captain James Cook (1776–1779).

THE COOK EXPEDITION

The British government and the leadership of the Hudson's Bay Company had long been concerned with the discovery of the Northwest Passage from the Atlantic to the Pacific Oceans. In 1762 Samuel Hearne traveled overland from Hudson Bay in search of copper deposits and the Northwest Passage. This expedition lasted several years and in 1772 the traveler went far enough north to demonstrate that the Northwest Passage could not be found south of 72° N latitude. Because of this the British Government made the search for the passage from the other side, the Pacific to the Atlantic Ocean. This was Captain Cook's main task.

In a secret instruction to Cook dated July 6, 1776, he is directed to sail directly to the northwest coast of America without stopping anywhere south of 45° N latitude, and from there to proceed along the coast, without exploring bays and rivers, until he reached 65° N latitude. It is not clear why he was directed to sail exactly to 45° N latitude. It is possible that the instruction was based on the notion that this latitude marked the boundary of "New Albion," to which Britain had vague pretensions.[129]

Cook's expedition was well supplied with literature and charts of the northern part of the Pacific Ocean.[130] The expedition's ships did not carry any publications in the Russian language devoted to the explorations in this region. They had a few translations into English of Russian geographic works and English editions of Russian maps. Among these was a small part of an article by the English geographer and essayist J. Campbell about the northeastern borderlands of the Russian Empire, based on the First Kamchatka Expedition.[131]

The information about Bering's overland trek (1725–1727) and his sea voyage of 1728 were based on Bering's brief communiqué of 1730 and the catalogue of the geographic coordinates for "the cities and famous Siberian places, put on the map."

Harris's publication reproduced one of the copies of the chart of Bering's voyage compiled by Pyotr Chaplin. This chart focuses on Bering's route from Tobolsk to Okhotsk. In the extreme northeast, the contours of the Chukotka Peninsula with Cape Shelagskym are shown extending to 72°30′, together with Chukotka Cape, with St. Lawrence Island shown to the northeast.

The main Russian cartographic materials among the navigational aids available on Cook's expedition were the maps by Müller and Truskot of 1754–1758, and that by Jacob von Staehlen of 1773, which, as has been shown above, did not reflect the most recent information known by Russia's Admiralty College. They had the potential to mislead any navigator who used them.[132]

Cook's expedition, sailing on the *Resolution* and the *Discovery*, approached the northwest coast of North America in the area of the modern state of Oregon. For three weeks Cook led his vessels along the American coast, but at a considerable distance from shore. Between March 29 and April 26 the expedition stood in Nootka Bay on the west coast of Vancouver Island and then sailed north-northwest along the American coast. Cook did not even suspect that the expedition was proceeding along the large Alexander Archipelago.

In Gulf of Alaska waters the expedition duplicated Bering's route, being guided here by Müller's map, which, in this particular area, was based on Sven Waxell's information. On May 1, 1778, Cook sailed where Chirikov had approached the American mainland, and on May 4 he reached the point from which Bering had sighted Mt. St. Elias on the horizon. Cook named it Bering's Bay; it is now known as Yakutat Bay.

On May 10 Cook was along the coast of Kayak Island, where in 1741 members of Bering's Second Kamchatka Expedition went ashore. Not being sure that Bering's men had been on this same island, Cook named it Kaye's Island. On May 12 the expedition vessels entered a large bay which Cook called Sandwich Sound and which later was renamed Prince William Sound by John Douglas, the editor of the first edition of Cook's journals. After a week in Prince William Sound, on May 19 the Expedition passed the end of the modern Kenai Peninsula (indicated on Müller's map as Cape Germogen), to the northwest of which a wide entrance was sighted. This deep inlet was taken by Cook to be the mouth of a river. From May 28 to June 6 the expedition explored its shores and formally registered them as possessions of the English Crown. It ought to be noted that Chirikov was exactly here, at the southern end of Kenai Peninsula, in 1741.

Cook then sailed along the islands of Shuyak, Afognak, and Kodiak, believing they formed part of the mainland. To the south of Kodiak he found a landmass which he named Trinity Island, but which was in reality a group of islands, comprising Sitkinak, Tugidak, and several small islands separated from Kodiak by narrow straits.

On June 17 one of the islands of the Evdokievskaya group was sighted, which by then was well known to Russian *promyshlenniki* (the modern-day Semidi islands). The next day the expedition approached the Shumagin Islands.[133] After the *Resolution* and *Discovery* passed the Shumagin Islands, Charles Clerke (captain of the *Discovery*) reported to Cook that his vessel was approached by Native kayaks, from which a note in the Russian language was passed. The dates 1778 and 1776 appeared—the English were able to understand only this, as no one among the expedition members understood Russian. Later it became clear that the note was a receipt for *iasak* and on this basis it was decided that here was the easternmost point of Russian penetration.

On June 21, at the islands of Sanak and Caton, which Cook named the "Halibut Islands," the expedition's vessels were once again approached by Native kayaks. The Natives greeted the voyagers in the Russian manner—bowing from the waist—and Cook drew the correct conclusion that Russians were frequent visitors there. During the time spent at Unalaska Island the Natives again brought the commander a note in the Russian language. Cook returned this note, adding to it a letter naming the vessels, their captains, and the dates of the expedition in English and Latin. The local inhabitants appeared to the expedition members to be rather "civilized": they presented many letters in Russian, while the ladies offered endearments for a few leaves of tobacco.

Cook anchored in Samgoonoodha Bay on Unalaska's northern coast until July 2, when he continued his voyage north.[134] Having passed the Krenitzin Islands, on July 3 the vessels found themselves at the western extremity of the Alaska Peninsula. The expedition continued exploring along the northern shore of the Alaska Peninsula within the waters of Bristol Bay. At the western end of this bay, at Cape Newenham, Cook changed course to the west-northwest and, not noticing Nunivak Island, on July 28 approached St. Matthew Island. On August 3 the expedition reached St. Lawrence Island, which Cook named Anderson Island, to honor the expedition's surgeon who had died that day. The vessels then proceeded northeast and on August 9 came to the westernmost point of the American mainland. Although Müller's map, as well as those derived from it, clearly showed that this cape was discovered by geodesist Gvozdev, Cook named it Cape Prince of Wales.[135]

The expedition then turned westward, toward the Chukotka coasts, and on August 10 entered St. Lawrence Bay. From here the expedition returned to the American coast and continued along it northeast up to the latitude of 70°30′, where they encountered impassable ice on August 17. At a coastal projection which they named Icy Cape (70°19′ N latitude and 161°41′ W longitude), Cook turned southwest, having lost hope of finding an entry to the route to the Atlantic Ocean. On August 29 the expedition's vessels once again approached the northern shore of Chukotka at the cape named by Cook North Cape (modern Cape Shmidt) and then sailed southeast toward the Bering Strait. Sailing in these waters along the route traversed by Bering's first expedition,[136] Cook checked the accuracy of Bering's cartography of the northeast coasts of Asia. On September 4, 1778, he made the following entry in his journal:

> In justice to *Behrings* Memory, I must say he has deleneated this Coast very well and fixed the latitude and longitude of the points better than could be expected from the Methods he had to go by.[137]

The next day, convinced that Bering had correctly charted the northwest coast of Asia, Cook made the following notation:

> Being now satisfied with the accuracy of that Gentleman's Discoveries in this part, I hauled to the Eastward.[138]

At that time, Cook was near the coast of St. Lawrence Island, not suspecting that this was the same island to which he gave the name of Anderson a month before.

From St. Lawrence Island the expedition sailed once again toward Alaska, and en route, between September 6 and 18, they charted the coasts of a large bay that they named Norton Sound. On September 18 the vessels were near large shoals, and Cook, quite correctly, deduced that a large river emptied into the sea here. Indeed, the expedition was then across from the mouth of the Yukon River, but Cook's men were unable to see it.[139]

On October 3 the expedition entered Captain's Harbor at Unalaska, where this time they discovered a Russian settlement and entered into a close relationship with the Russian sailors and *promyshlenniki*. Cook described his first meeting with three Russians:

> They were all three well behaved, intelligent men, and very ready to give me all the information I could desire, but for want of an interpretor we had some difficulty to understand each other. They seemed to have a thorough knowledge of the attempts that had been made by their Countrymen to Navigate the Frozen Sea, and the discoveries which had been made in this by *Behring*, *Tchirekoff* and *Spanburg*, but seemed to know no more of Lieutenant *Sindo*

or *Sind* than his name: Nor had they the least idea of what part of the World Mr. *Stæhlins* Map refered to when leaid before them; and when I pointed out Kamtschatka and some other known places, they asked if I had seen the islands laid down (on) the Chart and on my answering in the negative, one of them laid his finger upon the Chart where a number of islands are laid down and said he had been cruzing there for land and could never find any. . . . The Memory of few men is held in greater esteem than these Men do *Behrings*, probably from his being the occation of their fur trade being extended to the Eastward.[140]

The three Russians spent the whole night with Cook, and the next morning (October 12) visited Clerke. They then left, very pleased with the reception offered them by the Englishmen, and promising to return in a few days with a chart showing the islands between Unalaska and Kamchatka.

On October 14 Cook met with navigator Gerasim Grigorievich Izmailov, who was, as the English seaman rightly surmised, "the principal person amongst the Russians in this and the neighboring islands." Izmailov asked Cook to his tent and treated him to dried pink salmon and berries. The Russian sailor immediately favorably impressed the head of the English expedition. Cook says the following in his journal:

(H)e was a sencible intelligent man, and I felt no small Mortification in not being able to converse with him any other way then by signs assisted by figures and other Characters which however was a very great help . . . I found he was very well acquainted with the Geography of these parts and with all the discoveries the Russians had made and at once pointed out the errors of the Modern Maps.[141]

On October 16 Izmailov brought, and permitted the Englishmen to copy, two charts which, in Cook's words, were

both Manuscripts and had every mark of being Authentick. The first comprehended the Penschinskian Sea, the coast of Tartary as low as the latitude of 41°, the Kurilian Islands and the Peninsula of Kamtschatka.

The Second Chart was to me the most intresting as it comprehended all the discoveries made by the Russians to the Eastward of Kamtschatka towards America, which if we exclude the Voyage of *Behring* and *Tcherikoff*, will amount to little or nothing. The part of the America Coast where the latter fell in with, is placed in this Chart, between the latitude of 58° and 58½° and 75 degrees of longitude from *Okhotsk*, or 118½ from Greenwich. And the place where the former Anchored in 59½° of latitude and 63½° of longitude from *Okhotsk* or 207° from Greenwich. To say nothing of the longitude

> which may be erronious for many causes, the latitude of the coast discovered by these two Navigators; especially that by Tcherikoff differes considerable from the account published by Mr Muller, but it is hard to say whether Mr Mullers Chart or this is the Most erronious, nor indeed is it now a point worth discussing. But the islands lying dispersed between 52° and 55° of latitude, in the space between Kamtschatka and America deserve some notice, though according to Mr Ismyloffs account neither the number nor situation of these islands are well assertained. He struck out about one third of them, assuring me they had no existence, and altered the situation of others considerably, which he said was necessary from his own observations, and there was no reason to doubt it.[142]

We do not know which charts Izmailov showed to Cook, but to judge by the latitude span wherein the Aleutian Islands were located (52°–55° N latitude), they may have been the most recent cartographic sources, such as the chart by Timofei Shmalev (1775) and Vasily Krasilnikov (1777) discussed above.[143]

At first glance it seems surprising that Izmailov struck from this chart a third of the islands represented thereon, but if one considers that the role of all *promyshlenniki* charts was to serve as navigational aids for reaching the most fur-bearer-rich islands, it seems quite possible that the navigator and *promyshlennik* Izmailov was reluctant to show to the foreigners the islands that were rich in "soft gold."

The islands of St. Macarius, St. Theodore, St. Abraham, Seduction, and a few others, shown on the Müller-Truskot map, were absent from the chart shown to Cook by Izmailov. The Russians tried to convince Cook that they had searched for these islands in vain many times, but Cook had difficulty believing that "Mr. Muller could place them in his chart without some authority." In spite of that, Cook, based on the testimony of Izmailov and his comrades, "le(f)t them out of my Chart and made such corrections among the other islands as I was told was necessary."[144] Cook also entered Isanotsky Strait on his chart between the Alaska Peninsula and Unimak Island.

The charting of this strait was conducted, as we have seen above, by the Krenitzin/Levashov eexpedition, and the presence of this strait on Izmailov's chart serves as one more piece of evidence confirming this chart's reliability and modernity.[145] Cook moved the Andreanof Islands to the east on his chart, on Izmailov's advice.[146] The English also obtained from the Russians the Native names of all the islands, including that of Kodiak. As far as the American mainland was concerned, Cook wrote:

> I have already observed that the America Continent is here called by the Russians as well as Indians Alaska, which is the proper Indian name for it, and probably means no more than that part adjoining to

> Ooneemak, however the Indians as well as the Russians call the whole by that name and know very well that it is a great land.
>
> (T)hey assured me over and over again that they k(n)ew of no other islands but what were laid down on this chart and that no Russian had ever seen any part of the Continent to the northward, excepting that part lying opposite the Country of the Tchuktschis.[147]

New geographic information obtained by Cook from the Russians convinced him of the total unreliability of Staehlen's map, about which the English traveler says, with indignation, the following:

> If *Mr Stææhlin* was not greatly imposed upon what could induce him to publish so erroneous a Map? In which many of these islands are jumbled in regular confusion, without the least regard to truth and yet he is pleased to call it a very accurate little Map? A Map that the most illiterate of his illiterate Seafaring men would have been ashamed to put his name to.[148]

Gerasim Izmailov

Speaking of Russian seamen, at least one of them, Izmailov, impressed Cook very favorably, as Cook expressed it:

> This *Mr Ismyloff* seemed to have abilities to intitle him to a higher station in life than that in which he was employed, he was tolerably well versed in Astronomy and other necessary parts of the Mathematicks. I complemented him with an Hadlys Octant and altho it was the first he had perhaps ever seen, yet he made himself acquainted with most of the uses that Instrument is capable of in a very short time.[149]

This outstanding Russian sailor, Gerasim Izmailov (1745–1795?) was certainly worthy of such a high opinion by Cook. He was a representative of a new generation of Russian *promyshlenniki* who were taking possession of the oceanic approaches to Alaska. A graduate of the Irkutsk Navigational School, Izmailov took part in the expeditions of Sindt and Krenitzin-Levashov as an apprentice navigator. In April 1771 he was taken prisoner of war by the Polish mutineer Moritz Benyowski, who engineered an uprising on Kamchatka. The mutineers, having taken a merchant vessel in the mouth of the Bolshaya River, fled from Kamchatka on her.

Benyowski compelled Izmailov to sail with him, but after it became clear that Izmailov and Paranchin, a Kamchadal, were organizing a plot, after cruelly beating them, he put them ashore on an uninhabited shore of Simushir Island in the Kuriles.[150] While Benyowski sailed on to France, Izmailov and Paranchin managed to reach the coasts of Siberia, and in October 1772 they were brought to Irkutsk, where they were questioned at length as participants in Benyowski's mutiny. Once their personal lack of involvement had been established, they were freed.

In 1775 Izmailov conducted a survey of the coasts of Kamchatka, and in early 1776 he was appointed captain of the merchant vessel *St. Paul.* This vessel embarked on a five-year-long commercial voyage to the Fox Islands, and Unalaska became its base. Izmailov was an excellent navigator with much experience in Aleutian waters and in the Bering Sea.[151]

Cook entrusted to Izmailov's care a sealed package with a letter to the British Admiralty and a chart of northern coasts, which were delivered to Kamchatka the following year, and from there to England.[152]

Izmailov, in his official report, provided in detail all the information about the expedition he was able to obtain from Cook and his officers and stated that Captain Cook granted him "a spherical chart, [i.e., apparently in conical projection—AP] assuring me that it represented the true route of their voyage, and then indicated locations on the Mercator chart drawn by me from North latitude of 50°00′ to that of 71°00′, and from longitude 51°00′ to 77°00′ from Kurilskaya Lopatka eastward. As a sign of his charity, Mr. Commander presented me with a quadrant for the observation of the sun's altitude and also his own sword."

Izmailov appended to this report not only the "spherical" chart he received from Cook, but also the Mercator chart he himself had compiled using English data.[153]

Only one week after Izmailov's report was received in Petropavlovsk, the *Resolution* and *Discovery* entered that harbor. They were on their second voyage to the northern part of the Bering Sea, having spent the winter in the Hawaiian Islands, where on February 17, 1779, in a skirmish with Natives, Captain James Cook was killed.

On May 12 the English arrived at Bolsheretsk, where they were hospitably received by the "Chief Commander" of Kamchatka, Premier-Major Magnus Bem (1727–1806), a native of the city of Riga. The expedition's officers were assigned a fine house, servants, and a guard detail. The expedition was supplied, totally free of charge, with foodstuffs and ship's tackle. R. F. Vereshchagin, the priest of the village of Paratunka, regularly provided the deathly ill expedition leader, Captain Charles Clerke, with fresh milk. The English much enjoyed the Russian dances and songs with which they were entertained by Captain V. I. Shmalev, who was getting ready to assume the office of Chief Commander of Kamchatka that was being vacated by Bem.[154]

During the time the expedition's vessels spent in Petropavlovsk Harbor, the English and the Russians continued the exchange of geographic information which had begun on Unalaska Island.[155] Lieutenant James King, who visited Bem in May of 1779, on Clerke's instruction, gave the chief commander of Kamchatka a gift of "charts of several newly-discovered islands which he had brought here, and portraits of wild persons of male and female genders,"[156] and also showed him a

> small Chart of our discoveries this Voyage; this as was natural, was what above all things he wish'd to see, & he was too well bred after a Slight hint, to ask a second time to take a copy of it, or to be at all offended at a refusal.[157]

Nevertheless, according to the testimony of the expedition's assistant surgeon, the expedition's general chart fell into Russians hands after all. On May 26 the English learned, much to their dismay and indignation, that William Bayly, the expedition's astronomer, gave a chart of Cook's discoveries off American shores to the navigator of a Russian sloop then in Petropavlovsk Harbor in exchange for a chart of Kamchatka and adjacent localities.[158] The fate of Cook's chart remains unknown, but it is possible that it was utilized in Russian cartographic work dealing with the results of Cook's voyage (these will be discussed below).

Polutov

The navigator to whom Bayly delivered the general chart of the expedition was apparently Apprentice Navigator Dmitri Polutov, captain of the vessel *St. Nicholas*, which belonged to the Totma merchant Pyotr Panov. The *St. Nicholas* was then getting ready to quit Petropavlovsk Harbor and sail to the Aleutian Islands for fur procurement.[159] In the fall of 1779 Polutov arrived at Unalaska, and in the summer of 1780 he sailed to the American coasts, but, due to damage to the anchor, he was forced to stop at Unimak. From there the *promyshlenniki* traveled by baidaras to Sanak and Unga islands. In May of 1782 Polutov and Ocheredin joined the navigator Pankov and sailed for Kodiak, to trade and procure furs.

The Kodiak inhabitants did not permit the *promyshlenniki* to land on the island, however. Polutov was thus forced to turn back and he procured furs in the Fox Islands until 1785.[160] On the basis of this expedition a new map was compiled: *The Mercator Chart of the Voyage of the Vessel St. Nicholas under the Leadership of the Skipper and Foreman, Apprentice Navigator of the Okhotsk Command Dmitri Polutov, Dispatched from Petropavlovsk Harbor by Mssr. Companions Pyotr Panov and Arsenti Kuznetsov in '779 July Month to the Bay Sandwich Sound and from there back to the Mouth of the River Vorovskaya in the Year '785, 28th day of September and which Shows the Aleutian Islands and a Part of north-Eastern America, Compiled in the Year 1786, 25th Day of September.*[161]

This manuscript chart, executed in watercolors, with elegant drawings of fur seals, was apparently compiled from the chart Polutov obtained in 1779 from the English. The evidence for this is the presence of toponyms given by Cook to some geographic features of Alaska, particularly Sandwich Sound.

Setting out for the second time on the voyage north, the English entrusted to Bem, who was preparing to leave for St. Petersburg, a letter from Clerke to the British Admiralty, in which it was stated that the former chief commander in Kamchatka agreed "to take care of this letter and of the journal of our investigations, together with the chart of parts of America, Asia etc. which we have seen, and another one of the Sandwich Islands." The letter reached the Admiralty intact, but the charts which accompanied it disappeared.[162]

The Cook Expedition, now led by Clerke, left Petropavlovsk Harbor on June 12, 1779, and proceeded along Kamchatka's east coast north. "Upon their departure from Kamchatka, the Englishmen were issued, for their use, open orders to all Russian *promyshlenniki* to treat them in the Eastern Sea with all possible consideration and not to offer any aggravation."[163] The vessels passed through the Bering Strait and approached latitude 70°, where they encountered ice. For eight days Clerke attempted to push his way northward, but on July 27 he was forced to turn south. He died of consumption on August 22, approaching Avacha Bay. The command of the expedition passed to John Gore. Clerke was buried in Petropavlovsk,

FIGURE 20 The monument over Clerke's grave erected by the LaPerouse Expedition in 1787. Martin Sauer. *An Account of a geographical and astronomical expedition to the northern parts of Russia . . . .* 1802. pp. 146–147. UAF Rare Book A0114

where the grave of this remarkable explorer is carefully preserved to this day and respected by the Russian people (Fig. 20).

Once again the English were met in Petropavlovsk with hospitality, but this time they dealt not with Bem, who had departed for St. Petersburg, but with V. I. Shmalev, temporary commander of Kamchatka. On October 8, 1779, the expedition left for home and one year later, on October 7, 1780, the *Resolution* and the *Discovery* entered the mouth of the River Thames.[164]

Let us proceed to examine the cartographic results of Cook's third voyage pertaining to the north of the Pacific Ocean.[165] Facsimiles of cartographic materials received by the British Admiralty were published by Robert Skelton in 1955.[166] The summary chart of the northern Pacific Ocean included in these materials (*Chart of Part of the NW Coast of America. Explored by Captain Cook in 1778*) reproduces the document which was sent by Cook, along with his letter, to Philip Stephens, Secretary to the Admiralty, via Izmailov. This chart is in conical projection and probably formed the basis for the "spherical" chart which Cook "presented" to Izmailov at Unalaska.

The chart lacks any indication that Russian materials were utilized. Kodiak is represented as part of the mainland and the coasts of Unalaska are shown in a very fragmentary way. Included in the set are charts of larger scale, in Mercator projection, showing the Unalaska and Bering Strait regions (charts ## LII, LIII, LVI). In 1991 this author had the good fortune to locate a large-scale manuscript chart, hitherto unknown: *Chart of NW Coast of America and the Part of NE Asia with the Track of His Majesty's Sloops Resolution and Discovery from May to October 1778* (Fig. 21)[167] in the collection of the American Geographic Society, Golda Meir Library, University of Wisconsin, Milwaukee.

This chart was compiled by George Vancouver, who at the time of the expedition was a midshipman aboard the *Discovery*. This chart gives more detail (including data on the expedition's route with dates when geographic coordinates were determined) than the summary chart from the British Admiralty. It is entirely possible that this chart is a copy (or even the original?!) of the expedition's chart sent by Clerke with Magnus Bem across Russia to England. The chart lacks any indications that Russian sources were used.

The first chart to appear in print was *A Chart of the Discoveries Made by the Late Captain Cook & other European Navigators in the Great Pacific Ocean between Asia & America* published by Thomas Kitchin in the July issue of the *London Magazine* for the year 1780.[168] Other editions of charts reflecting the results of Cook's expedition appeared also,[169] but the official journals and charts were first published in 1784 in London, in the *Atlas* accompanying Captain Cook's *Voyage to the Pacific Ocean*. This atlas contains the *Chart of the North West Coast of America and the North East Coast of Asia Explored in the Years 1778 & 1779. The Unshaded Parts of the Coast of Asia are Taken from a MS Chart Received from the Russians.*[170] As far

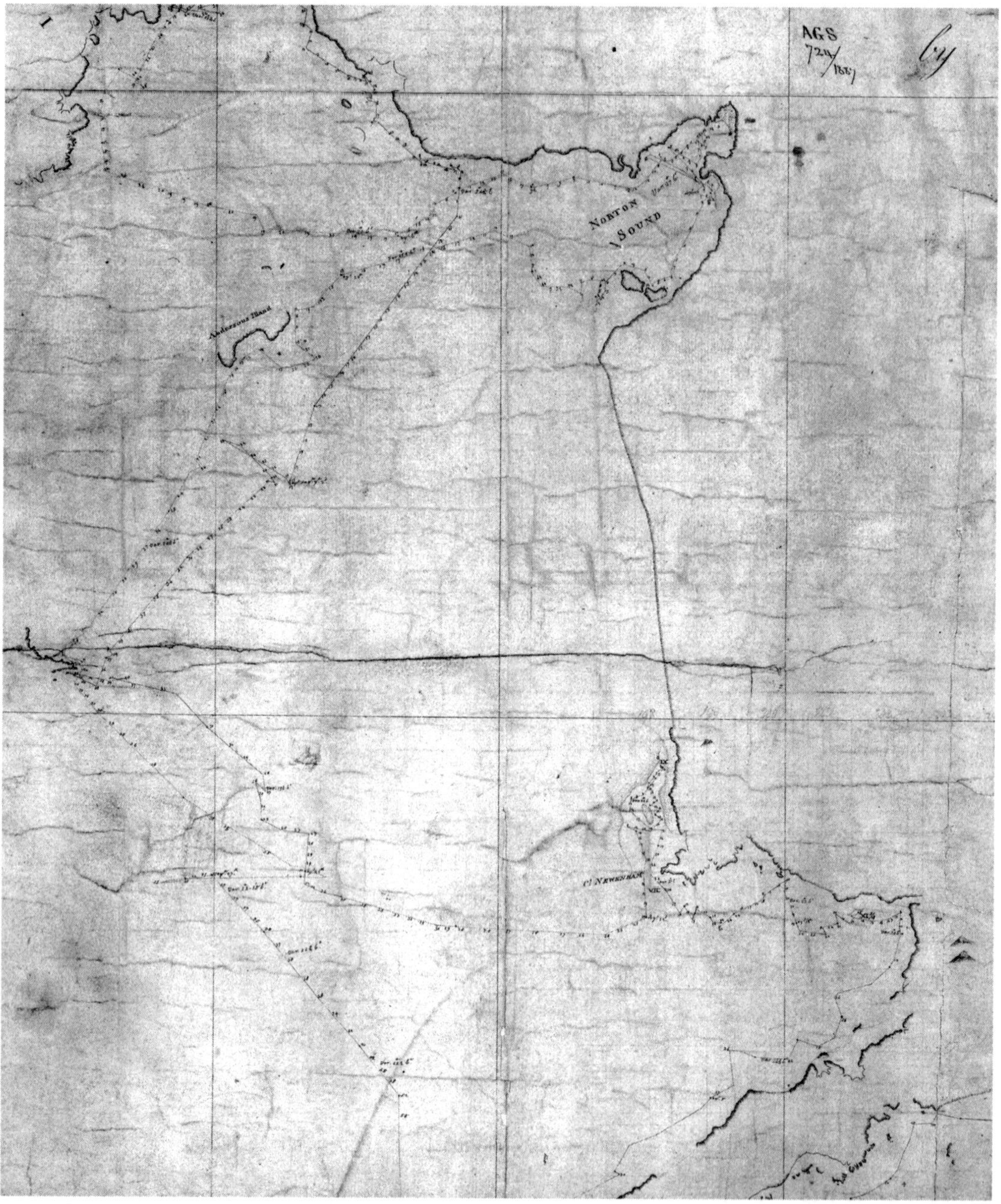

FIGURE 21 A portion of a manuscript map drawn by George Vancouver depicting the track of the Cook Expedition from May to October 1778. Courtesy of the Golda Meir Library, University of Wisconsin, Milwaukee. Manuscript AGS 724/1887

as representation of the coasts of the Aleutian Islands and the Alaska Peninsula are concerned, there is no reference to the use of Russian materials.

By 1778/1779 Russian seafarers already had copies of various charts and descriptions made by members of Cook's expedition. These materials were important not only to the navigators of merchant vessels, but also to the representatives of the Russian administration. For experienced seafarers sailing repeatedly in the waters of the Pacific Ocean along the Aleutian Islands and the coasts of Chukotka and Kamchatka, Cook's charts greatly clarified the geography of the ocean between Kamchatka and America. These charts established the general outline of the American shoreline between the Alaska Peninsula and Bering Strait. Russian charts compiled in the years 1779–1781 utilized English data. We now will proceed to examine them.

The first of these charts (Fig. 22) was created in 1779 by Apprentice Navigator Potap Kuzmich Zaikov, who participated in fur procurement off the shores of the distant Aleutian Islands as captain of the vessel *St. Vladimir.* He spent almost three years on Unimak and from there launched expeditions in the coastal waters of Alaska as far as Kodiak. On these voyages he conducted surveys and collected valuable information about the Natives.[171] Based on these explorations, Zaikov compiled a chart, *Mercator Chart which Shows the Aleutian Islands also Called the Northern Archipelago, According to the Description by Apprentice Navigator Potap Zaikov from the Year 1779; All of the Newly-Described Islands Indicated in Red Color Differ from Earlier Surveys in 4½ Degrees Longitude and from the English Observations by 2½ degrees . . . Copied from the Original Sergeant of Geodesy Ivan Rebrov.*[172]

The anchorage at Unalaska of the vessels of Cook's expedition is shown (with an erroneous date of 1748 instead of 1778) with the caption, "At this cape two English vessels stood at anchor in the year 1748." The presence on the chart of the northwestern coast of the Alaska Peninsula, which prior to Cook was totally unknown to the Russians, demonstrates the use of English data. The chart was delivered to the Irkutsk governor, Franz Klichka, and Zaikov was awarded a prize of 500 rubles for it and received the rank of Navigator.[173]

Pallas wrote: "I believe that Zaikov's chart which also shows the voyage of Dmitri Bragin . . . shown in green should be considered by mariners who sail these regions as the best and most accurate one of all those that exist at present in respect to the Islands; I have no basis to mistrust its details, though I have weighty reasons to doubt the extent and longitude of Aliaska."[174] A version of Zaikov's chart reflecting the voyages of 1772–1779 has also been preserved.

On Klichka's orders, Magnus Bem left all the Russian and English maps related to Cook in Irkutsk when he returned to St. Petersburg. A preliminary analysis of these materials and new charts summarizing the results of Russian and English explorations was done by Second-Major Mikhail Tatarinov at the Irkutsk Navigational School.[175]

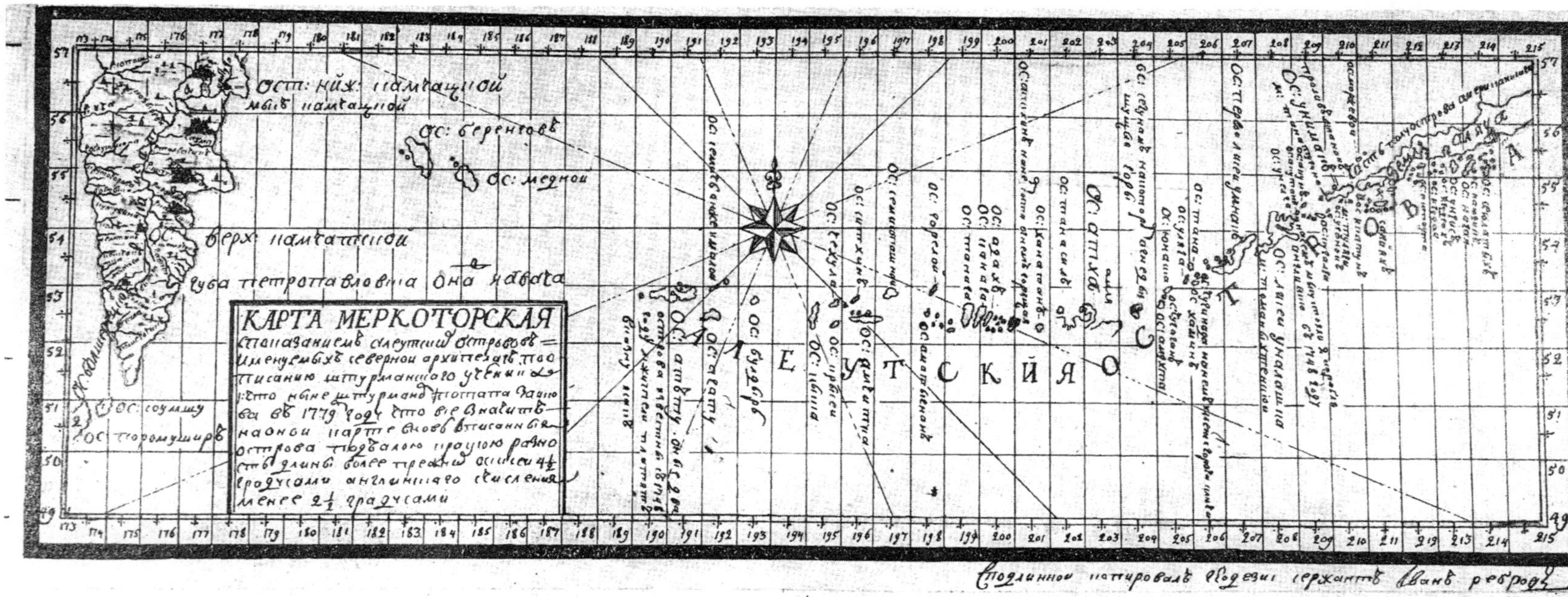

FIGURE 22 Mercator chart of the Aleutians drawn by Potap Zaikov in 1779. RGAVMF. F. 1331. Op. 4. No. 105

Tatarinov also prepared the journal of the expedition led by cossack commander Ivan Kobelev. Kobelev was dispatched to Chukotka in response to rumors about the appearance of foreign vessels at its shores. He sailed out of Penzhina Bay toward Anadyr on March 22, 1779, and reached the Diomede Islands on July 27.

An extract from Kobelev's journal prepared by Tatarinov mentions the meeting of the cossack commander on the island of Igellin with the "chief headman . . . who in conversation stated that on the American land, along the River Khevren, there is a fortified place called Kymgovei where Russian people live; they speak Russian, read books, write, venerate icons and so on; they differ in appearance from the Americans. Americans have sparse beards and then pluck even these, but the Russians who live there have thick and large beards."[176]

Thus Kobelev strengthened the rumors about an ancient Russian settlement on American shores, across from Chukotka. Nevertheless, during this expedition to the Diomede Islands, much real data about the geography and settlements of the northern part of Alaska were obtained. This was reflected in the schematic drawing compiled by Kobelev (Fig. 23) on the basis of information supplied by the headman of Igellin. It was used in the same year by Second-Major Tatarinov when compiling a chart of Chukotka and Alaska.[177]

On the latter chart there are clear indications that Cook's data, as well as data obtained from the Natives, were utilized. The inscription placed parallel to the eastern border of the chart states: "North American Land. The coasts are represented on the basis of Captain Cook's description, while rivers with American settlements, are derived from the sketch by Commander Kobelev, who visited the second island Igellin, where the local *toion* Kaimunia described these verbally and pointed them out on the sketch without distances . . . ." This inscription is one of the earliest instances of the use by Russians of Native cartographic images.

Research conducted by Svetlana Fedorova and Dorothy Ray proves that the Kobelev-Tatarinov chart is a clear step forward in the depiction of northern Alaska. Ray's ethnographic fieldwork demonstrated that the latter chart is unique not only because it shows several bays, including Kotzebue Sound first found by the Europeans in 1816, and rivers and islands "discovered," as it was believed much later, but because it also shows sixty-nine Eskimo settlements (sixty-one of which are shown with their toponyms).

The Eskimo settlements are represented on this chart in two groups about 400 km distant from each other: one on the Seward Peninsula, the second in the coastal delta of the Yukon and Kuskokwim rivers. In the process of mapping the present-day Native settlements in the Bering Strait region, Ray determined that all the settlements shown on the Kobelev-Tatarinov chart were located on the shores of the Seward Peninsula and Kotzebue Sound. In spite of the considerable time that has elapsed since the chart was compiled, linguistic changes, and differences in the methodology of obtaining information, the names of these settlements are recognizable, and they are shown in today's geographic sequence.[178]

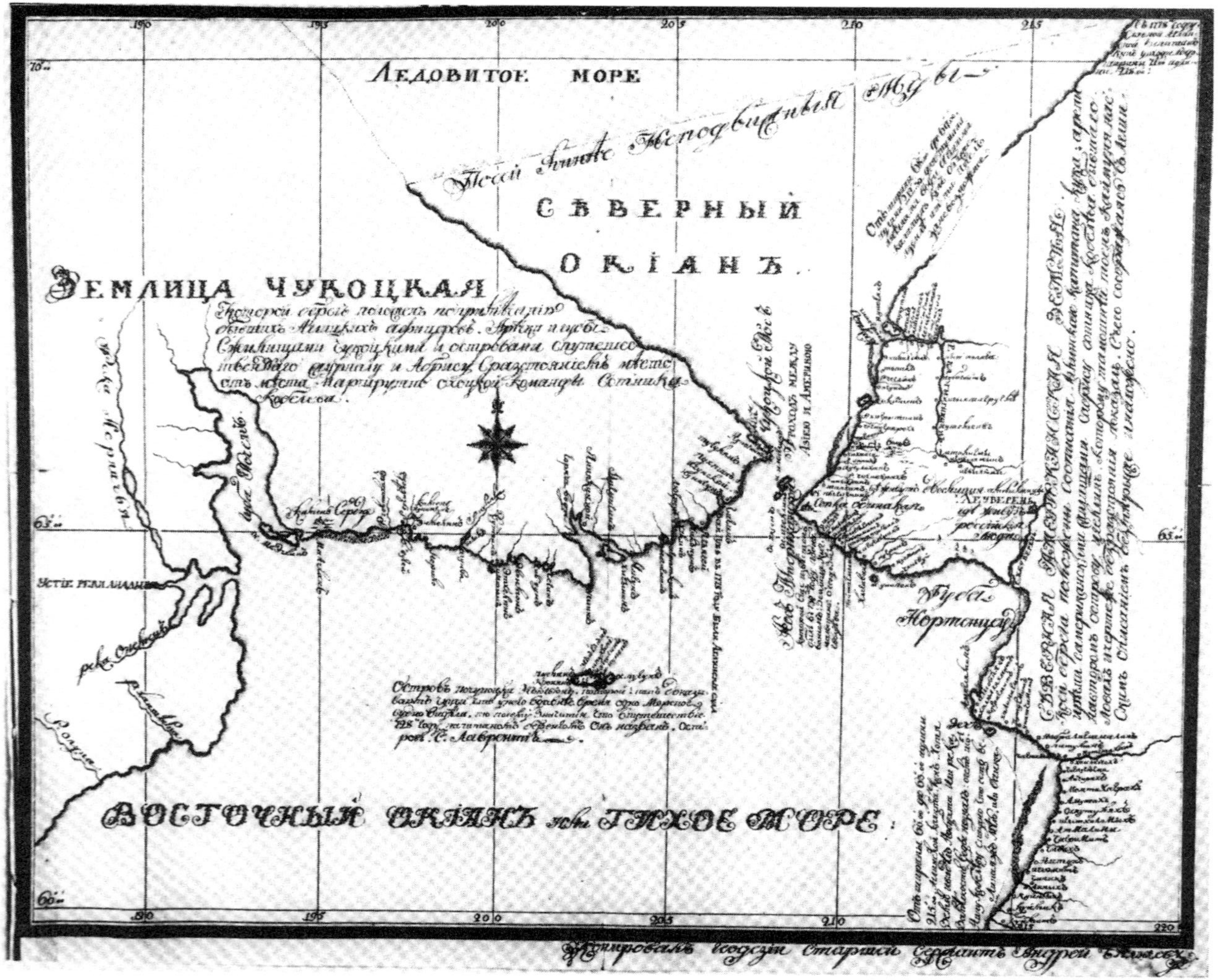

FIGURE 23 Kobelev's voyage 1779. RGADA. F. Gosarkiv, VII, 2539, l. 535

Toward the end of 1779 Tatarinov attempted to summarize all the data about the mainland and water expanses in the high latitudes, based on the descriptions and charts of domestic origin available to the administration of Siberia. He also used the materials of Cook's expedition delivered by Bem and some data about the Spanish voyages. The result was the *Chart of the Entire Northern Ocean near the North Pole Indicating the Searches in the Arctic Ocean and in the Pacific Sea of Mssrs. Naval Officers, Russian, English and Spanish.*[179]

This chart is drawn in azimuth projection with the center at the North Pole. North America is shown by a dotted line, which extends throughout the polar region and is joined to Greenland, but is separated from Asia by Bering Strait. A certain coastline extends to a point which is marked: "This point was reached in the year 1778 by the English Captain Commander Cook up to 71 degrees North latitude." The chart shows in detail the routes of the detachments of the Second Kamchatka Expedition, Bering's and Chirikov's, in the Pacific, and those along the coast of eastern Siberia (Muraviev, Malygin, Skuratov, Ovtsyn, Pronchishchev, and Laptev). Also represented are the routes of the expeditions of Krenitzin/Levashov, Sindt, and merchant Bakhov.

Great attention was paid to Cook's expedition. Various stages of his voyage are discussed in the chart's legend and compared with Russian data: "The observations published by Mr. Cook differ from the Russian ones in longitude by more than 10 degrees, specifically at Cape Aliaksa, under the letter H."[180]

Tatarinov compiled another chart in 1780 of the northern part of the Pacific Ocean that was first made known to the scholarly world by Svet and Fedorova.[181] Andrew David, having compared this chart with Cook's expedition materials, reached the well-founded conclusion that the coastline of Alaska followed Izmailov's Mercator chart, and the insert showing Avacha Bay followed the survey of the *Resolution*'s captain, William Bligh.[182]

There is another chart of the same type which was probably also connected to the work of Siberian cartographers, who analyzed the results of Cook's expedition and integrated them into the data of Russian investigations. This is the *Mercator Chart on Which Locations are Shown from Various Voyages Described by Sailors. . .* , preserved in the Russian State Archive of the Navy.[183] The author's name is not indicated, but, judging by the handwriting, it was the same Second-Major Tatarinov. Alaska's coast is probably drawn from the Mercator chart by Izmailov, as was the case with the preceding chart.

All the materials connected to Cook's expedition had arrived at the capital by 1780, when the Admiralty College and Academy of Sciences undertook their analysis and summation in order to publish a chart of the northern part of the Pacific Ocean intended to prove Russian priority in this region. In addition to the charts discussed above, several English charts summarizing the results of the 1778 and 1779 voyages were delivered to Izmailov, Bem, and Polutov.

At present, three copies of the summary chart of the voyages of Cook and Clerke (*Chart of the NW Coast of America and NE Coast of Asia Explored in the Years 1778 and 1779. The Unshaded Parts of the Coast of Asia are Taken from a MS Chart Received from Russians*) are preserved in Russian collections. One of these (Fig. 24) is a version that reproduces either the chart sent by Clerke in 1779 from Petropavlovsk appended to a letter to the (English) admiralty which was lost in Russia, or the chart obtained by the skipper of the *St. Nicholas*, Dmitri Polutov, from the expedition's astronomer, William Bayly. This chart is preserved in the Archive of the Foreign Policy of the Russian Empire in Moscow. The possibility should not be ruled out that it is an original document compiled on board one of Cook's vessels.[184]

Two others have the title *Copy of the Geographic Chart of the Voyage of Mr. Cook, Compiled by the London Meridian. It Encompasses the Northeastern Coasts of Asia Adjacent to the Arctic Ocean, as well as those of Kamchatka and North America Washed by the Eastern Ocean (Pacific); After his Voyage in the Years 1778 and 1779, his Routes Indicated by Dotted Lines in Black with the Addition of the Voyages of the Russian Sailors in the Eastern Sea in the Years 1728, 1741, 1766 and 1769, Each Marked by a Different Color Indicating also which Degrees of Longitude and Latitude were Reached.*

They are the same summary charts of the expedition in conical projection and in the English language, but with the addition of the routes of the First and Second Kamchatka Expeditions, Sindt's expedition, and the voyage of Krenitzin and Levashov.[185] The Aleutian Islands are shown without shading, that is, according to the explanation on the chart, from Russian manuscript charts obtained by members of Cook's expedition during their stay at Unalaska and in Petropavlovsk Harbor in Kamchatka. Except for the addition of the routes of Russian voyages, nothing was added. Even Kodiak Island is shown as linked to the mainland, though by this time the fact that Kodiak was an island was well known to the Russians.

Though the summary chart of Cook's expedition was indisputably more detailed and accomplished than the "spherical" one received by Izmailov at Unalaska, surprisingly, it was the latter which was utilized as the main material illustrating Cook's expedition data. It was published in Russia in 1781 and summarized the results of the English and Russian explorations in the Pacific Ocean.

The *Chart of the New Discoveries in the Eastern Ocean, 1781*[186] appeared as a separate publication in the Russian language and as a supplement to the work by Pallas examined above, *On the Russian Discoveries Between Asia and America*, published both in Russian and German.[187] There is no doubt that the navy and the Academy of Sciences entrusted the compilation of this chart to Pallas, historiographer of the Admiralty College. Pallas's authorship is not to be disputed, not only because the chart is appended to his work, but also because the original, in Pallas's handwriting, is preserved in the manuscript division of the Academy of Sciences in St. Petersburg. The content of it is absolutely identical to the chart published in 1781.[188]

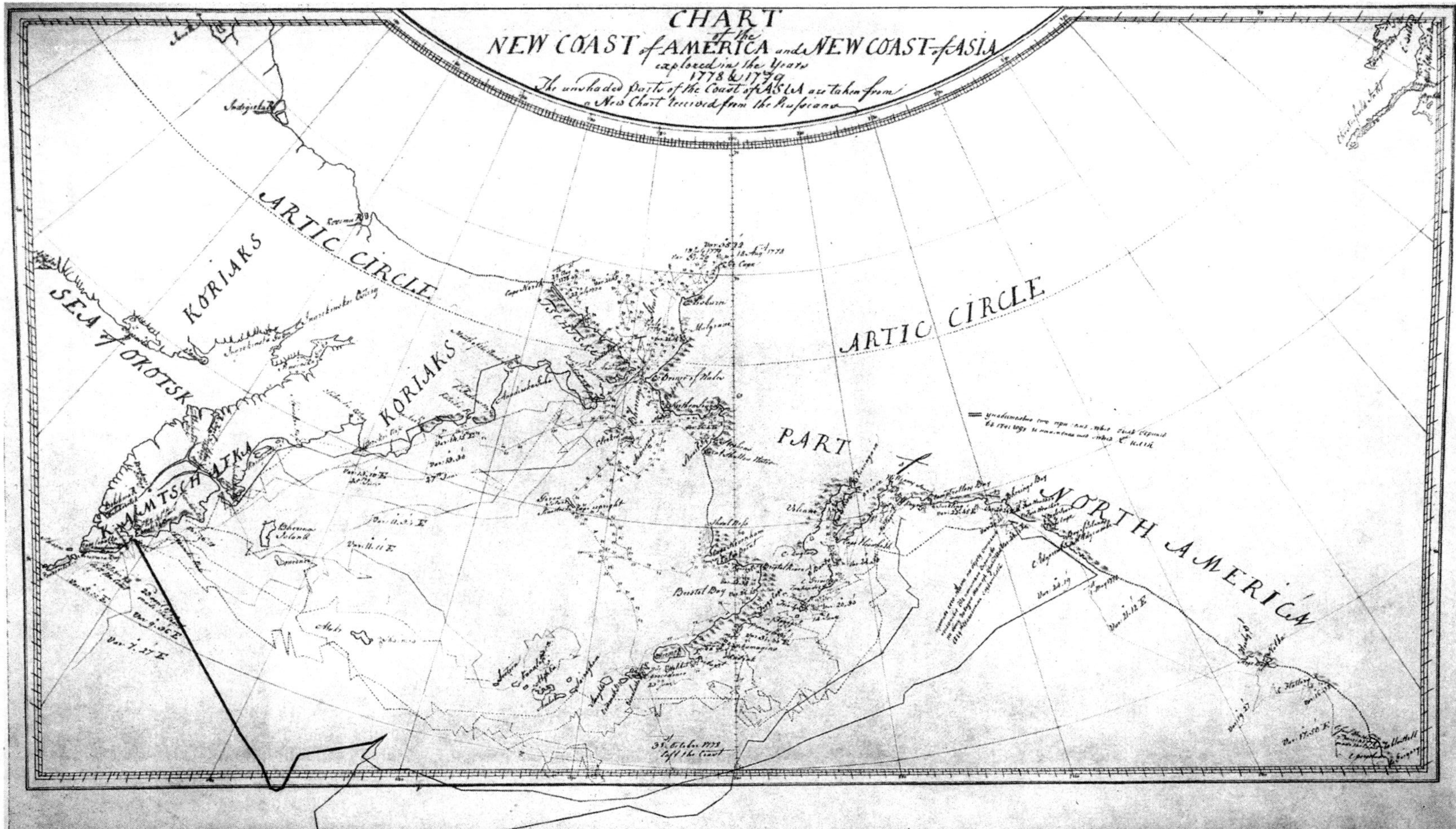

FIGURE 24 A copy of a Cook Expedition manuscript chart provided to the Russians in Kamchatka. RGAVMF. F. 1331. Op. 4. No. 107.

A Pallas manuscript map in the Academy of Sciences library is in German and only the title, *Chart of the New Discoveries in the Eastern Ocean*, is in Russian. Pallas himself devoted special commentary to this chart.[189] He states, among other things, that he found it more beneficial to illustrate his article not with the Krenitzin chart but to "present here a revised chart of all the islands known at present between Kamchatka and America, as well as the shores of America itself to the areas near the strait between Asia and America which have become known thanks to the voyage of the famous Cook therein. Therefore, the extremity of Asia and the entire American coast are represented in accordance with Captain Cook's observations, insofar as these are known to me."[190]

Overall, the chart of 1781 confirmed the theoretical model of the geography of the northern part of the Pacific Ocean created by Pallas, which was based on analysis of Russian voyages and stated two years before Cook's voyage. There is also no doubt that one of the aims in publishing this chart was subtle propaganda advocating Russian priority in the exploration of the Aleutian Islands and the coasts of America. Thus, in the region of the cape which Cook had named Cape Prince of Wales, the chart of 1781 bears the inscription: "American cape visited in the year 1732 by geodesist Gvozdev."

In the southern part of Alaska, the coast along which Chirikov sailed is indicated by a dotted line. In the area of the Bering Strait, islands "from the charts of Lieutenant Sindt" are indicated, also by means of a dotted line. This chart in the Central Archive of Ancient Documents of Russia (RGADA) emphasizes Russian priority by indicating the coastal segments and islands discovered and explored by Russians with edging in yellow, while the coasts explored by the Cook-Clerke Expedition are shown in pink watercolor.

The 1781 chart, like all eighteenth-century Russian charts which preceded it, did not reflect the true state of geographic knowledge of the northern part of the Pacific Ocean. We have noted that the basis for the 1781 chart was the chart compiled by Tatarinov in 1780 in Irkutsk, which in turn was based on Izmailov's Mercator chart. Pallas's chart did not reflect in full the data on Russian fur procuring and government expeditions and presented the data by Cook/Clerke in a rough manner. This was even though Russians had charts of the voyages of Cook's vessels in the Bering Sea in the years 1778–1779 that showed Alaska's coastline in much greater detail than the chart of 1781 under discussion here.

Apparently the leadership of the navy and the Academy of Sciences understood the shortcomings of Pallas's map and the need to further demonstrate the achievements of Russian sailors, because immediately after its publication the *Chart of the Voyage of Navigator Zaikov* (Fig. 26)was published in 1782, supplemented by data from other *promyshlenniki*. This chart by Potap Zaikov shows the coastline of the Alaska Peninsula and the adjacent territories in much greater detail than indicated by Pallas, but here, too, use of the summary charts from Cook's expedition is evident, though with significant corrections. Specifically, there is detailed

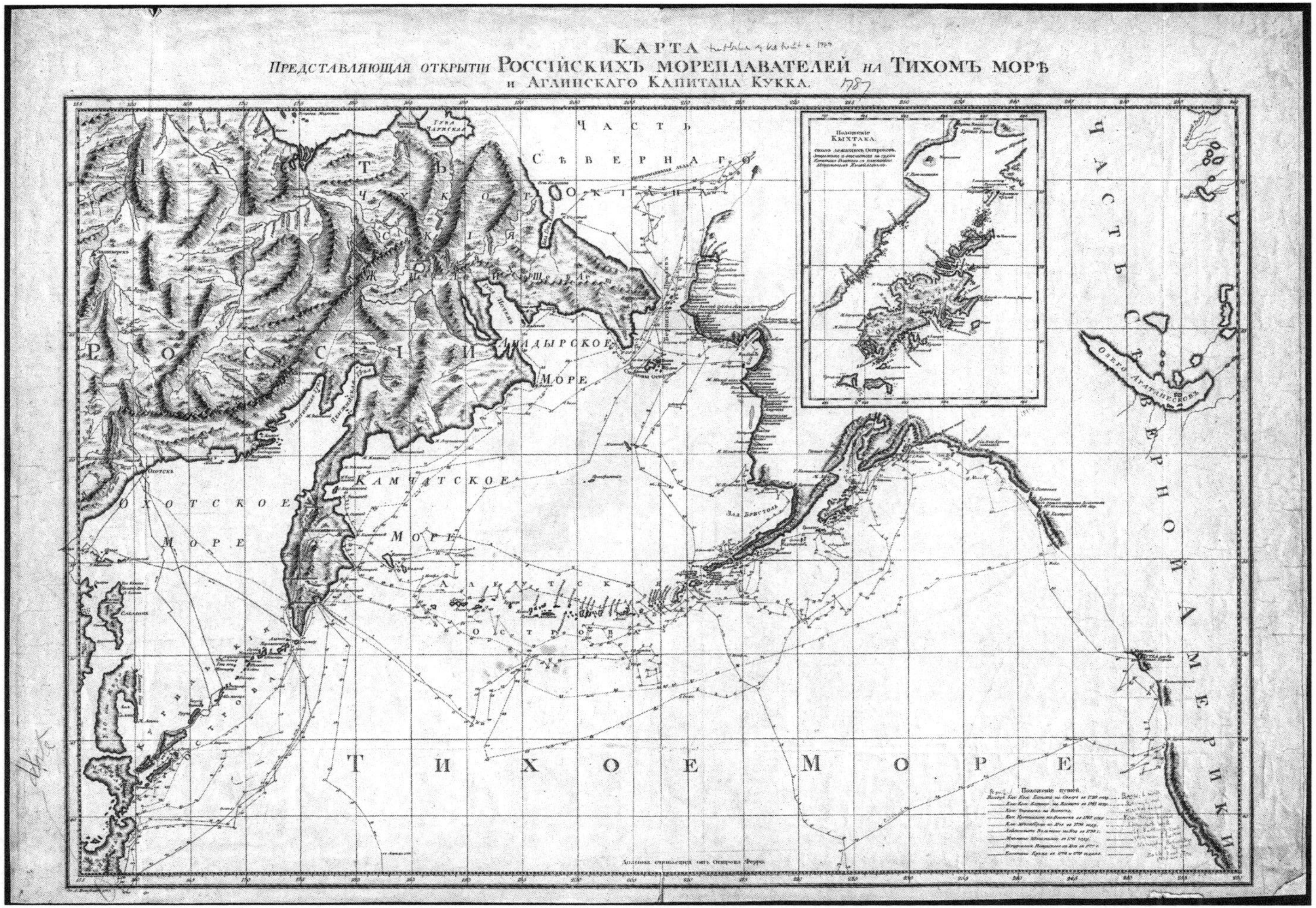

FIGURE 25 A. Vilbrekt, *Karta predatavliainschaia otkriti* . . . [Russian Discoveries and those of Captain Cook], 1779. UAF [1787 printing] Rare Map G9236 S12 1787 V5

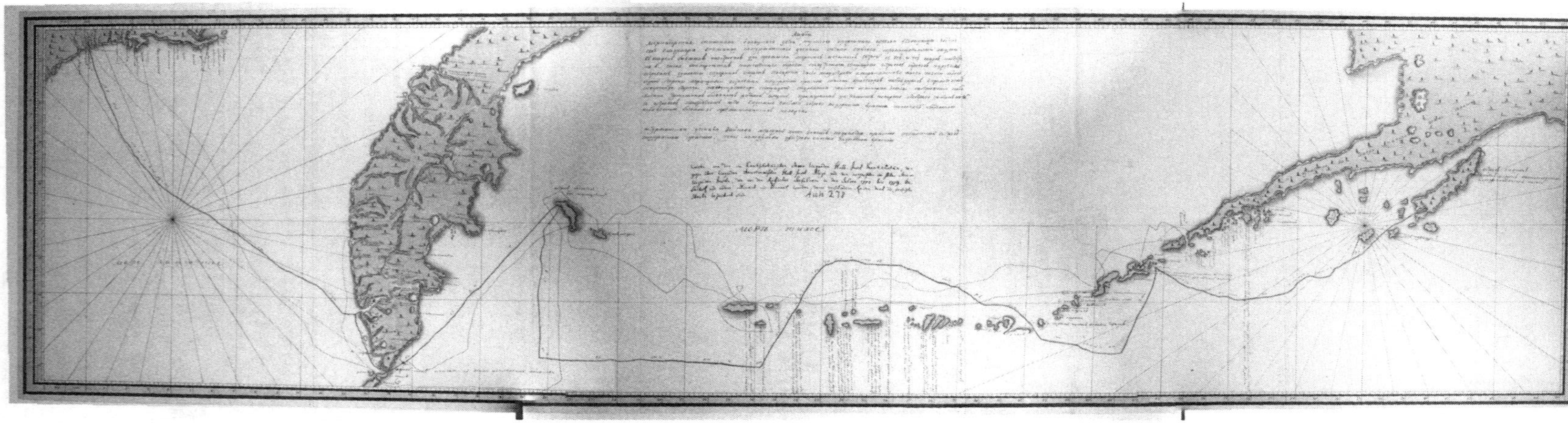

FIGURE 26 Manuscript chart of the voyage of Navigator Zaikov, 1782. Courtesy Göttingen University Von Asch Collection 278

representation of Kodiak Island, which in Cook's materials appeared to be linked to the mainland.

In Siberia and Kamchatka, cartographic work linked to the results of Cook's expeditionary vessels in the northern part of the Pacific Ocean continued. In 1781 a new version of the summary chart of the northern part of the Pacific Ocean was compiled, considerably enriched in comparison to Tatarinov's chart of 1780. This is the *Accurate Chart of Part of Kamchatka Showing the Okhotsk Coast and all the Kurile Islands and Part of Japan, as well as Northwestern America, with all the Aleutian Islands or the Northern Archipelago, Corrected in Irkutsk in the Year 1781 in Comparison to the Earlier Descriptions.*[191]

The chart's author is not named, but it is possible that it was once again Tatarinov, as it is clear that his 1780 chart served as a basis for the later work. Data from Russian *promyshlenniki*, among them Potap Zaikov and Dmitri Polutov, were utilized in the representation of the Aleutian Islands, islands off the Alaska Peninsula, and the coastline of the southeastern part of the peninsula. For this reason, the outlines of these coasts show much greater detail than the chart published in St. Petersburg.

Sindt's islands are not shown, but there is an inscription at the southern shore of Alaska which reads: "at this place in the year 1742 Captain Chirikov stood and in 1774 [apparently 1775 is meant—AP] the Royal Spanish frigate found white and blond Indians here." The apparent goal of this inscription is to underline Russian priority in the discovery of these regions and to point out the possibility of the presence of descendants of Russian sailors from Chirikov's command that were left on the American coast.

Summary maps were also developed independently in Kamchatka—the *Atlas of Navigator Lovtsov During the Winter in Bolsheretsk Ostrog in the Year 1782* was compiled there (Fig. 27).[192] Various surveys conducted in the Okhotsk and Bering seas are cited as sources for this atlas and were worked over and charted in Mercator projection by officers of the Okhotsk Port who had participated in earlier naval expeditions.

Sources noted for regions to the north of Kamchatka were "declarations by servicemen who were on these coasts, their reliability ascertained from Mssrs. Captains Clerke and Gore attached to the chief commander of the two English vessels under Commander Cook, from their Notes and declarations translated by the former commander in Kamchatka von Bem."[193]

The representations of the Aleutian Islands and the Alaska coast are approximate. It is clear that Lovtsov did not have the charts compiled in 1775–1777 at the Admiralty College. If the author had a chart available, its content differed little from the sources utilized by Tatarinov in 1780, as was pointed out by Andrew David.[194]

In 1782 the Academy of Sciences in St. Petersburg was preparing excerpts from four charts reflecting the results of Russian explorations in the northern part of the Pacific Ocean and comparing them to Cook's expedition data. The results of

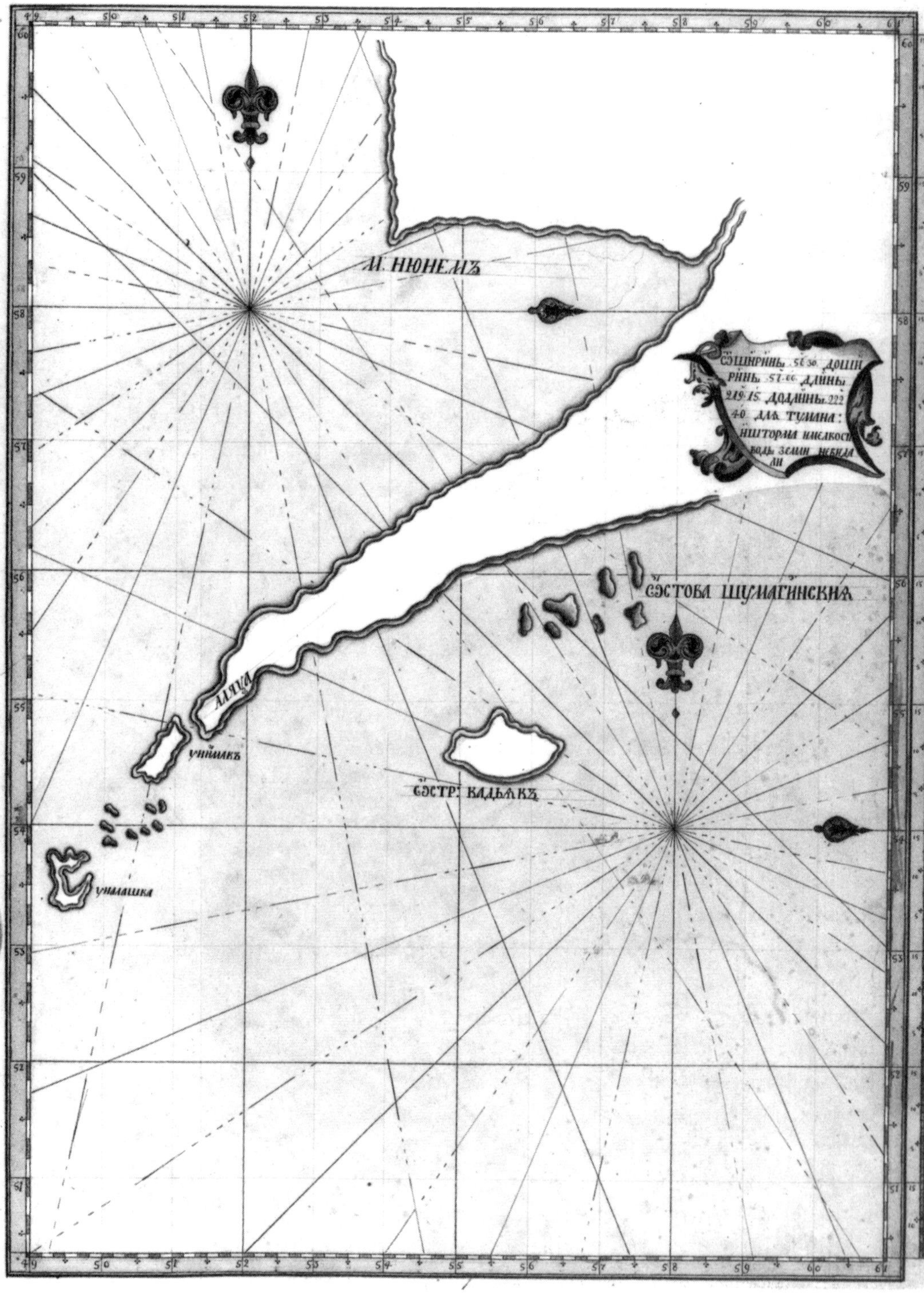

FIGURE 27 The Alaska Peninsula from Black, *Lovtsov Atlas*, Chart 10

this work survive in the form of two manuscript copies, one in the Russian State Archive of Military History, the other in the Archive of Foreign Policy of the Russian Empire. They are entitled *Compendium of various Charts Indicating the Routes of Russian Sailors to the Arctic Ocean and to North America. Printed at the Imperial Academy of Sciences.*[195] On one sheet are placed copies of the basic charts of the northern part of the Pacific Ocean published in Russia by that time: Müller-Truskott of 1758, Staehlen of 1774, Pallas of 1781, and Potap Zaikov of 1782.

It may be supposed that the academy made an effort to compare and generalize all the data available to that institution in order to finally evaluate the maps by Müller-Truskott and Staehlen. Many inscriptions on the charts provide evidence that the compendium was intended for such use, in particular the extended explanation of the chart of 1781. Here, apparently, Pallas himself cites arguments in support of the fact that Bering and Chirikov, homeward bound from the American coasts toward Kamchatka, sighted the Aleutian Islands. Above we have seen that the Russian Admiralty College had already decided this question by 1775–1777. Therefore the Academy's work of 1782 lost its justification. Voyage tracks were eventually published later in the 1780s. Vilbrekt's map shows the tracks of nine voyages, including those of Bering, Chirikov, Krenitzen and Cook (Fig. 25).

In 1779 the Spanish, who had no idea about the scope of Cook's expedition, continued their expeditions to the northern part of the Pacific Ocean.[196] The expedition under the command of Ignacio de Arteaga y Bazán on the vessel *Nuestra Senora del Rosario*, more widely known as the *Princessa*, and Bodega y Quadra on the *Nuestra Senora del Remedios* or the *Favorita*, sailed on February 11, 1779. A special chart to be used by the expedition was compiled by Mourelle and Canizares.[197] It was based on the maps by Bellin (1766), Bodega (1775), Müller-Truskott (1758), and Staehlen (1764). The expedition had orders to reach 70° N latitude.

Choosing it as their base, the Spanish surveyed Bucareli Bay in detail using boats. They also surveyed the coasts of Prince of Wales Island and the waters and islands adjacent to it.[198] On July 16 the expedition approached the coast in the vicinity of Mt. St. Elias (determining its coordinates to be 60° N latitude and 43° W longitude from San Blas), and then proceeded to the end of the Kenai Peninsula. Here the Spanish landed. Arteaga's explorations had no significance as far as priority was concerned, especially as a year earlier Cook's vessels had sailed here and to the west of Kayak Island, even if the results of the latter expedition were not yet known. The main achievement of the Spanish expedition was the surveying and description of Bucareli Bay.

In the course of Arteaga's expedition a general chart of the coasts between 58° and 61° N latitude and of Bucareli Bay was compiled. It is preserved in the National Archive of Mexico. This general chart shows the area from Cape St. Elias to the western extremity of the Kenai Peninsula. The latter is represented as a multitude of islands, only a few which are, indeed, separate from the mainland. This chart was published in 1796 by Tomas Mauricio Lopez.[199]

SHELIKHOV

The English and Spanish voyages caused great consternation among the Russian leadership in Siberia as well as in the capital. As early as 1766 the commander of the Okhotsk Port, Colonel Friedrich Plenisner, had firmly stated the need to colonize America down to California. Ten years later, Plenisner's successor, Timofei Shmalev, proposed establishing a colony in the Kuriles and dispatching three vessels with 180 men to conquer Kodiak.[200] Then in 1778, Cook's vessels appeared off the coasts of America and Kamchatka. The urgency of the Plenisner and Shmalev proposals became evident to the Siberian *promyshlenniki* and administrators. The officials were ready to assist private entrepreneurs willing to carry out these designs. Such *promyshlenniki* were on hand; the most outstanding among them was Grigory Ivanovich Shelikhov (Shelekhov) (1747/1748–1795).

A native of the city of Rylsk, in 1773 Shelikhov was a Siberian merchant serving as manager of the company of Ivan Golokov (1729–1805) based in Irkutsk. Between 1776 and 1781 Shelikhov, in partnership with various other entrepreneurs, annually outfitted vessels for fur procurement in the Aleutian Islands. On August 17, 1781, in St. Petersburg, Shelikhov, I. I. Golikov, and the latter's nephew, Captain Mikhail Golikov, concluded an agreement to form the North-Eastern Company for fur-gathering in the Aleutian Islands and on the coasts of North America.[201]

As has been shown by the research of Professor Richard A. Pierce, the idea of creating such a company and active colonization of America was not exclusively Shelikhov's achievement. In 1779 this idea was presented to Golikov by a certain Pyotr Kutyshkin, who entrusted the entire documentation he had prepared to Golikov, prior to the latter's departure for St. Petersburg in 1781. After the company had been created, Kutyshkin was not informed and spent the next twenty years trying to obtain compensation. In 1800 he even filed an official complaint before Tsar Alexander I, but apparently with no result.[202] Shelikhov never mentioned this fact, but always maintained the idea was purely his own.

Whatever the case may be, Shelikhov's leading role in the subsequent history of Russian America is not in doubt. The financial situation of the new company was strengthened considerably due to aid from one of the richest men in Russia, industrialist Nikita Demidov (1724–1789), who loaned Shelikhov 50,000 rubles.

In 1783, under Shelikhov's guidance, three galiots were outfitted and on August 16 Shelikhov sailed, together with his wife Natalia and two daughters, aboard the *Three Saints*. Sub-navigator Gerasim Izmailov, already known to us, was the captain of the flagship.

The vessels *Three Saints* and *Simeon* spent the winter of 1783/1784 on Bering Island. In addition to putting up meat stores and skins, Izmailov and Bocharov conducted repeated observations and found that "*along the line we laid down the magnetic needle deviated toward the east for 1/4 rhumb.*" From the start of the expedition Shelikhov proved to be a gifted leader for whom the health of the expedition members was as important as the fur-procurement and exploration goals of the expedition. One outstanding example is Shelikhov's order designed to fight scurvy, directing the sailors "to walk during snow storms by the sea and in clear weather in the mountains on skis for long distances."[203]

On July 13, 1784, the *Three Saints* and the *St. Simeon* arrived at Unalaska. The *St. Michael* arrived later, having spent the winter in the Kuriles (Shumshu Island). On August 3, 1784, Shelikhov reached the goal of his voyage—Kodiak Island—where he commenced to implement the plan for colonization of America that he and Golikov had worked out beforehand. The inhabitants of Kodiak Island, the Alutiiq, had stopped earlier Russian attempts (1761, 1776, and 1780).

Shelikhov named the bay in honor of his vessel, the *Three Saints*, on August 5, 1784. The Alutiiq were much impressed by the indifference of the Russians to the solar eclipse which took place that year. Despite this, resistance was offered to the intruders, but it was suppressed with determination and cruelty by the Russians.[204] The true scale of the military operations on Kodiak and the number of casualties among the Native population cannot be ascertained with certainty.

Shelikhov asserted that about 4,000 Natives assembled, from whom the Russians took about 1,000 prisoners. But Yuri Lisyansky, who visited there in 1804, wrote that there were only 400 Natives and cites *promyshlennik* Sapozhnikov, who maintained that there were no more than 350 Alutiiq. *Promyshlennik* Britiukov denies that a savage fight ever occurred as described by Shelikhov, but speaks only of several shots fired by the Russians, after which the Natives dispersed. He also writes that in the next battle more than 500 Alutiiq were killed. Later, on July 25, 1790, Izmailov reported these events to Billings [*sic*, Hall—LB] and denied that 150 to 200 local inhabitants were killed. He maintained that many cast themselves into the ocean out of fear and drowned. Izmailov also conceded that six to ten leaders were taken out onto the tundra and killed there. The Finnish mining specialist and naturalist Holmberg, who was in Kodiak in the 1850s, cited a tale by a very old Alutiiq about these events. According to the testimony of this Native about 300 of his tribesmen perished.[205] Having suppressed the aboriginals' resistance, Shelikhov founded a settlement and built a fort on Kodiak Island and then on the neighboring island of Afognak.

From the very first days of their sojourn on American shores, the expedition's navigators, Izmailov and Bocharov, began making surveys, descriptions, and charts of the shores of the new Russian colonies. That is, they were executing the assignments which were among the most important ones within the plan of creating widespread holdings by the Golikov-Shelikhov Company.

By the end of the summer a chart was completed, *Supplemental Plan of the Islands of Kykhtak and Others, as well as the Coasts of America, Showing Inlets, the Areas Discovered by the Merchant Company of Ivan Golikov and Grigory Shelikhov, under the Supervision of the Latter, in the Year 1784. These are Described by the Navigators Izmailov and Bocharov who Accompanied him* (Fig. 28).

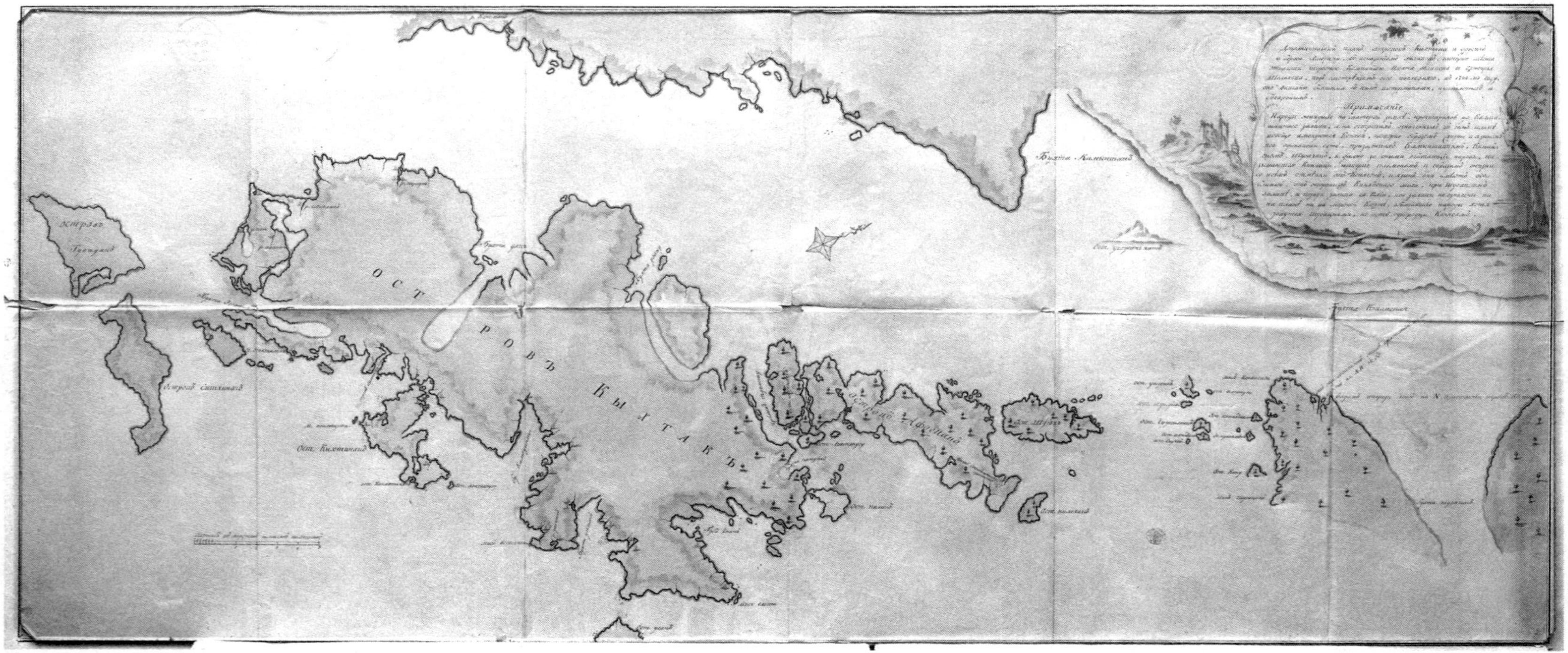

FIGURE 29 The Izmailov and Bocharov map of Kodiak. RGVIA F. VUA. No. 23441

This plan, like all the subsequent cartographic materials of Shelikhov's expedition, is remarkable for the great attention paid to representation of the aborigines who inhabited Kodiak and adjacent lands. In particular, the plan carries the following notation:

> The peoples, who live on the mainland up to Kamyshak Bay and on the islands shown on this plan are in general called Koniags (Alutiiq), all have the same way of life and language. The peoples living at the Kamyshak, Kenai and Shugachik bays and beyond are called Kenaitsy, absolutely different from the Koniags in their origin and way of life, and have a different language. But off the Kenai Cape, at Shugatsky Bay (Prince William Sound) and beyond the Cape of St. Elias, which are not marked on this plan nor on the sea chart, the peoples living there are called Shugatskie (Shugat) but are akin to the Koniag.

A plan of the same type was compiled by Izmailov in the years 1785–1786, but its supplementary data (the thematic data) pertain to the most economically valuable animals and to plants. This is the *Plan of the Island of Kadiak or Kykhtak Together with the Straits Around it, Inlets and Nearby Islands and Part of Mainland America where Mountain Ridges are Shown* [these are represented in perspective—AP] *as well as the Shores Described in the Year 1785 by the Skipper Under-Navigator Gerasim Izmailov made aboard the Vessel of the Rylsk Merchant Grigory Shelikhov The Three Church Teachers (Three Saints) Basil the Great, Gregory the Theologian, and John Chrysostom. The Products of Each Island, and in the Straits and Inlets Kinds of Fish are Indicated in the Plan's Explanation, Marked by Letters. Compiled in October 1786* (Fig. 29).[206]

Close attention is given to fish of various species, the character of their seasonal migrations, and Native modes of fishing. In particular are noted

> Streams which are entered by fish, rainbow trout, pink salmon, Siberian salmon, silver salmon, and also a fish called in the Koniag language *ty ilip* no less than five hand spans in size, in taste and appearance resembling the king salmon. It enters the streams from the sea in about the last days of May and is found up to mid-November. The people who live there put it up during the summer as dried food for winter. Near the islands and in the Kilyudin Bay they take the sea fish, halibut and cod, by means of fishing gear.

Alutiiq habitations are marked by special signs of red color, which are "square yurts of wood covered with sod and grass." The channel within the straits is indicated by a red line and marked: "safe passage for galiots." On the chart the sections charted on the basis of surveys and sections charted by an estimate from questioning of local inhabitants or superficial investigation are distinguished.

The large-scale plan of Kodiak and adjacent areas is an appendix to the first version of Shelikhov's summary chart, which was "*Created by Under-Navigator*

Gerasim Izmailov" in 1786. This original summary chart, preserved in the Russian State Archive of the Navy, is the *Mercator Chart of the Three Vessels Sent out of the Urak River in the Year '783, Companies of the Rylsk Merchant and Maritime Companion Grigory Shelikhov, to the Aleutian Islands. Charted From Okhotsk to Kurilskaya Lopatka on the Basis of Earlier Navigators' Data; from the Lopatka to the Fox Chain of Islands of Unalashka and Unimak on the Basis of Charts by Captains Krenitsyn and Levashov. From these to Unimak with Islands Adjacent to the Land Aliaksa from the Survey of Navigator Zaikov and Apprentice Navigator Polutov. From Unimak to Kodiak Island with Adjacent Capes, Shugat (Chugach) and Kenai Inlets, "Groze Fliuv" and Kamykshak, from Kamykshak Cape the Coast along the Lmr from the Survey by Under-Navigator Gerasim Izmailov. Appended to this Chart is a Plan Compiled in the Year 1786 of Kodiak Island and Part of the Mainland.*[207]

On this beautiful manuscript chart of many colors are shown the routes of Shelikhov's vessels and the results of surveys up to the year 1786. It is worth noting that Izmailov indicates his sources, including the surveys by Krenitsyn and Levashov, as well as charts by Potap Zaikov and Dmitri Polutov, which in turn had utilized the materials of James Cook's expedition. Besides this indication that Gerasim Izmailov used data from these charts when creating his own, other data confirm it. For example, such inscriptions as "Kenai (inlet) . . . named by the English Groze Fliuv," and islands "found by the English."

Prior to his return to Siberia, Shelikhov worked out detailed directions dated May 4, 1786, for K. A. Samoilov, the chief manager of the North-Eastern American Company. Besides the outline of Samoilov's administrative rights and duties, Shelikhov devoted great attention to the need for detailed study and mapping of new possessions in order to secure them as lands of the Russian Empire and also to serve further expansion.[208]

Samoilov was also instructed to inquire in detail from local inhabitants about the distribution of land and marine animals of economic value, and about useful minerals, specifically "mica, crystal, various pigments, copper ore, whetstone, limestone, and good clay; all is to be inventoried and a description appended, and samples of such, even of minerals little worthy of attention, are to be brought back."

Shelikhov also demanded that attention be paid to items of material culture and arts collected from various groups: "such as masks, headgear, wreaths, drums, handheld rattles and so on; also in good time to prepare young and good lads and girls from the Kenai and Chugach tribes for transport for their education and . . . instruct others in Russian literacy and speech for translation."[209]

The expeditions' skippers, Bocharov in particular, were instructed to continue detailed surveys, in the course of which

> everywhere are to be described bays, streams, harbors, capes, tide flats, underwater reefs and visible rocks. Usable lands, that is forests, meadows, their properties, their appearance and location; where and at what time of year,

what kind and in what quantity are found useful fish, by what means they are taken; useful plants are to be noted, they are not to be omitted. The main goal is the description of each settlement, where it is located and what the number of people is in each. A census is to be conducted, of male and female gender, the number of souls with a notation of their approximate ages. Each stream, lake, settlement, island—each useful location—is to be identified in the survey by a letter which would correspond precisely to those on the plans. The place names are to be written with the utmost effort as all places are named by the local inhabitants, *do not disfigure this with your own name so that locations can always be found by means of local names* [emphasis mine—AP].[210]

We see here that through his official directions Shelikhov confirms the long-established practice of the *promyshlenniks* in naming geographic locations.

In his instruction to the apprentice navigator Bocharov, Shelikhov stressed as a primary goal the voyage south mentioned in the directions to Samoilov. He ordered Bocharov to "sail down from Unalashka Island to the fiftieth degree North latitude and beyond and make every effort to search for new islands within the triangle between California and Unalashka meridian until the middle of September. Go on shore on Afognak Island for the winter. After the winter season, complete the unfinished search without fail."

Shelikhov ordered that a similar expedition "in search of islands" be dispatched in the northerly direction, where it was to "sail to the sixty-fifth degree North latitude and to make an utmost effort to survey, without omitting anything and with utmost care, any islands that are found as well as the mainland coasts. If on any of such islands animals would be found, take them for a year or longer, and having amassed as many skins as possible, proceed to the Island of Kyktak."[211] Shelikhov promised a reward of 1,000 rubles for any discovery of new islands in the north.

All these directives were repeated by Shelikhov on the very eve of his departure on the *Three Saints* for the return voyage to Kamchatka to "the Macedonian Greek Evstrat Ivanov Delarov," whom he appointed as chief manager of the "Company of Messrs Golikovs and Shelikhov which is in America" instead of Samoilov, ordering that the latter was "to be your assistant, and Malakhov is to be demoted from foreman to laborer because of his poor management of subordinates."[212]

In August of 1786 Shelikhov arrived in Kamchatka, and from there, at the beginning of 1787, he returned to Okhotsk. On April 13, 1787, in Irkutsk, he reported to Governor-General I. V. Yakobi about his voyage to America, and submitted to him the diary kept on the voyage, copies of instructions to chief managers, plans of *ostrogs* built on Kodiak and Afognak, a general chart of the voyage, and other materials.[213]

It is interesting that Shelikhov had managed to engrave and publish a general chart of the voyage and of the new Russian colonies by 1787 (Fig. 30). This

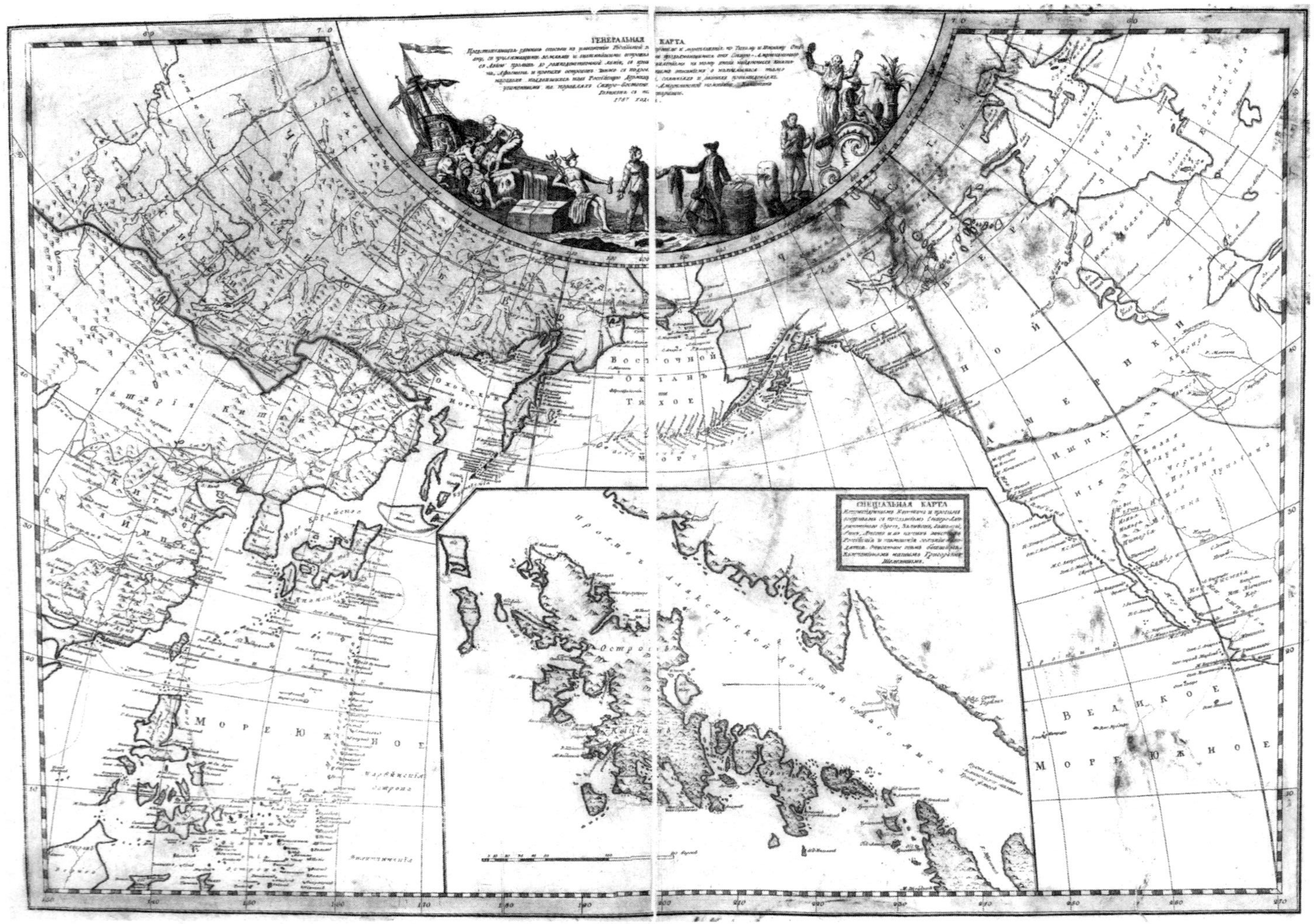

FIGURE 30 Map of the North Pacific showing the discoveries of the Golikov Company in 1787. RGADA. F. 192. Op. 1. Irkutskaia guberniia

is the *General Chart Presenting Convenient Means to Increase Russian Commerce and Sea Voyaging in the Pacific Sea and Southern Ocean with Adjacent Lands and Most Important Islands Stretching from the Strait between North America and Asia, with the Addition of the Newly-Acquired Islands of Kyktak and Afognak and Other Islands, also Including a Detailed Description of the Peoples who are There and who put Themselves under the Russian Scepter, their Settlements and Various Products, Compiled aboard the Vessels of the North-Eastern American Company of Captain Golikov and comrades in 1787.*

In addition to the main map there appears on the same sheet an additional *Special Map of the Newly-Found Kyktak and Other Islands, with an Indication of the Coast of the North American Mainland, Inlets, Harbors, Rivers, Forests and Locations where there are Russian and Local Settlements. Described by our Companion, Grigory Shelikhov, who was there.*[214]

The presentation of the chart corresponds to its obvious goal: to serve as a "cartographic declaration" of the establishment of a permanent Russian colony in America, as well as of the role in its formation of the North-Eastern American Company and personally of Grigory Shelikhov.

In Irkutsk Shelikhov met a member of Cook's expedition, the American John Ledyard, who had elected to walk across Siberia.[215] In a memorandum, undated and unsigned, which according to Professor Richard Pierce is in Shelikhov's hand, it is stated that Ledyard attempted to question Shelikhov about the history of the spread and the dates of establishment of Russian settlements. He received an answer that Russian villages were established "a long time ago" and that trade with the Indians was conducted along the coasts in the direction of California and in the continent's interior, that all the local tribes had become Russian subjects, and that any interference in these regions by other states would be decisively halted. Apparently Shelikhov considered Ledyard a threat to the Russian possessions in America and reported thus to Yakobi. On the latter's order, Ledyard was arrested in Irkutsk, returned to St. Petersburg, and expelled from the country.[216]

While Shelikhov and Golikov, with Yakobi's support, were busy in St. Petersburg seeking the establishment of a monopoly for fur procurement and development of Russian-American possessions, Evstrat Delarov and his subordinate navigators, Izmailov and Bocharov, actively commenced carrying out Shelikhov's instructions for the *promyshlenniki* and further exploration of the islands and coasts of Alaska to find new locations rich in fur-bearers.

Bocharov and Izmailov

In the spring of 1788 Delarov dispatched the vessel *Three Saints* to explore and chart islands off the American coast as well as the mainland. On the order of the governor-general of Siberia, Delarov instructed Izmailov and Bocharov to place copper crests and possession plates in all new lands they should visit.

In the course of this voyage, lasting from April 30 to July 15, 1788, the navigators Izmailov and Bocharov visited a number of small islands (among them Shelidak, Ugak, Suklia, Nikakhata, Khluk, Tkhalkha, Achaku, and Kaiak) and erected Russian possession plates on the islands of Khlikakh, St. Konstantin, and Elena. They discovered Yakutat and Lituya bays, leaving on their shores crests and plates which proclaimed them to be Russian territory.

The geographic results and history of this voyage are very well reflected in the charts compiled to summarize the results of the expedition. These are preserved in two versions in the Archive of Foreign Policy of the Russian Empire in Moscow.[217] One version is apparently unfinished. This is a multicolored manuscript chart, *Chart Illustrating the Voyage in the Year '788 out of the Island of Kodiak by a Vessel of the Company of Golikov and Shelikhov along the Coast of North America Eastward and South-Eastward. Also Areas Newly Discovered in the Course of this Voyage as well as the Locations where Russian Signs Entrusted by the Government to the Company were Placed.*[218]

The chart, however, does not show any details. Only a general coastal outline is given and the routes of the *Three Saints*. There are references to special plans, but they are absent from the archival file.

The second version of the summary cartographic materials from the expedition of 1788 is more interesting. This consists of a set of five manuscript charts and printed text, arranged on a single sheet.[219] One of the charts represented here is a finished version of the chart of the 1788 voyage discussed above. It is the *Sketch of Part of the American Coast. From the Island of Kykhtak to Lituya Bay Extending for 16 Degrees Difference in Longitude; Inhabited by Various Peoples, Many of Whom put Themselves under the Protection of Russia. This Coast has Four Bays of Sufficient Depth, Sheltered from all Winds and Suitable for Anchorage. A Cross was Erected at the Entrance to Lituya Bay. Also in That and Other Bays Inscribed Copper Plates were put into the Ground; the Inscription Signifies the Right of the Russian Empire to These Acquired Lands.*

Four large-scale charts are devoted to inlets and bays suitable as anchorages which the expedition discovered. On the sheet under present consideration there are these large-scale charts: *1. Sketch of Harbors within Chugach Bay. 2. Sketch of Harbor Located at the island of Tkhalka with Indication of Depths at Ebb Tide. 3. Sketch of Yakutat Bay Lying on the American Coast. 4. Sketch of Lituya Bay Located on the American Coast.*

It is clear that these sketches were created as navigational aids for future seafarers, which is what determined their content and the explanations appended to separate sections indicated on the charts by capital letters. For example, marked by such letters is a "Bay where ships are able to anchor, a river which may be entered at high tide by galiots, sandy locations where it is possible to stand at anchor for some time," and so on. Moreover, locations where the copper plates were placed are indicated, as well as places where the *Three Saints* stood at anchor, and the summer and winter dwellings of the Natives. In some places special characteristic features of streams entering the bays are given. For example, "a stream where the water smells of sulfur and saltpeter"; another is "a stream with red fish."

Of interest is the text explaining the content of the charts as well as the course of the voyage and methods used in exploration. The text contains a particular mention that Izmailov and Bocharov left Three Saints Harbor in Kodiak on May 1, 1788, carrying on board the galiot forty laborers, two interpreters from among the inhabitants of the Fox islands, and four Alutiiq, who it seems were hostages and used as guides. On May 5 the expedition arrived at the island of Suklia (modern Montague). Izmailov and Bocharov erroneously assumed that it was the place that Bering named Cape St. Elias.

On this island several copper plates with the inscription "Land Under Russian Rule" were buried, with detailed description of their locations. On May 11 Izmailov determined the latitude by observing the position of the sun at noon, 59°47′17″ N latitude, which is, indeed, close to the actual latitude of Montague Island, bisected more or less in the middle by latitude 60°.

The galiot then sailed to Nuchek Bay on the Island of Tkhali (Tkhalkha) (modern Hinchenbrook). Here the expedition was joined on May 20 by a Chugach who wanted to travel with the Russians. On May 21 the galiot *Three Saints* entered another harbor, where it was learned "that the inhabitants had no name for this harbor and it was therefore named for St. Konstantin and St. Elena." This harbor was in the wide Chugach Bay which Cook had named Sandwich Sound, and which today bears the name Prince William Sound, the name given by James Douglas, the editor of the first edition of Cook's journals.

Bocharov and Izmailov "learned from local inhabitants that Chugach Bay is very wide, has many bays and rivers that enter it, is filled with a multitude of islands which are rich in spruce and larch timber, alder, birch and poplar, berries such as raspberry, blueberry, crowberry, currants and sorrel; birds are geese, ducks, eagles, cranes, loons, magpies and ravens; animals are bears of two kinds, black and dark yellow, an animal called nuni with spiky bristles resembling bone (porcupine), three kinds of foxes, black, cross and red, marten, otters, wolverines, mink and beavers; in the interior there are wild rams with long white fleece, hares, caribou, squirrels, ermine and sables. In the rivers fish are taken, king salmon, other species of salmon and other sea fish. The language of the inhabitants resembles that of Kodiak."[220] The Russian travelers handed a Russian crest to the local *toion* Nekshum Atasha, to be worn on his chest and shown to foreigners.

On May 30 the Chugach led the vessel to the island of Ochok (Achaku, modern Middleton) where "Navigator Izmailov, taking along the voyager (the Chugach) and 17 laboring people, went ashore. He was met by dance and song." But conflict was not avoided. The "voyager," the Chugach, took off and the Russian who was guarding the baidaras was attacked by the local *toion* with a spear. A skirmish ensued, a *promyshlennik* was wounded, and the *toion* was killed. The Chugach did not return to the vessel. Therefore the Russians concluded that "It was nothing else but his mischievous intent to obtain the various items he was given by showing his good side and then to run away."

Izmailov returned to the vessel accompanied by a local inhabitant who "desired (?!) to travel aboard the galiot." This Native told the Russians that though this island was treeless, "it was rich in birds and off it seals and fur seals were taken." On June 1 the galiot approached Kayak Island. The expedition members described it in detail and recorded the basic navigational data. The Ochok Island Native pointed out, from aboard the vessel, a river mouth on the mainland shore. On June 3 Navigator Izmailov went to explore this river.

The Russians learned "from the Ochok Native that along that river and inside this bay (this was the area of the Icy River in Yakutat Bay) live the Ugulakhmiut, whose next neighbors to the east are the Koliuzhi, (Tlingit) with whom they quarrel constantly." To be better able to acquaint themselves with the locality, the Russians bought from the Natives a twelve-year-old boy named Naiakh Koin who spoke the Alutiiq and Tlingit languages. They paid for him one pound of iron, one big bead, and "three *sazhen* of yellow-colored beads." He was used as an interpreter in conversations with the local inhabitants. They also exchanged a Chugach for a "boy of the Chichkhan clan called Nakhusei Natsk, who knew the Chichkhan and Koliuzh languages."

Having thus acquired the means to socialize with representatives of newly encountered peoples, the Russians "learned that this bay is called Yakutat, the inhabitants are called Koliuzhi; they live on the mainland along various rivers. Near a large bay, indicated by the letter F, are their summer dwellings. They come here for fishing and hunting." They learned that the Chichkhan people lay to the south of the Koliuzhi, while toward the northwest, at the head of the bay, the Ygalakhmiut dwelled. "On June 11, during barter exchange with the Koliuzhi in Yakutat Bay, it was noticed that among them were persons whose faces were white and hair light brown." Members of the expedition concluded that "perhaps these persons were descendants of the Russians, the very same whom Captain Bering (of course, Chirikov is meant here) left on the American shore in the year 1741 somewhere to the south of Lituya Bay."

The Yakutat *toion* Ilkhak received a crest to be worn on his chest and a portrait of the heir to the throne, Tsarevich Paul. Continuing on their voyage, on July 3 the Russian travelers, with the aid of the boy-guide Naiakh Koin, discovered a new bay where "local inhabitants came to barter and we learned from them that this bay is called Lituya and around it live the Koliuzhi." The local *toion* Taikhnuk Takhtuyak also received a copper Russian crest, to be worn on the chest. It was discovered that he was subordinate to Ilkhak. The expedition returned to Kodiak on July 15.

In addition to the charts preserved in the Russian archive which were discussed here, one other chart by Izmailov and Bocharov is known. This is the 1788 chart of Latouche Strait and the island of Khlikakhlik in the Library of Congress of the

USA and published by Professor Richard Pierce.[221] Evaluating the cartographic work of this expedition, Pierce rightly noted that their authors demonstrated a fine feeling for the landscape, even though their work is not remarkable for geodetic exactness.[222]

The new governor-general of Irkutsk, Lieutenant-General Ivan Pil, in his report to Catherine II dated February 13, 1790, singled out the exploration of Yakutat and Lituya bays as the most significant result of the voyage by Izmailov and Bocharov. Pil also stressed that in these natural and convenient harbors, not occupied by other Europeans, the Natives, the Koliuzhi, and Chichkhan, dealt peacefully with the Russian *promyshlenniki*, as was the case in other localities which were visited by the voyagers in the summer of 1788.[223]

But the charts by Izmailov and Bocharov and the descriptions in the official reports by Shelikhov contained information which strongly disturbed the Russian authorities. This was the information about appearances of various foreign vessels along the coasts of the new Russian colonies in America. While they were in Lituya Bay on July 9, 1788, Bocharov and Izmailov learned from the local inhabitants that in the western part of that bay, in a small harbor, "a large foreign vessel stood two years earlier."

LA PEROUSE

This fact is also reflected in the *Sketch of Lituya Bay Situated on the American Coast*. Here, indicated under the capital letter F is the "place where in the year 1787 a three-masted foreign vessel stood at anchor." Clearly this is evidence of the sojourn in Lituya Bay of the expedition led by the French seafarer Jean-Francois de Galaup La Perouse (1741–1788?). This scientific expedition, under La Perouse, with the ships *Boussole* and *Astrolabe*, was to take possession of the point on the American shore, Puerto de Bucareli, which at that time was considered the northernmost extension of Spanish possessions. La Perouse had at his disposal the just-published charts of Cook's expedition and various works about Russian expeditions.

In June 1786, at the entry into Yakutat Bay, members of this expedition measured the height of Mt. St. Elias with a sextant. The expedition then apparently reached Cook Inlet, which was taken to be the mouth of a great river. They named it, as well as the inlet, Bering River. On July 2, 1786, La Perouse dropped anchor in a harbor. Its coordinates proved to be the same through many observations: 58°37′ N latitude and 139°50′ W longitude (from the Paris meridian).

Convinced that no one had visited this harbor; they declared it to be a possession of France and named the bay Port des Francais.[224]

Leaving Alaskan shores, La Perouse's expedition sailed to Monterey. After spending the winter in the southern part of the Pacific Ocean, they explored the Sea of Japan and eastern coasts of Asia. In September 1787 La Perouse called at

Petropavlovsk in Kamchatka, where he was warmly received. La Perouse reported to the French ambassador at the Russian court, Count Louis Philippe de Segur as follows: "It was a happy circumstance that I met in Petropavlovsk Mr. Kozlov-Ugrein, the governor of Okhotsk, who was touring the unembraceable expanse of territory which includes the Kamchatka Peninsula. At home, among my best friends, I would not have been able to find a better reception. He did not have any directives regarding myself, but he knew that sailors are citizens of the Universe and it could not have been possible to better the Russian welcome. We were offered all kinds of assistance that could be obtained in this poor land and any recompense for it was refused."[225]

La Perouse dispatched Lesseps, a member of his expedition, across Russia with materials to be presented to the Marshal de Castries about explorations conducted in the Pacific Ocean. Following their sojourn in Kamchatka, La Perouse's expedition once again set sail for the southern part of the Pacific Ocean. Following a visit to Australia in 1788, nothing was heard about this expedition until 1826, when remnants of La Perouse's vessel were discovered at an island to the north of New Hebrides. The materials of La Perouse's expedition (the ones he sent on to France) were published in 1797 in four volumes and an atlas. The charts for it were compiled by M. Bauche. Although La Perouse's astronomical observations were accurate and the charts were executed with great care, little was added to the areas visited by the Russian voyagers and Cook's expedition.[226]

MARTINEZ AND DE HARO

La Perouse's expedition was the trigger for the organization of another voyage, which was eventually reflected in Izmailov's and Bocharov's materials. The general chart delivered by La Perouse to the Spanish showed four Russian settlements in Alaska. Once again, the powers in Madrid were alarmed, as had happened in the wake of Cook's expedition. In haste an expedition was outfitted to be led by Esteban Jose Martinez in the frigate *Princesa* and the packet boat *San Carlos* (Captain Gonzalo Lopez de Haro). The vessels left San Blas on March 8, 1788, and headed for Prince William Sound, arriving at Montague Island (Suklia) on May 18.

After sailing for about two weeks in the vicinity of Montague Island, the vessels separated and were to rendezvous later at Unalaska. Lopez de Haro sailed along the Kenai coast. On June 26 he received incontrovertible evidence of a Russian presence in these areas, as the Natives, taking the Spanish for Russians, presented them with *iasak* receipts which are preserved to this day in Madrid in the expedition's files. De Haro then sailed to Kodiak, where he met Evstrat Delarov, who received him very hospitably. Delarov gave de Haro a chart of the coast with Russian settlements indicated, especially the strait between Kodiak and the mainland [modern Shelikof Strait—LB]. The Spanish sailors had no idea that such a strait existed, relying on Cook's data.

The vessels were reunited in the vicinity of the Trinity Islands and together sailed for Unalaska. There they were hospitably received by Potap Zaikov, who greatly enhanced the geographic knowledge of the Spanish about these areas. He gave Martinez three charts, which showed the Aleutian Islands and Alaska to the north at 59°50′. One of these charts is in the National Library of Mexico.[227] The main results of the expedition are reflected in the manuscript chart compiled by de Haro.[228] Data obtained on Martinez' expeditions were used in 1791 in a chart compiled by Bodega, who had been on earlier Spanish voyages. It shows the Aleutian Islands west of Unalaska, extending toward Kamchatka for 8° longitude.

It is clear that the expedition members never saw these islands, perhaps with the exception of those immediately adjacent to Unalaska. However, all but two were given Spanish names. Martinez probably recorded these islands from the charts given to him by Potap Zaikov, and later either he or Bodega gave them Spanish names.[229]

The Spanish vessels were in the vicinity of Suklia Island at practically the same time as Bocharov and Izmailov. The Russian sailors "learned from the Chugach that a foreign three-masted vessel stood at the Island of Takh'lia (Txal'kha), in Nuchek Bay." On the *Sketch of the Harbor Located at Tkhal'kha Island, with Indication of Depths at Ebb Tide* the "harbor and mooring" of the Spanish expedition's vessels are indicated.[230]

BILLINGS EXPEDITION

By the 1780s the coastal waters of Alaska were visited more and more frequently by English and Boston private merchant vessels. They were attracted there by the promise of easy gains, following the example of members of Cook's expedition who received large sums of money for furs sold in Canton in 1779. This increase in activities by foreign sailors could not leave Russia indifferent. By an edict dated August 8, 1785, Empress Catherine II ordered the Admiralty College to prepare the Northeastern Geographical and Astronomical Expedition under the command of Captain-Lieutenant Joseph Billings, an Englishman in Russian service, who had been a member of Cook's third voyage.

The goals of the expedition were "to determine the degrees of longitude and latitude of the mouth of the Kovyma (Kolyma) River, charting the entire coasts of the Chukotka Peninsula to the East Cape, as well as the many islands in the Eastern Ocean extending toward the American coast, and a complete study of the seas."[231] Thus, the mission of this expedition was analogous to the mission of Krenitzin and Levashov. Specifically, it was to provide exact coordinates for a chart of the northern part of the Pacific, the main outlines of which by this time had been well established by the government and private expeditions in conjunction with the materials of Cook's third voyage.

The Imperial Academy of Sciences in St. Petersburg took an active part in planning this expedition.[232] Academician Peter Simon Pallas drafted an instruction on the methods of collecting botanical, zoological, and ethnographic materials. Among the members of the expedition were the physician Karl Heinrich Merck, who later wrote on the basis of personal observations about the lifestyle of the Yakut, Chukchi, Eskimo, and Aleut; "assistants in natural history" I. Mein and K. Krebs; and staff artist Luka A. Voronin.[233]

The expedition's commander was issued fourteen charts of "earlier voyages in the Northeastern Ocean . . . with the necessary extracts from the journals of his predecessors from the year 1724 to the year 1779." Billings was also ordered to study in Yakutsk "descriptions of the most recent voyages in the Arctic and other seas and also all files pertaining to a route to the Kovyma (Kolyma) River and to the Land of the Chukchi. He is to copy all that is useful for him and draw the charts."[234]

One of the primary tasks of this expedition was to make one more effort to pass from the mouth of the Kolyma River along the Arctic coast into the Pacific Ocean in order "to discover if there are any islands or lands to the north of these coasts and of the so-called Bering Strait." In the Pacific, the expedition was to concentrate on "exploring the entire island chain stretching toward America . . . also the islands seldom visited and known incompletely which are situated along and under the lee of the coasts to the east of Unimak island and the Great Alyaksa Peninsula which is part of the mainland, such as for example the Island of Sanayak (Sanak), Kadiak and Lesnoi, the Shumagin Islands, Tumannoi (Chirikof Island) sighted by Bering and others."

Moreover, they were instructed "without losing much time" on the way to America "to attempt discovery of islands to the south and east of the Kamchatka meridian between 40° and 50° latitude . . . it is left to the discretion of the expedition's commander to conduct explorations of those parts of the American mainland which his predecessors were unable to examine."[235] It is interesting to note that the directives given to Billings's expedition for exploration in the waters of the Pacific Ocean repeat the instructions given by Shelikhov to his *promyshlenniki* and navigators.

Following the decree of the empress, the Admiralty College gave Billings detailed instructions regarding the goals of the Northeastern Geographical and Astronomical Expedition. As was the case with the Krenitzin/Levashov Expedition, the high command of the Russian Navy used it as an occasion to provide seagoing practice to the students of the Siberian navigation schools. For this reason Billings was in Irkutsk "to select, after examination, the five or six best students of the local navigation school and enroll them in your command so that under your supervision they will make observations and a chart. They are to remain with you to the end of your voyage on the same basis as the other lower ranks enrolled in Petersburg."[236]

The Admiralty instruction paid great attention to the oceanographic and cartographic work to be performed by the expedition. Besides the obvious need to make regular and accurate determinations from the astronomical observations of the geographic latitude, longitude, and magnetic deviation, they were to "describe and place on the map the first sightings of high elevations, the characteristic features of the coast, bays, locations of lakes, landing places and roads, noting if they are convenient or not to serve commerce, fisheries and other needs. You are also to note and record the time, strength, elevation and interval between the flood- and ebb-tides and water currents and to describe underwater and above-water rocks, shoals and other dangerous places; winds which are constant or lasting longer than others, those which are dominant, variable or steadily blowing in one direction, atmospheric events, and most of all the northern lights, specifically the electrical charge in the air when they are active and how this affects the compass; also the barometric and temperature readings are to be noted."[237]

Even though the St. Petersburg Academy of Sciences issued a special instruction to the scholars participating in the expedition, the Admiralty College stressed the necessity of gathering of scientific materials by all members of the expedition, but particularly of the expedition's corps of navigators. Thus, Billings was to constantly ensure the study "of the characteristics of the land, the products of the lands" to be visited by the expedition and "to diligently collect seeds, ripe fruit, and for the herbarium dried grasses, branches or pieces of trees worthy of note, their bark, resin, sap, sea plants, bones, shells, fish, reptiles, insects, birds, zoophytes and so on; some will have to be skinned and stuffed specimens prepared, while some, according to your judgment, can be preserved dried. You are also to collect pieces of minerals, fossils, salts, and colored and sulfuric soil. When and where each item was found and collected is to be recorded."[238] The expedition included the artist Luka Voronin, whose drawings were used in creating the engravings documenting the voyage (Fig. 32, 32, 33).

The Admiralty instruction gave great attention to the necessity of making a detailed study and description of the lifestyle, customs, and languages of the peoples inhabiting the areas the expedition was to study. The expedition was instructed "to inquire and learn exactly the power, numbers, characteristics, activities and customs of the peoples residing in those lands and also of those inhabiting lands which are as yet little known. You are to order a compilation of their vocabularies according to the standard issued to you.[239] You are as far as possible to put down on paper the pronunciation of their dialect in Russian and Latin script; to obtain, and were it impossible to do, to sketch and describe the things used by these peoples, their weapons, dress, handicrafts . . . inquire about their divine service, the things they love most, and make sure that everyone in your command leaves them untouched and not destroyed."[240]

The Admiralty singled out as the "main duty" of the expedition "compiling the most precise chart of these islands, determining their location by frequent observations, and most of all an effort to learn about the best landing places." In this

Admiralty instruction pertaining to science we encounter for the first time a recommendation to actively utilize Native geographic knowledge and a suggested method of how to incorporate such data in constructing the maps of the new lands. Members of the expedition were directed to inquire where and by what means local inhabitants traveled: "how that place whither they go is called, in what direction, how distant is that land or island? Should they indicate by gesture, by diligent art and unobtrusively, translate this into the compass setting and bearings and note this in the journal; for the distance, as it would not be possible to determine it, ask them how many days it takes to sail to these places and use this information to set your course if you should deem it advisable."[241]

In contrast to Shelikhov, the high command gave members of the Billings Expedition "the power to name the lands and islands acquired by you as you will, as long as these do not already have a name."[242]

By the time the Billings/Sarychev Expedition was active, a decision had been reached that it was imperative to officially declare Russian discoveries on the west coast of North America and to secure the defense of Russian colonial possessions by means of a squadron of naval ships. On December 22, 1786, Catherine II issued an order to prepare the first Russian circumnavigating expedition under the command of Captain Grigory Mulovsky "because of the intrusion by the English merchant *promyshlenniki* in the Eastern Sea for trading and fur-procuring" and to guarantee by means of naval ships "our right to the lands discovered by Russian sailors."

By the fall of 1787 a squadron consisting of five vessels was ready, but because of the outbreak of war with Turkey and a looming military conflict with Sweden, an order canceling the expedition was issued on October 28, 1787, and the vessels dispatched to the Mediterranean. The Mulovsky expedition, which never took place, was an effort to realize the circumnavigating project proposed in 1732 by N. F. Golovin. Not until 1803–1806 was this idea realized through the voyage of Ivan Krusenstern and Yuri Lisyansky.[243]

In the meantime, the Northeastern Expedition reached Yakutia, and in May of 1787 left the mouth of the Kolyma River with the specially built vessels *Pallas* and *Yasashna*. The aim was to attempt to replicate Dezhnev's voyage, sailing along the coasts of northeast Asia and passing from the Arctic into the Pacific Ocean. Almost solid ice forced the expedition to return to the mouth of the Kolyma River at the end of June. Following this failure, the decision was made to attempt rounding the northeastern extremity of Asia from the opposite direction: from the Pacific into the Arctic Ocean.[244]

Preparations for this voyage and hydrographic work along the coasts of the Sea of Okhotsk were already under way under the leadership of Captain-Lieutenant Robert Hall. Captain-Lieutenant Sarychev, the expedition's navigator S. Bronnikov, and geodesists A. Ocheredin and I. Britov surveyed the Okhotsk coast to the River Ulkan using local baidaras and baidarkas. On October 5, 1789, Billings's ship *Glory of Russia* entered Petropavlovsk Harbor.

In the spring of 1790 Billings clarified the details of some of the expedition's tasks in his personal instructions. For example, in the instruction given to shift master D. Gauz, dated April 23, 1790, there is an order to keep a journal during the expedition "with a topographic description of the locations which you might visit." In particular, the directions and extent of "the main ranges, their cliffs and slopes, their bases, their height, the composition of the rocks, clay and earth, strata worthy of note, the latter's location and other matters which constitute part of the art of mining" were to be noted and described. Special attention was to be paid to former and active volcanism. For this reason the instruction demanded that craters of extinct and active volcanoes be examined, lava samples collected, and samples collected "from test pits in rock, soil, clay and various other mineral ores and salts, and especially various fossils."

The instruction also prescribed that a general landscape description be compiled, specifically: "The surface of the ground is also to be described, the steep slopes and the various strata visible within them, the location of high and low places, forests, of which kinds of trees they consist, and the animals dwelling therein, swampy places, lakes, large and small rivers and their banks, islands and what kinds of fish and amphibians are there, and also the chemical properties of the waters."[245]

On May 9, 1790, *Glory of Russia* set sail for American shores. In the course of the voyage, surveys were conducted and descriptions made of the Aleutian Islands from Amlia to Unalashka, Sanak, the Shumagin Islands, the Semidi Islands, and Kodiak. On the American shore Sarychev completed a hydrographic survey of Three Saints Bay and determined astronomically the latitude and longitude, employing the method of observation of lunar distances. On July 19 the expedition approached Nuchek Bay and *Glory of Russia* dropped anchor at Montague Island. Merck and Billings were busy collecting various kinds of information and materials, while Sarychev, traveling with baidaras, surveyed Prince William Sound and Kayak Island.[246]

In the course of his surveys Sarychev skillfully combined reconnaissance survey data with interviewing the Natives. Members of the expedition became convinced about the very good geographical knowledge of the local inhabitants with respect to the islands and adjacent American coasts. As the expedition's clerk, Martin Sauer, testified, they had even created their own system of navigational markers, piles of stones on shore which were used like lighthouses, to which "each Native who passes must add one more rock."[247] Sarychev not only utilized the geographic data obtained from local inhabitants, he also tried his utmost to involve all members of the expedition in the hydrographic and cartographic work. He did not begrudge his time training them in elementary survey methods.

It appears that in compiling charts described in 1790, contributions by Izmailov and Bocharov were incorporated. However, acknowledgment is given on only one of the four manuscript charts made by hydrographers of the Northeastern Expedition in the year 1790, the previously mentioned *Plan of the First Offshore Entrance from the West into the Chugach Sound Showing Convenient Vessel Anchorages, Bays and*

Depth. Year 1790. The plan was "copied from the original by navigator of first lieutenant's rank Anton Batakov" and was signed "Fleet Captain of the First Rank Joseph Billings." There is a notation on the plan that it was obtained from Navigator Izmailov.[248]

Three other manuscript charts of 1790 that have survived deal with Three Saints Harbor on Kodiak. These are the *Chart of a Part of Kad'iak Island and Sikhlidakh* [*Sitkalidak*—AP], *Entrance into the Harbor of Three Saints, with Depth Measurements in Sazhen 6 Feet Measure in English Feet, Made in '790 by Fleet Captain Billings;*[249] *Flat Chart of the Entrance into the Harbor of Three Saints on the Island of Kad'iak, on General Chart Shown under No. 12. Composed in 1790 by Fleet Captain Gavriil Sarychev, on an Unadjusted Compass, Variation 26° 00' East;*[250] and *Plan of the Harbor of the Three Saints, Named in Aleut Manikaksik. Lies on SE side of Kad'iak Island.*[251]

None of these three charts, nor any of the subsequently compiled large-scale and general overview cartographic materials of the Billings/Sarychev Expedition,

FIGURE 31 The Billings Expedition at Three Saints Bay, Kodiak Island. *Gavriil Sarychev, Puteshestvie flota kapitana Sarycheva* . . . . 1802. *Atlas*, plate 36. UAF Rare Book C0015

makes any reference to the two expedition charts compiled by the navigators of the Golikov-Shelikhov Company. A possible explanation of this is that one of the government tasks entrusted to the expedition was the investigation of outrages perpetrated by the *promyshlenniki* against the local population. In this connection, the expedition's officers apparently did not choose to advertise their connections with Shelikhov's sailors in their documented reports.

To all appearances such contacts were not favored by Billings, who, according to Sauer's testimony, suspected the Russian clerk of the expedition, Vasily Dyakonov, of entering into secret disclosures of the expedition's goals, put him in irons and sent him off to Irkutsk.[252] Sauer himself did not evaluate the activity of the Russian *promyshlenniki* in a consistent way. On the one hand, he stated that many of them treated the local inhabitants like slaves, but, on the other hand, some of Shelikhov's men had his deep respect. Thus, for example, he wrote with unconcealed admiration about "a well-known Russian officer, Navigator Izmailov, whom members of Cook's expedition met at Unalashka in the year 1778." He also underscores the strict but just management of Delarov, meaning here of Russians as well as the Natives, for whom, for example, a school was organized where local inhabitants learned the Russian language and reading and writing.[253] (Sauer assumed that the school was organized by Delarov, while in reality the school was created by Shelikhov himself in 1786.)

Surveying Prince William Sound in the summer of 1790, Sarychev noted that the local inhabitants frequently used the Spanish and English words *amigo* and *plenty*. This was evidence that foreigners were visiting these areas. Indeed, the Chugach confirmed to Sarychev that "nowadays each year two- and three-masted vessels arrive and this year they were visited by two vessels which have sailed for Kenai (Cook) Inlet."[254]

Winter was spent in Petropavlovsk, but in the spring of 1791 *Glory of Russia* once again set sail for Unalaska. En route, surveys were conducted at the Commander, the Near, and the Andreanof Islands, as well as St. Matthew and St. Lawrence Islands. Furthermore, the expedition surveyed the Pribilov islands, recently discovered in 1786 by Gavrill Pribilov [a navy navigator seconded to the merchants, later commandeered by Billings—LB] who had sailed into these waters from Unalaska and found the island they named St. George.

Pribilov left a work crew there to fur hunt. The crew was able to see land in the distance and the foreman, E. Popov, went in 1787 by baidara to an island which was named St. Paul (modern Pribilof Islands). Here they found a copper sword hilt, a clay pipe, and a fire pit, evidence of earlier visits.[255]

After surveying the Pribilof Islands, St. Matthew Island, and St. Lawrence Island, the Billings/Sarychev Expedition proceeded farther north and reached St. Lawrence Bay in Chukotka at 65°40′ N latitude. Here Billings transferred the command of the *Glory of Russia* to Captain Sarychev while he himself set out on his land journey across Chukotka. Sarychev then returned to Unalaska Island and resumed his hydrographic survey of its coast. He also investigated the islands of Umnak, Akutan, and Unimak.[256]

Khudyakov

The main surveys in these regions were conducted by Sergeant of Geodesy Osip Khudyakov, who was aided in his task by "the three Unimak *toions* Galok Ek'iasnisanov, Chunniuk, and Tukkuiok, who eagerly undertook to conduct him to the indicated location, and also by an interpreter, baptized Ivan Galkin."[257] Sarychev gave Khudyakov detailed instructions with regard to surveying and describing the islands and the coast of the Alaska Peninsula. Here the steps for determining the true meridian and magnetic variation, recommendations on how to survey the coast from the sea and on the ground, and methods for checking the instruments (hour glass, lag lines, leads, and so on) are set out in great detail.

Because the surveys had to be conducted in the absence of any surveying grid on shore and were to be based only on a separate determination of latitude and longitude, in his instruction Sarychev devoted a great deal of attention to making sure that the angles between local points of orientation were measured several times and that such angles were checked by compass headings.

Khudyakov was particularly instructed to conduct the coastal survey with baidaras and baidarkas in this manner: "After everything is made ready you are to depart from the settlement of Illiuliuk, which is to be the initial point where you will take compass bearings; proceed near shore taking a rhumb forward, orienting yourself on some distinctive feature; distance from shore is to be frequently recorded, as well as how high the shores are, cliffy, mountainous, sloping or low. At the end of the rhumb bearings are to be taken on all visible places, but especially the points of capes, river mouths, above-water rocks, especially high mountains, their summits. Note that the bearing is to be repeatedly oriented to the same place, especially in regard to features at a distance, and the distance between settings also has to be great enough to avoid very sharp angles."[258]

Special attention was to be given to the straits between the Alaska Peninsula and the islands nearest to it:

> The strait between Aliaksa and these islands is to be described with utmost accuracy, and its width is to be secured by compass settings, traveling a certain distance along the shore by the rhumb and at the rhumb's end the compass setting is to be set and the bearing taken on a distinctive feature near the water on the opposite shore. You are also to measure it by sailing from shore to shore along a rhumb, set so that the angles between them do not exceed 45°; record with as much accuracy as possible the depths and the bottom.[259]

He was directed to "inquire about numbers of inhabitants of male gender, record in the journal how many are in each settlement, their subsistence activities, and ask also about the presence of noteworthy volcanoes, big rivers and lakes . . . make every effort, under any circumstances, not to give the people who live in those islands any offence; ask for assistance from them by making signs of friendship."[260]

The journals kept by Sergeant of Geodesy Khudyakov on the voyage survive in the Russian State Archive of the Navy. These testify to the precise, and at the same time creative, execution of Sarychev's instruction.[261] The geodesist worked practically without a break. If it happened that "by the *toions*' statement the turbulence at sea was great and a strong gale was blowing" he busied himself with the shore survey, using data obtained when questioning the local population. Khudyakov noted in his journal every instance of such use. Here he paid great attention to the accuracy of the information he presented, especially stating which distances were estimated instead of measured.

Khudyakov was forced to spend considerable time on Unimak Island because the *toion* Gallon Ek'iagiksanok and other Aleuts, informed the Russians that "It is dangerous to travel forward in November by baidara, because great storms occur both on the northern and southern sides near this island; there are no landing places where one could guide the baidara; there is no need to subject oneself to danger, but in December calm weather occurs . . ."[262] Khudyakov used this forced interlude to explore Unimak Island in detail with the help of an Aleut guide.

The winter of 1791–1792 was extremely hard for Khudyakov's party: the food supplies were gone and the rest of the men from his detachment were quartered in various Aleut settlements for food and shelter. Khudyakov himself, however, continued his hydrographic survey, traveling in a three-hatch baidarka. On January 7 Khudyakov, taking along the Unimak *toion* Gallon Ek'iagiksanok and the Aleut interpreter Fyodor Malinovsky, left the Chingangaliuk settlement in the three-hatch baidarka.

In order to continue the survey, Khudyakov's party had to pass through Isanotski Strait. Once again, the Natives provided the hydrometereological information pertaining to the strait. This time the informant was *toion* Pankov, who reported that "in Isanok impassable ice occurs, driven by the north wind. The Aleut who came from the Morzhovoe settlement maintained that at this time there is ice in Isanok. Ice floes are driven by the flood and ebb tides and it is not possible to pass in a large baidara." For this reason Khudyakov decided not to risk his people and sent them to Unalaska, while he himself sailed in a three-hatch baidarka on February 2 from Pogromnoe settlement "to the point designated in the instruction."[263]

According to information provided by the Aleuts, the northern coast of Unimak was still blocked by ice. The geodesist, accompanied by Aleuts in seven baidarkas, among who once again were Unimak *toions*, proceeded along the southern coast of Unimak. The Aleuts engaged in hunting and, to use modern forms of speech, meteorological reconnaissance.

Thus, on March 1, 1792, Khudyakov noted the following in his journal: "This day the absent *toions* returned and brought for food a seal they had killed. They declared that though the wind will abate somewhat, the (sea) turbulence will not permit us to sail from here soon. It would be better to cross to the north shore, from whence it would be easier, weather permitting, to cross the strait. Therefore I decided to portage the baidarkas to the north side of the island on the morrow, while I myself will go on foot doing the land survey."[264]

Khudyakov completed his plan rather well. Having spent the month of March surveying the Unimak shore of Isanotski Strait, he rejoined the Aleuts who were awaiting him on the north shore of Unimak. On April 6 he and the Aleuts crossed to the Alaska Peninsula. On the whole, Khudyakov's surveys considerably improved the maps of Unimak Island and Isanotski Strait, which had been investigated for the first time by the Krenitzin/Levashov Expedition in 1768–1769.

The surveys conducted in the course of this voyage fully satisfied the standards of exploratory hydrographic and geodesic work. Khudyakov took pains to measure the midday altitude of the sun whenever weather permitted in order to determine latitude and eventually achieve a precise linkage of his route surveys with the numerous determinations of latitude and longitude made by Sarychev in 1791–1792. Khudyakov's contributions were considerable. He completed the hydrographic survey of Captain's Harbor in Unalaska and investigated Amaknak Island, Unimak, Olenii (Deer) Island, and part of the Alaska Peninsula.[265]

The *Flat Chart According to Adjusted Compass from Isanotski Strait and in Part Island of Unimak*; and *Flat Chart of the Aleutian Island Naunalshka (Unalashka) and the Strait between it and the Islands of Spirkin (Biorka), Unauga (Unalga), Iakutak (Akutan) and Others 1792* are based on these surveys.[266]

In total, on the basis of the survey materials of the Billings Expedition of 1790–1792, fifty-seven charts were compiled of Kamchatka, the Aleutian Islands, Chukotka, and the coast of North America. These are reproduced in part, and in facsimile, in the *Atlas geograficheskikh otkrytii v Sibiri i v Severo-Zapadnoi Amerike XVII—XVIII vv.* (*Atlas of Geographic Discoveries in Siberia and in Northwest America in the seventeenth-eighteenth Centuries*), by A. V. Efimov.

Besides the tasks of completing surveys and providing descriptions, members of the Billings Expedition had a government mandate to regulate the relationship between the *promyshlenniki* and the local inhabitants. In September of 1791 Sarychev was joined by Captain Robert Hall, who arrived at Unalaska aboard the cutter *Black Eagle*. Hall had been surveying the Kuril Islands. In accordance with the directive of the Admiralty College, they organized a commission on *iasak* collection from the Aleut.[267] They also conducted a population census of the Andreanof and Fox Islands, which found a total of 1,178 persons of male gender listed from whom *iasak* was collected in 1791.[268]

Members of the Billings Expedition had the good fortune to obtain information about Tlingit (Koliuzhi) settlements in the areas frequented by Russian fur-procuring parties. This information is in a special *Note on the Estimated Number of Settlements of the Koliuzhi People*[269] included in Billings's report to Governor-General Ivan Pil. In this note it is stated that 7,750 Koliuzhi of male gender (including children) lived in the explored territory.[270] Members of the Billings Expedition tried by all possible means to regulate the relationships between the *promyshlenniki* and the local population, demanding that they conform to the law. They explained rights and duties to the Natives. This had an immediate effect on the *promyshlenniki*, who were used to complete Aleut submission.

As early as the spring of 1791 Aleksandr Baranov bitterly complained to Shelikhov, stating that "Mister Billings . . . left orders and inculcated the Aleuts not to obey the Russians, and not to deliver food without recompense by agreed prices; he instructed all to prepare foodstuffs for his return travel to Unalashka, promising pay, and for this reason, the Orekhov (*promyshlenniki*) suffer hunger, as the Aleuts do not obey in anything and do not feed them, but after a prolonged

FIGURE 32 The Billings Expedition at Unalaska. Gavriil Sarychev, *Puteshestvie flota kapitana Sarycheva*. . . . 1802. *Atlas*, plate 14. UAF Rare Book C0015

voyage they have nothing with which to pay."[271] Baranov had "to wean" the Aleuts away from Billings's "orders," which he managed to do rather rapidly.[272]

In the spring of 1792 the Pacific Detachment of Billings's expedition returned from Unalaska, arriving first at Kamchatka, and then at Okhotsk. Here they received Billings's order terminating the expedition. The expedition ended officially on September 7, 1795.[273]

The land detachment led by Billings traversed "the entire land of the Chukchi"[274] in 1791–1792 starting from St. Lawrence Bay. Valuable research on that region was done. Billings used the time spent in the Bering Strait region to search for the legendary settlement of bearded (Russian) people in Alaska. Siberians, long linked to this problem, were drafted for this effort: the "Yakut nobleman," Chukchi Nikolai Daurkin, and cossack commander Ivan Kobelev. In 1791 they crossed the Bering Strait in baidarkas, visited Diomede and King Islands, and on June 11 Kobelev was in the Inupiat settlement of Kigigmen (modern Wales). Thus he was the first European to land on the Seward Peninsula.[275]

FIGURE 33 The Chuckchee Peninsula. *Gavriil Sarychev, Puteshestvie flota kapitana Sarycheva* . . . . 1802. *Atlas*, plate 110. UAF Rare Book C0015

Kobelev and Daurkin did not find any Russian settlements, but, according to their data, "the Americans (Natives in the Bering Strait region) pleaded persuasively for the Russians to defend them from the attacks and pillaging by the Reindeer and Foot Chukchi, who arrive almost every year in baidaras in great numbers, kill the Americans, plunder their estate . . . and take wives and children as prisoners."[276]

Sarychev

Billings's Northeastern Geographical and Astronomical Expedition was a significant scientific achievement. Though its results were not fully known to contemporaries due to the traditional Russian policy of secrecy, they were discussed in scientific publications to a much larger extent than the results of previous Russian government expeditions. This was due mostly to the works by Sarychev.

The most striking demonstration of the expedition's achievements is the chart published in Petersburg: *Mercator Chart from 47 to 71 N. L(atitude) and from 125 to 231 Degrees Longitude from Greenwich. Represents the Northern Part of the Eastern Ocean, Lying between the Northeastern Side of Asia and the Northwestern Part of America, with Indication of the Voyages in this Sea in the Years 1789, 1790 and 1792 in the Time of the Former Sea Expedition under the Command of Fleet Captain Billings, with the Vessels Glory of Russia and Black Eagle, compiled by a Participant in the Expedition, Fleet Captain of the Second Rank Sarychev.*[277]

This chart summarizes the results of Russian investigations in the northern part of the Pacific Ocean. Sarychev showed the routes of the expedition's vessels in detail. In the explication, he indicates sources used in compiling the chart, among which is *The Shore from Okhotsk to Aldoma River, Harbors of Kamchatka and Petropavlovsk and Unalashka Island, Surveyed by Baidaras.*

Of interest is the fact that references to *promyshlenniki* data (including those by Bocharov and Izmailov) are totally absent. Also absent are the data and map by Kobelev showing Eskimo settlements in Alaska which by this time had been published in a revised version in *Mesiatseslov istoricheskii i geograficheskii* (*Historical and Geographical Almanac*).[278] The Greenwich Meridian was used in Russia for the first time on Sarychev's summary chart. This made its immediate use by foreign geographers and cartographers much simpler.

Several large-scale charts based on surveys by the Billings Expedition were first published in his *Atlas Pertaining to the Voyage of Fleet Captain Sarychev,* and as illustrative material for Sarychev's *The Journey of Fleet Captain Sarychev in the Northeastern Part of Siberia, the Arctic Ocean and the Eastern Ocean Lasting Eight Years during the Geographical and Astronomical Sea Expedition under the Leadership of Fleet Captain Billings from 1785 to 1793* (in two parts, St. Petersburg, 1802) (Fig. 34).

This atlas also contains drawings by the expedition's staff artist, Luka Voronin. These include views of the coasts as seen from the sea, settlements, and Natives. There are drawings of everyday items as well as cultural artifacts.

FIGURE 34 Billing Expedition chart. Gavriil Sarychev, *Atlas Sievernoi chasti Vostochnago Okeana* . . . 1802. UAF Rare Book C0039

The main content of the large-scale charts in the *Atlas to the Voyage of Captain Sarychev* is the coastline of the lands explored by the expedition and the depth measurements. The land is represented more schematically, especially in regard to relief. Relief is represented by shading, so that the drawing creates an impression that many regularly spaced ledges and terraces exist, which, of course, was not the case in reality. Thus the shading indicated only the existence of mountains, but did not represent their true forms.

On Sarychev's charts much attention was paid to Russian and Native settlements. This testifies to the development of Russian place names that were given to the new settlements of the *promyshlenniki* and the renaming of settlements in or alongside Native dwelling places. In such cases the local and Russian place names appear side by side. These charts were included, in revised form, in the *Atlas of the Northern Part of the Eastern Ocean* published by Sarychev in 1826. These charts and coastal drawings were used by sailors of many nations as late as the 1940s.

Sarychev's methods are built on his analysis of previously conducted hydrographic work in the northern part of the Pacific Ocean. But, like so many of his predecessors, the scholarly sailors, he shows considerable contempt for the explorations by the *promyshlenniki*. He declared: "Other voyages conducted by private persons do not deserve any attention; they have brought so little benefit, both in regard to discoveries and surveys of the coasts that they are not worth mentioning."[279] However, it becomes apparent from the text following that this outstanding Russian hydrographer clearly utilized the experience of those very same "private persons," at least in his survey methods.

Eight years of experience surveying in the northern part of the Pacific Ocean convinced Sarychev that many of the practices of oceanographic work were no longer relevant under the complex hydro-meteorological conditions of the coastal waters of Alaska. Analyzing the work of such an unchallenged authority as James Cook, Sarychev wrote:

> Who among sailors possessed more art, enterprise, and patience than Captain Cook. His fervor to represent the coasts in minute detail on charts was so great that he was not even deterred by immediate danger. However, while in local regions, he was compelled by strong gale force winds, fog and shoals to leave much of the coast unexplored; even the coasts that he sighted were in some areas not indicated in accord with their real outline, some bays were missed, islands were not shown as separate from the mainland, and finally he mistook fog for islands.[280]

To "obtain accurate charts of the local seas, the survey must be conducted by feel, so to speak. It must be conducted in large skin baidaras or on small oar-propelled vessels, which, because of their shallow draft, are suitable for travel in the immediate vicinity of the shore and which are able, when the gales blow, to find shelter in shallow river mouths and bays."[281]

When the weather was fine, it was possible to survey up to fifty versts of coastline per day in a baidara. Sarychev assumed (rather overoptimistically) that by this method "all the coasts of the northern part of the Eastern Ocean from 56° to the Bering Strait itself, as well as the Kuril and the Aleutian Islands, can be described in the course of two years." He cited his own survey of 1791 as evidence in support of the above, when "with a small Aleut baidara one hundred versts of the northern coast of Unalashka Island in a day and a half. I surveyed and made depth measurements in three rather large bays in not longer than a day. And in spite of this rapidity, the accuracy was complete."[282]

Besides using the Aleut vessels—baidaras and baidarkas—Sarychev used data in his explorations obtained from interviews with the local population. We can find solid material in his work on translating Native information into the accepted European cartographic language. For example, he points out that the nomadic peoples often indicated distances between points by the time necessary to traverse them. The unit they adopted for such a distance was the "dnishche," that is, "the distance the nomadic people carrying their entire kit can cover in one day. A large *dnishche* is equivalent to about 10 versts, a small one to about eight."[283]

The Aleut and other Natives of Russian America were Sarychev's guides. Thus, departing from Unalaska for the Alaska Peninsula, the Russian voyagers took along an Aleut man and an Aleut woman: "one of them spoke Russian rather well; both had visited all the islands to the east from Unalashka to the Shumagin Islands and knew their names." Later Delarov, the manager of the Russian colonies, put two interpreters at Sarychev's disposal. One of them was a Native Kodiak Islander, and "the other was from America's shore, somewhere south of Cape St. Elias. Both spoke Russian rather well." Still later, one "American" agreed to travel with him into the interior of Chugach Inlet (Prince William Sound) "and tell me the names of the islands and rivers."[284]

Due to their knowledge of the islands and of the hydro-meteorological conditions of the surrounding seas, the Natives frequently rescued Sarychev and his men from dangerous situations. In his book, Sarychev tells of an instance, when on March 20, 1792, the Makushin [Volcano Bay, not modern Makushin—LB] *toion*, observing Sarychev's concern with the stormy weather and "suffering from lack of food provisions," took pity on the travelers and told them that it was possible to reach the expedition's vessel "by portaging the baidarkas from Makushin Bay into Beaver Inlet, from whence it would be easy to get to the vessel."

Several Aleuts agreed to guide the Russian sailors and to carry the Russian and their own "baidarkas across the isthmus which divides the heads of these two bays for a distance of 3 ½ miles."[285] Not once, but many times, Sarychev had occasion to be convinced of the Aleuts' fine knowledge of the hydro-meteorological conditions of their sea territory. Thus, Aleuts from the Andreanof Islands informed him about seasonal currents at Unalaska Island, where, according to their observations of many years, in summer the currents flowed east along the northern coast and westward along the southern coast. In the fall, "the current on the island's northern side shifts to the west and along the southern side to the east." This information was confirmed by personal observations by members of the expedition and also through analysis by Sarychev of previous expeditions' logs.[286]

Besides this direct geographic information, Sarychev also obtained reliable historical and geographical data from the Natives about the presence of Bering's expedition vessel on the American coasts. On June 25, 1790, in Nuchek Bay, Sarychev tried to learn from the local inhabitants if anyone remembered a Russian ship, because he doubted that "Suklia was the same island as the one called Cape St. Elias, as it is shown on the Admiralty College charts issued to us." One of the Natives informed Sarychev "that he had heard from his father about this vessel and that it stood not off Suklia but off the island called Kayak which lies to the east from this place, a day and a half's travel, where the Americans take sea otters. People from this vessel came ashore and left some knives and beads in summer shelters."[287]

Having then visited Kayak Island, Sarychev found that the Native information fully corresponded to the data of Bering's expedition about where its members set foot for the first time in the Western Hemisphere. The location and concurrence in latitude in Commander Bering's journal confirm that the first vessel mentioned by the Americans in Chugach Sound and which their fathers saw by this island, is indeed the one on which Commander Bering sailed."[288]

Sarychev devoted a great deal of attention to analysis of the hydrographic surveys, which were the basis for the charts compiled by the expedition's members. Speaking about the uninterrupted surveying, he wrote that "on all the days that land was visible, or when something remarkable was encountered at sea" he always noted the latitude and longitude at noon. Moreover, "the distance to the visible land and the size of islands was recorded not by estimating by sight, but was determined by sailing and taking accurate bearings, so that from this information a chart of the entire voyage may be compiled, showing coasts and islands in the same way as . . . is shown by the route calculated by observations."

All bearings and points of the compass were used in the chart compilations "just as they were recorded in the journal, that is, by an unadjusted compass" [i.e., without taking into account magnetic delineations—AP]. This was a common practice of the times because it was impossible to determine precisely, under expeditionary conditions, the large number of azimuths for calculating the variation of the magnetic arrow.[289] The expedition's surveys, however, were based on the precisely determined coordinates of the initial and end points of a journey. Multiple astronomical observations to determine them were made by Billings and Sarychev by means of special field observatories.

Sarychev, "when compiling the General Mercator chart, corrected the longitude calculations every twenty-four hours . . . and placed the corrected values on the chart, in line with the errors found by observation from one longitude to the next."[290] As a result, the relative placement of main landmasses and islands in the northern part of the Pacific Ocean is represented on Sarychev's summarizing chart with the greatest degree of accuracy which could be achieved in those times.

The experience gained by conducting hydrographic and cartographic work with the Billings Expedition permitted Sarychev to create the first original instruction for marine surveying and cartography in the Russian scientific and technical literature, published in 1804: *Rules Pertaining to Marine Geodesy, which Serve as a Guide to Describing Seas, Coasts, Islands, Bays and Inlets, Harbors and Rivers, Traveling in Large Sailing Vessels and in Small Oar-Propelled ones, as well as Going on Land with a Measure; Includes Demonstrations of How to Compile Marine Charts and How to Place on the Same the Surveyed Locations.*

In the introduction Sarychev especially stressed the importance of his more than eight years of experience in the Billings Expedition, where he "was unceasingly engaged in surveying various localities and because of this had an opportunity to make many useful notes for improvements in geodesy."[291] This guide became the basic instruction manual in hydrography for officers of the Russian Navy. It was republished several times throughout the nineteenth century. His work determined the fundamental methods of hydrographic surveys in the Russian-American colonies.

Sarychev paid the greatest attention to precision in marine geodesic (hydrographic) work. Astronomical and geodesic instruments used in surveys are described in great detail. The following are examined: "azimuth compasses, hourglasses, log and log line, lead and lead lines, also astronomical instruments for use in determining compass variation, the checking of watches and the determining of latitude and longitude at sea [chronometers, sextants, telescopes, theodolites, and others—AP]."

He describes the methods of conducting an expeditionary survey of formerly unknown coasts:

> having approached the coast of the land to be surveyed to a distance which safety permits, the lead must be dropped with greater frequency and the course held parallel to the coast, attempting to follow its bends. Sailing between the islands, the course should be kept so as to be able to see them on every side, and bearings are to be taken on all visible features, such as the end of a coast, capes, islands, mountains, elevations, buildings and dry banks in the sea; the same applies to the middle of features noticed on shore, canyons in cliffs, gullies, mouths of rivers and streams, the summits of sharply-peaked high mountains, belfries, points [capes—AP], and submerged rocks in the sea. Bearings are to be taken more frequently, or less so, depending on the vessel's speed and the distance to locations on which bearings are being taken, in order to ensure that the angles when the bearings intersect are neither too sharp nor too wide.[292]

When bearings were to be taken on the points of capes and coasts, the person taking the bearings had to set his sights on "the visible line of their intersection with the water." Also, "the bearings must be taken no less than three times at each feature." As the survey progressed, the instruction directed that each location be observed with great attention and the description recorded in the journal: "the appearance of the coasts, cliffy or sloping, rocky or earth-covered, high or low, mountainous or level, also to be noted are the submerged rocks separated from the land.

"The alignment of capes, islands, offshore submerged rocks and distinctive features is to be noted, taking bearings simultaneously by azimuth compass and recording in the journals the rhumb and the exact time of the alignment." The height of the most distinctive mountains was to be measured geometrically using an octant.[293]

Sarychev addressed how to conduct a detailed coast survey by means of oar-propelled vessels (including baidaras and baidarkas). Instruments recommended for use in such a survey are three to four good azimuth compasses, leads "of different kinds, a good sextant, a precise pocket watch." It is directed that a journal of standard form be kept during a coastal survey. It had to contain nine obligatory columns: 1—rhumb numbers; 2 and 3—time in hours and minutes; 4—speed in knots; 5—direction of compass setting; 6—distance covered in sazhen; 7 and 8—approximate distances perpendicular to the route and to the right and left of the distinctive contours; and 9—verbal description of the locality and of various occurrences in the course of the survey along the route.

In describing the coasts it was directed that the same data be reflected as those prescribed for surveys from large sailing vessels. But the character of the shoreline was to be described in greater detail: "is it rocky, earthy, sandy or clay, what kind of timber grows thereon, noting if it is suitable for what kind of construction."[294] To make the future reworking of the materials of such a detailed survey and the compiling of large-scale charts on their basis easier, sketches were to be made.

The geodesic basis of the survey (determining the geographic latitude of a location) was ensured, when the weather was good, by measuring the elevation of the sun at midday by means of an octant. A simplified method was permitted in reconnaissance surveys if various circumstances (enmity on the part of the Natives, lack of time, lack of an octant or sextant) made the precise determination of the elevation of the sun impossible. This involved using a water level or mercury poured into a box, which permits an approximate determination of the angle of the rays reflected from the surface of the mercury.[295]

In the third part of this methodological work Sarychev lays down the directions for land surveying. The instruments recommended for use are astrolabes or azimuth compasses, measuring chains or lines with spikes fifty, thirty, and fifteen sazhen in length. The results of the measurements are to be recorded in the survey journal with six columns:

1—number of the main line;

2—rhumbs;

3—distances along the main line;

4 and 5—distances perpendicular to the main line on the right and on the left toward various objects and the contours of the landscape;

6—brief description of the characteristics of the locality and occurrences in the course of the survey.

The main lines were to be set along the shores, in straight segments, by means of alignments, noting in the journal "what kind of terrain the main line crosses, woods or swamp, meadows, tilled ground or fields, recording at which measurement this terrain commenced; it is also necessary to record if the locality is low and subject to flooding or is elevated. The characteristic features of the ground, the quality of the wood and its suitability for construction are to be recorded."[296]

The results of the daily survey were "to be entered on a map" each evening. A great deal of attention was devoted to the study of rivers. A particular directive stated that "when measuring rivers, the local inhabitants are to be asked how far the headwater is and if any other remarkable rivers enter it, how much higher or lower the water level may be than the usual; beyond this, it is to be determined, by examining the shores, how high the water rises in the spring; the local inhabitants are to be asked when the river ice breaks and when the river freezes; what kind of fish are found in the river, is it navigable, and what kind of structural timber is found on its shores. All of this has to be recorded in the journal."[297]

When surveying particular rivers, "pilots knowledgeable about the course of the river and the localities on it are to be engaged, from one settlement to another, and the place names" were to be recorded. The geodesic basis for land surveys, according to Sarychev's instructions, was to be provided by points of reference with latitude and longitude established by astronomical observations. Also the latitude and magnetic compass variation were to be determined at every stop.[298]

In the fourth and last part of Sarychev's instruction the subject is: "On compiling marine charts. Flat and Mercator, how to draw the surveyed areas on each and how to transfer the data from one chart to another." In this section the author, using the generalized experience of his countrymen in compiling marine charts, presents data that were absent in the French textbooks (originals and translations) by Pierre Bouguer (1698–1758) and Etienne Bézout (1730–1783), used at that time by St. Petersburg's Noble Cadet Corps.[299]

Sarychev, like his French predecessors, discusses only the mathematical aspect of compiling marine charts. In particular, he sets out detailed methods for transferring hydrographic data from the originating charts (which were called in Russia "flat" or "particular") to the Mercator ones. These were compiled in the conical cylindrical projection using directions and distances corrected for magnetic variation, and utilizing all astronomically determined geographic coordinates.

In that period, such compilation was accomplished in two stages. First, the images compiled without taking into account magnetic variation were transferred to a general flat chart with corrections in accordance with averaged data for determining magnetic pole variation. The author recommended the use of the vessel's course for a twenty-four-hour period as the basis for constructions on the general chart. On it a mesh of squares was laid and the survey data were transferred from the corresponding squares on the originating chart. In the second stage the Mercator chart was compiled. A cartographic mesh in Mercator projection was constructed using special tables "of latitude increase;" the content of the corrected flat charts was then transferred onto the auxiliary mesh of parallelograms in the Mercator from the squares of the flat charts.[300]

Though Sarychev's work was of great significance in evaluating the expedition's results, works by other members of the expedition were also very important. The editing of these was also entrusted to Sarychev.

In 1811, using the journals of Billings and Hall, and some other materials of the expedition, he published *The Voyage of Captain Billings across Chukotka Land from the Bering Strait to the Nizhne-Kolymsk Ostrog. The Voyage of Captain Hall on the Vessel Black Eagle in the Northeastern Ocean in the Year 1791 with Appended Vocabulary in Twelve Dialects of Uncivilized Peoples, Observations of the Cold in Verkhne-Kolymsk Ostrog and the Instruction Issued to Billings from the State Admiralty College.*

The expedition's German naturalist, Carl Heinrich Merck (1761–1799), was engaged in collecting botanical, zoological, and ethnographic materials in accordance with the instruction by Academician Peter Pallas. These materials were delivered by him in 1794 to the St. Petersburg Academy of Sciences, where they were incorporated into the Academy's collections. Merck also brought the diary he kept during the expedition to Petersburg. Part of this material was used by Pallas in his main work, *Zoology of the Arctic*.[301] Merck's diary was published for the first time in English in 1980.[302]

In contrast to Merck, who submitted all of his materials to the St. Petersburg Academy of Sciences and to Pallas, was another foreign participant in the expedition, Martin Sauer, who spirited away some of the expedition's materials. He violated the instruction that specified that members of the expedition were to surrender materials in their entirety. On Billings's initiative, Sauer's papers were searched on September 4, 1794, and the undelivered materials were found. He was arrested and kept under guard at the Admiralty, but was set free on Billings's intervention.[303] Sauer managed to save some of the documents, however, and they somehow got out of Russia. Sauer's book was published in 1802, practically simultaneously with Sarychev's *The Voyage*.[304]

The results of the Billings Expedition were of great interest to the *promyshlenniki* of the Golikov-Shelikhov Company, who continued to actively explore the coasts and seas of Russian-American possessions. Extensive surveys of the Alaska Peninsula coasts were completed in 1786 and 1791 by Navigator Dmitri Bocharov.

This work resulted in the compilation, *Plan of the American Cape called Aliaksa with Survey of its South Side and Offshore Islands to the Island of Unga, from there to Isanotski Strait, Correcting the Previous Surveys, Conducted During the Sea Voyage of Navigator Bacherov* [*sic*] *by Means of a Three-Hatch Aleut Baidarka. From the Island of Kad'iak to Unalashka Island and Back to Kad'iak in the Year '786, from Isanotski Strait toward Northeast up to the Mainland of America. With the Survey of the Coasts, Bays, and, in Part, Rivers and Lakes Conducted during the Voyage of the Selfsame Bacherov with Two Baidaras along the North Side in the Year '791, then Portaging toward Kad'iak Island. Compiled in the Same Year in the Month of November.*[305]

The Alaska Peninsula is shown on this chart in much greater detail than on any compiled by previous expeditions, including the general chart by Sarychev. The investigations conducted in the 1780s–1790s by the company navigators and the data from James Cook's and Billings's expeditions were summarized in 1797 by James Shields. An Englishman, Shields spent about six years in the service of the Shelikhov-Golikov Company engaged in shipbuilding at Voskresenskaya Harbor [near present-day Seward—MF] on the Kenai Peninsula. Besides shipbuilding, he completed five sea voyages out of Kodiak, Yakutat Bay, and Sitka Island.

Based on the materials of these surveys, a manuscript chart was compiled: *Mercator Chart with Position of the Route in the Golikov-Shelikhov Company in the Post of Sailor-Navigator James Shields on the Company Vessels Named Below.* On this chart are shown the routes of the *Northeastern Eagle* in 1793, the frigate *Phoenix* in 1796–1797, and the cutter *Dolphin* in 1795.[306] An interesting characteristic of this chart is the fact that on two (out of three) copies, two versions of the longitude determinations appear: one following Cook (from the island of Ferro), the other following Izmailov (from Okhotsk).

The Value of Shelikhov's Maps

The summarizing small-scale charts compiled by the Shelikhov and Golikov Company are distinguished by their lesser detail, and at times, gross errors. Nevertheless, they are of interest as historical sources. They also reflect Shelikhov's colonial ambitions, as they are also a cartographic declaration of Russian possessions in America. Shelikhov's charts are important as documents reflecting subject matter which did not find a place on the geographically more precise maps, such as the Mercator chart by Sarychev.

Thus, the rather schematic chart appended to Shelikhov's book[307] differs from the summarizing Mercator chart by Sarychev in that Eskimo settlements along the

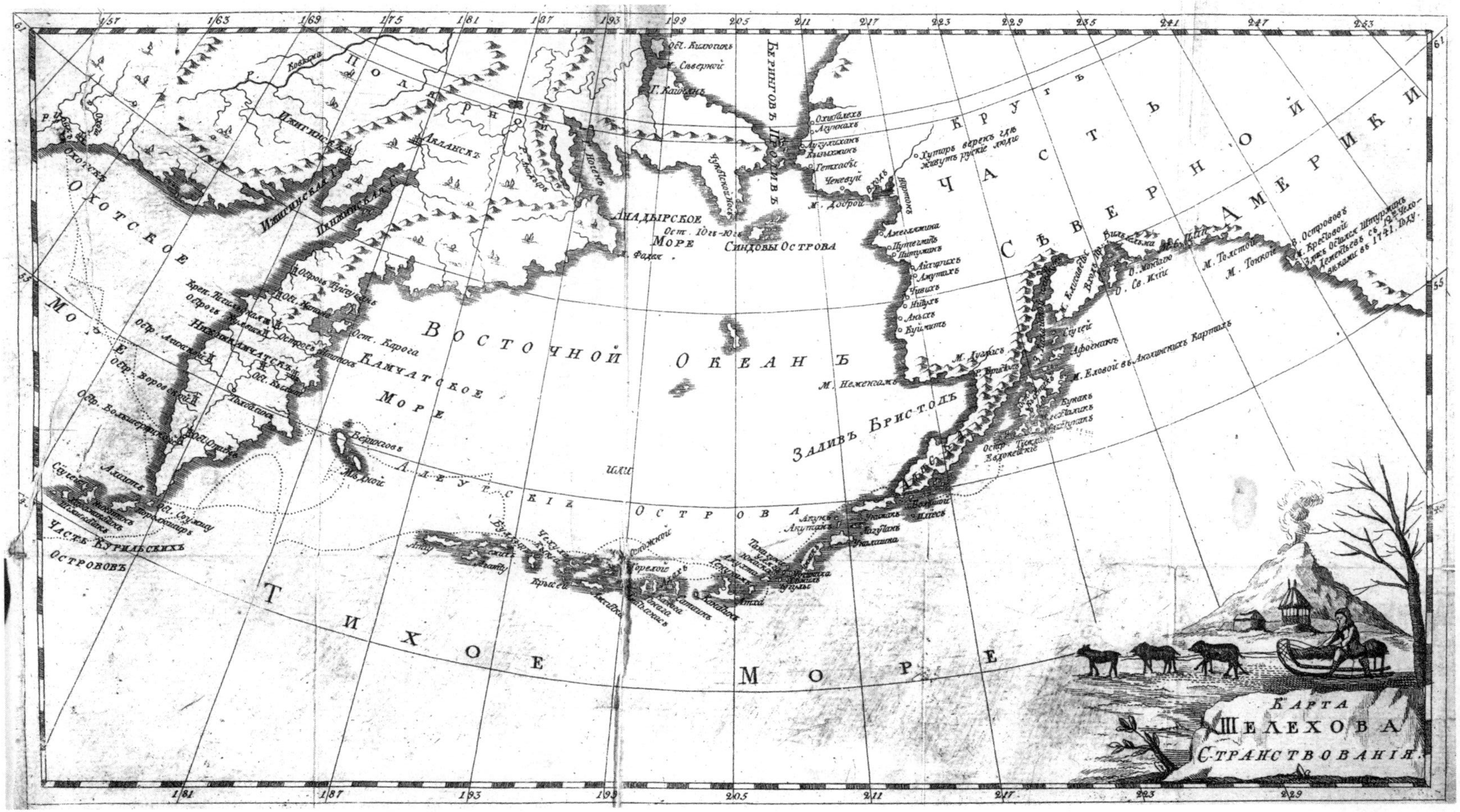

FIGURE 35 Shelikhov's voyage of 1783. Printed 1791, Courtesy Newberry Library Ayer 124. R98, S54, 1791

northern coast of Alaska are shown (Fig. 35). It is apparently based on the chart by Kobelev-Tatarinov of 1779 discussed earlier. Fully in accord with Shelikhov's drive to assert a Russian presence in Alaska as early as possible, this chart shows a settlement on the river emptying into Norton Sound with the inscription, "Homestead believed to be inhabited by Russian people," and on the southern part of the Alaskan shores there is a notation, "Cross Cape. Here Navigator Dementiev with 12 men stayed in the year 1741."

The same characteristics appear on the summary chart, compiled under Shelikhov's guidance but finished after his death in 1795. This is the *Navigational Chart of the Northeastern Part of Asia and Northwestern Part of America, from Okhotsk to Nootka, Including the Western Part of America from Cape Aliaksa to Cape Chirikov. Joined to the Russian Empire by the Honorable Citizen Grigory Shelikhov in the Years 1782 and 1788. The Most Remarkable Places, Numbered Variously, are Shown in Detail at the Bottom of this Map.*[308]

The legend on this chart states: "The northeastern part of Asia, with Kamchatka, and the Kuril and Aleutian Islands, were taken from Russian charts. The northwestern part of America from Cape Aliaksa to the Arctic Sea, from Cook's chart; from Cape Aliaksa to Lituya Bay from Shelikhov's chart; from Lituya Bay to Cape Chirikov, from Russian and the newest French charts; from Cape Chirikov to Nootka, also from French charts."

A careful examination of the chart leads one to conclude that the sources mentioned in the legend above are not wholly accurate; at least not in the area to the north of the Alaska Peninsula, where, in contrast to Cook's chart, numerous Eskimo settlements are shown and the coastline also represents details taken from the 1779 chart by Kobelev-Tatarinov. Again in contrast to all the summary charts which preceded this one, Shelikhov's map shows the first attempt to cartographically indicate the distribution of the Indian tribes of the northwest coast of America, their names, and the approximate borders of their territories. Tribes are indicated which by then were known to the Russians: Kenaitsy, Koniuag, Chugach, Tkhalkha, Ugalakhmiut (Eyak), and Koliuzhi (Tlingit).

Of historic cartographic interest are the eight inserts shown on these summary charts by Shelikhov. Seven of them represent materials from the surveys by company navigators (in the main by Izmailov and Bocharov) of various sectors of the Russian-American possessions (Kodiak Island, and on it St. Paul Harbor, the Khlikakhlik Islands with their strait, Yakutat Bay, Chugach Sound and Lituya Bay, the Tkhalkhe Islands) discussed by us earlier.

On these inserts on Shelikhov's summary chart, Native geographic place names predominate, though by this time Shelikhov, apparently under the influence of his drive to assert Russian priority by all possible means, had reversed his initial instruction to preserve the local toponyms. This is evident from the "Instruction by G. I. Shelikhov and A. E. Polevoi to manager I. F. Popov . . . in regard to the acquisition of new hunting grounds and the assigning of Russian place names to them" sent to the colonies on July 31, 1794. This instruction stated, in part: "We are also ordering you when discovering new locations, regardless of whether they are shown on foreign charts or not, to give them Russian names, honoring our ruling Sovereign, that is, using the names of the Imperial family, such as Paul Island, Constantine island, Alexander Island, Elena Strait, Olga Bay and so on, or in honor of the saints, as you deem fit. You must order the sailors to record these names on the charts and you are also obliged to give Russian names to localities which are already known. This is to counteract foreign pretensions."[309]

Judging from the entire corpus, it appears that practically no materials of the Billings Expedition were utilized when Shelikhov's summary chart was compiled. That the expedition took place is indicated by a point under the number 4, with the explanation in the legend that "this point was reached by Captain Billings in the year 1791."

The chart contains a detailed description, rich in inaccuracies, exaggerations and fantasies, about how Shelikhov "united" to Russia the western Aleutian Islands and Alaska. This description is expanded in Shelikhov's book. The chart has a whole array of geographic errors and inaccuracies that evidence the use of dated materials. This was pointed out by the famous Russian seafarer Vasily Golovnin.[310] The most obvious of the errors is the drawing near the southern part of the North American continent of a fantastical sea with the inscription: "An attempt to trace the route of Washington Olap, an American, in the year 1739."

MORE NON-RUSSIANS ENTER THE FIELD

Toward the end of the eighteenth century, the contributions of foreign sailors to the cartography of the coasts of Russian America grew. These sailors continued to be intrigued by the question of a sea passage from the Pacific into the Atlantic Ocean, but they were even more attracted by the prospect of gaining easy profits from the furs obtained in the waters off the northwest coast of America. The vast majority of the private expeditions seeking "soft gold" were English. The most ambitious one was the expedition by Nathaniel Portlock and George Dixon, with the ships *King George* and *Queen Charlotte*, organized by the King George's Sound Company.

This expedition entered Cook Inlet (or as the members of the expedition called it, King George's Sound) in March 1786 and hunted marine mammals until September with little success. In March of 1787 Portlock and Dixon again entered Cook Inlet, finding the English fur trader John Meares with his vessel, the *Nootka*, frozen in the ice of the inlet. Captain Meares was engaged in sea mammal hunting off the northwest coast of America, principally in Cook Inlet and Prince William Sound in 1786–1789. In 1790 he published a description of his voyages with three charts and seven plans of separate sectors of the coasts.[311]

Portlock remained near the Alaska coasts, while Dixon sailed south, with a stopover at Yakutat Bay. He called this harbor Port Mulgrave. Continuing south, he made "many discoveries," according to his own declaration. Dixon returned to London in September 1788, and by 1789 he had published a book that included a chart of the northwest coast of America and three cartographic sketches.[312] In 1789 Portlock also published a description of his voyage with his own version of the summary chart and five large-scale plans.[313]

On September 30, 1787, the first two American vessels, the *Columbia* (Captain John Kendrick) and the *Lady Washington* (Captain Robert Gray), sailed toward the northwest coast and arrived at Nootka, whence, after wintering, the *Lady Washington* sailed for the Queen Charlotte Islands, and then on to China, arriving there in January 1790. The *Columbia* had arrived somewhat earlier (toward the end of 1789). The voyage of Captain Gray provided the most cartographic results; among them, considered the most important, the passage by the *Lady Washington* through the Strait of Juan de Fuca. The expedition was unable to reach beyond 55°43′ N latitude.

In September of 1790 the *Columbia*, this time under Gray's command, left Boston for the northwest coast, where she was engaged in procuring marine mammal furs until the year 1792. In the course of this voyage Gray discovered the mouth of a river he named in honor of his vessel (the Columbia River). Some of this expedition's materials were published only relatively recently.[314]

In August of 1791 the northwest coast of America was visited by the French fur-procuring expedition of Etienne Marchand, with the ship *Solide*, out of Marseilles. They hunted near Krusof Island and Cape Edgecumbe. Marchand, having Dixon's book on board, determined that his own lunar determination of longitude agreed well with the observations by the English sailor. The results of this expedition were described in a book by Count Fleurien based on the notes by Captain Chanal and the ship's surgeon Roblet.[315] In 1800 an atlas was published as a supplement to the above-mentioned work. It included a hydrographic chart of the known world between 60° S and 70° N latitudes, part of Meares's chart, a copy of the 1752 map by Delisle, a Spanish chart, and two plans of the bays on the northwest coast of America.[316]

To a certain degree this atlas is typical of the cartographic publications of commercial voyages. The basis of the general charts in the atlas was Cook, much dated by this time, along with some large-scale plans of particular segments of the sea from the fanciful map of Delisle-Bauche.

Cartographic results from the expeditions of foreign fur traders, especially the books and charts by Portlock, Dixon, and Meares, were used by Aaron Arrowsmith in his general *Chart of the World*, compiled in Mercator projection and published in 1790.[317] The only Russian expeditions shown were the routes of Bering and Chirikov. Several plans of the harbors on the northwest coast of America, based on sketches by various fur traders, were published by the hydrographer of the East India Company, Alexander Dalrymple.[318]

The last of the large foreign expeditions of the eighteenth century were the voyages of the Spanish explorer Alessandro Malaspina in 1791 and the English explorer George Vancouver in the years 1792–1794.

The Royal Government of Spain decided to add a visit to the seas off Alaska to the program of the scientific round-the-world voyage of Malaspina and Jose Bustamante y Guerra. There was renewed hope of finding a passage from the Pacific to the Atlantic Ocean in the area of the newly discovered inlets of Prince William Sound and Yakutat. Among expedition personnel were naturalist Antonio Pineda, botanist Luis Nee, scientific assistant and artist Jose Quio, artist Tomas de Suria, and cartographer Felipe Bauza. In Chile the expedition was joined by the German naturalist and botanist Thaddeus Haenke.[319] In 1791 Bodega y Quadra, in Mexico, prepared a chart especially for this expedition in which he utilized Russian materials, specifically those by Potap Zaikov.[320]

In June 1791 the corvettes of the Malaspina expedition, the *Descubierta* and the *Atrevida*, reached 58° N latitude in the area of Mount Edgecumbe. Malaspina then sailed for Yakutat Bay, or Port Mulgrave as it was then called, where he hoped to find a passage east. Malaspina and members of his expedition spent over a month here, making valuable observations on the customs and way of life of the Tlingit. They conducted numerous scientific experiments and gathered flora and fauna specimens. Materials from these explorations have great value as historical and ethnographic sources.

Having failed to find a passage to the Atlantic Ocean at Yakutat, Malaspina's expedition returned to San Blas. This voyage did not provide any new cartographic results, though when Bodega y Quadra was compiling his summary chart of the Spanish explorations on America's northwest coast (1792) he used the precise determinations of longitudes made by Malaspina.[321] Malaspina's scientific expedition was the last Spanish voyage in Alaska's waters.

GEORGE VANCOUVER

The concluding chapter of the scientific study and cartography of the Alaska coast in the eighteenth century was the English government expedition under the leadership of George Vancouver. This expedition's mandate was to implement a treaty between Spain and Britain concluded in 1790, ending the Northwest Coast rivalry between these two sovereign states (the Nootka Controversy). Spain ceded their claims to territories that would eventually become British Columbia to Britain. Vancouver was also to study the northwest coast of America from 30° to 60° N latitude.

The expedition's vessels, the *Discovery* and the *Chatham*, approached American shores on April 17, 1792. They explored Puget Sound in detail and then, after sailing north to Safety Cove, turned south. On August 28 they arrived at Nootka, where Bodega y Quadra, representing Spain, was waiting to conclude official negotiations.

Here Vancouver made a series of precise astronomical observations determining the geographic coordinates. The expedition's vessels then sailed south, although the *Chatham* was delayed while making a survey of the Columbia River. In 1793, after a winter spent in Hawaii, the expedition reached Prince of Wales Island (at Port Protection), after conducting detailed examinations and surveys on the way. In the process of conducting this work, Vancouver named "Cape Decision," as he believed that his voyage had once and for all decided the question about the mythological tales of Fuca, Fonte, and others like them. The winter, once again, was spent in Hawaii.[322]

Vancouver reached his northernmost point in the navigational season of 1794. On March 15 the *Discovery* and the *Chatham* left Hawaii together, but soon the ships separated. Vancouver planned to examine Cook Inlet in detail and then sail eastward to where his surveys ended the previous year. Having sailed with Cook, he was familiar with the general outline of the northwestern coasts. He set his course directly north. On April 3, 1794, the sailors saw a mountainous land that proved to be an island. The coordinates were determined (55°48′ N latitude and 209°16′ E longitude) by astronomical observations on April 4.

Vancouver believed this island to be the one first sighted by Bering, who named it *Tumanniy* (Foggy). However, this name was assigned to a different island by Cook. Therefore, Vancouver decided to name it Chirikov because the English sailor considered that Chirikov's achievements were not sufficiently represented on the map of North America. In a bay near the Trinity Islands two Natives came aboard the *Discovery*. They said that three Russians lived on shore. From that day on Russia's presence on American shores was evident to the expedition on a daily basis.[323]

One of the main tasks set for this expedition was to investigate Cook Inlet and determine if there were any passage to the east. Entering the inlet and then going ashore, members of Vancouver's expedition found abandoned winter dwellings made of logs. Vancouver quite rightly concluded that they had been built by Russians. On April 30 the *Discovery* was visited by ten Russians and twenty Natives. Being in a difficult situation because of ice masses and a strong current in the inlet which made sailing difficult, they found it hard to obtain the information they needed. But the *promyshlenniki* explained simply that Vancouver was not on a river as Cook had believed the inlet to be, but in a large bay, deeply cutting into the land, and that its head was about fifteen versts from where the British vessel was anchored.

Furthermore, they told the English sailors about the bay and river of Turnagain Arm, and that the river, with a short portage farther on, could be used to cross the Kenai Peninsula and for travel between the Russian settlements of Cook Inlet and Prince William Sound. In the course of the conversation the Russians continually stressed that the American mainland with all the adjacent islands was the exclusive domain of the Russian Empire. Besides this meeting, the English encountered the *promyshlenniki* again more than once and even visited one of their settlements, the Aleksandrovsky Redoubt, situated at the entry into the bay, in a harbor Vancouver named "Port Chatham."[324] Vancouver stated: "So far as any conclusion could be drawn from this short interview, the Russians seemed to live upon the most intimate terms of friendship with the Indians of all descriptions, who appeared to be perfectly satisfied in being subjected to the Russian authority."[325]

Tangible evidence of the Russian presence was confirmed by Captain Puget, who arrived with the *Chatham* in Cook Inlet on May 7, 1794. Specifically, when he was off Cape Banks, he was visited by two Russians, one of whom was Egor Purtov. The latter said that the *Chatham* had missed a very good harbor where there was a Russian settlement (Aleksandrovsky Redoubt) and that an eight-gun sloop stood there, commanded by the chief manager of the Russian colonies, Aleksandr Baranov, who would have been pleased to visit Puget. Purtov also stated that forty Russians lived there and as many also in two redoubts, one group in Prince William Sound, the other in several small settlements in Cook Inlet.

Puget attempted to learn about Billings's expedition from the Russians and also how far in along the inlet one could sail, but Purtov did not respond, either not understanding the question or claiming that he did not. Puget met several times more with Russians and Natives who used the Russian language. Like Vancouver, he had many occasions to be convinced of the Russians' significant power and influence over the Natives. With the help of the Natives he eventually found the *Discovery*. After reuniting, both vessels stood at anchor at Turnagain Island. Here they were visited by two Russians, who then proceeded, together with the Englishmen, to Nikolaevsky Redoubt,[326] situated deep within the bay.

The British vessels arrived there on May 10, 1794, and received a polite invitation to visit the Russian settlement. Their arrival was greeted by a two-gun salute. The redoubt's commander, Stepan Zaikov, was away in Prince William Sound at that time. The English learned that the settlement had existed for twelve years. At the time of the visit by the Vancouver Expedition there was a large building housing thirty-six Russians, and a commandant's house. On shore, near the water line, a two-masted vessel of sixty to seventy tons' displacement was beached. The redoubt's armament consisted of four small cannons, about twelve muskets and two or three pistols. The English perceived this small number of guns to be a sign of the *promyshlennikis*' confidence in their safety.

Vancouver's investigations in Cook Inlet confirmed the information provided by the Russians that this, indeed, was an inlet and not a river (Fig. 36). Vancouver determined by astronomical observations that the northern head of the inlet had the coordinates of 61°29′ N latitude and 211°17′ E longitude.[327]

The expedition then started to sail to Prince William Sound. At the entry of Cook Inlet, near Cape Bede, the expedition's vessels were once again visited by Egor Purtov with a request that they not sail, as Baranov desired a meeting with Vancouver. The English, however, did not wait for Baranov's arrival. Having received from the Russians a large quantity of fish as provisions on their return trip, they left Cook Inlet on May 15.

On May 25 both vessels dropped anchor at Port Chalmers (at Montague Island). From there two of the ship's boats were dispatched to survey Prince William Sound. On June 7 the English were once again visited by three Russians, one of whom, "Ammus Niconnervitch Ballufian" (apparently Amos Balushin), proved to be more knowledgeable about the geography and charts than his companions. The English explorers asked Balushin to show them a Russian chart of the region. Balushin replied that such a chart was recently brought from Kamchatka to Port Etches (Fort Konstantin and Elena in Nuchek Bay at the island of Tkhalkha) and promised to bring it to the English.

The crew in the ship's boats, headed by Mr. Johnstone, received all kinds of support from the Russian *promyshlenniki*, who shared with them not only their geographic knowledge but also good provisions delivered by the Natives. Johnstone was cordially received by Pyotr Kolomin (Colomenee), from whom they learned that the English shipwright James Shields was serving with the Baranov's company. Though Kolomin had no Russian charts, he informed the English that the galiot's captain Izmailov (Smyloff) would be able to give the English detailed information. However, he was at sea.

On June 8 Russians from Fort Konstantin and Elena again came to the vessels. Among them was Balushin, who brought the promised chart and gave permission to copy it. The chart was compiled in the years 1789–1790, but the English were unable to translate its heading. The Russians either could not or did not want to add anything about the chart's authors and the methods of its compilation. The chart represented the northern part of the Pacific Ocean from 50° to 63° N latitude.

Vancouver gave the following characterization of this chart:

> On comparing this chart with such parts of the American coast as had been surveyed by Captain Cook, or seen by ourselves, though it had not any great apparent pretensions to accuracy, yet it was satisfactory in certain particulars, with respect to which we had entertained some doubts; especially with regard to the coast lying between Cape Trinity and Point Banks, which was represented in this chart as an extensive group of islands under the denomination of Kodiak. The largest of these extends from Cape Trinity to Cape Greville, where a separation takes place between it and the land to the westward of St. Hermogenes Island, which land the Russians call Fogniak; and forms with the land to the southward what Captain Cook called Whitsuntide bay, but which is a passage leading westward into those straits that separate Kodiak from the continent, to the south-westward of Cape Douglas. These straits are delineated as being in general upwards of 10 leagues wide, free from interruption; the land about Cape Trinity forming their south-west point of entrance. The parts thus described, that had not been seen by us, I have incorporated into our charts, but I cannot take upon me to be answerable in any respect

> for their accuracy. This survey in many particulars varies very materially from that published by Mr. Meares.[328]

Apparently, according to this description, Vancouver utilized a chart used by the *promyshlenniki* with the surveys we considered earlier by Bocharov and Izmailov. Vancouver compiled all of the available information in the creation of his chart (Fig. 37).

On June 12 the *Chatham* parted from the *Discovery*, which was under repair, and engaged in a survey of Kayak Island and then Yakutat Bay, now full of ice. After its repairs had been completed the *Discovery* arrived in Nuchek Bay on Tkhalkha Island (Port Etches) and dropped anchor not far from Fort Konstantin and Elena. Vancouver and his officers were cordially received on shore by Kolomin and Balushin. This time the English met with Izmailov, who gave the expedition's members new data on the geography and population of the Russian-American colonies. He mentioned that in the territory extending from Unalaska and Tkhalkha Island there were 400 Russian *promyshlenniki*.

Using Izmailov's information, Vancouver came to the conclusion that the population estimates made by Cook and Meares were overstated, as at the time of his visit there were half the number of Indians. He did not consider this fact to be a result of the Russian colonization because he observed only "old burials." In the English explorer's opinion, the Russian intrusion was relatively painless. Specifically Vancouver wrote:

> Although we could not gratify our curiosity to the extent I could have wished respecting the situation of the Russians, yet I could not avoid feeling a degree of satisfaction in observing the comfortable manner in which they seem to live amongst these untutored children of nature; having gained them over to be obedient to their wishes, they appear to maintain their influence not by fear, as their conquerors, but by having found the way to their hearts, and by securing an affectionate regard.[329]

This was especially apparent in Cook Inlet and Kodiak, though in Prince William Sound, according to Vancouver, the Natives were not as benevolently inclined toward the Russians. Vancouver explained this as being due to the fact that the *promyshlenniki* had appeared there more recently.[330]

On July 1 the *Discovery* entered Admiralty Bay, so named by Dixon (modern Yakutat Bay), renamed by Vancouver "Bering Bay," because Vancouver believed, erroneously, that Bering had stood in this bay. Vancouver's expedition continued to investigate Alaska's coasts until August 22. In this time interval Puget again met with Egor Purtov, who led a hunting party of 9 Russians and about 400 Natives from Kodiak and Cook Inlet. On Purtov's orders, the Natives abundantly supplied the British voyagers with fish. Purtov gave the Englishmen additional information

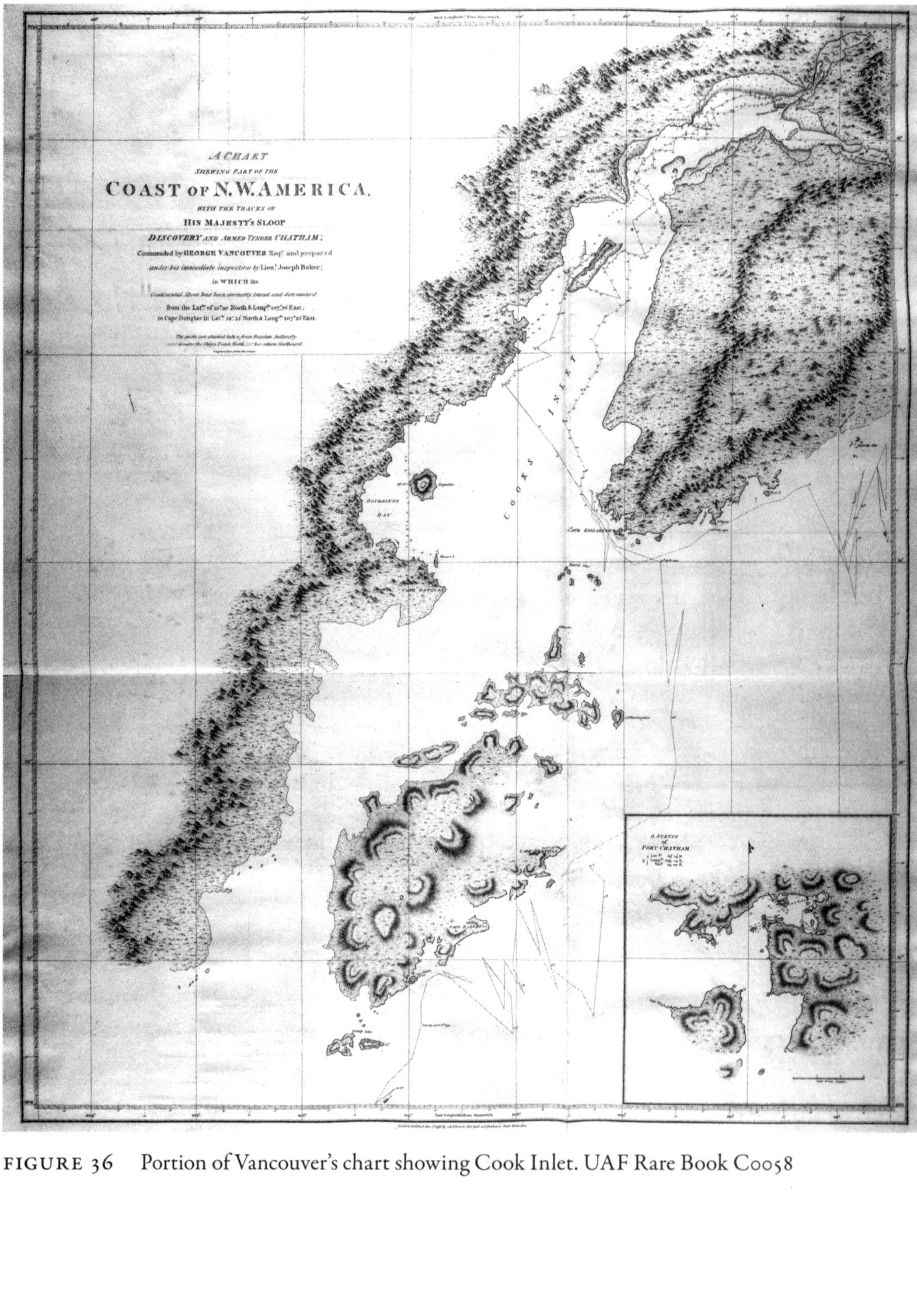

FIGURE 36 Portion of Vancouver's chart showing Cook Inlet. UAF Rare Book C0058

about the geography of the regions where his party hunted. He named the rivers situated near Cape Sakling [Suckling?—LB] (Controller Bay). He also warned the travelers that in the sea, in the southwestern direction from the locality called Icy Bay, there was a dangerous rocky bank about fifteen miles long. Vancouver and Puget noted that the Russians, who seldom used compasses, oriented themselves by the shoreline and specifically they were never wrong in determining the location of this bank.[331]

Vancouver's expedition arrived at Nootka on September 2, 1794, and by September 1795 they were back in Britain. Vancouver's health began to fail and he died in 1798 at the age of 40, having prepared for publication five volumes of the expedition's materials and had worked on the sixth, concluding volume. This was completed by his brother, John, and Peter Puget.[332] Among the works of the expedition is the geographic atlas that became one of the most authoritative sources for creating navigational charts for the areas investigated by Vancouver and Puget. In this atlas Vancouver, in contrast to many other foreign researchers, indicated on the charts of Alaska (Nos. 8–12) sections of the coast and submerged banks that were depicted on the basis of Russian sources, and placed Russian "factories" (forts) on them.

Even though Malaspina determined the coordinates (especially the longitude) with greater precision than Vancouver, the achievements of the English sailor became known four years earlier than those of the Spanish. They were widely advertised by the charts of Arrowsmith's company and the British Admiralty. However, in France the Spanish charts were used. Thus, at the beginning of the nineteenth century, there developed, outside of Russia's borders, two schools of cartography of these regions, the English and the Spanish (or the French).[333] Neither had the details which by that time were characteristic of the Russian large-scale charts in their depictions of the coasts and islands of Alaska. However, thanks to Vancouver, the English charts did consider the Russian investigations to a greater degree than the Spanish or the French.

All charts compiled by non-Russian expeditions, or by the fur-trading sailors, differ considerably from the Russian ones in that they carried no local toponymy. Each explorer considered it his duty to give new geographic names to the "discovered" localities regardless of the fact that they may have been named by an earlier sailor—even from the same country, not to mention Native place names. The English sailors made one exception to this rule: they kept the names given by Cook. As a result, there were a variety of names which were applied to one and the same place (Spanish, English, French, and Russian, though the latter mostly reflected the Native toponymy).

Thus, toward the end of the eighteenth century, through the common efforts of Russian, British, and to a lesser extent Spanish sailors, the general outline of the geography of the northern part of the Pacific Ocean became established. On the whole, the charts adequately represented the distribution of dry land and sea in this part of the globe. The greatest role was played by Russian explorers and

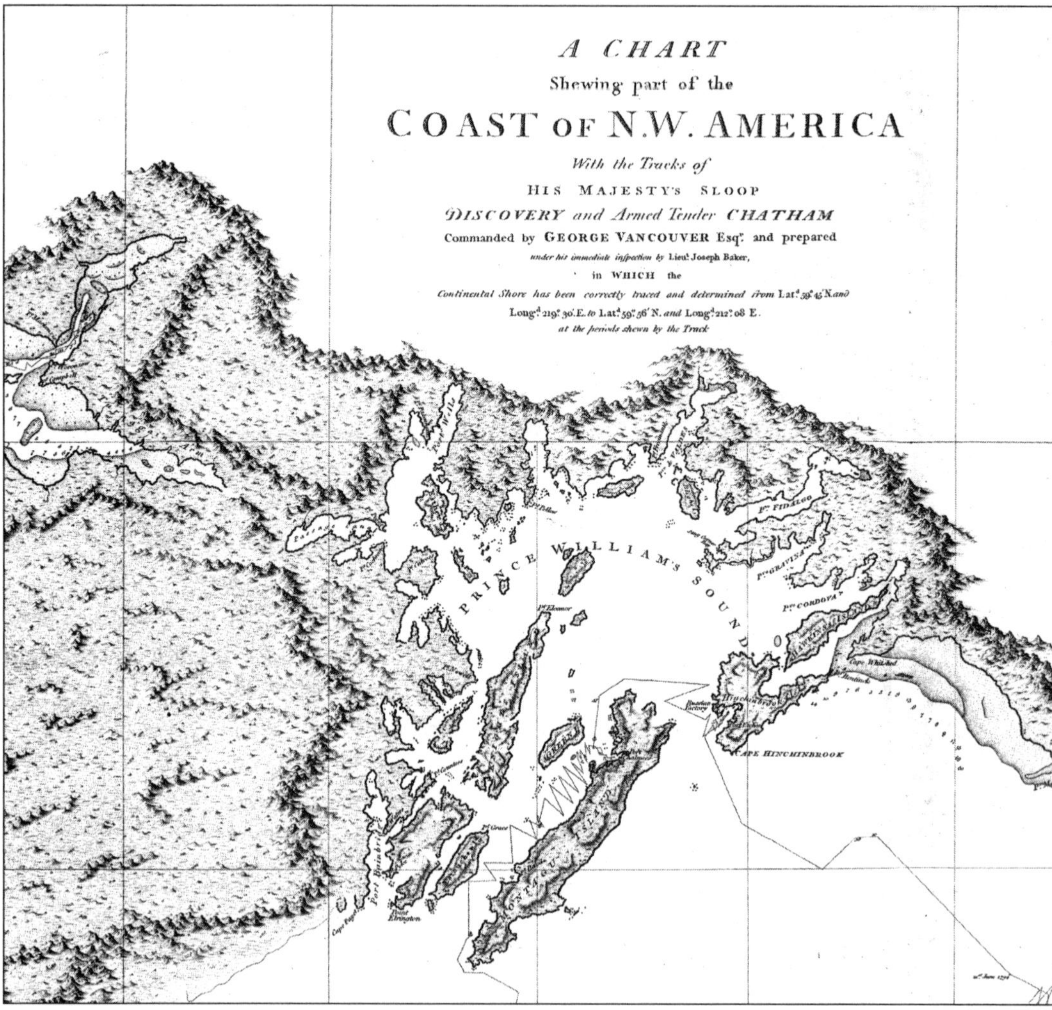

FIGURE 37 Portion of a Vancouver Chart showing Prince William Sound. UAF Rare Book C0058

promyshlenniki. The fur-procuring parties aided in the compilation of geographic information about the Aleutian Islands and Alaska's coasts, which were reflected in not-so-perfect cartographic sketches.

The greatest shortcoming of these charts was the lack (at least up to the 1770s) of precisely determined coordinates, the latitudes and longitudes. Their merit is the preservation of the Native geographic nomenclature and the relatively true rendering of the general outlines and orientation of the lands and seas made by the Russian *promyshlenniki*. The government scientific expeditions made it possible to improve the detailed, but geometrically imprecise, *promyshlenniki* charts by adding astronomically determined coordinates and including them in general geographic charts of the northern part of the Pacific Ocean.

Maps were further refined by including data from foreign expeditions, especially those of Cook and Vancouver. The establishment of navigational schools in Siberia supplemented the ranks of the *promyshlenniki* with qualified navigators. The latter's hydrographic surveys, by the 1780s, differed little from the works of government expeditions. This is evident from the work of such remarkable navigators as Bocharov and Izmailov.

The hydrographers of the government expeditions learned much from the accumulated knowledge gained by the *promyshlenniki* from exploring the Aleutian Islands and coastal Alaska. This exploration was made possible through the use of Native baidaras and the baidarkas. They learned from the customary methods of orientation and the knowledge of Aleuts, Eskimos, and other local tribes about their environment

Toward the end of the eighteenth century, synthesis is evident in the surveys and cartography of the Russian possessions in America, in the methods utilized and in the precision of the professional hydrographers of naval expeditions and the navigators of fur-procuring vessels. This synthesis formed the foundation of the future of Alaska cartography in the nineteenth century.

The results of the geographic study and cartography of the northern part of the Pacific Ocean toward the end of the eighteenth century are shown in the *Map of the Marine Discoveries of Russian Voyagers in the Pacific and Arctic Oceans Made in Various Years. In His Imperial Majesty's Personal Cartographic Service and According to the New Observations of Foreign Seafarers Corrected and Engraved in 1802.*[334]

On this map, for the first time, information from Russian and leading foreign researchers of the northern part of the Pacific Ocean is analyzed and unified. This is reflected in both the maps and in the text, where its sources are listed and a "Description of the main areas lying on the American coast from Kenai Bay to Queen Charlotte Island showing their foreign names" is given. Data from the expeditions of Cook and Billings were used to depict the most northern part of the Alaska coast (to "Cape Malovodny").

From "Cape Malovodny to Kamishak Bay the coast is fixed according to the descriptions of Captain Cook, Geodesist Gvozdev and Commander Kobelev," and here for the first time on an official Russian map numerous Eskimo settlements are shown from Kobelev's information. In the description of the sources used for compiling the remaining parts of the map the use of highly accurate data from scientific expeditions and information from *promyshlenniki* is in evidence.

In particular it is indicated that the

> Aleutian, Andreanof, Fox and other islands lying in the north, the Alaska Peninsula, and Kykhtak (Kodiak) Island received their main positions from the descriptions of Captains Cook and Billings, and the details from private maps obtained from the American Company . . . and from Kenai Bay (Cook River) the remainder of the American coast received its main position from

the descriptions of the English Captain Vancouver made in 1792, '93 and '95, and individual and detailed positions were obtained from the American Company.

NOTES

1. I think not Basov, but his in-law, Dmitry Nakvasin. [LB]
2. *Russkie ekspeditsii po izucheniiu severnoi chasti Tikhogo okeana vo vtoroi polovine XVIII v.* Sbornik dokumentov. Moscow: Nauka 1989:39 [hereafter—*Russkie ekspeditsii . . . vt. pol. XVIII v.* 1989.
3. *Russkie ekspeditsii . . . vt. pol. XVIII v.*, 1989:31–33.
4. See Lydia Black, *Russians in Alaska, 1732–1867*. Fairbanks, 2004:59–77 for a brief explanation of the business side of these early voyages. [MF]
5. A. Polonsky. *Perechen' puteshestvii russkikh promyshlennykh v Vostochnom okeane s 1743 po 1800 god. Arkhiv Russkogo geograficheskogo obshchestava* (Archive of the Russian Geographical Society) [further—ARGO], razriad 60, opis 1, no. 3, folio 27.
6. *Russkie ekspeditsii . . . vt. pol. XVIII v.*, 1989:35.
7. R. V. Makarova, *Russkie na Tikhom okeane vo vtoroi polovine XVIII v.*, Moscow: Nauka 1968:47. On Nevodchikov's expedition, see also H. H. Bancroft, *History of Alaska*, San Francisco 1886:102, 103, 107.
8. RGAVMF, fond 216, opis 1, no. 82, folios 1162, 1163, 1165, 1165v. AVPRI, fond RAK, no. 7, 281 v.
9. AVPRI, fond RAK, no. 7, folio 280. *Russkie ekspeditsii . . . vt. pol. XVIII v.*, 1989:345.
10. *Russkie ekspeditsii . . . vt. Po. XVIII v.*, 1989:38.
11. Report of Timofei Perevalov to Elizaveta Petrovna dated June 7, 1750, RGAVMF, fond 216, opis 1, no. 72, folios 40–46.
12. *Russkie ekspeditsii . . . vt. pol. XVIII v.*, 1989:39.
13. *Russkie ekspeditsii . . . vt. pol. XVIII v.*, 1989:39.
14. RGADA, fond 192, opis 1, Irkutskaia guberniia, no. 10, parts 1 and 2. RGVIA, fond VUA, no. 20289, 20290.
15. RGVIA, fond VUA, no. 20289.
16. RGADA, fond 192, opis 1, Irkutskaia guberniia, no. 10.
17. *Russkie ekspeditsii . . . vt. pol. XVIII v.*, 1989:45.
18. RGAVMF, fond 216, opis 1, no. 73, folios 48, 71–78. A. I. Andreev assumed, apparently erroneously, that fifty-three charts were sent. See A. I. Andreev, "Russkie otkritiia v Tikhom okeane v pervoi polovine XVIII veka." In *Izvestiia VGO*, vol. 75, issue 3, 1943:18.
19. *Russkie ekspeditsii . . . vt. pol. XVIII v.*, 1989:346–347.
20. Barratt, *Russia in Pacific Waters*. 1981:48. P. I. Kabanov, *Amurskie voprosy*, Blagoveshchensk, 1959:27.
21. E. L. Besprozvannykh, *Priamur'ie v sisteme russko-kitaiskikh otnoshenii*. Moscow, 1983:99–114.
22. L. A. Goldenberg, *Katorzhanin-sibirskii gubernator*. Magadan, 1979:239.

23. For more details on F. I. Soimonov, see Goldenberg, ibid., 1979:111–153.
24. AVPRI, fond RAK, no. 5, folio 78, no. 8, folio 75s and 75v.
25. H. H. Bancroft, *History of Alaska, 1730–1885*. The Works of Hubert Howe Bancroft, vol. XXXIII, San Francisco, 1886:115.
26. AVPRI, fond RAK, no. 8, folio 75. *Russkie ekspeditsii . . . vt. pol. XVIII v.*, 1989:347.
27. *Russkie ekspeditsii . . . vt. pol. XVIII v.*, 1989:51–52. AVPRI, fond RAK, no. 8, folio 82.
28. *Russkie ekspeditsii . . . vt. pol. XVIII v.*, 1989:60–65. RGAVMF, fond 216, opis 1, no. 74, folios 5–12. The document was published earlier in somewhat abridged form and with variant wording in *Shchukinskii sbornik*, issue 5, Moscow 1906:150–153, 172–173; and in Andreev, *Russkie otkrytiia v Tikhom okeane i v severnoi Amerike v XVIII–XIX vekakh*, Moscow 1944:23–29.
29. AVPRI, fond 339, RAK, opis 888, no. 923/3. (Formerly in RGADA, fond 1922, Irkutskaia guberniia, no. 54.) *Atlas . . .* 1964, no. 149: 99–100.
30. *Russkie ekspeditsii . . . vt. pol. XVIII v.*, 1989:8, 93.
31. *Russkie ekspeditsii . . . vt. pol. XVIII v.*, 1989:101–109. AVPRI, fond RAK, no. 20, folios 95–116v.
32. *Atlas . . .* 1964:100. In the published materials on the history of Russian expeditions in the Pacific Ocean in the second half of the eighteenth century this chart is dated apparently erroneously to 1764, see *Russkie ekspeditsii . . . vt. pol. XVIII v.*, 1989:347.
33. O. M. Medushevskaia, "Kartograficheskie istochniki po istorii russkikh geograficheskikh otkrytii na Tikhom okeane vo vtoroi polovine XVIII veka. *Trudy Moskovskogo gosudarstvennogo istoriko-arkhivnogo instituta*. Vol. 7, Moscow, 1954. *Atlas . . .* 1964, no. 150, p. 100. Variant of this chart, preserved in the Russian State Archive of Antique Documents (RGADA, fond 192, Irkutskaia guberniia, no. 54) was published by A. I. Andreev, ibid., 1948.
34. RGAVMF, fond 1331, opis 4, no. 88.
35. *Russkie ekspeditsii . . . vt. pol. XVIII v.*, 1989:52.
36. *Russkie ekspeditsii . . . vt. pol. XVIII v.*, 1989:54.
37. RGAVMF, fond 216, opis 1, no. 77, folios 388–395. G. A. Sarychev, *Puteshestvie flota kapitana Sarycheva po severo-vostochnoi chasti Sibiri, Ledovitomu moriu i Vostochnomu okeanu . . .* St. Petersburg, 1802, Parts 1 and 2:296, 300.
38. RGAVMF, fond 216, opis 1, no. 77, folio 394.
39. This chart was first published in 1946 by V. F. Gnucheva, see V. F. Gnucheva, *Geograficheskii departament Akademii Nauk XVIII veka* (Trudy Arkhiva AN SSSR), issue 6, Moscow/Leningrad 1946:126, p. 292. See also *Atlas . . .* 1964, 138: 92–93. Copy in RGAVMF, fond 1331, opis 4, no. 89, *Certified by Captain of the Fleet Petr Fondezin*. In *Russkie ekspeditsii . . . vt. pol. XVIII v.*, 1989:347–348.
40. Barrat, ibid., 1981:58.
41. *Russkie ekspeditsii . . . vt. pol. XVIII v.*, 1989:76. Original in RGAVMF, fond 228, opis 1, no. 21, folios 1–2. First published in *Russkaia Tikhookeanskaia epopeia*, Khabarovsk, 1979:333–334.
42. *Russkie ekspeditsii . . . vt. pol. XVIII v.*, 1989:79–83. Original in RGAVMF, fond 216, opis 1, no. 77, folios 399–408. Published in part in *Russkaia Tikhokeanskaia epopeia . . .*, 1979:337–343.
43. *Russkie ekspeditsii . . . vt. pol. XVIII v.*, 1989:83; Goldberg, ibid., 1982:529.

44. *Russkie ekspeditsii . . . vt. pol. XVIII v.*, 1989:10.
45. *Russkie ekspeditsii . . . vt. pol. XVIII v.*, 1989:97–100. Original in RGAVMF, fond 216, opis 1, no. 77, folios 415–424, dated March 4, 1765.
46. *Russkie ekspeditsii . . . vt. pol. XVIII v.*, 1989:99–100.
47. RGAVMF, fond 913, opis 1, no. 131, F. 1; fond 216, opis 1, no. 84, F. 461.
48. Barratt, ibid., 1981:63–64.
49. *Russkie ekspeditsii . . . vt. pol. XVIII v.*, 1989:355.
50. RGVIA, fond VUA, opis 16, nos. 23437, 23438. See also RGAVMF, fond 1331, opis 4, no. 98.
51. RGADA, fond 192, Irkutskaia guberniia, no. 39, 41; RGVIA, fond VUA, no. 23752–23754; AVPRI, fond RAK, no. 925/5; RGAVMF, fond 1331, opis 4, nos. 91, 92, 95–98. Two of the charts in the holdings of RGAVMF have been published by A. V. Efimov, *Atlas . . .*, 1964:151, 152; 101–103.
52. RGAVMF, fond 1331, opis 4, no. 95.
53. *Atlas . . .*, 1964:101.
54. RGAVMF, fond 1331, opis 4, no. 95.
55. *Russkie ekspeditsii . . . vt. pol. XVIII v.*, 1989:132. Original in RGAVMF, fond 913, opis 1, no. 131, folios 121–123.
56. *Atlas vidov Kamchatskikh i Aleutskikh ostrovov. Xzarisovki kapitan-leitenanta M. Levashova 1767–1768 gg.*, RGAVMF, fond 1331, opis 4, no. 702. *Vidy iz zhurnala Levashova i Krenitsyna*. RGVIA, fond VUA, no. 23503.
57. For greater details on the ethnographic significance of these materials, see R. G. Liapunova, "Etnogrficheskoe znachenie ekspeditsii kapitanov P. K. Krenitsyna i M. D. Levashova na Aleutskie ostrova 1764–1769 gg." in *Sovetskaia etnografiia* no. 6 (1971). R. G. Liapunova, Ocherki po etnografii aleutov (konets XVIII—prvaia polovina XIX v), Leningrad, 1975 [Roza G. Liapunova, *Essays on the Ethnography of the Aleuts (At the End of the Eighteenth and the First Half of the Nineteenth Century)*. Translated by Jerry Shelest with the editorial assistance of William B. Workman and Lydia T. Black. Fairbanks: University of Alaska Press, 1996]. Richard A. Pierce "Exploration in North America." In *Exploration in Alaska. Captain Cook commemorative lectures June–November 1978*. Anchorage, Alaska, 1980:115–116.
58. RGAVMF, fond 179, opis 1, no. 131, folio 289. *Russkie ekspeditsii . . . vt. pol. XVIII v.*, 1989:355.
59. For more detail on foreign publications depicting Krenitsyn expedition, see the foreword by A. I. Andreev to his collection of materials, *Russkie otkrytiia v Tikhom okeane i Severnoi Amerike v XVIII–XIX vekakh*, Moscow-Leningrad, 1944:10–11; and R. V. Makarova, *Russkie na Tikhom okeane vo vtoroi polovine XVIII v.*, Moscow, 1968:13–15; Raisa V. Makarova, *Russians on the Pacific 1743–1799*. Translated and edited by Richard A. Pierce and Alton S. Donnelly. Kingston, Ontario, 1975:8–10.
60. Part IV, Sankt-Peterburg, 1790:263–392.
61. A. N. Kristofovich, *Geologia Tikhii okean*. Leningrad, 1929:40–41. Barratt, ibid., 1981:64.
62. A. P. Sokolov, "Ekspeditsiia k Aleutskim ostrovam kapitanov Krenitsyna i Levashova 1764–1769 gg. *Zapiski Gidrograficheskago departamenta . . .* part 10, Sankt-Peterburg 1852, 70–103.

63. I. V. Glushankov, *Sekretnaia ekspeditsiia*. Magadan, 1972. *Russkie ekspeditsii . . . vt. pol. XVIII v.*,1989:351, 355.
64. *Zapiska kapitana shliupa "Kamchatka" direktoram Rossiisko-Amerikanskoi kompanii o neobkhodimosti zashchity russkikh severo-amerikanskikh i drugikh kolonii ot posiagatel'stva inostrantsev*. September 10, 1819 goda. Rossiiskii gosudarstvennyi istoricheskii arkhiv, St. Petersburg [further—RGIA], fond 1643, opis 1, no. 41, 3 v.
65. S. E. Mostakhov, *Uchastie mestnogo naseleniia v geograficheskom izuchenii severo-vostochnoi Sibiri (XVII–nachalo XX v.)*. Candidate's dissertation in geographic sciences, Moscow, 1965:71. About Daurkin, see also A. I. Alekseev, *Uchenyi-Chukcha Nikolai Daurkin*, Magadan, 1961.
66. For copies (of Daurkin's map), see Grekov, ibid., 1960:209; *Atlas* . . . 87–89, 128, 129, 130, 131.
67. D. J. Ray, *The Eskimos of Bering Strait, 1650–1898*, Seattle and London, 1975:30–31.
68. M. B. Chernenko, "Puteshestvie po Chukotskoi zemle i plavaniia na Aliasku kazach'ego sotnika Ivana Kobeleva v 1779 i 1789–1791 gg. *Letopis' Severa*, vol. 2, Moscow, 1957:121–141.
69. Ray, ibid., 1975:29–30.
70. RGAVMF, fond 216, opis 1, no. 77, folio 169; *Russkie ekspeditsii . . . vt. pol. XVIII v.*, 1989:353. There are two variants of this chart. One of these was formerly preserved in RGADA (at present it is located in AVPRI, fond 339, RAK, opis 888, no. 926/6). A copy of this variant has been published by A. I. Andreev in the appendicies to the collection *Russkie otkrytiia v Tikhom okeane i Severnoi Amerike v XVIII veke*, Moscow, 1948. The chart is executed in watercolor. The second variant is in black India ink. It is preserved in RGAVMF, fond 1331, opis 4, no. 90. A copy of this variant was published by Efimov in *Atlas* . . . 1964, 103–104, 153, with the incorrect reference to the Archive on Ancient Acts (that is to the first variant). For details, see A. I. Alekseev, *Beregovaia cherta*. Magadan, 1987:104–105.
71. *Russkie ekspeditsii . . . vt. pol. XVIII v.*, 1989:113–115. AVPRI, fond RAK, no. 8, folios 254–257. Published in *Russkie otkrytiia v Tikhom okeane v XVIII veke*, 1948:172–177.
72. *Russkie otkrytiia . . . vt. pol. XVIII v.*, 1948:171.
73. V. N. Berkh, *Khronologicheskaia istoriia otkrytiia Aleutskikh ostrovov ili podvigi rossiskago kupechestva*. St. Petersburg, 1823:71.
74. RGADA, fond 192, Irkutskaya guberniya, no. 58. Published in *Atlas* . . . 1964:145, p. 95.
75. This and following citations are after *Russkie ekspeditsii . . . vt. pol. XVIII v.*, 1989:116–118, Minutes of the Admiralty College session of April 27, 1767. Original in AVPRI, fond 339, RAK, opis 888, no. 21.
76. See RGAVMF, fond 216, opis 1, no. 77, folios 202–211.
77. *Russkie ekspeditsii . . . vt. pol. XVIII v.*, 1989:117–118.
78. The watch logs of the voyages of the galiot *St. Catherine* and the hooker *St. Paul*, as well as tables of calculations of geographic coordinates, are preserved in St. Petersburg. See RGAVMF, fond 913, opis 1, no. 116, folios 21–141. Also nos. 124–125, no. 126, folios 1–91, 134–169. Nos. 127–128, no. 136, folios 1–58 no. 137, folios 1–32, no. 138, folios 1–65, nos. 130–132, 139, no. 122, folios 93–146, 148–189, no. 138, folios 66–89.
79. RGAVMF, fond 212, opis 11, no. 2112, folios 65–66; fond 913, opis 1, no. 4, folio 77.

80. V. Berkh, *Zhizneopisanie admirala Alekseia Ivanovicha Nagaeva* [*Biography of Admiral Alexei Ivanovich Nagaev*], St. Petersburg, 1831:62.
81. *Atlas* . . . 1964:97, 98.
82. AVPRI, fond Geograficheskie karty, opis 504g, no. 720.
83. There is more with the following inscription: "Island of Alyaksa or mainland, on this the Russian people are in ignorance and the local inhabitants do not know where it ends and if it should be called an island or the mainland; no one can represent this until a true exploration is conducted; people say the land is low, tundra, and the peaks are far distant about 30 or 50 versts." [LB]
84. AVPRI, fond 339, RAK, opis 888, no. 928/8.
85. AVPRI, fond 339, RAK, opis 888, no. 30, folios 1–8v. Citation is from folio 5 and 5v.
86. Ibid., 8v.
87. RGADA, fond 192, Irkutskaya guberniya, no. 62.
88. RGAVMF, fond 1331, opis 4, no. 667 . . . *Gosudarstvennoi Admiralteiskoi kollegii sochinena v Chertezhnoi 1777 goda.*
89. RGAVMF, fond 1331, opis 4, no. 101. Published in *Atlas* . . . 1964, 155, p. 104.
90. RGAVMF, fond 1331, opis 4, no. 100.
91. RGAVMF, fond 1331, opis 4, no. 102. Published in *Atlas* . . . 1964:154, p. 104.
92. *Karta na chast Tikho-Vostochnogo okeana s novoobretennymi dalnymi ostrovami Unimak i Alyaska, mezhdu kotorymi v prolive podle berega Unimaka s 768 na 1769 god flota kapitan Krenitsyn zimoval s galetom Sv. Ekateriny* (*Map of the Part of the Pacific-Eastern Ocean with the Newly-Discovered Remote Islands of Unimak and Alaska, between which in the Strait beside the Coast of Unimak from '768 to 1769 the Fleet of Captain Krenitsyn Wintered with the Galiot St. Catherine*). RGAVMF, fond 1331, opis 4, no. 103. Published in *Atlas* . . . 1964:156, p. 105.
93. Jacob von Staehlen, *An Account of the Northern Archipelago, Lately Discovered by the Russians in the Seas of Kamchatka and Anadir.* Translated by C. Heydinger, London, 1774. Also published in German and French editions in the same year.
94. *Atlas* . . . 1964:157, p. 105.
95. *Russkie ekspeditsii* . . . *vt. pol. XVIII v.*, 1989:115.
96. AVPRI, fond 339, RAK, opis 888, no. 22: *Raporty i kontrakty kuptsov raznykh gorodov, strantsvuyushchikh po Kamchatke. (1748–1775)* [*Reports and Contracts of the Merchants of Various Cities Who Traveled in Kamchatka (1748–1775)*]. 80v.
97. For the most thorough modern analysis of Müller's contributions, see Gerhard Freiedrich Müller, *Bering's Voyages: The Reports from Russia.* Translated, with commentary by Carol Urness. Fairbanks, 1986.
98. J. L. S., Neue Nachrichten von denen neuentdekten Insuln in der See zwischen Asien und Amerika; aus mitgetheilten Urkunden und Auszügen verfasset von J. L. S. Hamburg and Lepzig, 1776. James R. Masterson and Helen Brower, eds., *Bering's Successors, 1745–1780: Contributions of Peter Simon Pallas to the History of Russian Exploration toward Alaska.* Seattle: University of Washington Press, 1948.
99. William Coxe, *An Account of the Russian Discoveries between Asia and America to which are added the Conquest of Siberia and the History of the Transactions and Commerce between Russia and China*, London, 1780.

100. W. Coxe, *Travels in the Northern Countries of Europe*. Five vols., London, 1802 (Reprint Philadelphia, 1812). See vol. II, p. 295.
101. RGADA, fond 199.
102. S. Engel, *Memoires et Observations. Anecdotes russes*. 106. Lausanne 1765: 234–235.
103. Ant. F. Buesching, ed., *Woechentliche Nachrichten von neuen Landcharten, geographischen, statischen und historischen Buechern und Sachen*, 14 vols., no. 50 (December 13, 1773), Berlin, 1773–1786.
104. G. W. Steller, *Beschreibung von dem Lande Kamtschatka, dessen Einwohnern, deren Sitten, Nahmen, Lebensart und verschiedenen Gewohnheiten, hrsg. Von J. B. S.* Frankfurt und Leipzig, 1774. Published in English as *Steller's History of Kamchatka*. Edited by Marvin Falk, translated by Margritt Engel and Karen Willmore. Fairbanks, 2003. Stepan P. Krashininnikov's own *Opisaine zemli Kamchatki* . . . was published in St. Petersburg in 1755, shortly after his death. He had been asked to prepare a volume on Kamchatka using both his own and Steller's research. It was published in translation by the Oregon Historical Society in 1972 as *Explorations of Kamchatka: North Pacific Scimitar*. [MF]
105. J. E. Fischer, *Sibirische Geschichte*, St. Petersburg, 1768; Russian edition, 1774.
106. *Woechentliche Nachrichten von neuen landcharten, geographischen, statischen und historischen Buechern und sachen*, ed. Anton Friedrich Buesching, 14 vols., Berlin 1773–1786, vol. II: 28, 110. Masterson and Brower, ibid., 1948:15–16.
107. *Russkie ekspeditsii . . . vt. pol. XVIII v.*, 1989:377.
108. P. S. Pallas, *Neue Nordische Beyträge* 1783, 1:273–313. In the current work, the author used the English translation of Pallas's article by Masterson and Brower, ibid., 1944:19–46.
109. Masterson and Brower, ibid., 1944:24. Pallas, ibid., 1781:275–276.
110. Buffon, *Histoire naturelle*. Supplement, v. V, 591–592.
111. Masteson and Brower, ibid., 1944:26–27. Pallas, ibid., 1781:283.
112. Pallas, ibid., 1781:286–287.
113. Pallas, ibid., 1781:294.
114. Pallas, ibid., 1781:299–300.
115. Pallas has the greatest respect for Müller and defends him throughout this work. [LB]
116. Pallas, NNB 1781, vol. 1:300–301.
117. Masterson and Brower, ibid., 39–40.
118. Masterson and Brower, ibid., 41–42.
119. Masterson and Brower, ibid., 35, 43.
120. José Torrubia, *I Moscoviti Nella California*, Roma, 1759.
121. Miguel Venegas, *Noticia de la California, y de su conquista temporal*. Vols, I–IV, Madrid, 1757.
122. Stuart R. Tompkins and Max L. Moorehead, "Russia's Approach to America. Part 2: From Spanish Sources, 1761–75," *British Columbia Historical Quarterly*, 13 (1949), nos. 3–4:231–255.
123. The Marqués de Almodóvar's reports have been published in the document collection: *Colección de documentos inéditos para la historia de España*. Vol. 108, Madrid, 1898:1–324. See also E. Vila Vilar, *Los Rusos en America*, Sevilla, 1966:19–26; and

the Introduction in Y. M. Svet, *Tretie plavanie kapitana Dzheimsa Kuka. Plavanie v Tikhom okeane v 1776–1780 gg.* (*The Third Voyage of Captain James Cook. The Voyage in the Pacific Ocean in 1776–1780).* Moscow, 1971:13–15; Barrat, ibid., 52–53.

124. Svet, ibid. (1971):13–14.

125. Vila Vilar, ibid. (19665):37–45. E. Volkl, *Russland und Lateinamerika, 1741–1841*, Wiesbaden, 1968:53–54. Henry R. Wagner, *The Cartography of the Northwest Coast of America to the Year 1800*, Berkeley, 1937:172 [vol. I]. *La Administración de D. Frey Antonio María de Bucareli y Ursúa, cuadragésimo sexto Virrey de México*, vol. 2, Mexico City, 1936:221–224. Svet, ibid., 1971:14.

126. For the most recent English translation of Spanish exploration accounts, see Wallace M. Olson, *Through Spanish Eyes: Spanish Voyages to Alaska, 1774–1792*, Auke Bay, Alaska, 2002. [MF]

127. Wagner, ibid., 1937:175–177.

128. Wagner, ibid., 1937:177–178. F. A. Maurelle, "Tagebuch einer im Jahr 1775 zu Untersuchung der nordlich von Californien fortgesetzten Kusten, gechehenen Reise (mit Tabellen)," in Pallas, ibid., 3: 198–273.

129. Wagner, ibid., 1937:183–184.

130. A detailed analysis of these sources is found in James Beaglehole's introduction to the scholarly edition of the materials of Cook's third voyage, and also in the introductory article by Y. M. Svet to the translation into the Russian language of [Beaglehole's] publication. See J. C. Beaglehole, ed., *The Journals of Captain James Cook on his Voyages of Discovery: III. The Voyage of the Resolution and Discovery, 1776–1780*. Cambridge, 1967, Part One, Introduction: lvi–lxviii; Svet, ibid., 1971:15–24.

131. J. Campbell, "A distinct account of part of the north-East frontier of Russian Empire, commonly called the country of Kamchatka or Kamschatska." In J. Harris, *Navigantium at que itinerarium bibliotheca of voyages and travels.* second edition, vol. II, London, 1748; third ed., London, 1764:1016–1041.

132. Analysis of the Russian materials available to Cook's expedition appears in the introductory essay by Y. M. Svet to the [Russian language—LB] publication of the journals of Cook's third voyage. See *Tret'ie plavanie* . . . 1971:16–24.

133. Y. M. Svet, *Plavanie v Tikhom okeane v 1776–1780 gg.*, 1971:31–32.

134. J. C. Beaglehole, ed., *The Journals of Captain James Cook on his Voyages of Discovery: III. The Voyage of the Resolution and Discovery, 1776–1780*. Part One (Cambridge, 1967). [hereafter—*The Journals* . . .] 1967:384–393.

135. Svet, ibid., 1971:32.

136. Svet, ibid., 1971:32.

137. Beaglehole, 1967:433.

138. Beaglehole, 1967:434, fn 1 (quoting from the ship's log).

139. Svet, ibid., 1971:377–378.

140. *The Journals* . . . , 1967:449.

141. Ibid., 1967:450.

142. *The Journals* . . . , 1967:453[–454].

143. Those were Izmailov's own charts, he was a Naval Service navigator serving the trading company by permission seconded to them. [LB]

144. *The Journals* . . . , 1967:453[–454].

145. Bechevin's ship was the first to winter in Isanotski in the 1760s; after that many vessels passed through and made camp there. Isanotski appeared on many *promyshlenniki* charts and Izmailov probably sailed the strait himself. [LB]
146. Ibid., 1967:455.
147. Ibid., 1967:456.
148. Ibid., 1967:456.
149. Ibid., 1967:457.
150. The persona of Maurice Beniovski, and his tales about his adventures and "discoveries" in the course of his voyage from Kamchatka to France, originally published in London in 1790 (Maurice Benyowski, *Memoirs and Travels of Mauritius Augustus Count de Beniowski*) continue to be an object of heated discussions in the world of scholarship. Benyowski maintained that he, besides and prior to visiting the Kurile Islands and [regions] farther south, spent some time in the Bering Sea, in the vicinity of St. Lawrence Island and then was on the coasts of Alaska, even down to Kodiak Island. A majority of scholars have doubt in regard to the speed with which Benyowski covered these enormous distances. Thus, for example, Benyowski claims that having left Bering Island on May 25, he was at St. Lawrence Island by June 1, and, after a ten-day stay near this island (from June 5 to 10) Benyowski's vessel on June 11 was sailing along the Alaska coast in the region of Cape Rumiantsev [on modern US maps, Cape Romanzof]. Discussion of his voyages has been renewed by Edward Kajdanski who examined Benyowski's origtinal manuscript preserved in the British Library. See E. Kajdanski, "The authenticity of Maurice Benyowski's account of His Voyage through the Bering Sea: The earliest Description and the Earliest Drawings of St. Lawrence Island." *Terrae Incognitae. The Journal for the History of Discoveries*. Vol. XXIII (1991), 51–79. Despite the detailed and deep historico-geographical analysis conducted by Kajdanski, doubts remain. Kajdanski's main argument rests on Benyowski's reliable description of St. Lawrence Island. But, this island had been visited [literally, in ms., "Seen"—LB] by many others earlier, among them Izmailov and other participants in I. Sindt's expedition, whom Benyowski might have met on Kamchatka. There are no cartographic materials among Benyowski's papers. We have seen that Izmailov himself, in his conversations with James Cook never mentioned such a voyage.
151. E. Markova-Dvoichenko, "Shturman Gerasim Izmailov." *Morskie zapiski* vol. 13, no. 4, 14–27. New York 1955. "Izmailov . . ." in R. A. Pierce, Russian America: a Biographical Dictionary. Kingston, Ontario, Fairbanks, 1990: 205–207.
152. A. David, "Russian Charts and Captain Cook's third voyage." *The Map Collector*, Autumn 1990, no. 52: 2–6.
153. *Russkie ekspeditsii . . . vt. pol. XVIII v.* (1989):181–182.
154. *The Journals . . .*, (1967):660–669.
155. The question of the exchange of geographic information between members of Cook's expedition and the Russians has been examined by Y. M. Svet and S. G. Fedorova. See Y. M. Svet and S. G. Fedorova, "Captain Cook and Russians." *Pacific Studies*, vol. II, no. 1 (Fall 1978):1–19; and also Andrew David, ibid. (1990).
156. *Russkie ekspeditsii . . . vt. pol. XVIII v.* (1989):185.
157. *The Journals . . .*, (1967):667.
158. *The Journals . . .*, (1967):1248.

159. *Russkie ekspeditsii . . . vt. pol. XVIII v.*, (1989):183.
160. Raisa V. Makarova, *Russians on the Pacific, 1743–1799* (Kingston, 1975) 74–75.
161. RGADA, fond 192, opis 1, Irkutskaia guberniia, no. 66.
162. David, ibid. (1990):5.
163. RGADA, fond Gosarkhiv, no. 2529, part 1, verso 42, quoted after Cook, ibid. (1971):624.
164. Y. M. Svet, ibid., (1971):35.
165. The cartographic results of Cook's voyage to the coasts of Alaska and northeastern Asia were studied by Y. M. Svet, S. G. Fedorova, Andrew David, and Glyndwr Williams (see Glyndwr Williams, *The British Search for the Northwest Passage in the 18th Century*, London, 1962; [and others—LB]; G. Williams, "Alaska Revealed: Cook's Exploration in 1778." In *Exploration in Alaska. Captain Cook commemorative lectures*, (Anchorage, 1980) 69–87.
166. R. A. Skelton, ed., *The Journals of Captain Cook on his Voyages of Discovery. Charts and Views by Cook and his Officers reproduced from the original manuscripts.* (Cambridge, 1955). For a more recent compilation and reproduction of the original charts, see Andrew David, *The Voyages of the Resolution and Discovery, 1776–1780.* Vol. 3 of *The Charts & Coastal Views and the original Engravings Assoctiated with them . . .* London, 1997. The front matter includes an essay on Russian charts of the North Pacific by A. Postnikov.
167. American Geographical Society Collection. Vall Number AGS 724/1887 (Gift of J. Carson). MS Chart of the NW Coast of America and the part of the NE Asia with the track of His majesty's sloops Resolution and Discovery from May to October 1778 by George Vancouver. (Scale ~ 1: 2,000,000). Size: 34 3/4 × 48 inches.
168. Thomas Kitchin, *A chart of the discoveries made by the late Capt. Cook & other European navigators in the great Pacific Ocean between Asia & America* by. Printed for R. Baldwin in *London Magazine* (July 1780). An improved version of this chart was published in the December 1780 issue of the same journal.
169. The first account of Cook's third voyage with maps was an unauthorized edition published in 1781: *Journal of Captain Cook's last voyage to the Pacific Ocean . . . Illustrated with cuts and a chart.* London, 1781. [MF]
170. "Chart of the North West coast of America and the North East coast of Asia explored in the years 1778 & 1779. The unshaded parts of the coast of Asia are taken from the MS Chart received from the Russians." In Cook, *Voyage to the Pacific Ocean. Atlas.* London, 1784.
171. "Perechen' puteshestviia shturmana Zaikova k ostrovam mezhdu Azieiu i Amerikoiu nakhodiashchimsia na bote 'Sv. Vladimir." *Sobranie sochinenii vybrannykh iz mesiatseslovov na raznye gody.* Part V. St. Petersburg, 1790.
172. "Karta merkatorskaia s pokazaniem Aleutskikh ostrovov imenuemykh Severnyi arkhipelag po opisaniu shturmanskago uchenika /chto nyne shturmanom/ Potapa Zaikova v 1779 godu, chto vse znachit na onoi karte vnov' opisannyia ostrova pod aloiu kraskoiu, raznost' dlinny bolee prezhnikh opisei 4½ gradusa, angliskago schicleniia menee 2½ gradusami. . . . S podlinnoi kopiroval geodezii serzhant Ivan Rebrov." RGAVMF, fond 1331, opis 4, no. 105. Published *Atlas* . . . 1964:161, 108.
173. Raisa V. Makarova, *Russians on the Pacific. 1743–1799.* Kingston, 1975:69–70, 245.
174. Masterson and Bower, ibid., 1948:86.

175. Svet and Fedorova, ibid., 1978: 11–13
176. *Russkie ekspeditsii . . . vt. pol. XVIII v.*, (1989: 188–189. RGADA, fond 7, opis 2, no. 2539, folios 62–64. Published: A. V. Efimov, *Is Istorii Velikikh Russkikh Geograficheskikh Otkrytii v Severnom Ledovitsom i Tikhom Okeane (From the biblio).* Moscow, 1950:264–268.
177. Svetlana G. Fedorova, *The Russian Population in Alaska and California. Late 18th Century–1867.* Kingston, 1973:250. RGADA, fond Gosarkhiv, VII, no. 2539, 1, 535. The map was first published by Fedorova, see S. G. Fedorova, *Russkoe naselenie Aliaski i Kalifornii. Konets XVIII v.—1867.* Moscow, 1971, Fig. 1. The map reproduced here is in German, published from Peter Simon Pallas, *Neue nordische Beyträge*, vol. 4, 1783, endpapers. It reproduces the place names, but lacks the annotations.
178. Dorothy J. Ray, "Eskimo Place-Names in Bering Strait and Vicinity," *Names* 19 (1) 1974:4–7. Ray, ibid., 1975:45.
179. RGADA, fond 192, opis 1, Arkhangel'skaia guberniia, no. 16, folio 1. Published by Fedorova, "Issledovatel' Chukotki i Aliaski kazachii sotnik Ivan Kobelev," *letopis' Severa*, vol. 5, Moscow, 1971.
180. RGADA, fond 192, opis 1, Arkhangel'skaia guberniia, no. 16, folio 1.
181. Svet and Fedorova, ibid., 1978.
182. David, ibid., 1990:5.
183. RGAVMF, fond 1331, opis 4, no. 106. Published in *Atlas . . .* (1964), 165, 109–111.
184. AVPRI, fond 339, RAK, opis 888, no. 929/9.
185. RGAVMF, fond 1331, opis 4, no. 7. Gosudarstvennyi istoricheskii muzei [further—GIM], Otdel kartografii, G. O., 1882. Published in *Atlas . . .* 1964, 163, 109.
186. RGAVMF, fond 1331, opis 4, no. 109. RGADA, fond 192, opis 1, Irkutskaia guberniia, no. 93, folios 1, 2, 3. Folio 3 is a working compilation, original, which lacks inscriptions and legend.
187. Pallas, ibid., vol. 1 (1781): Carte der Endekungen zwischen Siberien und America, bis auf das Jahr 1780. The chart is appended to the end of the volume. P. S. Pallas, O rossiiskikh otkrytiiakh na moriakh mezhdu Aziei i Amerikoi. *Mesiatseslov istoricheskii i geogrficheskii na 1781 g.* St. Petersburg, 1781.
188. This chart was first published by V. F. Gnucheva, see V. F. Gnucheva, *Geograficheskii departament Akademii mauk XVIII v.* Leningrad, 1946:292, no. 124. A copy of this map also appears in Efimov's *Atlas*—no. 162:109. Both Gnucheva and Efimov noted only that this chart was compiled by a person involved in cartographic work of the Academy of Sciences, but they missed the fact that it is identical to the chart of 1781 by the academician Pallas. Pallas was linked to the work summarizing the results of the Russian explorations in the northern part of the Pacific Ocean and their comparison with Cook's expedition data.
189. S. P. Pallas, Erkundigungen, die beygefuegte Karte betreffend. *Neue Nordische Beitraege 1.* 1781: 265–269.
190. Masterson and Bower, ibid., 1948:47.
191. RGADA, fond 192, opis 1, Irkutskaia guberniia, no. 88.
192. This atlas was recently discovered in the provincial archives of British Columbia in Victoria and published as *The Lovtsov Atlas of the North Pacific Ocean, compiled at Bol'sheretsk, Kamchatka in 1782 from Discoveries made by Russian Mariners and*

Captain James Cook and his officers by Vasilii Fedorovich Lovtsov. Translated with an introduction and notes by Lydia T. Black. Edited by Richard A. Pierce, Kingston, Ontario/Fairbanks, Alaska, 1991.

193. *The Lovtsov Atlas* . . . 1991:40.
194. David, ibid., 1990:5.
195. AVPRI, fond 339, RAK, opis 888, no. 930/10, fond VUA, no. 23741.
196. Christon A. Archer, "Russians, Indians, and Passages. Spanish Voyages to Alaska in the Eighteenth Century." *Exploration in Alaska*. Anchorage, 1980: 134.
197. Carta reducida de las costas y mares septentrrionales de Californias formada hasta el grado 58 de latitud por las observaciones hechas por el teniente de navio Don Juan Francisco de la Bodega y Quadra, y el alferez de fregata Don Francisco Antonio Maurelle, cuia se representa por medio de sombura de tinta, y quanto se manifiesta por sombra encarnada pertenece a le de monsieur Bellin impreza al ano de 1766. Described in Wagner, ibid., 1937:344–345.
198. Wagner, ibid., 1937:191–192.
199. Wagner, ibid., 1937:192–195.
200. A. I. Alekseev, "Brat'ia Shmalevy i ikh deiatel'nost." *letopis' severa*, vol. 3 1962:103.
201. Makarova, ibid., 1974:122.
202. Shelikhov, *A Voyage to America, 1783–1786*. Translated by M. Ramsey, edited with an Introduction by Richard A. Pierce, Kingston, 1981:7–8.
203. AVPRI, fond 339, RAK, opis 888, no. 39. *Zapiska Shelikhova o ego puteshestvii v Vostochnoe more. 1783 god*. Verso 1.
204. Shelikhov, ibid., 1981:10–11.
205. H. J. Holmberg, *Ethnographische Skizzen ueber die Voelker des Russischen Amerika*, Erste Abteilung [Part One], Acta Societatus Scuentarum Fennicae, vol. 4, Helsingfors [Helsinki] 1856:410–417. Published in English as *Holmberg's Ethnographic Sketches*. Edited by Marvin Falk and translated by Fritz Jaensch. Fairbanks, 1985. Shelikhov, ibid., 1981:11–12.
206. AVPRI, fond 339, RAK, opis 888, no. 935/15. RGVIA, fond VUA, no. 23443.
207. RGAVMF, fond 1331, opis 4, no. 110.
208. AVPRI, fond 339, RAK, opis 888, no. 46, F and verso 1, F 5. This is the *Nastavelnie Konstantinu Alekseevichu Samoilovu* first published in *Russkie otkrytiia* . . . 1948:185–199). Part of this document was also published in *Russkie ekspeditsii . . . vt. pol. XVIII v.*, 1989:220–224. A copy is preserved in the Archive of the St. Petersburg sector of the Institute of History, Russian Academy of Sciences [further—LOII], fond 36, 1, no. 476, folios 385–391.
209. AVPRI, fond 339, RAK, opis 888, no. 46, verso 1, folio and verso 6.
210. AVPRI, fond 339, RAK, opis 888, no. 46, folio and verso 8.
211. Ibid., folio 9–10 verso.
212. AVPRI, fond 339, RAK, opis 888, no. 49, folio 2–3 verso.
213. Makarova, ibid., 1974:124–125.
214. Makarova, ibid., 1974:124–125.
215. See Stephen D. Watrous, ed., *John Ledyard's journey through Russian and Siberia. 1787–1788. The journal and selected letters*. Madison, Wisconsin, 1966.

216. Shelikhov, 1981:19–20. Pierce used a document housed in the Iudin Collection, Box 2, Folder 29.
217. AVPRI, fond 339, RAK, opis 888, no. 936/16, 937/17.
218. Ibid., no. 936/16.
219. AVPRI, fond 339, RAK, opis 888, no. 937/17.
220. AVPRI, fond 339, RAK, opis 888, no. 937/17, text.
221. (Latouche Passage by Gerasim Izmailov and Dmitrii Bocharov, 1788. Library of Congress). In 1790, this plan was copied for the Billings/Sarychev Expedition. The copy was executed by *Navigator of first lieutenant's rank Anton Batakov, certified by the Fleet captain of the First Rank Iosif Billings.* This *Plan of the first passage [strait] to Chugach Inlet at the American coast with indication of convenient anchorages, bays, and depths* is in RGAVMF, fond 1331, opis 4, no. 113.
222. Shelikhov, ibid., 1981:117.
223. Makarova, ibid., 1974:128–129.
224. Wagner, ibid., 1937:199–200.
225. Letter dated September 23, 1787. *Russkie ekspeditsii . . . vt. pol. XVIII v.*, 1989:245.
226. Wagner, ibid., 1937:200.
227. Wagner, ibid., 1937:204–205. The chart Mez. BN 15/1/10, folio 280; LC.
228. Wagner, ibid., 1937:205. Gonzalo Lopez de Haro. *Carta reducia qua contiene la costa septentrional de la California asta los 62 grados de latitud Norte correjida y emndada en la expedicion que hicieron los dos buques de S. M. la fregata Princesa y el paquebot San carlos desde el Puerte de San Blas asta la latitud Norte de 60 grados 11 minutos en el año de 1788.*
229. Wagner, ibid., 1937:205. Juan Francisco de la Bodega y Quadra, *Carta general de quanto asta Noy se ha descubierto y examinado por los Espãnoles en la Costa Septentional de California, formada . . . por D. Juan Fancisco de la Bodega y Quadra.* Ãno de 1791.
230. AVPRI, fond 339, RAK, opis 888, no. 937/17.
231. *Russkie ekspeditsii . . . vt. pol. XVIII v.*, 1989:204.
232. D. A. Shirina, *Letopis' ekspeditsii Akademii nauk na severo-vostok Azii v dorevoliutsionnyi period.* Novosibirsk, 1983:36–48.
233. *Russkie ekspeditsii . . . vt. pol. XVIII v.* 1989:14–15.
234. *Russkie ekspeditsiii . . . vt. pol. XVIII v.* 1989:205 RGAVMF, fond 227, opis 1, no. 48, folios 32–38. Original. Published in G. A. Sarychev, *Puteshestvie po severo-vostochnoi chasti Sibiri, Ledovitomu moriu i Vostochnomu okeany.* Moscow, 1952:31–35 [and in] Divin, *Tusskaia tikhookeanskaia epopeia.* Khabarovsk, 1979:363–367.
235. *Russkie ekspeditsii . . . vt. pol. XVIII v.* 1989:206.
236. *Russkie ekspeditsii . . . vt. pol. XVIII v.* ,1989:209. RGAVMF, fond 214, opis 1, no. 1, folios 11–13, originals. Published in Sarychev, ibid., 2:279–294.
237. *Russkie ekspeditsii . . . vt. pol. XVIII v.*, 1989:210.
238. Ibid.
239. The standard questionnaire is found in RGAVMF, fond 214, opis 1, no. 1, folios 86–92 and 94–97.
240. *Russkie ekspeditsii . . . vt. pol. XVIII v.* 1989:214.
241. *Russkie ekspeditsii . . . vt. pol. XVIII v.* 1989:214.

242. *Russkie ekspeditsii . . . vt. pol. XVIII v.* 19891984:214.
243. *Russkie ekspeditsii . . . vt. pol. XVIII v.* 1989:15.
244. A. I. Alekseev, *Gavriil Andreevich Sarychev*, Moscow, 1966:48–50.
245. RGAVMF, fond 214, opis 1, no. 60, folio 1. Published in *Russkie ekspeditsii . . . vt. pol. XVIII v.* 1989:265.
246. G. A. Sarychev, *Puteshestvie po severo-vostochnoi Sibiri, Ledovitomu moriu i Vostochnomu okeanu*. Moscow, 1952:105–106. *Russkie ekspeditsii . . . vt. pol. XVIII v.* 1989:16.
247. Martin Sauer, *An Account of a Geographical and astronomical Expedition to the Northern parts of Russia . . . .* London, 1802:160–161.
248. RGAVMF, fond 1331, opis 4, no. 113.
249. AVPRI, fond 339, RAK, opis 888, no. 938/18.
250. AVPRI, fond 339, RAK, opis 888, no. 941/21.
251. This chart is in the voyage journal of G. A. Sarychev, RGAVMF, fond 913, opis 1, no. 171, folios 94–104, 107, and 108 verso–117. Published in *Russkie ekspeditsii . . . vt. pol. XVIII v.* 1989:271.
252. Sauer, ibid., 1802:213.
253. Sauer, ibid., 1802:171–172.
254. Sarychev, ibid., 1952:158.
255. A. S. Polonsky: *Perechen' puteshestvii russkikh promyshlennikov v Vostochnom okeane s 1743 po 1800 god*. ARGO, razriad 60, opis 1, no. 3, folios 155–157.
256. *Russkie ekspeditsii . . . vt. pol. XVIII v.* 1989:16.
257. *Russkie ekspeditsii . . . vt. pol. XVIII v.* 1989:301.
258. *Russkie ekspeditsii . . . vt. pol. XVIII v.* 1989:302.
259. Ibid.
260. *Russkie ekspeditsii . . . vt. pol. XVIII v.* 1989:302–303.
261. RGAVMF, fond 913, opis 1, no. 279. *Zhurnal, vedennyi geodezii sezhantom Osipom Khudiakovym vo vremia sledovaniia na amerikanskoi baidare do ostrova Unimaka, ot onago* v troeliuchnoi baidarke do Olen'ia ostrova s opisaiem ostrovov, lezhashchikh u mysa Aliaksy c 7 sentiabria 1791 goda ot seleniia Iliuliuk do 22 aprelia 1792 goda / on 157 folios/.
262. RGAVMF, fond 813, opis 1, no. 279, 2 verso, 3 verso, 38 verso, 39 verso.
263. RGAVMF, fond 913, opis 1, no. 279, folios: 65 verso–70.
264. Ibid., folios 72, 81–82 verso. The citations are from a clerk's copy of Khudiakov's journal. Original journal is in RGAVMF, fond 913, opis 1, no. 280.
265. RGAVMF, fond 913, opis 1, no. 199, folios 278–280.
266. RGAVMF, fond VUA, nos. 23446, 23458.
267. *Russkie ekspeditsii . . . vt. pol. XVIII v.* 1989:305, Document no. 107.
268. Ibid., 1989:17, 363. G. A. Sarychev, *Puteshestvie flota-kapitana Sarycheva . . . .* St. Peterburg, 1802:174. The results of the census are presented by Sarychev in his journal and also in the table he compiled: *Tabel' Aleutskim ostrovam i na onykh seleniiam chislu muzheskago polu liudei, kak iasashnykh tak i ne iasashnykh, i skol'ko v kotorom selenii obiazalis' ezhegodno platit' iasak, ravno i chislo vnesennykh noniche za 1791 g.* RGAVMF, fond 913, opis. 1, no. 251, folio 188.
269. AVPRI, fond Spb, Glavnyi arkhiv, 1–13, opis 10, 1789–1791, no. 1, folio 7–8 verso.
270. Ibid., folio 7.

271. AVPRI, fond 339, opis 888, no. 78. *Letter of A. A. Baranov to G. I. Shelikhov from captain's Bay, Unalashka island*, folio 2.
272. Ibid., 5 verso.
273. Ibid.
274. Ibid., p. 311.
275. Ray, ibid., 1975:52–53. Kobelev's journal has been published. See M. V. Chernenko, Puteshestvie po Chukotskoi zemle i plavanie na Aliasku kazach'iego sotnika Ivana Kobeleva v 1779 i 1789–1791 gg. *Letopis' Severa*, 2, 1957:121–141. "Abstract of the Jouranal of the Cossack Sotnik Ivan Kobelov Concerning the Chukchi Country and the islands and Corner of America Lying Opposite to it." In Masterson and Brower, *Bering's Successors*, Seattle, 1948:93–96.
276. AVPRI, fond 339, RAK, opis 888, no. 42. *Kratkii ekstrakt iz donesenei Billingsa o sekretnoi ekspeditsii 1785–1794 gg.*, folio 20.
277. Today, this chart is a great bibliographic rarity. Several copies are in Russian archives and libraries, see, for example, AVPRI, fond 339, RAK, opis 888, no. 934/14. The manuscript compiler's original is apparently the *Merkatorskaia karta, predstavliaiushchaia severnuiu chast' Vostochnago okeana, bklichennuiu mezhdu severo-vostochnoii storonoiu Azii i severo-zapadnoiu chast'iu Ameriki, s naznacheniem plavaniia po semu moriu v 1789–92 gg., vo vremia byvshei sekretnoi ekspeditsiie pod nachal'stvom flota kapitana Billingsa*. RGAVMF, fond VUA, 23762. An interesting example of Sarychev's printed chart, with handwritten notations in German, in the eighteenth- and early nineteenth-century hand, is in Newberry Library (Chicago), call number Ayer P 133; S 25; 179. The summary chart by Sarychev has been published by Gnucheva, ibid., 1946:265–266, no. 318; *Atlas . . .* 1964:115–116, no. 177.
278. *Atlas . . .* 1964:114, no. 174.
279. Sarychev, *Puteshestvie flota kapitana Sarycheva . . .* 1802:iii.
280. Ibid., 1802:vi.
281. Ibid., 1802:vi.
282. Ibid., 1802:vi–vii.
283. Ibid., Part 1, 1802: 25.
284. Ibid., Part 2, 1802:39–40; 47–48.
285. Ibid., Part 2, 1802:149–151.
286. Ibid., Part 2, 1802:182.
287. Ibid., Part 2, 1802:52–53.
288. Ibid., Part 2, 1802:60.
289. Ibid., Part 2, 1802:x.
290. Ibid., Part 2, 1802:xi.
291. RAN *Pravila prinadlezhashchie k Morskoi Geodezii . . . Contra-Admiral Sarychev.* Petersburg, 1804:iii.
292. Ibid., 1804:13–14.
293. Ibid., 1804:14.
294. Ibid., 1804:23–24.
295. Ibid., 1804:23–24.
296. Ibid., 1804:58–60.
297. Ibid., 1804:62.

298. Ibid., 1804:64–66.
299. Pierre Bouguer, *traite' complet de la navigation* . . . . Paris, 1706.
300. Sarychev, ibid., 1804:88–100.
301. P. S. Pallas, *Zoographia rosso-asiatica*. 3 vol. St. Petersburg. Petropoli, in officina Caes. Academiae scientiarum impress, 1811, edit. 1831.
302. H. Merck, *Siberia and Northwestern America 1788–1792. The Journal of Carl Heinrich Merck, Naturlist with the Russian scientific Expedition led by Captain Joseph Billings and Gavriil Sarychev*. Translated by Fritz Jaensch. Edited with introduction by Richard A. Pierce. Kingston, 1980. Z. D. Titova, *Etnograficheskie materialy Severo-Vostochnoi geograficheskoi ekspeditsii, 1785–1795 gg.* Magadan, 1978. Iu. Bronshtein and N. Shnakenburg, "Zapiski doktora K. Merka uchastnika ekspeditsii Billings-sarycheva v 1785–1792 gg." *Sovetskaia Arktika*, no. 4, 1941:76–88.
303. RGAVMF, fond 179, opis 1, no. 131, folios 126–128. *Russkie ekspeditsii . . . vt. pol. XVIII v.*, 1989:364.
304. Martin Sauer, *An Account of a Geographical and Astronomical Expedition to the Northern Parts of Russia* . . . London, 1802.
305. RGVIA, fond (kollektsiia) 419, no. 240. Published in *Atlas* . . . 1964:117, mo. 180.
306. Three copies of this chart are preserved. Two are in AVPRI, fond 339, RAK, opis 888, no. 944, folios 1 and 2 (the latter in somewhat larger scale with the addition of two baidara parties, the *eastern* and *western* ones in 1797 and one in RGAVMF, fond 1331, opis 4, no. 102.
307. G. I. Shelikhov, *Rossiskago kuptsa Grigoriia Shelikhova stranstvovanie v 1783 godu iz Okhotska po Vostochnomu okeanu k ameikanskim beregam, Izhdiveniem V. S.*, St. Petersburg, 1791. This first edition of the book by the founder of Russian-American Company is a bibliographic rarity. More widely known are the editions of 1792 and 1973 supplemented by materials from earlier voyages. See G. Shelikhov, *Rossiiskago kuptsa Grigoriia Shelikhova prodolzhenie stranstvovaniia po Vostochnomu okeanu k Amrikanskim beregam v 1788 godu*, St. Petersburg, 1792 [and] G. Shelikhov, *Rossiskago kuptsa imenitago ryl'skago grazhdanina Grigoriia Shelikhova pervoe strnstvovanie s 1783 po 1787 god iz Okhotska po Vostochnomu okeanu k Amerikanskim beregam i vozvrashchenie ego v Rossiiu*. St. Petersburg, 1793. The *Chart of Shelikhov's voyage*, taken from the 1793, edition is reproduced in Efimov's *Atlas* . . . 1964:118, no. 182. During the Soviet era, Shelikhov's book was reissued, edited, and with an introductory essay by B. P. Polevoi. See G. Shelikhov, *Rossiskogo kuptsa Grigoriia Shelikhova prodolzhenie stranstvovaniia po Vostochnomu okeanu k amerikanskim beregam v 1788 gg.*, Khabarovsk, 1971. For the English edition, see Shelikhov, A Voyage to America, 1783–1787. Translated by Marina Ramsay and edited by Richard A. Pierce. Kingston, 1981.
308. AVPRI, fond 339, RAK, opis 888, no. 93/11, folios 1, 2, 3 which represent variations of this chart, differing from each other in minor details. RGVIA, fond VUA, no. 23, 461. RGAVMF, fond 1331, opis 4, no. 124. Copy of this chart was reproduced by Efimov in *Atlas* . . . , 1964:119, no. 184.
309. Richard Pierce, ed. *Documents on the History of the Russian-American Company*. Translated by Marina Ramsey. Kingston, 1976:60–61. V. Divin, comp. and ed., *Russkaia Tikhookeanskaia epopeia*, Khabarovsk, 1979:413–416. *K istorii Rossisko-Amerikanskoi Kompanii*, Krasnoyarsk, 1957:52–60.

310. V. M. Golovnin, *Sochineniia i perevody*, vol. 5:127. St. Petersburg, 1864.
311. John Meares, *Voyages made in the years 1788 and 1789 from China to the North West coast of Amrica. To which are prefixed, an introductory narrative of a voyage performed in 1786, from Bengal in the ship* Nootka. . . . London MDCCXC. Wagner, ibid., 1937:210–211. Maps ## 758, 758a, 759; Plans ## 760–766.
312. George Dixon, *A voyage round the World but more particularly to the north west coast of America; performed in 1785, 1786, 1787 and 1788, in "King George" and "Queen Charlotte," captaines Portlock and Dixon*. London, 1789. Wagner, ibid., 1937:207, maps 732–735.
313. Nathaniel Portlock, *A voyage round the world, but more particularly to the north-west coast of America, performed in 1785, 1786, 1787 and 1788, in the King George and Queen Charlotte* . . . London, 1789. Wagner, ibid., 1937:207, maps 738–743.
314. Robert Haswell, "Journal of Robert Gray's voyage in 1788–9 and 1790–2." An Appendix to vol. 1, Bancroft, Hubert Howe, *History of the Northwest Coast*. 2 vols. San Francisco, 1886. Frederick William Howay, ed., *Voyages of the "Columbia" to the Northwest coast 1787–1790 and 1790–1793*. Boston, 1941.
315. Comte Charles Pierre de Claret Fleurieu, *Voyage atour du Monde, pendant les anee's 1790, 1791 et 1792 par E'tienne Marchand*. . . 4 volumes. Paris, An VI–VIII 1798–1800).
316. Wagner, ibid., 1937:212, maps 849, 571, 852, 850–851.
317. Wagner, ibid., 1937:213, maps 745–748.
318. Ibid.
319. For the journal of the voyage, see Alessandro Malaspina, *The Malaspina Expedition 1789–1794: The Journal of the Voyage*, vol. 2. *Panama to the Philippines*. Edited by Andrew David, Felipe Fernandez-Armesto, Carlos Novi, and Glyndwr Williams. Hakluyt Society, Third Series, no. 11. London and Madrid, 2003. The Alaska portion of the voyage was only one component of this massive scientific enterprise and it is covered in this volume. For a narrative of the voyage, see Thomas Vaughan, *Voyages of Enlightenment: Malaspina on the Northwest Coast*, 1791–1792. Portland, 1977.
320. Wagner, ibid., 1937:223, map 772.
321. Wagner, ibid., 1937:225–228, map 800. Archer, ibid., 1980:139–141.
322. Wagner, ibid., 1937:239–245.
323. George Vancouver, *A Voyage of "Discovery" to the North pacific Ocean, and round the World, in which the coast of North-West America has been carefully examined and accurately surveyed*. Vol. III, London 1798:83–89.
324. Fedorova, ibid., 1973:123.
325. Vancouver, ibid., 1798:122–123.
326. Fedorova, ibid., 1973:123.
327. Vancouver, ibid., 1798:125–145.
328. Ibid., 1798:175.
329. Ibid., 1798:175.
330. Ibid., 1798:200–201.
331. Ibid., 1798:225–231.
332. Wagner, ibid., 1937:245–249.
333. Wagner, ibid., 1937:250–254.
334. RGVIA, fond VUA no. 23784.

CHAPTER IV

The Exploration and Cartography of Russian America (1799–1867)

THE FORMATION OF THE RUSSIAN-AMERICAN COMPANY

Shelikhov and Golikov lobbied for a monopoly to exploit the fur riches of the new colonies, but the imperial government was not yet ready to approve it. First, there was Russia's international situation, specifically the beginning of wars against Sweden (Russo-Swedish War, 1788–1790) and Turkey (Russo-Turkish War, 1787–1792). Second, the Empress Catherine was more inclined to give maximum support to the military operations in the south, initiated by her all-powerful favorite Potemkin, than to spend money to develop overseas colonies. And third, she disliked A. R. Vorontsov of the Commission on Commerce, who was supportive of a monopoly.

It is also possible that Catherine, considering the recent independence of the British North American colonies, feared the possibility of a Russian version of the same. Moreover, waging wars on two fronts, the empress did not wish for any new military or political complications. Catherine instead awarded silver swords and gold medals with her portrait on them to Shelikhov and Golikov, and they continued on without a monopoly.

In view of the empress's hostility toward the idea of a monopoly, Shelikhov began to organize new "independent" companies. In 1790 he organized the Northeastern American Company and the Predtechenskaya Company. Somewhat later Shelikhov created the Unalaska Company on the island of the same name where he decided to establish a permanent settlement.

Also in 1790 Shelikhov made one of his most fortuitous decisions. He sent Aleksandr A. Baranov (Fig. 38) to Kodiak as the new manager of the Northeastern American Company. A merchant from Kargopol, Baranov remained at his post for twenty-eight years and decidedly influenced the entire subsequent course of Russian America's history.[1]

Following Shelikhov's sudden death in the summer of 1795, his wife Natalia Shelikhova continued the drive to establish strong Russian colonies in North America, supported by her son-in-law, the very influential courtier, Nikolai P. Rezanov (1764–1807).[2] The internecine rivalry between the various fur-procuring companies increased considerably after Shelikhov's death, sometimes leading to armed conflict.

FIGURE 38 Portrait of Alexander Baranov. Lithograph based on a drawing by M. Tikhanov, 1818. GIM

Nevertheless, there was an increasing need to unify because of foreign competition and the urge to secure the imperial government's protection and support.

In the summer of 1797 the Golikov-Shelikhov Company, the Northeastern, the Northern, and the Kuril companies, united with a company of Irkutsk merchants headed by N. P. Mylnikov to form the United American Company. This merger did not extinguish the existing conflicts of interest, it merely masked them. The sharp internal dissent within the United American Company was expressed through frequent mutual complaints and accusations sent to Irkutsk and St. Petersburg.

In the meantime, Catherine II died in 1796 and her son Paul I ascended the Russian throne. Paul strived to reverse the programs of his powerful mother in all aspects of foreign and domestic policy. Consequently, the new emperor was open to arguments for a monopoly in Russia's American possessions. The government desired to "establish order out of the chaos brought about by merchant rivalry in Irkutsk following the death of Shelikhov."[3]

On July 9, 1799, Emperor Paul I issued a royal order to the ruling Senate regarding the establishment of the Russian-American Company, "under Our Most High Protection."[4] On the basis of the proposals from the Commerce Collegium submitted in August 1797 and January 1799, Paul confirmed the final version of "the rules and privileges" of the company "for twenty years."[5] The geographic boundaries of the territory were defined in the first paragraph of this act as follows:

> On the basis of the discovery since ancient times of the northeastern part of America beginning with 55° North latitude and of the chains of islands that stretch from Kamchatka northward toward America and southward toward Japan and of the right by occupation of the same by Russia, the Company may use all the industries and establishments currently situated on the northeastern coast of America from the above-mentioned 55° to the Bering Strait and beyond, and also in the Aleutian, Kuril and other islands lying in the northeastern ocean.

The company also had the right to explore and annex to its possessions lands to the south of 55° N latitude, "provided the same were not occupied by any other nation and were not subjects of the latter."

Representatives of other companies active at that time in Russian America continued their "litigation and slanders" for a while, attempting to protect their commercial interests, but their fight against Shelikhov's heirs was uneven in view of the latter's connections at the court. By royal order, the main office of the Russian-American Company was transferred to St. Petersburg.

The close ties of the company to the government were strengthened in the spring of 1802 when members of the imperial family—Alexander I (who had become emperor in 1801 after Paul was murdered), the Dowager Empress Maria

Fyodorovna, and others—became shareholders [Income from the shares was designated for charity—LB].

Thus, from its very founding, the Russian-American Company had a quasi-governmental character and was charged with the task of securing and expanding Russia's colonial possessions in the Western Hemisphere. In support of this, a role was assigned to the Russian Orthodox Church. Expanding upon earlier lay efforts, the activities of its missionaries left persistent traces, preserved to this day in the culture and language of the local peoples. The work of Russian missionaries and priests in the study of languages, customs, beliefs, and material cultures of the Natives has been well researched by later ethnographers. However, their geographic observations have not received due attention. For this reason, in the next section we shall leave the topic of the history of the cartography of Russian America for a time, in order to assess the contribution of the Russian Orthodox Church to the study of the environmental and social geography of Russia's American colonies.

THE GEOGRAPHIC INVESTIGATIONS OF RUSSIAN ORTHODOX CHURCH MISSIONARIES

Lay missionary activities in the Aleutian Islands were first undertaken by the *promyshlenniki*. In 1759 foreman and skipper Stepan Glotov baptized several Aleuts. He brought one of these, the teenager Mushkal, to Kamchatka. Later, Mushkal, known by his Christian name Ivan Glotov, in his capacity as interpreter and then as *toion* (headman) of Umnak Island, spread Christianity by his example. Shelikhov devoted much effort to conversion of the Natives to Orthodoxy. In 1784 he founded the first school in Kodiak. A similar school for Creoles and Natives existed in the late eighteenth to early nineteenth centuries in Unalaska. In both, instruction was in the subjects of "Russian Language Arts, God's Law, Arithmetic, and Navigation."[6]

Navigation was taught in these schools by the company's navigators, for which they received "a reward of special compensation." Other subjects were taught by "those of the *promyshlenniki*, who have some skills in the area and they, too, receive special recompense." An institution for girls where handicrafts and domestic skills were taught was opened. Shelikhov noted the aptitude of the Natives and Creoles in music. In consideration of the need to provide choral singing in the church services he took twelve Creoles to Irkutsk in 1787, specifically for musical training. "Half of them perished due to the change in climate, but the rest returned as fine musicians."[7]

The first official Orthodox spiritual mission was dispatched to the American colonies in 1792. It consisted of eight monks (six from the Valaam Monastery and two from the Konev Monastery). The head of mission, Archimandrite Iosaf—born Iosaf Bolotov (1761–1799)—was educated at Tver with additional education at Rostov seminary. He was called back to Irkutsk to be consecrated Bishop of Kodiak in April 1799 but perished in the wreck of the *Phoenix* while returning to American shores along with two monks from the Konev Monastery (Hieromonk Makarii and Hierodeacon Stefan).

The fate of other members of the mission varied. Hieromonk Juvenalii, who had been an artillery officer prior to being tonsured, boldly went to preach the faith of Christ to the islands and American mainland. He was killed by the Natives at Queenhagak on the Bering Sea in 1796. Hierodeacon Nektarii left the American colonies for Irkutsk in 1806. He died in 1814 in the Kirensky Monastery. Monk Iosaf died in Kodiak in 1823, while Hieromonk Afanasy served there as a priest until 1825, when he safely returned to Russia. Monk Herman lived in Russian America for the longest time of all. He died in 1836 on Spruce Island near Kodiak. With his death the Kodiak Mission ceased to exist. Besides those permanent members of the mission, Cathedral Hieromonk Gideon of the Alexander Nevsky Monastery in St. Petersburg was in Kodiak from 1804 to 1807.[8]

Archimandrite Iosaf

The Holy Synod of the Russian Orthodox Church issued guidelines to Archimandrite Iosaf for describing the lands where the missionaries would be active. Iosaf was intelligent and, for his time, well educated. He was the first to conduct a serious study of the Kodiak Americans and of the natural environment of Kodiak Island and adjacent territories. In 1797 he brought his description of the colonies to Irkutsk.

The results of Iosaf's observations were published in the journal *Drug prosvieshcheniya* (*Friend of Enlightenment*) in 1805 under the title *Topograficheskoe, esteticheskoe i moralnoe opisanie ostrova Kadiaka* (A topographic, aesthetic and moral description of Kodiak Island), also published as a separate brochure.[9] These publications apparently were arranged by Count Rumiantsev, who had a lively interest in Russia's American colonies and who had received Iosaf's text from Amvrosii, the metropolitan of Novgorod, in 1803.[10] Iosaf's published materials are widely used by ethnographers and historians as a valuable source for the study of the material and spiritual culture of the Native people of Kodiak and adjacent islands.

We shall consider this description, compiled by the head of the Russian Orthodox Mission, in greater detail in the context of our own study. We shall use the complete text sent by Iosaf to the synod that is now in the archive of foreign policy of the Russian empire in Moscow.

The guidelines for the description, composed by someone at the Holy Synod and issued to Iosaf, are themselves of interest. They continue the Russian tradition of using the formal questionnaires used in the geographic explorations of the eighteenth century. The questionnaire was worked out in minute detail and makes it clear that the author was acquainted with previous exploration questionnaires of the St. Petersburg Academy of Sciences, the Noble Cadet Corps, and other institutions (that is, the questionnaires by V. N. Tatishchev, G. F. Müller, M. V. Lomonosov, etc.).[11] It is also obvious that the compiler of this questionnaire knew the history and geographic specifics of the Russian colonies in America. The questionnaire is divided into four main parts:

1. Topographic (physical geography)
2. Climatic
3. Statistical (social, geographical, and ethnographic)
4. Aesthetic (beliefs, attitude toward Christian gospel, aesthetic and intellectual capabilities of the Natives)

The questionnaire opens with six questions about the geographic location of the islands, including latitude and longitude, distance from the mainland, population figures for the local and Russian populations, and the number of settlements under their control.[12]

The next five questions elicit a general description of the islands near Kodiak and of the coast of America. The guidelines for the topographic description end with questions about the flora ("is there on Kodiak as well as on other islands, timber, fruits and other plants, what Russian trees and fruits grow there and that are adapted to the local climate and which predominates . . . also, is there any timber specifically suited for shipbuilding?"). Rivers and lakes were to be described indicating "their extent and depth, how their water tastes, are there fish, what kind, and are they suited for food."[13]

In the second (climatic) section of the description the questions deal with information about "when spring, summer, autumn and winter commence, how long the seasons last and when they end, what winds prevail and are there any that are harmful to the people or livestock?" Information was requested about fauna, useful minerals, and environmental conditions that would permit the development of agriculture. There are questions of a medical and geographical nature ("which sicknesses harmful to humans or to livestock occur most frequently? What are their causes and are they the same everywhere?").

The section with the most detail deals with the social description of the Russian colonies. This section provided the future basis for all subsequent programs for the study and description of the Natives in the Russian colonies, their social organization, material and spiritual cultures, and their morals and customs. In essence, this is one of the earliest ethnographic questionnaires for the study of Alaska Natives. In this respect, it is of considerable historical and scientific interest.

Much attention was paid to Native social order and governance. For this reason the questionnaire required that information be provided on dispute settlement, the role of elders, and how the population is governed. How do the Natives treat crimes and punishments?

Detailed information was to be provided about Native clothing, dwellings, and food. Any changes that came about after the Russians arrived were to be noted.

In particular, all items used as food were to be listed, with special attention paid to edible animal and fish species. In regard to living conditions, information was sought about "in what locations the inhabitants concentrate their settlements, how many families live in each settlement, what are the distances between settlements, do they have houses and of what materials are they made, what is their appearance and size, interior arrangements, do they have any decorations, and how many families live in a dwelling."[14] There were questions about family life, rites, festivals, and games.

Considerable space in the next section is devoted to fishing and hunting, trade, and vessels used locally. The tools and methods of catching mammals, fish, and birds, the use of furs and skins, and the nature of barter and trade were to be described. What constitutes the inhabitants' wealth? What kind of personal and home adornments do they use, do they use money and, if so, how is it counted?

Regional differences among the Natives should be documented. Do "the inhabitants of all the islands have the same lifestyle, facial appearance, stature, language, mores, customs, food, clothing, handicrafts, and hunting and fishing, or are they different on each island, one submitting to the other or despising some group, and if so, why?"

The topic of war was addressed. Were there firearms among the Natives and where did they come from? Were there any fortified structures? What about wars among the Natives? Why do they go to war and how is war conducted? What is the command structure? What armaments are involved, what kind of battles do they engage in, and do they engage in sea battles? Do they arm themselves, how do they proceed, do their wives and children accompany them, or are they left behind; what kind of battles do they engage in, are there battles at sea, and what kind of vessels are used in such? How do they treat prisoners of war?[15]

Naturally, the compilers of the synod questionnaire were interested not only in the secular matters discussed above, but also much concerned with problems directly concerning missionary activities. Members of the mission were to report if "there are many Americans now subject to Russia and converted to the Christian faith, and estimate how many remain unconverted." They were also to report "if there are in any islands missionaries from other states, or do any such come aboard the ships or the frigates (naval vessels), and to whom are the islanders attracted most."

The last section of the questionnaire (aesthetic) is devoted in its entirety to the spiritual life of the Natives, their faiths, their attitude toward conversion to the faith of Christ, the needs of Orthodox education, and the preparation of local inhabitants as clergy. First of all, it was necessary to establish "if the inhabitants, besides those who have been converted to the Christian faith, have any concept of God, how do they call the Deity, do they have any divine services or do they venerate idols? What kind of idols? What are these called? What kind of offerings are made and with what intent? Are there any sects among them? Do they have a concept of virtue and vice, and what is the substance of such concepts? What are their thoughts on future life and how is such conceptualized? Are they hospitable and kind, are they humane? In their mores and actions, what is considered a vice? Do they willingly accept Christian faith and virtues? Are they constant in these and are they reverent in respect to divine service? Is there any resentment on the part of the unconverted against Christians, and if so, what are the reasons for it?"

Naturally, the leadership of the Russian Orthodox Church was interested in the spread of their religion in the colonies and the outlook for missionary activity. For this reason, the concluding section of the questionnaire contains questions about the number of churches and the areas that were reached by their influence, questions about the need for icons, books, candle wax, incense, and so on. This document provides clear evidence that the Holy Synod actively supported the widespread practice of preparing Natives to be clergy. For this reason, several questions deal with Natives' intellectual endowment and if "it is possible to instill in them the necessary enlightenment" for engaging them in future missionary activity.

The Russian Orthodox Church became a constant and influential advocate for the spread of education among the Native population. Another humanitarian aspect of the Christian Mission's activity—defending the Natives against the arbitrariness of the *promyshlenniki*—was shown in questions about humane treatment of the local inhabitants by the Russians: "Do they burden them either by labor or taxation for the *iasak* . . . do they aim for enlightenment of the Natives or for their own greed?"[16]

The head of the Russian Orthodox Mission on Kodiak Island, Archimandrite Iosaf, not only answered the questions fully, but approached his task creatively, striving to include all the information he obtained in the colonies that he deemed reliable on the basis of his own experience. He provided a detailed description of individual islands and the part of the mainland known to the Russian *promyshlenniki*. He limited the geographic boundaries of the description to the territory actually controlled by the Russian-American Company.

By Iosaf's calculation, there were no more than 225 Russian *promyshlenniki*, including settlers. The Natives numbered "in Kodiak alone over 8,000 of both sexes," but he did not attempt to estimate the number of (Alaska) Natives as a whole "because the further one proceeds toward Nootka or King George Sound, the more populous is the area; and on the north side of Alaska, though there are many inhabitants, only those who are close to the Alaska Peninsula are acquainted with the Russians." In Iosaf's opinion, those who "accepted the holy baptism and subjected themselves to the Russian scepter, number with the Unalashka Aleuts of both genders, up to twelve thousand."[17]

Iosaf describes the center of the Russian-American colonies on Kodiak as follows:

> In the main Russian establishment in St. Paul Harbor, on the rocky shore a wooden fort has been built in a semicircular form. Within its walls are large

> barracks for the *promyshlenniki* and quarters for the office, school, and navigators, a tackle shop, hostage quarters, stores with cellars and various craft shops such as a copper smithy, a metal shop, a blacksmith shop and a shed. On a bluff separated from the fortification by a creek stands a wooden church with a belfry. On the other side of the fortification on the open shore is the manager's house and quarters for the Spiritual Mission; further behind the fortification are guest houses and a hospital. Along the shore, going from the church, are private houses of various *promyshlenniki* who are family men. Also on Kodiak, in two places, are barracks for *promyshlenniki* serving in the *artels* (fur hunting camps).[18]

The archimandrite composed a description of the environment, taking into account the *promyshlenniki* need for natural resources and attempting to understand the Natives' traditions of utilizing the environment. Iosaf characterizes Kodiak's climate succinctly and correctly: "Almost through the entire year there is no real warmth, which is usual in localities that lie in the same degrees of latitude as Kodiak. Nor is there frost. Mostly the atmosphere is filled with thick sea fog. No winds harmful to humans or animals have been recorded. It is known, however, that when the northern and western winds blow, the weather is always clear, otherwise, when the winds are from NE to SW it rains in all seasons."

The general characterization of the climate may be considered to be one of the earliest phenological observations made in the Russian-American colonies. For example, Iosaf describes seasonal change in Kodiak and adjacent lands: "It is not possible to state precisely when spring begins, as throughout the year snow is seldom seen at the foot of the mountains, while in the mountains it also does not disappear through the entire year. It is possible to judge by the appearance of the greenery, which usually appears in force by the middle of May, but in places warmed by the sun one often finds it in January. New snow appears in the mountains in the first days of September, but at the mountain's foot snow almost never appears until November, and even then the snowfall is only when north and west winds are blowing; when winds are from other directions, the snow is always washed away by rain."

He describes the animal kingdom, now the wealth of the new Russian colonies, thus: "The forest animals in Kodiak are bears, red foxes, cross foxes and silver foxes, otters and ermines; birds are forest eagles, two kinds of crow, large and small, magpies, other birds of prey and several species of small birds similar to European ones. On the American mainland, on Alaska (Peninsula), Kenai and in the Chugach (Prince William Sound) there are brown and black bears, wolves, foxes of different varieties, wolverines, lynx, caribou, porcupine, wild sheep, ground squirrels, tree squirrels, porcupine or American hedgehogs, otters, river beavers, mink and marten; the birds are the same as in Kodiak."

Iosaf testifies that on Kodiak Island iron ore was found in many places and that the inhabitants "used it in burned form to paint their faces." Russian experiments in making iron were unsuccessful: "iron was obtained in a small quantity but it turned out to be brittle." On the possibility of obtaining copper, coal, and sulfur: "On the American mainland, beyond Cape St. Elias on the Copper River, copper is found in chunks. In this raw form it is easily worked, but when raw copper is smelted it becomes brittle. Where the veins are and whence the inhabitants obtain such chunks of copper, time will tell. On the American mainland, in the Kenai Gulf, there are sufficient quantities of anthracite coal, and near the volcanoes, burning sulfur, pure and transparent like amber, is collected. Amber is also washed up by the sea at the western end of Kodiak. The inhabitants used it as adornment, hanging it in their ears. They deem it to be a precious stone and do not believe that it is resin."[19]

Being an Orthodox missionary, Iosaf quite naturally paid a great deal of attention to the description of Natives within the Russian colonies. He did not solve the question of the origin of local tribes, assuming that "Alaska Peninsula inhabitants, Kenai people, and the Chugach derive their names from the names of their places of settlement."

He reports that:

> Though the Kodiak inhabitants believe that all peoples surrounding them are their descendants, the similarity of their language with that of the Alaskans and those who are farther away, up to across from the Chukchi Peninsula and between the latter and the Chukchi (Siberian Yup'ik) suggests that their origin is in Asia, across the Chukchi Peninsula. The Kenai Nation apparently came here from the mainland interior, from America, as their speech is closer to that of the inhabitants of the interior along the Mednaya (Copper) River; their lifestyle and rituals also are similar . . . but the Kenaitsy, (Dena'ina), Mednovtsy (Ahtna) and the Koliuzhi (Tlingit) together with the Ugalentsy (Eyak) have no similarities in their tongues.

Discussing the Natives' social order Iosaf notes that:

> The advice of elders and experienced men, but most of all of men who gained fame in wars, does sway the hearts of the people, but they have no concept of subordination. However, their *toions*, though they have the title of a superior, due only to the fact that they are always family men, became famous in past wars, and as a consequence have many slaves or prisoners. The latter are treated arbitrarily, they are free to give the slaves away as a gift to anyone they choose, or to sell them, or bequeath them to their heirs. But they are not entitled to rule over their neighbors as they wish, nor to punish crimes.[20]

In contrast to earlier Russian descriptions that allege a total absence of religious notions among the Natives, Iosaf maintains that they:

> Acknowledge a Supreme Being whom they call Ayugan, or a spirit of good, who teaches people to make baidarkas, assists them in hunting and in all other human needs; they make no sacrificial offerings to him because they know that he will never work evil. The evil spirit they call iyatk. This one always attempts to do harm to humans. For this reason, to pacify it, they dance dressed in strange costumes, some even shamanize, but they have no idols whatsoever. Everything that is dictated by custom they deem to be a virtue, what is contrary to custom they deem to be a vice. They acknowledge a future life, but offerings are based on what they sense here on earth. They believe that in the future life good people will have abundant hunting, while bad people will be beggars and have eye trouble. For this reason, before the enlightenment, they put into the graves of the best people all their weapons and hunting gear; moreover, a prisoner who was a favorite in life was killed and buried with the man to serve him in the future life.[21]

In the description much space is given to the Natives' everyday life, hunting and fishing, clothing, and food. The local inhabitants' remarkable adaptation to their environment is stressed, as is their natural aptitude for seafaring. In Iosaf's opinion, "they possess aptitudes for language arts as well as fine arts, but those Americans educated by the Company and taught the language and grammar prior to the Mission's arrival are more inclined to take up positions associated with navigation and therefore are entrusted with important positions, according to each man's aptitude."[22]

The archimandrite notes the traditional features of material culture and lifestyle and also changes resulting from the arrival of the Russians. Formerly beads and the possession of slaves were the signs of wealth, but "nowadays they prefer European dress, shawls, and iron tools." Another European introduction was venereal disease which rapidly spread among the Natives. Iosaf explains this by the "lack of tidiness" among the Natives. Local illnesses ("fits, chest pains and an unusual kind of large boil of which not only the Native inhabitants but also Russians often die, and eye diseases") are linked in the description to the environment of the Native lifestyle. Iosaf singles out the cause of eye diseases as "the reflection of sun's rays from the mountains, and the soot from the burning oil used for light."

Ioasaf reports an intriguing Native method of curing eye disease. He testifies that the Natives "rather successfully remove the film from the eye by an operation: a live louse, of which there is abundance in hair and clothing, is tethered on a hair. The louse is allowed to crawl on the film. When it is noticeable that the louse has made itself truly fast, it is yanked away. Through several repetitions, the film is removed."[23]

The archimandrite ends his description by presenting a set of recommendations for improving colonial management and for the development of missionary activity. He notes that:

> all (Natives) accept the holy baptism, but it is remarkable that the old ones, firm in their error, do so only to keep up with the young, considering it a dishonor not to be baptized. Reverence toward the sacred services and constancy in Christian virtues is noticed more among the younger ones, especially among those who frequently interact with the Russians.

> He concludes that:

> The ability of Russians, scattered through such great expanse, among such numerous peoples, to keep Natives in the expected submission depends not on force but solely on good order established within the Company and indulgences, kindnesses and various political means that restrain their urges toward barbarity and internecine fighting, the constant wars to which they were much given before the Russians. Nobody can lord it over them despotically. Feeling oppression, they usually decide to overthrow and kill the oppressor. If that is impossible to do, they go into the land's interior, gather strength, and then make frequent attacks.

Archimandrite Iosaf understood well that "without assistance from laborers who are Natives or hostages, the Company is not able to function in Kodiak or on the American shore. The hostages are taken to ensure fidelity. They learn Russian customs and restructure their wild and natural manners for future benefit. Through them their fathers and kinsmen are moved to visit our Russian harbors and settled points frequently and thus mutual links are firmed up more and more."

He supported the long-established practice of hostage-taking, specifically the desirability of attracting the offspring of the Natives and "prisoners ransomed by the Russians from their conquerors" to company work, to attempt to convert them to the Christian faith and to give them a basic education. Thus the supremacy of the Russians and their expansion would be guaranteed by two main, self-evident factors. The first would be the fact that the Natives "love their children fervently, but neither father nor mother may chastise them except verbally."

In this connection it becomes obvious that the Native parents of children taken as hostages would be concerned for their children's safety and would be loyal to the Russians. The loyalty of prisoners ransomed by the Russians was ensured because of their gratitude for their liberation from an uncertain and dangerous future under threat of enemy tribes. Secondly, the use of such prisoners aided the expansion of the colonial boundaries. Iosaf notes equality of rights between ordinary Russian laborers and the Natives, including hostages who were employed in hunting and other enterprises. In particular, he notes that "all these, the hostages as well as the Native employees, are supported by the Company with food and are issued clothing and footwear. The quick, efficient, and diligent among them are often placed in the same positions as Russian *promyshlennye*, and those who travel alongside the Russians receive the same benefits. Those hostages who have been noted for their intelligence are accepted into the school in Kodiak, but they are not forced to attend."[24] Archimandrite Iosaf urged that such practices be continued in the future.

He drew a map of the *Northeastern American Shore Occupied by the Russians and a Plan of the Main Establishment in Kodiak* that is absent from the archival file.

Hieromonk Gedeon

Following the death of Archimdrite Iosaf, another Orthodox observer arrived in Kodiak: a hieromonk of Alexander Nevsky Monastery named Gedeon, who also wrote a description of the Russian-American possessions.

Amvrosii, Metropolitan of Novgorod [and St. Petersburg—LB] entrusted Gedeon with the inspection of the clergy of Kamchatka and Russian America. He was to continue missionary activity and in particular, he was entrusted with the directorship of the school in Kodiak. Gedeon arrived aboard the frigate *Neva* in June 1804 as a passenger on the first Russian circumnavigation.[25]

Although Gideon's instructions specified that he was "under no circumstances to enter into civil matters," the curious missionary, like his predecessor Iosaf, compiled a description of his voyage and of Russia's American colonies that far exceeded the need to propagate the Christian faith. Following Iosaf's format, Hieromonk Gedeon greatly expanded and enriched the data of his predecessor, especially in characterizing the material and spiritual culture, lifestyle, and mores of local inhabitants. The Natives' seasonal festivities (games), rituals, and customs linked to marriage, childbirth, and funerals, methods of hunting and fishing, and most of all of warfare are described in this work.

Hieromonk Gedeon managed to collect detailed information on the cosmographic notions of Kodiak inhabitants, as well as their past history and migrations, based on oral tradition. The following is about the origin of the Kodiak Islanders:

> The Kadi'iak folk, according to the legends recounted by the old people, moved to Kad'iak from Aliaksa (the Alaska Peninsula). Their ancestors formerly lived on the north side of Aliaksa near a great river, the Kvignak (unidentified). They had an anayugak (leader, angayuqaq) named Atlivatu. He had an only, well beloved daughter who disappeared without a trace. He set out with his command to search for the girl, and also persuaded another angayuqaq, Iakunak, to join him. They wandered over many places for a long time. Nearing the southern side of Aliaksa, they sighted land and called it Kigikhtak (Qikertaq), which means in their language 'an island'. This name was applied to Kokiak until the arrival of the Russians.
>
> Later on, Atlivatu and Iakunak became curious, investigated, and the advantages persuaded them with their families to resettle on Qikertaq.

The affinity of the languages of Kad'iak and Aliaksa inhabitants lends support to the story.[26]

The statement quoted here testifies to the considerable geographic knowledge of Kodiak Islanders. Moreover, it contains possibly one of the earliest testimonies about the Yukon River, "the great river Kvignat," that later appeared in Russian sources under the name of Kvikpak or Kvikhpaq. [Kvikhpak is the Yupik Eskimo name for the Yukon; the latter derives from the Athabascan Yukana.—LB]

Gedeon also recorded that Kodiak Islanders had:

> the following notions about the creation of the world: there was once upon a time a Kashshiakhliuk, a sage, a trickster. At that time there was neither day nor night. He began to blow through a straw, causing land to emerge gradually from the water and imperceptibly to spread. Then, while he continued to blow, the sky opened, the sun appeared; in the evening the stars came out and the moon rose. Finally, they saw animals and people.

It appears to us that such a rich description of the world's creation is closely linked to the subsistence activities of the Kodiak Natives—the seafarers and sea mammals for whom the view of gradually increasing land rising from beyond the horizon out of the sea is practically an everyday occurrence. As Gedeon states, the experience of sea voyages led the Kodiak Islanders to the conclusion that the earth was round. In this connection he notes: "They have concluded that the earth is round on the basis of the following occurrence: their ancestors sent out two baidarkas manned by young men, who returned as old men but failed to reach the end of the earth. Moreover, in their opinion the Kuzhikagmiut inhabit the last place in the world, beyond them nobody lives. By this they mean the farthest inhabitants of the Fox Aleutians."[27]

Besides such general observations, Gedeon notes that the Natives of Kodiak had knowledge, transmitted from generation to generation, about local sailing conditions in Kodiak and Alaska Peninsula waters and particularly the ability to forecast the weather according to signs, so that "A diligent hunter goes outside frequently during the night to observe the clouds, not the weather, and, leaves to hunt on the basis of his observations."[28]

Gedeon expands upon Iosaf's information about Native methods of curing various illnesses, including names of healing herbs, the locations where they grow, and how they are used. He notes that one of the universal remedies against all diseases was fasting. "Their rules for maintaining their health are as follows: those who wish to live in good health and for a long time must, first, never eat to satiety; second, they must never oversleep the dawn; third, they must abstain from intercourse with women; and, lastly, they should almost always be in motion."

These health maintenance rules were supplemented by exceptional care about the spiritual peace of the growing generation. "The education of children consists of preventing the child from crying, not just the infants but even seven-year-olds. From this one may conclude that mothers and fathers love their children dearly. They never inflict any kind of corporal punishment, neither on minors nor adults, resorting only to admonition. If it is noticed that there is no improvement, they scold and shame (the culprit) before the assembly of old men. If this last method does not prove effective, the person, an object of contempt, is deemed to be stupid or savage."[29]

Such a healthy atmosphere within the family was guaranteed by the fact that "Husbands treat their wives with great affection. In the presence of others, they may not speak harshly to each other nor show a stern countenance; were one of them to do so, the other would consider this an unbearable insult."[30]

Father Gedeon's activity in Russian America left a noticeable effect. Baranov states that Gedeon was "with the Kodiak school as its chief director from the month of March of the year '805, where he demonstrated his abilities, diligence and effort in its organization as well as in education of the local youth; he excelled in his morality and kind disposition towards and treatment of all persons, showing goodwill in his behavior, refraining from every instance of conflict and discord; rather he reconciled various feuds among others throughout his stay on Kodiak."[31]

The first festive graduation examination of the Kodiak School took place in the presence of Rezanov in 1805 and the second in 1807, in Baranov's presence. Baranov reported later that "from among the fosterlings of the Russian-American school some have already taken teaching positions." And, on Rezanov's order, the compilation of a dictionary and grammar of the language of Kodiak Natives[32] had begun "under the direction of the interpreter Paramon Chumovitsky." Hieromonk Gedeon brought Prokopii Lavrov, who had started his education in the colonies, with him on the return to Russia. Lavrov was consecrated and served in Siberia.[33]

Gideon's successor as head of the Russian Orthodox Mission, Monk Herman, later named Patron Saint of Alaska, was also well educated. The memory of this hermit, a devotee who spent the last twenty years of his life alone on Spruce Island, is revered to this day by Orthodox inhabitants of Alaska and the Aleutian Islands.[34]

Ioann (Ivan) Veniaminov

The scientific and educational activity of Russian missionaries reached its apogee in the creative work of Father Ioann Veniaminov (Ivan Evseevich Popov), who later became Innocent, Metropolitan of Moscow (1797–1879).

Ivan Evseevich Popov was born in August 1797 in the village of Anginskoe, on the left bank of the Angara River in Irkutsk Guberniya. At the age of nine he entered the Irkutsk Seminary, where he spent eleven years of his life. Besides devoting himself most diligently to his studies and to active self-education, he learned many crafts from his uncle, Hieromonk David. He received the new surname Veniaminov at his consecration as priest.

In 1817, before graduation, Veniaminov married Ekaterina Ivanovna Sharina, the daughter of a priest. After graduating from the seminary in 1820 and being consecrated as a priest in May 1821, Veniaminov served two years in Siberia, and in May 1823, in accordance with the decision of the church to open an Orthodox parish at Unalaska, he left Irkutsk with his family for Russian America. The young priest's decision to accept the post was influenced by colorful tales of the *promyshlennik* Kryukov, who spent forty years of his life in the Aleutian Islands.

In September of the same year, Veniaminov, his elderly mother, his wife Ekaterina, a small son and Veniaminov's younger brother Stepan arrived on Baranov Island (Sitka), where they spent the winter. This enabled Veniaminov, with the help of a company employee, to study the Aleut language and also to use the rich library that had been left fifteen years earlier by Rezanov. Veniaminov arrived at Unalaska on July 28, 1824, where he undertook the building of a church and the arrangement of his own house. Unalaska District Manager Rodion Yakovlevich Petrovsky assigned Native laborers to him, to whom he rather quickly taught various building skills. Veniaminov himself took an active part in construction and in particular made furniture for his house.

Veniaminov opened an elementary school for boys on March 12, 1825, and the new Unalaska Island church was consecrated on June 29, 1826. Besides holding church services and his travels throughout the islands by baidarka and baidara, the Unalaska Island priest engaged in linguistic, geographic, and ethnographic studies. He created the Aleut language alphabet, and translated the Russian Short Catechism and Gospel of St. Matthew into the Aleut language.

Veniaminov kept a daily record of weather for seven of his ten years in Unalaska, the earliest for Alaska. Visiting remote corners of his parish, traveling by baidarka, he learned about the regional peculiarities of the environment and local populations, accurately recording it all in his diaries.[35] In 1823 the church of St. Nicholas was ordered to be built on Atka, and in 1828 the Creole Yakov Netsvetov was assigned to this church as priest [to the parish of the Western or Atka District of the Russian-American Company—LB]. In 1827 Veniaminov paid his first pastoral visit to the Pribilof Islands. In 1828 he traveled in a three-hatch kayak to Akun Island in the Krenitzin Island group. In 1829 and 1832 he visited the new post of the Russian-American Company that had been established in Nushagak on the American mainland.

In 1834 Ferdinand Petrovich von Wrangell, who admired Veniaminov as a scholar as well as a priest, insisted on his transfer to the capital of the colony, Novoarkhangelsk (Sitka) on Baranov Island. Veniaminov arrived there in November of the same year. Veniaminov created a workshop in Sitka, where he busied himself with furniture construction and repair. He also made and repaired clocks, and even built an organ. Though the inhabitants of Sitka and adjacent American coasts, the Kolosh (Tlingit) rather strongly resisted baptism, Veniaminov began to study their language.

In 1836 Veniaminov visited Fort Ross, the Russian colony in California, and also the Spanish mission of San Rafael and the small settlement of San Francisco.

In 1838 Veniaminov sailed aboard the frigate *Nicholas I*, arriving in St. Petersburg in 1839. There he appealed to the government that his daughters, Fekla, Olga, and Paraskeva, be admitted to a special royal educational institution for girls from noble but not affluent families. An agreement was reached and the girls were maintained there at the imperial family's expense.

In 1840 the St. Petersburg Academy of Sciences published Veniaminov's *Notes on the Islands of the Unalashka District*,[36] at the Russian-American Company's expense, and his grammar of the Aleut language was published in 1846.

While he was in St. Petersburg, Veniaminov learned that his wife had died in Irkutsk on November 25, 1839. In 1840 Veniaminov accepted the monastic tonsure and received the monastic rank of archimandrite with the name of Innokenty (Innocent). Nicholas I granted him an audience. On December 15, 1840, Veniaminov was consecrated Bishop of the new diocese of Kamchatka, the Kuril, and Aleutian Islands. Nothing kept him in Russia anymore, as his two sons, Gavril and Innokenty, had begun their seminary education in the capital.

On September 26, 1841, Veniaminov once again arrived in Russian America, where, as in the past, he undertook strong missionary and educational activities as well as travel. In 1842 he visited Kodiak, Unga, Unalaska, and Atka.

On August 18, 1842, Veniaminov arrived in Kamchatka. There, in the winter, he accomplished an amazing trip of 5,000 versts across the peninsula, traveling by horse, reindeer, and dog team, reaching Okhotsk in April 1843. From there, in the fall of the same year, he again set out for Russian America. He visited Kamchatka more than once before 1850, completed yet another overland journey to Ayan (1847), and sailed aboard the brig *Great Prince Konstantin* to the Pribilof Islands, Unalaska, St. Michael Redoubt, to Michigmen Bay on the Asiatic coast, and to Kodiak.

He received the rank of archbishop with the see at Ayan in 1850. Veniaminov traveled for almost ten years, beginning in 1852, through Siberia, Chukotka, Kamchatka, and Amur. In 1858 he accompanied Governor-General Muraviev on a voyage along the Amur River for negotiations with China that resulted in the Treaty of Aigun in May, which returned the Amur lands to Russia. On May 15, 1859, Veniaminov celebrated the first service in the Yakut language in the Trinity Cathedral in Yakutsk. In September 1862 he returned to Blagoveshchensk, ending in a shipwreck that he survived.

In 1865 Archbishop Innokenty was appointed a member of the Holy Synod, and on January 5, 1868, the Emperor appointed him to the position of Metropolitan of Moscow, the head of the Russian Orthodox Church. Metropolitan Innokenty died on March 31, 1879, and was buried in the Trinity-St. Sergius Monastery next to his friend and predecessor, Metropolitan Filaret.

In 1849 Petersburg's geologist Konstantin Grewingk named a volcano on the Alaska Peninsula after Veniaminov, and in 1979, on the 100th anniversary of his death, Father Innokenty, Metropolitan of Moscow, was canonized as a saint of the Russian Orthodox Church.

Veniaminov's contributions are widely used as a respected source on the ethnohistory of the Native people of Russian America,[37] but his work has not been studied sufficiently by geographers. Therefore we shall focus on this aspect of his creative endeavor in greater detail.

The most detailed natural history of Russia's American possessions is found in Veniaminov's *Notes on the Islands of the Unalashka District*. In presenting and analyzing these data, the scholarly missionary extensively used the work of natural scientists who had conducted studies in Russian America (geologist Postels, botanists Mertens, Kastalsky, and Chamisso, and sailors such as Litke and Voronkovsky) as well as his own observations and information amassed by the Natives about their environment.

He was very modest in evaluating his own work, saying that "although my notes are as complete as possible and in every aspect as accurate as possible, they are nothing more than materials arranged by categories without any kind of revisions, and some of it, such as topographic descriptions of each island separately, is not needed by ordinary readers."[38]

But we must point out that it was these topographic descriptions that gave us the first complex characterization of the natural environment of Russia's American possessions.

Of the five main parts or districts—Sitka, Kad'iak, Unalashka, Atkha, and Northern—the Unalashka District was investigated in the most detail. It encompassed all the islands and part of the American mainland situated between 53° and 59° N latitude and 158° and 170° W longitude from Greenwich meridian, that is, from Amukhta Island to the (outer or western) half of the Alaska Peninsula and the Pribilof Islands.

For this area Veniaminov provides detailed descriptions of the relief, hydrography, meteorology, fauna, and flora. One of the general questions discussed at the time by Russians was the reason for the treelessness of these islands. The missionary scholar suggested a very original solution. He postulated that the reason was the volcanic origin and relative geological youth of the islands so that the mainland flora had not spread there yet. In support of this notion Veniaminov cites the fact that spruces imported from Sitka took hold and even produced cones.

While maintaining the importance of volcanic activity in the formation of the islands, Veniaminov notes that the presence of granite permitted him to state that they were "remnants of a primeval continent" built upon by volcanoes and once constituted a single whole. He postulates that "on the expanse of the Bering Sea there was once a continent hurled into the abyss of the sea by the terrifying force of the subterranean fire and the islands in the Bering Sea and rocky cliffs on the low shores are remnants of this continent."[39]

Earlier we saw that Peter Pallas expressed similar views; modern science supports the correctness of Veniaminov's thesis that the bottom of the Bering Sea is of a continental nature. The difference is that nowadays the creation of the sea over this continent is linked not with geological cataclysms but with the post glacial rise of the oceans caused by waters from the melting continental ice. Arguing for the presence of a land connection between Asia and America, Veniaminov also cites the Aleut oral tradition that they came to the islands from the west.[40]

It is natural that Veniaminov, like all his predecessors, pays the greatest attention to the natural wealth of the colonies, the animal world. When describing particular animals Veniaminov cites their Native names.

After describing the colonies, Veniaminov provides a detailed description of all the islands from east to west.[41] In discussing the possibility of the existence of islands not yet discovered, Veniaminov cites the Aleut stories about an island called Aklyun (Haklyun) supposedly lying south of Samalga. Subsequent investigations proved that such an island does not exist.

Veniaminov describes how the Aleut navigated on a voyage beyond sight of land:

> They steered first by the Samalga and Four Mountain Islands, which kept the first island always behind the stern and the second over the right shoulder . . . They traveled in this direction until the Four Mountain Islands disappeared from view. Then the Aleuts dropped buoys or signs, customarily using white inflated sea lion stomachs for this. In order to prevent the wind from carrying these away they attached a stone on a long rope to each buoy. From the first buoy, noting the direction of the sea turbulence, they traveled further until the buoy began to disappear. Then they dropped the second and the third, after which that island of Aklyun came into view.[42]

From this we discern that the main method of orientation at sea for the Aleuts was their knowledge of the location of islands relative to each other and conditions of sea turbulence along the coasts (the direction of the waves and, possibly, their refraction and interference when they hit the shore). Such methods were known to exist among many other sea coast dwellers whose subsistence was based on fishing and sea mammal hunting.

Veniaminov, who many times underscored in his work the depth of Aleut knowledge of natural characteristics of the seas with which their lives were inseparably linked, pointed out that "when the seas are difficult they always gauge the falling and the velocity of the wave and always distinguish ordinary turbulence from rough waters in the shallows and underwater reefs."[43] Veniaminov's admired the exceptional endurance of the Aleuts on sea voyages, when they were able to paddle for fourteen to twenty hours without rest, and their keen sight, which the Aleuts themselves explained by the fact that they did not use salt with their food.[44]

Like all his predecessors, Veniaminov noted the Aleut ability to learn the scientific methods of navigation. He wrote: "Among those who had the opportunity to study navigation here, several were reputed to be expert in their work (I am not referring to the Creoles). For instance, a certain Ustyugov, an Aleut, knew navigation very well. His *Marine Chart of the Nushagak River* (the first one) is even now considered most accurate."[45]

Besides orientation by shore outlines and wave direction, the Aleut apparently had concepts about the compass directions based on observations of the sun and the moon. "Their means for observation of either were the lookout points near a village on the surrounding mountains. In regard to the sun they noted that at the time of its solstices, it stays in one position for two and a half days, while before and after that it moves slowly. With respect to the moon they said that it can be seen on the third day after its birth. Their astronomers could point out on the horizon or in the sky the place or point of each new moon, where it must set or rise, for the entire year."[46]

The Aleuts also possessed knowledge about the link between phases of the moon and ocean tides. Besides the usual division into four basic seasons, there was a "division by months, beginning with March, and they predicted the new and full moons of March and of other months by the flow, ebb, and swiftness of the sea tides." Their entire calendar of twelve months was predicated on natural occurrences.

Veniaminov specifically stressed Aleut knowledge of hydro-meteorological conditions and their synoptic habits. In this connection he wrote: "The old-time Aleuts knew how to predict the weather very well, especially the winds, and even nowadays there are such people among them. In foretelling the weather they relied on the sunset as the surest sign and the next morning's dawn, by which experts could unerringly know what the weather would be for the coming day. So carefully did they examine the changes of the colors of the dawn that they called this 'speaking' with the sun and the dawn."[47]

While engaged in the compilation of an Aleut dictionary, Veniaminov collected data about this people's autonyms and their past history. He specifically determined that "The inhabitants of these islands, called 'Aleuts' by the Russians and all Europeans, call themselves Unangan. This word has no meaning in Russian and cannot be derived from any other Aleut word" [The word translates into English as "people."—LB].

> It is said that their forefathers originated in and originally lived in the west, in some great land which was also called Alaskha, that is, continent. There were no storms, no winters, but constant pleasant breezes, and the people lived peacefully and in tranquilly. But enmities and civil wars forced them to move

> right up to the sea. But they could not remain in peace for long here, either. Pressed by other peoples, they were compelled to seek refuge in the islands and then, moving from island to island, they populated the local ones.

Besides this gradual movement to the east, the Aleut also had a tradition of distant baidarka voyages in the Pacific Ocean. Veniaminov writes the following:

> Prior to the outbreak of fighting and internecine wars, they customarily traveled. Moved by the glory to be gained by voyaging, they traveled to the east and west, in order to acquire knowledge of other peoples and their customs, and to show themselves. One of these voyagers reached the northernmost cape of America, which they called then Kigaditigan Kamga, that is, Northern Head. When they returned to their homeland they reported that everything there was of ice.[48]

In contrast, the information about the Kolosh of Southeast Alaska in Veniaminov's works is much more laconic, but he was one of the first to note that "The Kolosh call themselves Tlingit with the addition of the word aptukuan i.e. people of everywhere or people of all settlements. How they received the appellation Kolosh or Koliuni is not known."[49] In Veniaminov's opinion, the Tlingit were the most gifted of all the North American Indians known to the Russians. However, such singling out did not mean that he believed the Aleut to be incapable of occupying the same posts as were customarily occupied by the Russians. On the contrary, Veniaminov strove to provide them with opportunities in the same way that he did for the Russian *promyshlenniki* and the Creoles.

Specifically, on his petition in 1833, the Aleut were permitted to officially enter the clerical estate. From then on it was not necessary to request "the special permission of the highest authorities" for such a move.[50] In theory, this decision opened the way to Russian citizenship for the Aleut, as those of them who occupied responsible posts entered the Creole estate, while a Creole who had Russian subordinates became a "man"; that is, he received all the rights of the free taxable population of Russia.[51]

THE GEOGRAPHIC STUDY AND CARTOGRAPHY OF RUSSIAN AMERICA 1800–1840

In a preceding chapter we noted that by the beginning of the nineteenth century a map had been created that reflected the relationship between sea and land of the northern part of the Pacific Ocean. However, the outline of continental Alaska, especially of the areas north of the Alaska Peninsula, was only approximate and information about Alaska's interior was vague.

Vasily Ivanov

The search for more fur led to the exploitation of new regions, specifically on the mainland. In the 1790s the fur party of Vasily Ivanov, an employee of the Lebedev-Lastochkin Company, was the first to go into the Interior of Alaska. Ivanov and his party set out from Lake Iliamna at Christmas and traveled until Easter, a period of about four months, covering some 500 versts, according to Ivanov's calculations. The travelers passed rivers and lakes and encountered about ten different tribes.

Information about this expedition came to Russia through officers of the Imperial Navy, Nikolai Khvostov and Gavriil Davydov, who were in Russian America in 1802 and 1803. Davydov obtained "In the Kad'iak office of the Russian-American Company . . . a report about the journey of *promyshlenniki* Ivanov, who had been in charge of all his companions" that he included in *The Two Journeys to America by Naval Officers Khvostov and Davydov, Written by the Latter*, published in 1810–1812 in St. Petersburg. Davydov paraphrases Ivanov's report in the following manner:

> Toward Christmas we left Lake Iliamna on snowshoes and followed the River Nugulna which flows into that lake. Then we crossed Lake Klichikh, out of which this river flows. From there we traveled north for seven days, passing four settlements where we observed from 40 to 60 men, to the River Bamadna that flows out of high mountains. From this river we spent an entire day crossing the ridges. In four days we reached the River Khakhlitna [Holitna River—LB], having passed a populous settlement. From Khakhlitna we traveled eight days northward through level plains to the River Kagylna that empties into the sea. On it is a settlement where we saw up to 70 or more men. From there it takes eight days to the River Galtsana, passing a small settlement of the Rabbit chief. (It is not known why the leader was called the Rabbit chief.) From the Rabbit to the Tutna, a pretty large river, it is six days' travel, past the Mandychatskoe settlement that musters up to 90 fighters (adults, able to do battle). Along the Tutna it is two travel days to the settlement of Unagmak, where there are more than 200 inhabitants. From it, along the same river, it is one day's travel to the settlement of Tulyuka, where there are about 150 people. Then we came traveling across a portage (a narrow place where in the summer boats are portaged from one river to another) to the mouth of the Balsanda, where there is a large settlement called Izyntylkhuk containing more than 700 inhabitants.
>
> From Izyltylkhuk we traveled along the coast for a day to the village of Ilgeetsko, situated on a river of the same name. There are up to 70 inhabitants there. Thus it can be seen that Tutna and Balsanda both empty into the sea and are 24 hours' travel time from each other. But Ivanov says later that the mouth of the Tutna is north of the Balsanda and that the width of the Tutna (I surmise at the mouth) is from four to six versts, and that in that river is a multitude of islands. He was told that along the Tutna there are 44 settlements

> in which there are up to 7,000 inhabitants. All the rivers mentioned here, with the exception of the Balsanda, enter the Tutna.
>
> Everywhere Ivanov saw many freshwater fish, such as burbot, pike, and others, but no sturgeon. When the rivers are covered with ice, fish are taken with traps. Dried salmon are stored for spring only, as in winter they take many caribou and they eat meat. There are plenty of other animals, such as beaver, river otters, all kinds of foxes, marten, wolverines, grey squirrels, muskrats, and hares. In the forests there are plenty of fowl, partridges, grouse, and many others. In summer there are multitudes of geese, swans, and ducks.
>
> In winter the people travel by dog teams on large sleds. They always live in a permanent settlement; their houses are made of logs, round and partly covered with sod. In their houses he saw large bladed spears with copper inlays, axes, narrow and tall, but mostly blunt ones [used to cut up fish, ice, and wood—LB]. They have pots of clay, vessels of bent wood and bark, large stone knives, and arrow blades of walrus ivory. Ivanov also saw several iron broadswords resembling Yakut spears. He offered little seed beads and the inhabitants bought them eagerly, especially white ones, as well as large white beads and various iron items.
>
> The forest growth in those regions is the following: spruce, straight and thick, small larch-like growth, poplar, bird cherry, mountain ash and pretty thick birch. It is said that along the Balsadna there is mica, blockstone, and whetstone. Ivanov returned to Iliamna the day before Easter. On the way he passed many lakes, large and small.[52]

The information recorded by Davydov was supplemented in a small way by the Russian naval historian Vasily Berkh, who met Ivanov in Kodiak in 1804–1805 and asked him about his trek into Alaska's interior. It is unfortunate that the geographic nomenclature recorded by Ivanov, written down from second hand verbal accounts by others, do not resemble any of the terms recorded just a few years later. Even today they cannot be deciphered or identified by means of a modern map.

Russian scholars surmise that Ivanov's route went from Lake Iliamna to the upper reaches of the Mulchatna River via Lake Clark and from there along the Holitna or Stony River to the middle valley of the Kuskokwim. In his report Ivanov mentioned two large rivers: the Tutna and the Balsanda. However, his information is not clear as to which one is situated farther north. Just the same, it is accepted that Ivanov's party crossed from the Kuskokwim River valley to that of the Yukon to the Eskimo village of Okhogamiut [Akahamiut, southwest of Russian Mission on the Yukon—LB]. The two rivers come closest to each other precisely at this place, which accords with Ivanov's information. Professor Richard A. Pierce, however, identifies the northernmost river reached by Ivanov's party as the Kuskokwim.[53]

For the first time, reliable information was obtained by Russian eyewitnesses about parts of Alaska's interior. Valuable data collected about the lifestyles of

Natives in the Interior contrasted with very different customs of Natives on the islands.

Davydov also presents his own observations about those sectors of Russia's American colonies that he and Khvostov were able to visit. They traveled from Kodiak to Yakutat and they made a special effort to collect information about the Tlingit, with whom the Russians entered into active and very complicated relationships. Davydov's work includes a short Russian *Vocabulary of the Dialects of the Peoples Called Koliuzhi who Inhabit the Area Between Chugach Sound and Yakutat*. Khvostov and Davydov were interested in cosmographic concepts and the astronomical and geographical knowledge of the aborigines inhabiting the Russian colonies. Some of the information presented in Davydov's book, specifically the notion among the Kodiak Islanders that the earth has no end, was later utilized in Veniaminov's work discussed above.

Davydov especially admired the navigational skills of the local inhabitants, whose abilities he observed. This is what he wrote about it:

> Since they use all their senses always to their maximum they have developed them to a point of perfection unknown to Europeans. They travel on the sea with the same assurance as on dry land. On several occasions we crossed rather wide straits in stormy and foggy as well as snowy weather and directly found the settlement that we were aiming for. Many of them forecast fine or stormy weather accurately. One who is getting ready for hunting further out to sea rises before the sun comes up, sits on top of his habitation or upon a hilltop and, observes how it rises on the horizon, and decides to go or not to go. This has become their habit, or better to say way of life, so that they are compelled to follow it. It is a rare morning that passes without the savages meeting the sun with their full attention engaged as if they wanted to perfect their experience in regard to weather forecasting. Some may deduce that the reason for this is some kind of beliefs, but I am convinced otherwise.[54]

Earlier we noted that Veniaminov observed similar remarkable forecasting abilities among the Unalaska inhabitants. Davydov specifically stressed the remarkable habits of the Natives in the field of orientation and safe watercraft piloting under the complex weather and navigational conditions in the coastal waters of Alaska. He says that "if you are aboard a ship in dangerous waters at night, or in the fog, you may rest assured when there is an American in the forecastle, he will at all times notice rocks at a considerable distance. It is amazing how they are able to discern them in stormy weather when the entire sea is just as white and turbulent near the submerged rocks." The Russian professional navigator was astonished at how quickly the Natives absorbed the conditions of navigation in areas that were new to them, so that "in an inlet a Koniag has passed through once he knows all the rocks right away and one can rely on him as if he were the most excellent pilot."[55]

Davydov stresses that seasonal weather variations played a great role in Native life and that ten of the twelve months that composed their calendar, the year beginning with August, were characterized accordingly. Thus, the month of October was the month when "hoarfrost falls on the grass," November—"snow appears on the mountains," December—"streams and lakes freeze up," January—"sixth month after the autumn," February—"dried salmon is cut into pieces," March—"ice breaks up," April—"ravens lay eggs," May—"waterfowl lay eggs," June—"seals give birth," July—"porpoises give birth." The remaining two months are characterized by astronomical signs: August—"the Pleiades begin to rise," September—"Orion rises."[56]

A New Way to Supply the Colonies

The first Russian round-the-world voyage under the command of Johann von Krusenstern and Yuri Lisyansky (1803–1806) provided crucial government support to Aleksandr Baranov. One of the most practical tasks of the expedition was the resupplying of the Russian-American colonies. The old method of transporting supplies across Siberia with subsequent transport across the Pacific Ocean was exceptionally complex and expensive. For example, it required the deployment of 4,000 horses annually.

A detailed proposal for supplying the colonies by sea was presented to the Russian government by the Minister of Commerce, Count Nikolai Rumiantsev (Fig. 39). Rumiantsev was an enlightened person, a well-known statesman in the time of Tsar Alexander I. He founded the Rumiantsev Museum and an enormous library in Moscow (now Russia's State Library). Deeply interested in geographic studies, in the early nineteenth century Rumiantsev spent a considerable amount of his own funds to support the study of the colonies. He shared the costs for one ship with the Russian-American Company, while the second ship was paid for by the Russian government. A major player in the organization of this circumnavigating voyage was Nikolai Rezanov, who was closely tied to the interests of the Russian colonies in North America.

On June 10, 1803, Rezanov was given the title of chamberlain and appointed ambassador to Japan. On Rumiantsev's insistence, Tsar Alexander I decided to dispatch the ambassador to Japan with the same ships that were to deliver goods to the Russian colonies in North America. This enhanced the international importance of the expedition. It was also decided to dispatch a scientific team to study the coasts of the Pacific Ocean. Naval officers Krusenstern (1770–1846) and Lisyansky (1773–1837) (Fig. 40) were selected to lead this first Russian circumnavigation. Besides being graduates of the Naval Academy in St. Petersburg, both of them had trained and served in Britain's Royal Navy, taking part in many sea battles. The general proposal for the organization of the circumnavigating expedition was presented to the Naval Ministry by Krusenstern, who was appointed the expedition's leader.

FIGURE 39 Portrait of Count Rumiantsev from Choris, *Voyage Pittoresque autor du Mond*. . . 1822, frontispiece. UAF Rare Book B0083

YURI LISYANSKY

The physical, scientific, and technical parts of the preparations were entrusted to Lisyansky. In 1803 he purchased two vessels, named the *Neva* and the *Nadezhda*, in London, for the price of 22,000 pounds sterling. He also acquired navigational instruments, including several sextants, azimuth compasses, barometers, a hydrometer, Taunton's artificial magnet, and chronometers made by Arnold and Pettington. Special importance was placed on the precise determination of astronomical coordinates, and for this reason all instruments, especially the chronometers, were checked at the Pulkovo Observatory in Russia by Academician F. I. Shubert. The *Neva*,

FIGURE 40 Portrait of Captain Lisyansky from his *A voyage around the world in the years 1803, 4, 5 & 6. . . .* 1814. UAF Rare Book A0914

commanded by Lisyansky, was to sail to Russian America, while the *Nadezhda* under Krusenstern would sail to Japan.

The *Neva* arrived at Kodiak during a very complicated period for Russia's North American colonies. Baranov had been energetically pursing an advance along the coasts of Alaska southward but was checked by a Tlingit uprising on Sitka (now Baranov) Island. In 1802 the Tlingit had taken the fortress by storm and burned it along with a merchant vessel in the harbor. English sailors from the vessel of Henry Barber actively participated in this attack.[57] Baranov prepared long and hard for military operations against the Tlingit. The arrival of the Russian expeditionary vessel *Neva* in St. Paul Harbor at Kodiak was most fortuitous. Toward the end of August 1804, Lisyansky approached Sitka, where he was joined on September 19 by Baranov with the vessel *Ermak*. By September 25 the entire force of Russian *promyshlenniki* and the Kodiak Islanders were assembled at Cross Harbor in Sitka. It included 800 men with 350 baidarkas.[58]

On September 28 the *Neva*, the *Ermak*, and Baranov's baidarka fleet approached the Tlingit settlement. There, on September 29, "Baranov, having gone ashore accompanied by a number of armed men, raised a flag atop a rather high mountain in the middle of the abandoned settlement." According to Lisyansky, "six cannons, four of which were copper and two cast iron," were placed into this fortification that "by itself would be considered unassailable." A landing party from the *Neva*, led by Captain-Lieutenant Lisyansky himself, arrived at this fortification at noon "when it was named Novoarkhangelskaya, while several salvos were fired from all cannons." Thus was established the settlement that soon became the center of Russia's American colonies, the modern city of Sitka.[59]

On October 1 a Russian detachment approached the nearby Tlingit fortification. The first attempt to capture it by storm was unsuccessful, ending in the Russians' retreat under cover of the ship's artillery. In the course of this assault two sailors were killed and many were wounded, among them Baranov. On October 7 it was discovered that the Tlingits had abandoned their fortress, leaving only an old man and a boy there. On October 8 Lisyansky visited the abandoned Tlingit fortification and recorded his impressions as follows: "Coming ashore, I observed a most barbaric sight that could bring even the most hardened heart to tremble and recoil. Assuming that we could trace them in the woods by the voices of infants and dogs, the Sitkans put them all to death. . . . The entire set of circumstances led us to conclude that the fortress had contained no fewer than 800 persons of male gender."[60]

On November 10, 1804, the *Neva* left Sitka and on November 15 she arrived in St. Paul Harbor at Kodiak. Here the members of the expedition came on shore for the winter. Lisyansky rendered a painting which was later reproduced in his account of the voyage (Fig. 41). Despite taking part in the battle against the Tlingit, Lisyansky organized a hydrographic party to survey regions that had been insufficiently explored. As soon as August 1804 Lisyansky "sent out his navigator into Chiniak Bay to describe it." The winter was spent on Kodiak observing natural history and ethnography. The *Neva*'s captain extensively utilized information obtained from local inhabitants, in particular, about the appearance of a new volcanic island in the Distant Aleutian (Fox) Islands. Lisyansky tells the story:

> The island appeared suddenly around the middle of April 1797. The first news about this miracle was brought by the Aleuts who, coming in from the sea, assured everyone in Captain's Harbor (on Unalashka) that they had seen fire over the sea surface not far away. The fire-breathing mountain, emitting

> flame, was emerging from the depths of the sea little by little so that in May 1798 the newly-emerged island could be seen from the Makushin settlement on Unalashka, though it was situated not less than 70 versts to the northwest. This island nowadays, so they say, resembles a hat, is relatively high and has a circumference of about 20 versts. It has been noted that it has not grown since 1797. The molten matter that ruptured the surface of some peaks scattered the mountain rocks of which they were composed. It is asserted that this new work of nature could be seen from the very beginning of its emergence from the Island of Umnak.[61]

Using Native information in his descriptions, this enlightened Russian sailor did not always evidence the requisite sensitivity when encountering the aboriginal beliefs and rituals practiced by the local inhabitants, lacking the tact of missionaries such as Iosaf, Gedeon, and Veniaminov. Lisyansky, describing his encounter with one of the Kodiak *toions*, Minak, says that:

> this old man, about 80 years of age, is the greatest shaman or sorcerer on the island. Perhaps trying to impress us with his fables, he insisted that he often meets an unclean spirit with whose help he foretells the future for the folk. We laughed heartily at his tales and the poor old man got so angry that he left us without saying farewell to anyone.[62]

Just the same, and in spite of his relatively short stay in the colonies, Lisyansky repeatedly noticed the depth of Native knowledge about the natural environment of their islands and in particular their ability to forecast the weather. Despite this, in April 1805, during a trek with a baidarka to Ugak island, Lisyansky disregarded the advice of the local *toion* to stay on the island for the night, because, the *toion* "insisted, a northeast wind will begin to blow soon." On the return trip to St. Paul Harbor he got drenched because, just as the *toion* had predicted, "in the afternoon the wind began to blow, first from the east creating such waves that they constantly washed over the baidarkas."[63] Speaking of the seagoing abilities of the Natives, Lisyansky particularly notes that the Kodiak Islanders "sail without any kind of danger over a thousand versts" in their baidaras and baidarkas.[64]

Describing Russian possessions on the American coasts, Lisyansky remarks with concern about the exhaustion of the fur resources in the colonies, noting that "a few years ago sea otters were observed nearby, but nowadays they have retreated rather far into the sea, while gray fur seals are not found at all." Insisting on the necessity of instituting measures to protect the colonies' fauna, perfect fur procurement methods, and improve the living conditions of the Aleut and Kodiak Islanders, he specifically proposed that Native fur hunters be transported to remote hunting grounds with their baidarkas aboard large Company vessels, which later became a steady practice of the Russian-American Company.[65]

Having spent the winter on Kodiak, the *Neva* arrived off Sitka Island on June 22, 1805, and dropped anchor at Novoarkhangelsk (Fig. 43). Lisyansky was struck by Baranov's progress in enhancing amenities in the new settlement. He writes (on June 23): "About ten o'clock I went ashore and with the greatest of pleasure saw the remarkable fruits of Baranov's tireless labors. During our short absence he managed to put up eight new buildings, which in their size and appearance might be considered beautiful even in Europe. Also, he developed 15 vegetable gardens near the settlement. He now has four cows, two heifers, three bulls, a sheep and a ram, three goats, and a considerable quantity of swine and chickens."[66]

While anchored off Sitka Island, Lisyansky and his officers continued their hydrographic surveys and description of Alaska waters and land. The *Neva*'s navigator, Danilo Kalinin, conducted the survey and description of the strait leading to Cross Harbor "and separated the island on which stands Echkom" (Cape Edgecombe). Lisyansky named it Kruzov Island in memory of Admiral Aleksandr Ivanovich Kruz (1731–1799) to whom the seafarer was greatly indebted.

Lisyansky notes that from the time of the Chirikov voyage to that of Vancouver it was not known if Sitka was an island or part of the mainland. Vancouver discovered

FIGURE 41 Saint Paul (now city of Kodiak) drawn by Lisyansky, in Lisyansky, *A voyage around the world in the years 1803, 4, 5 & 6....* 1814. UAF Rare Book A0914

HARBOUR of NEW ARCHANGEL in SITCA or NORFOLK SOUND.

FIGURE 42 The new settlement of New Archangel, in Lisyansky, *A voyage around the world in the years 1803, 4, 5 & 6. . . .* 1814. UAF Rare Book A0914

Chatham Strait and thus determined that Sitka was an island. The investigations by Lisyansky and his officers proved that among the multitude of islands off Sitka there are four main ones: Iakobiev, Kruzov, Baranov, and Chichagov. We see that in his plan for naming islands Lisyansky bestows new geographic names without paying due attention to the local toponyms.

On September 2, 1805, the *Neva* was ready to commence her homeward voyage. Taking leave from the Russian-American colonies, Lisyansky remarked in his notes: "At this time Baranov came to us from the fort and I took my leave from him, but not without regret. He deserves all sorts of respect commensurate with his abilities. In my opinion, the Russian-American Company cannot have a better head in America. Besides his knowledge, he is accustomed to perform various tasks and does not spare his own estate for the communal benefit."[67]

The main cartographic results of the circumnavigating expedition to the shores of Russian America are reflected in the *Collection of Maps and Drawings Belonging to the Voyage of Fleet Captain of the 1st Rank and Cavalier Yuri Lisyansky Aboard the Ship Neva*.[68] This atlas contains the general *Chart of the Russian Possessions in the Northwestern Part of America, Based on the Newest Descriptions and Confirmed by Observations of Captain of the Fleet and Cavalier Yuri Lisyansky, Conducted on Kodiak Island and in the Port of Novoarkhangelsk, 1805* (Folio 4) and also four large-scale navigational charts:

1. *Chiniak Bay and St. Paul Harbor, Described in 1805 by Navigator Kalinin under Special Supervision by Captain of the Fleet and Cavalier Yuri Lisyansky* (6 Verso);
2. *Kodiak with the Islands Surrounding it, Described by Captain of the Fleet and Cavalier Yuri Lisyansky in 1805* (Folio 7);
3. *Sitka Bay Described under the Supervision of Captain of the Fleet and Cavalier Yuri Lisyansky. 1805. Novoarkhangelsk* (Folio 8);
4. *Three Saints Harbor* (Folio 9).

These charts were compiled with the highest level of scientific hydrography of the times. Their astronomical and geodesic accuracy was guaranteed by numerous determinations of coordinates, latitude, and longitude, using meticulously checked instruments and field observatories that Lisyansky conducted in Kodiak and Sitka. Besides the depth measurements indicated on the charts, they include an indication of the nature of the sea bed: shell, rock, sand, silt, silt with sand, etc.

When compiling his charts, Lisyansky utilized all the sources that were available to him. This is evidenced, for example, by an inscription on the general chart in the area of Yakutat Bay: "shoal shown on Spanish charts, unknown to the Russians." The charts comprising the atlas are complemented by sketches of the coasts as seen from the sea in the approaches to the main harbors and coastal segments where navigation was difficult, a practice by then obligatory in all the publications of Russia's Admiralty. We must note that such sketches convey much more accurate representations of the character of Alaska's coastal relief than by means of hatching (parallel lines used to indicate topographic relief).

Besides hydrographic and navigational content, Lisyansky's general chart includes the location of tribes. The Kenaitsy, the Chugachi, the Ugalakhmiut (Eyak), and the Koliuzhi (Tlingit) are indicated by their tribal names. Lisyansky's book contains a detailed and precise description of the lifestyle, mores, and material and spiritual culture of the Natives of Russia's colonies. His atlas includes drawings of objects of everyday use, including weapons and clothing. Examples of Indian artifacts were brought to St. Petersburg and enriched the collections of the Academy of Sciences. These were collected by members of the expedition in Russian America and some were presented by the colonial leadership.

Lisyansky's book, first published in St. Petersburg in 1812,[69] (Fig. 43) appeared in an English-language edition translated by Lisyansky himself in 1814. In this edition maps from Lisyansky's *Collection of Maps and Drawings . . .* were included, reworked by Aaron Arrowsmith.[70]

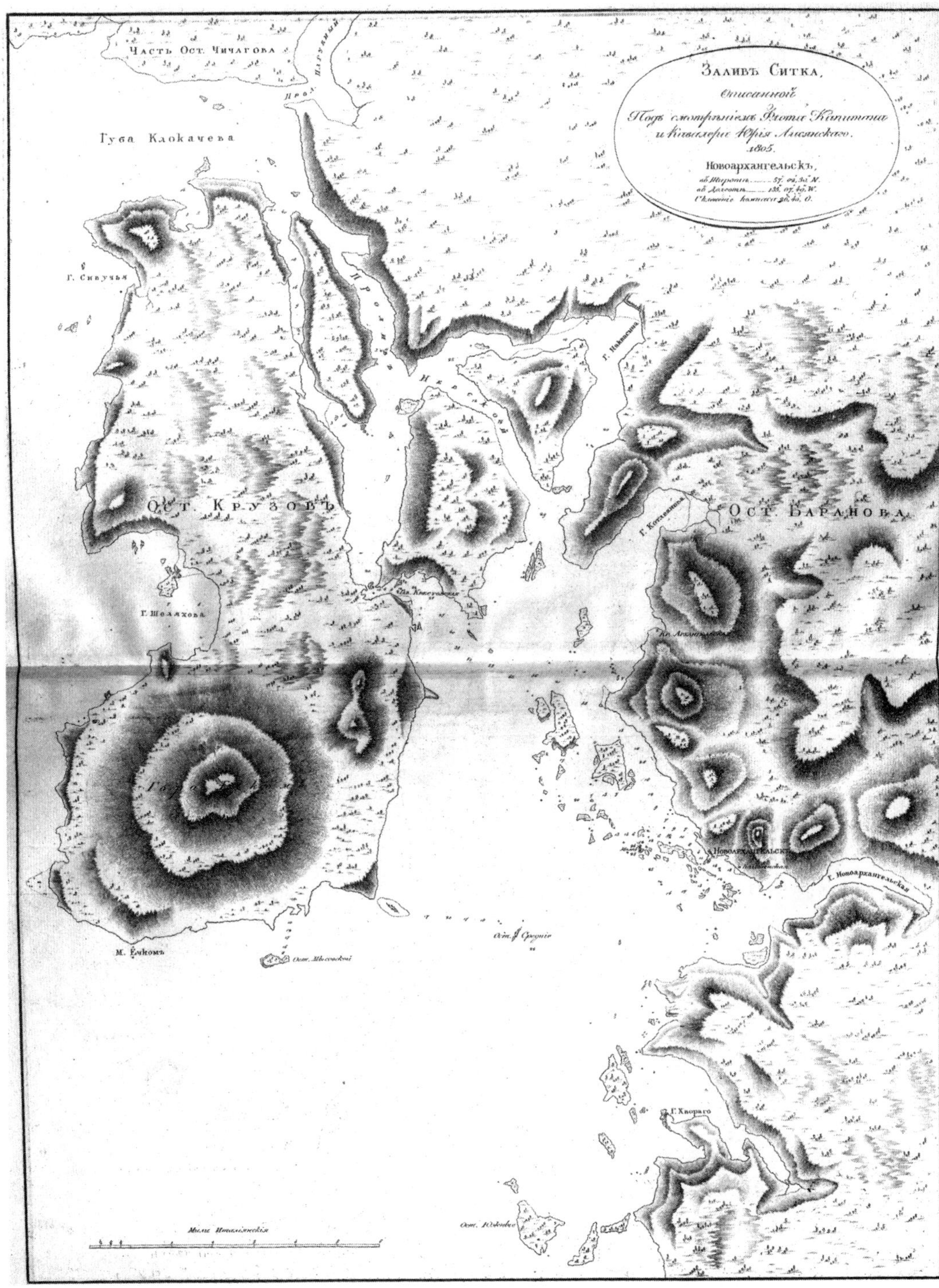

FIGURE 43 Lisyanski's map of Sitka. Lisyansky, *A voyage around the world in the years 1803, 4, 5 & 6 . . . . Atlas* 1812. UAF Rare Atlas C0088

GEORG HEINRICH LANGSDORFF

After the *Neva*'s departure from the colonies, the brig *Mariya* arrived there, bringing some of the Krusenstern contingent that had been in Japan: Nikolai Rezanov, Nikolai Khvostov, Gavriil Davydov, and the German naturalist and member of the St. Petersburg Academy of Sciences Georg Heinrich (Grigory Ivanovich) Langsdorff (1774–1852). The publication of Langsdorff's bio-geographical investigations conducted in 1805–1806 in Russian America, with his description of Alaska and of the Russian settlements, gained widespread attention through publication in Germany in 1812 of materials from his travels.[71] In 1814 Langsdorff's work was published in English.[72]

In his work Langsdorff included ethnographic descriptions of the Native peoples of America and also their relationships with the *promyshlenniki*. Mentioning many times the arbitrary acts of colonists against the aborigines, the German naturalist also stressed that ordinary Russian laborers suffered no less than the local inhabitants. Langsdorff wrote with indignation that "not only the Aleuts but many Russian *promyshlenniki* who are not employed as skilled craftsmen or clerks, but serve as sailors, perform agricultural tasks, or transport timber, find themselves in scarcely a better situation. They are treated exceptionally poorly and are kept at work until they are totally exhausted. . . . The food of such people consists for the most part of fish, whale and walrus meat."[73]

NIKOLAI REZANOV

The sojourn of Nikolai Petrovich Rezanov (1764–1807) in the colonies was of great significance for the future development of Russian America. He visited Unalaska, Kodiak, Sitka, and California from 1805 to 1806. During this time he maintained an active correspondence with the directors of the Russian-American Company, part of which pertained to his plans to expand and strengthen Russia's colonies in America. These plans were published in the second part of Pyotr Tikhmenev's work devoted to the company's history, while a general account of his voyage appeared in several issues of the journal *Otechestvennye Zapiski* in the years 1822–1825.[74] Rezanov wanted to expand the boundaries of Russia's colonies southward, to the fortieth parallel, as was previously championed by Shelikhov.

For this reason, in 1806 Rezanov sailed to California aboard the vessel *Yunona*. Arriving in San Francisco Bay, he entered into trade negotiations with the Spanish for foodstuffs needed in the colonies and ascertained that the Spanish had no permanent settlements to the north. This was immediately conveyed to the chief manager, Baranov, who in turn notified the Imperial Russian government.[75] The information obtained by Rezanov was immediately utilized to expand the fur hunt. By 1806 the Russians visited Bodega Bay, naming it Rumiantsev Bay, and established sea otter hunts there.[76]

In November 1811 Baranov sent his assistant, Ivan Aleksandrovich Kuskov (1765–1823) there, aboard the schooner *Chirikov*, accompanied by twenty-five Russian laborers and an Aleut party with forty baidarkas. They determined that the bay was not suitable for settlements as there was no timber. For this reason the new Russian settlement site was chosen on a steep forested shore to the north of the harbor in Bodega Bay. Having prepared a sufficient quantity of timber, Kuskov founded Fort Ross in June 1812. It became the southernmost Russian outpost in America, if we do not count the camp in the Farallon Islands established somewhat later, at Drake Cape, where the Russians hunted sea lions and fur seals.[77]

Among the materials sent by Rezanov to the main office of the Russian-American Company in St. Petersburg was a grand proposal for colonial reorganization. The main idea was the necessity of creating a large Russian population on the coasts of Alaska. In order to attract settlers it was proposed that various incentives be offered, such as land and all necessary equipment. Rezanov also called for an improvement in the living conditions for the Russians, Creoles, and the Native inhabitants of the colonies. One of the first measures proposed was an improvement of education in Russian America, to which Rezanov himself paid great attention in the course of his sojourn in the colonies.

In 1805 he reformed the school in Kodiak, transforming it into the central colonial school, part of which was later transferred to Sitka. Students were accepted at this school without regard to their parentage, social origins, or nationality. Its graduates were used in the colonies in the most varied colonial tasks. They worked as sailors, craftsmen, and clerks. Rezanov devoted great attention to the Creoles, assuming quite reasonably that children born to Russian *promyshlenniki* and Native mothers would become most pliable servants of the company. In this connection, on January 15, 1818, the company's management petitioned the Interior Ministry of Russia to liberate the Creoles from taxation and provide them with the opportunity to serve in the colonies "and gain through their talents and their service to the State ranked estates and honors."[78]

The Council of the Interior Ministry, chaired by Minister of the Interior Kozodavlev, discussed this petition and stated that "the Council considers it beneficial to leave them as free Citizens of those localities where they were born, either occupying positions with the Company or having their own establishments, as has been suggested by the late Mr. Chamberlain Rezanov."[79] This decision was incredibly liberal for the Russia of those years, as, perhaps only in theory, it offered an opportunity to attain, after having reached a certain rank, such as lieutenant in the navy, the honor of personal nobility, and even, if attaining a yet higher rank, of a hereditary Russian nobleman.

It should be noted that even before this decision, the Russian-American Company provided opportunities for such careers when, in 1805, it dispatched, on Rezanov's initiative, eight Creole boys to continue their education in Russia, of whom only four survived, "three having studied at the Kronstadt Navigational School and one at shipbuilding school; after completing their studies they were sent back, one with the rank of navigator apprentice, and another as ship's carpenter. They are to receive from the Company salaries from 500 to 1,200 rubles per annum."[80]

According to data for the year 1818, more than sixty children were being educated at the colonial school. They were taught Russian language and literature, basic mathematics, art, navigation, and English language.[81] Baranov labored to expand the curriculum, believing "that the subjects of instruction should be, besides Russian language arts, arithmetic, geometry, mathematics, physics, navigation, mechanics, chemistry and laboratory arts, drawing and painting, artillery, fortification, geodesy, and English and Spanish languages."[82] This high level of instruction would be achieved by the 1830s.

In 1816 it was decided that Creoles should also be educated at the St. Petersburg Academy of Medicine and Surgery in order to satisfy the demand for well-prepared medical personnel in the colonies. It is difficult to determine the exact number of Creoles who were educated in Russia, but data in the annual reports of the Russian-American Company indicate that each year between five and twelve people were being educated in the metropolis not only in naval, business, and medical disciplines, but also in the various skilled trades needed in the colonies. Among the cadets of the Kronstadt Navigation School were Creoles who later became well-known explorers of Alaska: Andrei Klimovsky, Aleksandr Kashevarov, Fyodor Kolmakov, and Pyotr Malakhov.[83]

The colonial library, begun by Shelikhov in 1794 with "classical, historical, mathematical, moral, and economics" books,[84] was enlarged by the first circumnavigating expedition. The *Neva*, commanded by Lisyansky, brought to Kodiak a significant book collection donated by Russian literary authors and foremost representatives of Russian society. The books were collected after a call by Rezanov. According to Khlebnikov, in 1835 the Sitka library contained 1,200 volumes, of which more than 600 were in Russian, 300 in French, 130 in German, 35 in English, 30 in Latin, and the rest in Swedish, Dutch, Spanish, and Italian. The collection was valued at 7,500 rubles.[85]

An important consequence of the first Russian circumnavigation was the expansion and perfection of the geographic investigations and cartography of Alaska. Although Rezanov was not a professional sailor, his correspondence with the directors of the Russian-American Company demonstrates that he clearly understood how the colonies should develop. He foresaw the improvement of the vessels used by *promyshlenniki* by studying the American (or as they were called in Russian America, Bostonian) ships. In his letters sent from Sitka Port to St. Petersburg (July 2, 1806) he noted the remarkable sailing capabilities, speed, and longevity of these ships. He recommended following their example, by skin-plating the vessels with copper and oiling with drying oil (linseed oil or similar) and ocher. According to

American observations, this would increase the longevity of service by more than two years in the complex hydro-meteorological conditions in Alaska.[86]

IVAN VASILIEV

Following the 1805 Russian circumnavigation, regular shipping between the Russian-American colonies and the Baltic Sea ports of Russia began. By September 13, 1807, the *Neva*, sailing out of Kronshtadt under Captain Ludwig von Hagemeister (1780–1834), again arrived in Sitka with navigator Ivan Filipovich Vasiliev (1776–1812). Vasiliev immediately commenced a survey of St. Paul Harbor and the Russian settlement there, and of Sitka Island and Sitka. The charts he compiled in 1807–1809 are outstanding in their precision and as a rule were accompanied by expressive drawings of the settlements.

Among the hydrographers who valued Vasiliev's cartographic work was Vice-Admiral Vasily Mikhailovich Golovnin (1776–1831), who checked the accuracy of the chart of Sitka Bay compiled by Vasiliev in 1810. Golovnin concluded that Vasiliev's survey was of such accuracy that he included the chart in his own (unpublished) atlas with a few very minor corrections.[87] Vasiliev's charts and drawings, like many other materials pertaining to Russian America, remained unpublished.

Fedorova discovered two of Vasiliev's charts in Golovnin's unpublished atlas that is at the Museum of Agriculture of Moscow State University.[88] It seems that these are the two charts that Golovnin checked in 1810.

Two other charts by Vasiliev exist in three versions preserved in the Russian State Archive of the Navy. The first of these is *Map of St. Paul Harbor at Kad'iak Island where there is the Main Settlement of the Russian-American Company, Compiled on the Basis of the Survey Made in the Year 1808*.[89] The chart was drawn from a detailed survey of the harbor's shores and rather detailed depth measurements on straight-line tacks drawn between shores. Two isobars are shown that delineate offshore shoals and banks. The land relief is shown by means of a wash that indicates only the presence and general shape of separate elevated landforms. As is the case with the charts of other Russian hydrographers discussed earlier, the wash creates the impression of the relief having a step character. We must, however, note that in spite of this stylized representation of the land formations, Vasiliev devoted serious attention to specific features of the landscape by explaining these in extensive annotations pertaining to the chart, such as, for example, "swamp that dries out in summer." He also represented rather vividly and realistically the general view of the landscape surrounding the Russian settlement in wonderfully executed drawings. One of these is a "View of the main Russian American settlement on the island of Kad'iak at St. Paul Harbor to NwtN, taken from across the strait from the island at a distance of about 100 sazhen."

Vasiliev paid even greater attention to the landscape when drawing the chart representing the region of the new Russian settlement on Sitka (Baranov) Island.

This is the *Chart of Sitka Bay together with all the Islands within it and Partial Measurements of Depth Taken at Ebb Tide in Sazhen; Surveyed and Compiled by Assistant Navigator 14th Class Vasiliev in the Year 1809.*[90] The chart shows in detail the shores of Sitka. The cartographic representation is supplemented by inscriptions containing not only geographic but also historical information.

The inscription appearing along the river emptying into Glubokoe Lake contains the following: "Stream 20 sazhen wide, explored on foot upstream for about 5 versts but because of thorny growth and thick forest the source was not reached. Shallow with many rapids, current up to 4 knots per hour." Also noted are: "Stream with a waterfall of 50 sazhen, stream where in summer fish and hay are put up" and other comments.

Besides the main chart, which is drawn to a scale of one verst to an inch, there is a large-scale (twenty sazhen to an inch) map of Sitka Fort with a detailed "Explanation of the plan." The chart is accompanied by drawings and profiles of importance to sailors: "View of the lighthouse on Mayachny Island (Signal Island), 26 feet in height, elevation of the island 60 feet; View on Sitka Cape of St. Lazarus Mount (Mt. Edgecumbe) . . . " and "View from Sitka to the NW." Two drawings of Sitka that complement Vasiliev's chart are of great geographic and historic interest: "View . . . of the Russian American settlement at Sitka, the Sitka Fort, taken from the E side from the ship *Neva* September 25, 1807 by Assistant Navigator 14th class Vasiliev," and the somewhat simplified "View of the Russian American settlement in Sitka Bay from the eastern side, America," signed by Chernyshev.

Taking advantage of the *Neva*'s arrival, Baranov began the search for new islands to the south of the Aleutian chain, a search that proved fruitless. In November of 1808 he instructed Lt. Hagemeister to add a search to his voyage with *Neva* from the Kuril Islands to Kamchatka and to sail on the forty-fifth parallel N latitude from Urup Island to the east for about 150–200 miles, tacking back and forth within one to two degrees latitude, in search of a large island. This was to follow up on information provided by the Japanese, who had suffered shipwrecks in the Rat and Andreanof Islands.[91] Moreover, Lebedev's *promyshlenniki* told Baranov that when they were sailing for Unalaska in 1786 they saw land to the right, but a storm and fog prevented them from investigating it.[92]

A stimulus for investigating northern Alaska was the persistent rumor about Russian people on the Kheuveren River. This rumor arose in the 1760s. The search for undiscovered Russian settlements played a conspicuous role in the planning of government expeditions and also in Russian-American Company management plans for northward expansion. In 1811 Russia's Admiralty Department presented a plan to the government for a new expedition into the northern part of the Pacific Ocean. The project had been formulated by Vice-Admiral Gavriil Sarychev, who proposed that the leadership of this expedition be entrusted to Captain-Lieutenant Golovnin.[93] It was proposed that special attention be paid to investigation of the western coast of North America in the sector "between Cape Stephens and Catme Point."

Sarychev "testified that during his sojourn on the Chukotka Peninsula he heard from the Chukchi that on the coast directly across there is a fine bay and Russians live in its vicinity; it would be useful to use the occasion to check the Chukchi testimony, especially as a large river empties into it, judging by the shallows between Cape Stephens and Catme Point and by the fact that driftwood is constantly carried to Gor Island (St. Matthew Island) that lies directly across from it. This has to be clarified by a coastal survey."[94] In order to answer these questions, the expedition planned to sail directly to the Bering Strait and then, approaching the American coast about thirty to forty miles to the south of Cape Prince of Wales, search for "the harbor and the river" and collect information from local inhabitants.

Golovnin was to proceed along the coast to Norton Sound and "from Cape Stephens to Cape Catme Point dispatch baidaras to sail close to shore for a coastal survey; he himself is to determine exactly the shoal across from it and rejoin the baidaras at Catme Point. Should he complete the survey sooner, or for other reasons, he was to commence the survey from Catme point and then proceed to Norton Sound and rejoin the baidaras; from there he is to sail toward Cape Prince of Wales and on the return trip to Kamchatka touch upon Preobrazheniya Island." This expedition was never realized, but a part of the Admiralty program would be carried out by another expedition, the 1815–1818 voyage of the *Rurik*, commanded by Captain Otto von Kotzebue (1787–1846).

OTTO VON KOTZEBUE

Kotzebue's expedition was financed by Count Nikolai Rumiantsev, who considered the expedition's main goal to be finding a free passage from the Atlantic to the Pacific along the American Arctic coast. Krusenstern took an active part in the planning. Kotzebue's expedition sailed from Petropavlovsk in Kamchatka for the Bering Strait in June 1816. On July 27 it reached St. Lawrence Island. The Native islanders stated that their island was called Tshiboki and the land to the east (America) was called Kililak. Members of the expedition surveyed the western shore of St. Lawrence Island and the expedition then continued on its course for the Bering Strait. On July 30, after passing King Island, it reached the western extremity of Cape Prince of Wales. Kotzebue placed four islands in Bering Strait on his charts, instead of the three Diomede Islands, and named the largest of them for Captain Ratmanov under whom Kotzebue had served with Krusenstern's expedition. Then the expedition proceeded east-northeast along the American coast, conducting surveys aided by rather fine weather and the twenty-four-hour daylight.

On May 31 the *Rurik* dropped anchor in a bay they named for Shishmarev. A group led by Kotzebue himself explored the bay and found that there was an island at its mouth, which they named Sarychev Island. More detailed exploration of the bay was prevented by the alleged hostility of the Natives (Eskimos). The *Rurik* continued eastward along the American coast and on August 1 the expedition reached

dess. et Lith. par Choris

Lith. de Langlumé

Ouluk principal établissem.t sur l'île d'Ounalachka

FIGURE 44 The settlement on Unalaska. Choris, *Voyage pittoresque autour du monde . . .* 1822. UAF Rare Book B0083, plate VII

an entry into a large water expanse that "tore" the mainland apart. At first, members of the expedition decided that they were at the entry of the long-sought Northeast Passage to the Atlantic. However, a detailed exploration of this water body proved that this was not so.

This was the expedition's most important discovery, which they named Kotzebue Sound, "on the expedition commander's insistence." Chamisso Island was put on the chart, as well as Eschholtz Bay and other coastal features. On the return sail, proceeding along the sound's western coast, they discovered an entry leading west. Kotzebue attempted to find out from an Eskimo how far west this bay extended. The Native engaged in pantomime, which the members of the expedition interpreted to mean as paddling in a baidarka for nine segments with stops for the night between. Kotzebue concluded, erroneously as it was proved later, that the water body he had discovered continued to the ocean in the area of Shishmaref Bay or Norton Sound. Nine days of baidarka travel were necessary to cover the distance. An attempt to penetrate this waterway they called Good Hope Bay proved futile because of numerous shoals.

Members of Kotzebue's expedition had an opportunity to appreciate the cartographic abilities of the Natives. One of them drew the general outline of the northwestern headland of Kotzebue Sound and indicated the locations of Eskimo camps on it. He called these *Kegi*. On the expedition's chart this segment is shown by dotted lines on the basis of the Eskimo drawing. On August 14 the *Rurik* reached the northeastern headland of the sound, which they named Cape Krusenstern. The expedition had been instructed to find good anchorage in Norton Sound that could be used as a point of departure for further investigations of northern Alaska. However, because of the discovery of Kotzebue Sound, which was more suitable for that purpose, the commander concluded that he had carried out his assignment, and he turned the ship southwest.

The *Rurik* reached the Asian shore of Bering Strait, put in at St. Lawrence Bay and then, on September 7, 1816, dropped anchor in Captain's Harbor at Unalaska Island.[95] After spending the winter in the Sandwich Islands (Hawaii), in the summer of 1817 the expedition once again sailed to the northern part of the Pacific. Because of serious damage suffered by the *Rurik* during a hurricane, considerable time was spent on repairs while at Unalaska. Several seamen were wounded and Kotzebue received a heavy blow to his chest, which troubled him for the rest of his life. During their stay a detailed survey of Captain's Harbor was conducted (Fig. 44). In July the expedition set sail again for the north. After surveying the Fox Islands en route, the expedition approached St. Lawrence Island. Here they charted the island's southeastern coast. The attempt to sail farther, toward Bering Strait, was halted by solid ice.

Because of Kotzebue's worsening state of health, it was decided to return.[96] The *Rurik* had failed to penetrate farther north than James Cook. Even though the expedition was not able to carry out its main assignment, Kotzebue's voyage considerably advanced the geographic knowledge and cartography of the northern coasts of Alaska. The expedition charted Shishmaref Bay, Sarychev Island, Kotzebue Sound, Escholtz and Spafariev Bays, the Choris Peninsula, Chamisso Island, and other features. Highly sophisticated observations and investigations were conducted in the course of the voyage by naturalist Johann Friedrich Escholtz, the poet and scholarly zoologist Adelbert von Chamisso, and by the expedition's artist, Louis Choris. The expedition returned to Russia carrying a wealth of natural science and ethnographic collections. Hundreds of sketches and drawings of Natives executed by Choris are being studied to this day.[97]

The expedition's achievements were described in Kotzebue's three-volume work published in 1821 in German and immediately translated into English and Russian. The third volume of this work was devoted to the notes by the naturalists (Fig. 45), who also published their own works separately.[98] The main cartographic results are reflected in the *Atlas of the Voyage of Lt. Kotzebue Aboard the Ship Rurik into the South Sea and to the Bering Strait*, published in 1823 as a supplement. The following charts deal with the northern part of the Pacific Ocean:

Folio 6: *Flat Chart of the W Part of Bering Island, July 1816.*

Folio 7: *Mercator Chart of the Bering Strait, August 1816* (Fig. 46).

Folio 9: *Mercator Chart of a Part of North America from Cape Prince of Wales to Kotzebue Sound.*

FIGURE 45 Members of the Kotzebue Expedition examine ground ice. Choris *Voyage pittoresque autour du monde . . .* 1822. UAF Rare Book B0083, plate IX

Verso 10—Folio 11: *Mercator Chart of the Fox Chain of the Aleutian Islands, August 1817. St. Lawrence Bay, September 1816. Flat Chart of Escholtz Bay, August 1816.*

Verso 12—Folio 13: *Plan of Captain's Harbor on Unalaska Island, 1817.*[99]

All the charts are executed with great thoroughness and to the highest standards expected of navigational aids of that time. Attention was paid to the coastlines and the navigational approaches to them. Coastlines surveyed by members of the expedition personally are represented by solid lines with shading; coastlines drawn on the basis of charts prepared for the Kotzebue Expedition are shown by thin solid lines; coastlines drawn on the basis of approximation, specifically from information elicited from the Natives, are shown by dotted lines.

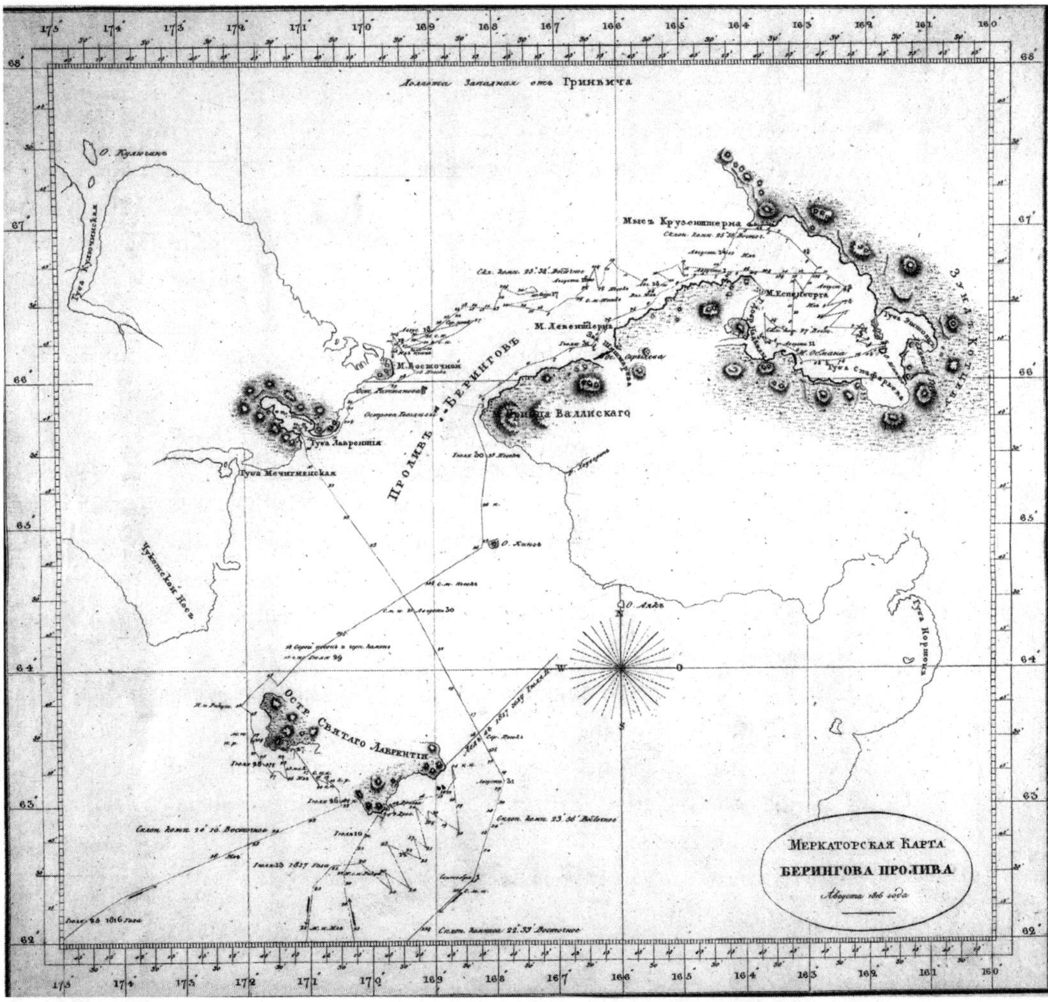

FIGURE 46 Bering Strait with newly discovered Kotzebue Sound, Otto von Kotzebue. *Atlas . . .* 1823. UAF Rare Book C0020

The *Rurik*'s routes, with dates and determination of coordinates, are given, and this, too, permits us to evaluate the relative precision of these charts. The charts' hydrographic content is detailed and expressive. Large-scale plans, in particular the plan of Captain's Harbor, show in detail the extent of dry land at ebb tide, the rocks both under and protruding from the water. Shore cliffs are represented realistically.

Native toponyms that had become long established through usage by *promyshlenniki* appear on the Kotzebue charts. But in the regions where this expedition pioneered the investigations, Native terms are practically absent, despite the fact that members of the expedition elicited local place names from the Natives and recorded them in their notes. The names of expedition members (Kotzebue Sound, Escholtz Bay, Shishmaref Bay, Cape Espenberg, and Chamisso Island) became a part of the chart.

Petr Karsakovsky Explores Southwest Alaska

The continued search for a settlement of long-lost bearded people in the early nineteenth century corresponded with the drive by the Russian-American Company to expand its operations north of the Alaska Peninsula. The company needed to expand into new areas for fur-gathering and counter the activities of foreign fur traders in Russian America. Although Baranov had long been planning an expedition to the north it became a reality only in 1818, when Ludwig von Hagemeister replaced him as Chief Manager.

Pyotr Korsakovsky, about whom little is known, was appointed to head this expedition. Tikhmenev notes that Korsakovsky was an employee of the Russian-American Company and had "some experience in compiling descriptions and was acquainted with the trade relations among the mainland Natives."[100] The expedition was dispatched from Kodiak to the north of Bristol Bay. Information about his voyage is based on instructions from the chief manager in Sitka addressed to the manager of the Kodiak office, sent in April of 1818, and also on the expedition journal that was discovered by M. B. Chernenko in 1956 in the Manuscript Division of the State Library of Russia in Moscow.[101]

Count Rumiantsev once again participated in financing an expedition by putting 3,533 rubles in silver at Hagemeister's disposal. In accordance with the proposal sent to the Kodiak office, the expedition was planned as a continuation of the investigations conducted by Kotzebue. It was proposed that there should include among the members of the expedition "one of the interpreters who was aboard the *Rurik* on its voyage north and he is to be offered, like the rest, a reward." The expedition was instructed:

> to endeavor . . . to reach those people that were seen by those on the *Rurik* and even farther north, where flows the River Kiveren, along which live bearded people, according to various traditions. These are believed to be descendants

of Russians driven there at an unknown time by storms. Various rumors from the Chukchi who travel to the islands in the Bering Strait and trade with those people support these traditions.[102]

The expedition party was advised to utilize local inhabitants in order to orient themselves and to study the localities. With this aim in mind, they were advised "to hire guides from one settlement to another; upon reaching the locality known to them, the guides should be discharged and others in the new settlement attracted by recompense to the same service." Hagemeister planned to launch a parallel expedition by sea. Native interpreters who participated in the Kotzebue voyage were to be included in that one also. In his letter to Kodiak, the chief manager formulated the following instruction to the expedition's leader pertaining to the study of new lands and peoples:

> A compass is enclosed. The voyager is to note daily the compass direction close to the route taken, the course of the rivers, and the estimated width and a description of their shores, whether sandy, rocky, or earthy, are to be included, as well as forest growth near rivers, peoples, at what distance and in what settlement, their numbers, their name, the name by which they are known to their neighbors and the name by which they call themselves. (Endeavor to hire guides from one people to the other; for gifts and exchange of small scale take some goods that are to be expended sparingly; do not reveal all your treasure so as not to arouse a wish among the wild ones to possess it; it is projected that only an acquaintance will be established, *hostages are not to be taken* [emphasis by AP]. It is to be expected that through such kind intercourse and reasonable demonstration of our strength not used for the destruction of the peoples, they, too, would be kindly inclined.
>
> Their way of life, methods, occupations, hunting and fishing, travel along the rivers and by what means, and whether they reach the sea and in what direction are to be noted. Endeavour to note minerals, are there transparent ones or those appearing like metal, of exceptional colors or weight, and take samples, recording the location where they were found. Observe what the people hunt, as the Chukchi trade red and black foxes and a large number of marten parkas from the bearded men. Should you reach the sea, place a marker with an inscription and ask the people if they have any news of the arrival of the vessel that we plan to dispatch in the summer. A copy of the voyage of Hearne and Mackenzie is enclosed[103] for the brave seeker of Fame with the wish that he should add to the information for which the world is indebted and thus make his name famous.[104]

From the instruction cited here it is evident that Hagemeister wanted to ensure the expedition would provide information as complete as possible about the areas visited, their natural resources, and especially their populations. The geographic aim of the expedition is self-evident: to link the Russian explorations in Alaska with the work of the Hudson Bay Company. During the early decades of the nineteenth century, an important component of exploration was the secure engagement of Natives, with knowledge of their own lands.

The goodwill of the Natives was also needed to ensure the economic success of the Russian-American Company as well as the safety of the company's employees. Hagemeister stressed this:

> In conclusion, I wish God's Blessing for the faithful sons of the Fatherland whose love for it will sustain them in their labors; deal with the wild ones with the utmost modesty, not offering any offense, as they are our fellow men, and make the name of Russians pleasant for them to hear, which will also serve the safety of those who form the community of travelers; do not engage in revenge for minor insults with punishment that is not known to them and for which they in return may find an occasion to destroy all; suffer these with indifference and ascribe them to their ignorance. Show that it is not cowardice that restrains the use of weapons, demonstrating the strength thereof on inanimate objects (but never on people). Respect their property and even stupid superstitions; leave it to time to work on those through good examples. Along with every possible demonstration of meekness, take utmost precautions against attacks. Should those happen (God preserve us), act with courage, but see them as enemies only as long as they fight. In the defeated, see your fellow man.[105]

Hagemeister believed that it was rational for the expedition to winter on the mainland. In his supplemental instruction, sent to Kodiak on April 11, 1818, he stressed the "necessity to proceed from Ilemna (Iliamna Lake) to N or NNW; old rumors about two large rivers promise much of interest, but not having detailed knowledge of what was done before nor the methods employed, nor people able to explain the latter, I have to leave it to the (Kodiak) Office to act in accordance with information collected from those who have been to the interior and during the journey itself."[106] As is evident from the *Journal of the Voyage from St. Paul Harbor of P. Korsakovsky and the Kolmakov Expedition*,[107] Pyotr Korsakovsky and his companions visited the Russian missionary Hieromonk Afanasii on Afognak. Here the organization of the expedition was finalized. Besides Korsakovsky, three more Russians were included: Fyodor Kolmakov, Pyotr Gorokhov, and Gavriil Patyukov;[108] Anderi Klimovsky, a Creole, and twenty "Americans" (Natives of the Aleutian Islands, Kodiak, and the Alaska Peninsula). Among them was an "Indian, who had been Kolmakov's guide to the settlement of the Aglogliut Indians and who was used to transport fur goods to Katmai and has been faithful to the Company for a long time."[109]

On May 19 Korsakovsky's party set out across the Alaska Peninsula, utilizing the rivers Kiyulik and Igegik. The Agligmiut Natives (Central Alaskan Yupik) and their *toion* Kipalek received the exploring party very well. Kipalek himself, "also Tuliman and interpreter Nikita Potorochin agreed to join and were issued a three-hatch baidarka."[110] Having spent about two weeks among the Agligmiut, Korsakovsky was able to give a detailed description of their way of life and mores.[111]

On June 14 the expedition arrived at the mouth of the Nushagak River. After passing it, Kolmakov led the detachment to the mouth of the Igushik River. From there, following the advice of *toions*, the travelers reached Kulukak Bay, having used the streams of the Interior to get there. Here they investigated the islands "Kaiashik," "Ugumkhtu," "Naliukak," "Unilgaik," "Ingikvan," and "Kilkin." On one of these they erected a cross with the inscription: "R.A.C. The land expedition was here on the 30th day of June 1818."[112]

The Russian exploratory expedition reached the mouth of the Togiak River and ascended it for about fifteen versts. Then members of the expedition sailed north by baidara to Cape Newenham, from where they could see "Kuskokhkin" (Kuskokwim) Bay.

Between July 12 and 17 the travelers stayed in the camp of the Kukokvigmiut. Here they learned that "they buy items of European manufacture, that is, iron items, copper kettles, and also seed beads from the Kiyatinskie (Kiiatagmiut Eskimo), and others obtain them in trade from the Kuikhpakskie" (inhabitants of the Kvikhpak or Yukon Valley).[113] The Natives also knew about Aleksandrovsky Redoubt on the Kenai Peninsula. On July 17 the expedition set out on its return voyage. After rounding Cape Tikhy (modern Cape Constantine), they reached Nushagak Bay on July 21. Here, in the Aglegmiut settlement, the travelers were handed Hagemeister's orders, left by ship, that they should travel northward in the direction of the two great rivers. The expedition divided. On July 26 Kolmakov set out to "investigate the forests," while Korsakovsky, accompanied by Klimovsky, Gorokhov, and Patyukov, left Bristol Bay on July 27 for Iliamna Lake. There they were joined by the supervisor of the "old settlement" (Staraya Iliamna village), Eremei Rodionov, a man well acquainted with local territory.

In Iliamna Korsakovsky received "from the local *toion* the news that allegedly an Indian from the Kukhtkhokhninskoe settlement visited the Russians in the north three times . . . he went to live in his own settlement and he is said to have something in his possession from these people."[114] This information confirmed one more of the long-known rumors about an ancient Russian settlement in the north. Ever more determined, Korsakovsky headed north on August 9 with twenty-four men, "along the trail" from Iliamna Lake. "Kenaitsy who know these regions well" safely guided the team to Mulchatna River or the Nushagak River, where on August 16 the travelers reached a Kenaitsy settlement. Here the expedition's leader decided to send Eremei Rodionov north with seven Natives selected by the *toion* to act as guides. The rest of the expedition remained in the settlement because the *toion* warned Korsakovsky that at that time of year it would be difficult to find food for a large detachment: "because, he said, the fish enter our river very early and then in small numbers."[115]

Even though there is no information about Rodionov's further progress, Professor VanStone proposes that he could have reached the Kuskokwim River.[116] In any case, Rodionov's party returned to Korsakovsky's camp in the upper reaches of the Mulchatna River by the beginning of September. During his sojourn on the Mulchatna, the expedition's leader recorded another Native testimonial about Russian settlement in Alaska's north, as follows: "The Old man Kylyshbak, who made two journeys northward along the sea coast, saw bearded people. They had a copper barrel [as in barrel of a cannon—LB], one end wider, the other narrower, resembling a blunderbuss. Another man had a copper barrel resembling that of a gun, decorated with black shadows and white lines . . . Kylyshbak compares their clothing to ours, he says it is cut the same."[117] On September 4 Korsakovsky's expedition set out on the return trip. Arriving at Iliamna Lake, at the spot where they had left their baidarkas, they went by way of Iliamna Lake, crossed to Cook Inlet by Kamyshak Portage, and on October 4 arrived safely at St. Paul Harbor in Kodiak.[118]

Initially the expedition's results were evaluated by Semyon Yanovsky (1789–1876) who had replaced Hagemeister as the chief manager of Russia's colonies in America, in a rather restrained manner as follows:

> The expedition dispatched last year to the north has returned, having traversed Iliamna Lake, the Nushagak River and up to Black Cape. As it appears to me, they reached the cape which is named on Arrowsmith's map Cape Newenham. According to the memorandum from the Kodiak office based on Korsakovsky's descriptions, it would appear that he discovered the Mountains of Gold, but it is not possible to know the reality. Journals have not been sent, and the maps are drawn without any scale and do not correspond to the description. Though Korsakovsky did indicate degrees, the extent of Iliamna Lake does not agree at all with the description; the rocks, clay and various items he brought have been retained by the Kodiak office. What for? Unknown. Many localities are marked according to rumors obtained from inhabitants. Though this expedition did not achieve major discoveries, there is hope for the future. The effort by Korsakovsky, Klimovsky and the other participants is deserving of notice.[119]

Yanovsky's note indicates that schematic maps were compiled with the longitude and latitude net and that these maps and the other reported materials were presented to the chief manager. Originals of these cartographic documents have not been located so far, but it is evident that they were used at the time in compiling generalized summary maps of Russia's American possessions. Korsakovsky himself did not conceal from his superiors that he considered his task not fully carried out and stood ready to travel north again at the first opportunity. Yanovsky considered that "he should not be deprived of this fine intent and should be given an opportunity to achieve glorious deeds and be of service to the fatherland." For this reason, he ordered the Kodiak office to organize another expedition in 1819 and assign to it "the same Korsakovsky as expedition leader and all those who were with him and to whom this voyage is familiar and in accordance with Korsakovsky's request, five two-hatch baidarkas, that is, 10 Aleut hunters, are to be added."

The main tasks of the expedition were formulated by Yanovsky in his instruction as follows:

> He, Korsakovsky, is to depart, together with all the people entrusted to him, on the 10th day of April, to the portage in Cook Inlet at the Burned Mountain. From there he is to portage to Iliamna Lake, timing it so that his arrival would coincide with the ice breakup on the lake. He is to traverse the lake and then to proceed to the sea along the River Kuchak (Kvichak). In his note Korsakovsky remarks that it would be better for the expedition to proceed by sea. I agree to this, he is to travel by sea from the mouth of the Kuchak River along the coast to Pushigik Island and further to Black Cape, where they found the Aglagmiut Indian.

According to Korsakovsky's description, in this area there are good bays and ships' anchorages. In this connection, Yanovsky directed the expedition to reach the area by early May and stay there until June, engaging in walrus hunting and trade with the Indians there. Yanovsky planned to dispatch the Russian-American Company vessel *Konstantin* to this area as the expedition's support. The vessel was to collect walrus ivory and items bartered from the Natives.

Upon receiving "support from the *Konstantin*," in the form of food supplies and necessary items, the expedition was to proceed north, "either portaging across Black Cape or rounding the same as circumstances might dictate. Then endeavor to reach that river about which you heard rumors. As it appears to be the Kheuveren River, put out the utmost effort to reach it." In setting the by-now traditional final task of the expedition in Russian America, Yanovsky stressed the necessity of investigating the still unknown coasts and rivers in detail in the course of the voyage.

> Should you reach the River Kheuveren, you need to penetrate it as far as possible and measure and describe it. You should not lose any time unnecessarily nor remain in a locality on the way; you should calculate wintering in such a way as to get to the mouth of the Nushagak River, where a settlement will be established this summer. Should you find rocks that are translucent or contain shining particles on the way, collect some for delivery to the office. As far as minerals are concerned, you should remember that the heavier the rock the more valuable it may be; also, should you see any clays or earth of an unusual

> type, endeavor to take samples, but note the locality and record it, so that it would be impossible to err.[120]

We observe that in his instruction Yanovsky was compelled to recognize the insufficient education of the voyagers, and in order to obtain materials of good quality he was forced to employ detailed explanations that at times were naive from the scientific point of view. It was directed that while proceeding north the expedition should leave "signs" for the *Konstantin*, which was to be dispatched to the same area. Korsakovsky was to return for the winter to the mouth of the Nushagak River, where by the time of his arrival it was planned to build "a small house and all that is necessary for peace and safety." Taking into account the insufficient quality of the cartographic materials produced in the course of Korsakovsky's first trek, the new chief manager assigned Apprentice Navigator Andrei Ustyugov to the expedition, who, as Yanovsky states in his instruction, "knows how to take observations and calculate the latitude and who has received full instruction as to determination of place and keeping a journal, and survey of coasts, rivers, and bays; for this he will be issued all means . . . Korsakovsky should not interfere with him in any way, on the contrary he is to offer support in everything. Ustyugov, like the rest, will be under Korsakovsky. His journals are to be delivered to me directly."[121] Yanovsky appended to his instruction a detailed "guideline for journal-keeping."[122]

One could expect some new data to be gained from the journal, reflecting the results of Korsakovsky's second expedition. Unfortunately this document has not been located so far, and information about this journey is found only in the work of Vasily Berkh, in which he presented descriptions of both voyages of Pyotr Korsakovsky on the basis of testimonies by Fyodor Kolmakov, Pyotr Korsakovsky, the Aleut "apprentice navigator" Andrei Ustyugov, and also of the "Uglakhmiut old man."[123]

In accordance with the plan, the detachment proceeded from Cook Inlet to Iliamna Lake and then down the Kvichak River to the sea. Following along the coast of the Alaska Peninsula they reached Togiak Bay and Hagemeister Island, where the *Konstantin* awaited them with provisions delivered from Kodiak. Then the expedition proceeded toward Kuskokwim Bay, but reached only as far as Good News Cape. An Eskimo whom Korsakovsky met advised them not to ascend the river, pointing out that serious obstacles would be encountered by the Russians, the main being the lack of food.

Korsakovsky also learned from the Eskimos that a large island lay to the northwest of Kuskokwim Bay, between the mouth of the Kuskokwim and that of the Yukon. This was Nunivak Island, which was destined to play a significant role in further explorations of coastal Alaska. The expedition turned back from Good News Bay. The main achievement of Korsakovsky's second expedition was the founding of the (New) Aleksandrovsky Redoubt at the mouth of the Nushagak River.[124]

The management of the Russian-American Company had first planned to move this redoubt to the Kuskokwim River, but on the basis of information provided by Korsakovsky it was concluded, in 1821, that such a step would be premature in view of:

> A. Neither the land nor the specific features of this river are yet known. B. It is rumored that at its mouth the coast is low and lacks forest growth and though upstream there is forest, the wild ones dwell there in large numbers. Consequently procurement of timber will be difficult and even possibly dangerous. The wild ones will have no difficulty in attacking us with united forces. C. This river can be reached only by sea, with a vessel.[125]

The 1819 march once again awakened the hope among the leaders of the Russian colonies of finding a settlement by bearded fellow countrymen among the Natives. The tales told to Korsakovsky by the Ugalakhmiut old man fed this hope, as he said that white people, reminding him of Russians, had formerly lived farther north but then, allegedly, resettled on a large island separated from the American continent by a wide strait. According to the old man, these people lived in wood-and-stone houses with roofs of thin copper that shone brightly on clear days. Berkh interpreted this to mean that all the tales about white men were really about the English settlements of Hudson Bay and in Quebec and various cities in the United States where houses were covered with copper.[126] It is also possible to suggest that the rumors about resettlement to a large island were a reflection of information received by the Eskimos about the Kotzebue Expedition's sojourn at St. Lawrence Island.

Apprentice Navigator Andrei Ustyugov, a member of Korsakovsky's expedition, left a significant trail in the history of the cartography of Alaska. He faithfully followed Yanovsky's instruction and thus the charts he compiled answered all the demands of that time. Veniaminov specifically noted that "His navigational chart of the Nushagak River (the first one) to this day (the year 1840) is considered very accurate."[127] Besides this specific chart of the Nushagak River, Ustyugov compiled a summary *Chart of the Northwest Coast of America from Bristol Bay to the Port of Good News*.[128] This chart apparently reflected not only the 1819 surveys conducted by Ustyugov himself, but also the cartographic materials we mentioned earlier from the 1818 expedition.

Along with Ustyugov's surveys made during the 1819 expedition, hydrographic measurements were taken from the sloop *Konstantin*. The result was the *Flat Chart of the NW Coast of America from the Nushagak River to Cape Newenham. Compiled by Adjusted Compass by Skipper Dmitry Pometilov in the Year 1819 Aboard the RAC Sloop Konstantin*.[129] The chart shows the expedition vessel's route, depth measurements, and sea bottom data, as well as the Aleksandrovskaya fort founded by Korsakovsky at the mouth of the Nushagak River in 1819 and an "Old Aleut settlement."

Mikhail Vasiliev and Gleb Shishmareu

Immediately following Korsakovsky's expeditions, the Russian Navy and also the Russian-American Company embarked on an exploration of the coasts north of the Alaska Peninsula. In 1819–1821 Russia's Admiralty outfitted a circumnavigating expedition whose goal was to investigate the possibility of a sea passage from the Pacific to the Atlantic Oceans along the northern coast of America. Experienced captains Mikhail Vasiliev on the *Otkrytie*, and Gleb Shishmarev on the *Blagonamerennyi*, explored the northern coasts of Alaska in 1819 and 1820 up to the vicinity of Icy Cape. Vasiliev reached 71°06′ N latitude in June of 1820, traveling thirty miles farther than Cook. Shishmarev reached 69°01′ N latitude.

The expedition spent the winter of 1820–1821 in San Francisco. While there, navigator Mikhail Rudalev made the first accurate Russian description of San Francisco Bay.[130] In July 1822 the *Otkrytie* and the *Blagonamerennyi* once again attempted to go north, but were stopped by ice earlier than in the preceding season.[131] At that time Lieutenant Aleksandr Avinov, senior officer on the *Otkrytie*, was dispatched in a longboat to search for a Russian settlement on the coast between Cape Newenham and Norton Sound. Avinov and his crew came to the firm conclusion that such a settlement did not exist, but their work near Alaska's coasts produced much new hydrographic data.[132] In the course of the Vasiliev and Shishmarev Expedition detailed oceanographic surveys were conducted and navigational charts drawn. However, these were not published; they are preserved in the Navy Archive in St. Petersburg. In particular, while visiting Sitka, members of the expedition made new surveys of the coasts and the sea in the region.

This chart shows detailed depth measurements that were made on two courses perpendicular to each other, an indication of the character of the bottom, a detailed outline of the coastline, and an accurate plan of Novoarkhangelsk. *Blagonamerennyi* conducted detailed surveys of the St. Matthew Island region, and specifically of the Sarychev Strait that separates that island from the modern Hall Island.

The northernmost surveys by the expedition were reflected in two charts, copies of which are reproduced in the *Atlas* by Efimov. They are *Chart of the American Coast between Cape Lisburn and Rikord, Surveyed from the Sloop Blagonamerenny, Depth in Sazhen, Year 1820*, also compiled by Vladimir Petrov,[133] and *Mercator Chart from the Bering Strait to Icy Cape with the Inclusion of the Northwest Coast of America and Northeast Part of Asia, Compiled from Surveys Conducted from the Sloops Otkrytie and Blagonamerennyi in the Years 1820 and 1821*.[134] The latter chart shows not only the routes of the two expedition vessels, but also those of the coastal surveys by Lieutenant Avinov in the longboat.[135]

Vasily Khromchenko

In tandem with the Vasiliev and Shishmarev Expedition launched by the Admiralty, the Northern Expedition outfitted by the Russian-American Company worked in the same waters: the vessels *Golovnin*, commanded by Vasily Khromchenko (Khramchenko) (1792–1849) and *Baranov* commanded by Arvid Etholen (1790–1876), a Finlander by origin. The initiative for this expedition came from Count Rumiantsev who recommended appointing Khromchenko, a former member of the Kotzebue Expedition, as the expedition's commander.

The management of the Russian-American Company accepted Rumiantsev's advice. However, at the same time it viewed Rumiantsev's active participation with reserve. The basis for this stance was clearly spelled out in a letter from the Main Office of the Russian-American Company to Muraviev:

> Dear Sir, Matvei Ivanovich!
>
> It is known to us that Count Nikolai Petrovich Rumiantsev wishes that Kotzebue's companion, Mr. Khromchenko, should be dispatched by you in the wake of the Rurik northward for exploration and discoveries. Such a patriotic wish is by itself deserving of praise, but it is harmful to our Company, as Kotzebue's tales have inflamed the foreigners to visit, one of the most harmful of their goals. For this reason we wish that discoveries and descriptions should be solely by our Company and we beg you to mention in instructions to anyone whom you might dispatch to the North that he is entrusted to act in the name of the Company and that **report must be made only to the Company** [emphasis by AP]. The Company will assume the cost of gifts assigned by the Count as rewards for service in discoveries in the North, as long as its goals and interests are not damaged.[136]

In a directive issued on October 6, 1821, the leaders of the Russian-American Company demanded even more firmly the maintenance of secrecy with respect to the conduct of the expedition to the north and caution in relations with Count Rumiantsev, stating specifically:

> Anything to the contrary and side issues have no place here. For this reason the participation in the discoveries in the North by Count Rumiantsev is not always pleasing to the Management. He looks out for his own goals and seeks to rob the Company of its fame acquired though its efforts and to acquaint the foreign public with his discoveries. For this reason the Management informs you that if the Count is acquainted with the successes of the expedition to the North, this information should be scant, not to its fullest extent. The good of the Company demands it. The basis for this is its privileges, according to which no one may act within the territory belonging to the Company except for Company Agents acting for its benefit.[137]

In these documents we find a recurrence of Russian traditions of secrecy about new geographic information and also the intent of the Russian-American Company to monopolize such information. This led, as we shall see below, to rather inadequate reporting of the results of the expedition in the press. The Khromchenko and Etholen Expedition left Sitka in 1821 and spent nearly two years in northern waters. They made a significant contribution to the exploration of Alaska.

In the first stage of the voyage the travelers sailed to Bristol Bay, where, in accordance with the chief manager's instructions, they surveyed and described Hagemeister Island and the strait of the same name between it and the mainland. Etholen surveyed Good News Bay.

After finishing these tasks the sailors set their course to the north, where, independently of each other, they located Nunivak Island. As we saw above, the first information about this island was obtained from the Eskimos by Korsakovsky's expedition two years earlier. Khromchenko was the first to approach Nunivak, though later his primacy was contested on the basis that three days earlier Avinov raised the flag of St. Andrew of Russia when he was conducting surveys in a longboat as part of Vasilev's expedition.[138]

Thus, in 1821 four Russian vessels conducted hydrographic work in the area between Bristol Bay and Norton Sound. Etholen and Khromchenko were not able to make a complete survey of Nunivak's coast, but Etholen completed the cartography of a part of the strait that separates this island from the mainland, the strait that to this day bears his name, the Etolin Strait. He also entered the rivers Nushagak and Kuskokwim, while Khromchenko, continuing the survey of Norton Sound where Cook left off, found an inlet he named Golovnin Bay to the west of Cape Darby.

In 1822 Khromchenko obtained detailed information from the Eskimos inhabiting the shores of this bay about the southwestern part of the Seward Peninsula. On this basis Khromchenko concluded that from this bay, utilizing the interior waterways and portages, it was possible to reach Shishmaref Bay and the Arctic Ocean.[139] More detailed information about this region was not obtained until the end of the nineteenth century. Khromchenko learned some facts about the shores of the Seward Peninsula to the north of Golovnin Bay from Tungan, an experienced Eskimo who had traveled to St. Lawrence Island and the Siberian coasts.

The Khromchenko and Etholen Expedition found that no additional undiscovered islands existed in the Pribilof Islands group. New information was obtained about Walrus Island (near St. Paul Island), and detailed surveys were taken near Cape Mendenhall on Nunivak Island. However, one of the main purposes of the expedition was to survey the Alaska coast between Cape Vancouver and the Seward Peninsula and to obtain information about the inhabitants of these areas. The sailors were not able to achieve full success because of foul weather and shallow coastal waters. No information was obtained about Yukon Delta Natives.

The failure to conduct a survey here possibly delayed the final discovery of this great river until the journey of Andrei Glazunov, who came to it by land out of St. Michael Redoubt in 1834. Khromchenko's geographic data about the regions in the upper reaches of the Nushagak River were obtained by questioning the Natives.

Field exploration of these regions only began seven years later. Just the same, he correctly evaluated the importance of the Nushagak River system and concluded that it could be utilized to penetrate the Alaska interior and the basins of other important rivers flowing into the Bering Sea. The Eskimos reported an abundance of beaver in these regions and expressed their readiness to trade.

The expedition's results stimulated the decision by the Russian-American Company to organize subsequent explorations of the Interior by Ivan Vasiliev, Fyodor Kolmakov, and others. These explorations led to intensive exploration of the basins of the Nushagak, Kuskokwim, and Yukon, accomplished in a timespan of less than twenty-five years, and also to the establishment of the Kolmakov Redoubt on the Kuskokwim and the Mikhailov Redoubt northeast of the mouth of the Yukon.

Though the ethnographic information collected by Khromchenko that is reflected in his journals is not remarkable for its volume or detail, it is the earliest recorded about the Eskimo population of southwest Alaska and their migrations. In particular, the 135-word vocabulary reported by Khromchenko, the earliest of those for this region, is of great value for Eskimo linguistics. Khromchenko's information about the inhabitants of Nunivak Island is of even greater value for ethnographic studies. It is clear that these people did not encounter any Europeans before the expedition's arrival.[140] Excerpts from Khromchenko's journal were originally published in St. Petersburg in the fall of 1824, in seven issues of the journal *Severny Arkhiv* (*Northern Archive*).[141] The role of the Khromchenko and Etholen Expedition was evaluated very highly, as is seen from the letter from the State Admiralty Department to the management of the Russian-American Company dated March 13, 1824, # 264:

> having paid just praise to these sailors who delivered detailed information about the coasts, islands and other localities of Northeastern America, and noting that with the acquisition of this information *there is no longer any need to dispatch a special officer for the survey of the American coast from Norton Sound to Bristol Bay, as was earlier proposed* [emphasis by AP], I saw it as my duty to renew the presentation made in 1822 to the Lord Head of the Naval Staff of His Imperial Majesty about the merits of Messrs. Khromchenko and Etholen and to ask His Excellency for the second time to request Royal awards for them. To the first, the order of St. Vladimir, 4th degree, to the second-named the rank of Navigator of 12th class so that these worthy sailors, realizing that their efforts for the benefit of the fatherland have not gone unnoticed, will renew their efforts to achieve more successes in the future.

Sarychev signed the letter.

The Admiralty petition was not responded to in full measure. The "voluntarily serving skipper" Etholen received the rank of Navigator 14th class, while Khromchenko received the Order of St. Anna, third degree.[142] Further, in September 1824 Khromchenko was promoted to the rank of lieutenant.[143]

The original charts and surveys, the results of Khromchenko's and Etholen's expedition, were not published, as was the case for the works of their colleagues Vasiliev and Shishmarev. Moreover, in contrast to the Admiralty Expedition, whose manuscript charts are known and have been analyzed above, the cartographic materials of this Russian-American Company expedition have not been located so far. For this reason, the general manuscript chart that summarizes the results of all the Russian expeditions up to 1822 is of special interest. This is the *Mercator Chart, Representing the Arctic Ocean, Bering Strait and Part of the Eastern Ocean with the Coasts of Chukotka Land and North America*, preserved in two copies.[144]

The explanatory remarks on this chart state the role played by various expeditions in exploring the northern part of the Pacific Ocean as follows:

> On this chart the outlines of the coastlines are corrected from the surveys by Vasiliev of 1820 and 1821 from Kotzebue Sound to Icy Cape. Northern coast of St. Lawrence Island from the survey by Shishmarev of 1821. The Arctic Ocean coast from Kolyma River to Shelagsky Cape from L. Wrangell of 1821. Southern part of Nunivak Island discovered by K. Vasiliev in 1821. Golovnin Sound. Kuskokwim River, eastern and northeastern parts of Nunivak Island discovered and surveyed by M. Khromchenko and Skipper Etholen in 1821.

Khromchenko's information about the possibility of a passage from Golovnin Bay to Shishmaref Bay is shown on this chart, quite erroneously representing it as an unbroken water passage between these two bays. In published form, the results of the expeditions of the years 1818–1822 are reflected in the *Atlas of the Northern Part of the Eastern Ocean*,[145] a summary of all the previous cartographic work of the Russian hydrographer Vice-Admiral Gavriil Sarychev. This atlas contains twenty-six charts and coastal plans of Asia and North America. Besides the materials of the Billings/Sarychev Expedition discussed earlier, it also contains charts by Kotzebue, Golovnin, Hagemeister, M. N. Vasiliev, and others. This fundamental work clearly demonstrated the gaps in the cartography of the northwestern coasts of America.

In this connection, in June of 1829, Sarychev issued an instruction to the main office of the Russian-American Company to order the captains of its vessels to engage in surveys and descriptions of those coasts of Russian possessions in America that were not shown in the *Atlas of the Eastern Ocean*.[146] In this way Sarychev's atlas came to serve as the base for further development, to be constantly upgraded by employees of the Russian-American Company and by members of government naval scientific expeditions.

The results of the 1820–1822 expeditions were analyzed in detail and discussed by the leadership of the Russian-American Company as well as by the State Admiralty Department. There, starting in 1821, V. N. Berkh conducted large-scale work summarizing geographic information about the northern regions of

America and Asia and compiled general geographic maps. The manuscript *Map of the Russian Possessions in North America Published at the Expense of the Russian-American Company Composed by Captain Lieutenant Berkh and Drawn in the Draftsmen's Office of the State Admiralty Department by Student of the 1st Class A. Tokarev. 1821, May 31*[147] was apparently compiled for the specific purpose of reflecting the collected information about Alaska and adjacent regions of the American continent in its entirety.

This work is possibly the first effort after Chirikov's of 1742 to represent the lands explored and known well by the Russians on a general map of North America that included its eastern part to Hudson Bay and the Baffin Strait. The link between this work and the problem of finding a water passage from the Pacific to the Atlantic Oceans is obvious: information is included that was obtained from Natives and from explorers who traveled by land. There is specific information about the rivers that empty into the Arctic Ocean. The routes followed by the land expeditions of Samuel Hearne and Alexander Mackenzie are shown.

Inscriptions such as these are given: *The Sea Seen by Hearne in the Year 1777* and *The Sea Seen by Mackenzie in the Year 1789*. Hearne's route is shown crossing a river with the following inscription: "According to the Indians, this river debouches into the sea." To the east of Russia's American possessions a hypothetical river is shown, with the explanation that according to the American (Indians), it flows from the south and empties into the sea.

The Mercator chart of northern polar regions appended to the second volume of Berkh's 1823 *Chronological History of All Voyages into the Northern Polar Regions with the Inclusion of a Survey of this Area's Physical Characteristics* is remarkable for its broad view. This map shows the eastern coasts of Greenland, northern coasts of Europe and Siberia, and the coast of Alaska between 71° and 59° N latitude on a single geographic canvas for the first time.[148]

The results of such analyses of materials on the geography of polar regions and of the experiences of preceding voyages led the leadership of the Russian naval department to the idea of launching a new overland expedition. The program for this trek was outlined in a special memorandum by the State Admiralty Department, *On a New Expedition to the North to Search for a Passage Eastward* . . . directed to M. I. Muraviev via the head office of the Russian-American Company. Specifically, it included the following statement:

> The Chancellor (Rumiantsev) had not given up on the idea of exploring the shores of the Arctic Ocean by land since the Rurik's return (that is, the Kotzebue expeditions). Since the time when it became known that the English captain Franklin had set out from England to descend the Mackenzie River and then proceed toward the Bering Strait, His Excellency's intention to carry out this idea has firmed up. Such an expedition may solve four important geographic problems.

1. Whether the coast where Icy Cape is located is part of the mainland or whether it constitutes an island; if the latter is the case, is it possible to effect a passage to the east between Icy Cape and Cape Lisburne, as we have no knowledge of the area between these two capes.
2. Is there a coast to the north of Icy Cape, and if so what is the trend of this coast to the east?
3. To determine the situation and trend of the mountains known as Kamenistye (Rocky) that ought to terminate somewhere at the very shore of the Arctic Ocean. The answer to this question is also important politically, as if these mountains lie to the west of the Mackenzie River they could rather conveniently be accepted as a natural boundary between the possessions of our Russian-American Company and England.
4. To determine by means of astronomical observations the situation of the mouth of the Mackenzie River, not known to this day."[149]

Careful preparations were made. Very little exploration by the Russian-American Company had been done utilizing dog sleds. This new expedition would need that, so the Admiralty Department's proposal was accompanied by "remarks compiled by Mr. Matyushkin, who traveled together with Baron Wrangell for three years over ice and amassed great experience in such undertakings (dog sled travel)."[150]

Count Rumiantsev provided 20,000 rubles in silver for the organization of this expedition, while the head office of the Russian-American Company authorized an additional 10,000 to be expended, "but no more, except for the leader's salary and that of his subordinates, paid by the Company, which will continue." The Admiralty Department considered that Khromchenko and Etholen were the most suitable persons to carry out the proposed exploration program. However, the Russian-American Company management responded that "The Head Office considers it difficult to dispatch Messrs. Khromchenko and Etholen at the same time, as in such a case the colonies will be deprived all at once and for an undetermined period of two of their best navigators."[151]

Much attention was paid to the scientific side of the expedition. For this reason, the application of the Imperial Botanical Garden to the Russian-American Company asking that the latter provide rare plants and other "natural history objects" was quite opportune. This request by the Botanical Garden was accompanied by detailed guidelines on specimen collection. Later on these were used as a standard by all the exploratory expeditions in Russian America.[152]

Although two-thirds of the cost of the expedition was to be financed by Count Rumiantsev, the management of the Russian-American Company apparently had no intention of letting the sponsor have unrestricted freedom in the use of any obtained materials. The company management's hardening position in respect to the use of any results of geographic explorations is best evidenced by the *Obligations* signed by Pyotr Egorovich Chistyakov (1790–1862) on January 29, 1824, in preparation for assuming his post as chief manager of the Russian colonies in America (he occupied this post from 1825 to 1830). In these *Obligations* Chistyakov wrote, in particular, "should I decide to publish something to the world, I can do it only with the consent of the Company. . . . As follows from the preceding point, I undertake, upon my return to St. Petersburg or even earlier via the Head Office, to deliver to the Company all charts, plans, views and journals."[153]

In spite of efforts by the Russian-American Company leadership and that of the State Admiralty Department, information about Russia's geographic explorations in the northern part of the Pacific Ocean, the Arctic Basin, and in Alaska spread beyond the borders of the empire. The chief "guilty person" was Otto von Kotzebue, whose works were published in foreign languages. This information disturbed Britain above all, because its Hudson's Bay Company was the immediate neighbor of the Russian-American Company in Alaska. But this new large-scale Russian expedition to the north was canceled in 1827 because of the death of its chief sponsor, Count Rumiantsev, and the refusal of his heir, S. P. Rumiantsev, to confirm expenditure of the funds allocated for it by his father.[154]

GREAT BRITAIN EXPLORES ARCTIC ALASKA

Great Britain viewed the Russian successes in the exploration of America's polar regions with considerable jealousy. The secretary of England's Admiralty (1804–1845) and founder of the Royal Geographic Society (in 1830), Sir John Barrow, wrote in 1817: "It would be somewhat mortifying if a naval power but of yesterday should complete a discovery in the nineteenth century, which was so happily commenced by Englishmen in the sixteenth."[155] In 1818 the British parliament announced a considerable reward to the voyager who should discover the Northwest Passage from the Atlantic to the Pacific Ocean.

In the same year the British vessels *Dorothea* and *Trent*, under captains David Buchan and John Franklin (who had served as a lieutenant aboard the *Trent* under Beechey), attempted to sail from the Atlantic to the Bering Strait via the North Pole, but were stopped at Spitsbergen by ice. In 1819 John Franklin explored the Arctic coasts of North America between the mouth of the Coppermine River and the eastern end of the continent.

In 1821 the British expedition of William Parry and George F. Lyon, with the *Fury* and the *Hecla*, was dispatched to the northeastern part of Hudson Bay to search for a passage to the Pacific Ocean, traveling westward in the late spring of 1822. The members of the expedition believed that the local inhabitants—the Eskimos—were a fairly reliable source of information. In particular they were excited by a map drawn for them by an Eskimo woman. Her information was that at a distance of three-days' travel west of the Melville Peninsula there was open sea. Further talks with the Eskimos strengthened Parry's and Lyon's conviction that the passage was very near, and they wrote: "This little North-West Passage set us all castle-building, and we already fancied the worst part of our voyage over."[156] In June of 1822 the expedition reached the Igloolik settlement on the island of the same name.

At the beginning of August another Eskimo once again drew a map that confirmed the presence of a passage to the northwest. The expedition's vessels were halted by ice in a strait the Eskimos called Closed. Having spent the winter on Igloolik Island, the British were forced to turn back in August 1823.[157] The cartographic results of this expedition are reflected in the *General Chart Showing the Track of H.M. Ships Fury and Hecla, on a Voyage for Discovery of a North-West Passage, A.D. 1821–22–23. The Shaded Parts of the Coast Show the Discoveries of this and the Preceding Expedition*, first published on January 7, 1824, by John Murray in London.[158]

Canadian scholar Barry M. Gough, who studied the correspondence between Krusenstern and John Barrow and between Franklin and Barrow, established that Krusenstern proposed to borrow Franklin's plan for exploration of the Northwest Passage: a combination of an overland voyage out of the mouth of the Mackenzie River and a sea voyage out of the Bering Strait eastward.[159] Franklin was especially disturbed by the possibility that exploration by the Russians might lead to an expansion of Russian possessions to the fur-rich areas of the Mackenzie River basin.

Beechey Expedition

To counter such a development the British Admiralty organized a new expedition. The execution of the maritime portion in Alaskan waters was entrusted to Commander Frederick William Beechey (1796–1856) on the ship *Blossom*. The *Blossom* was magnificently outfitted with instruments to conduct hydrographic surveys and precisely determine coordinates; among them were latitudinal and azimuth circles, a large theodolite, a latitudinal instrument, a redundantly reflective instrument, two azimuth compasses, four chronometers (three box and one pocket), four barometers, twelve thermometers, four hydrometers, steel survey chains, tapes, and stakes.[160]

Although the only scientist aboard the *Blossom* was George Tradescant Lay, the other officers were apparently selected for their competence in conducting scientific observations. George Peard had solid geological knowledge. Lieutenant John Wainwright conducted astronomical observations, kept the records of various surveys, and dispatched useful notes to the Hydrographic Department. Edward Belcher, who, together with Thomas Elson, was Beechey's main assistant in conducting hydrographic and topographic surveys, gathered a rich collection of geological specimens. The first officer, James Wolfe, did much to compile charts and conducted astronomical observations in the field observatory. William Smith provided drawings of collected specimens. The expedition's surgeon, Alexander Collie, and his assistant, Thomas Nelson, also took an active part in the expedition's scientific activities, especially in specimen collection. Collie was interested in natural science, meteorology, and geology.

Aboard the *Blossom* were the newest hydrographic data and charts, including detailed materials from Kotzebue's circumnavigating voyage. Despite the Anglo-Russian competition, these were delivered to the British Admiralty by Admiral Krusenstern in 1823.[161]

The twenty-six-gun British Navy fighting schooner, the *Blossom* (on the voyage she carried only sixteen guns) set out on January 12, 1825, for Bering Strait. She was to rendezvous with the expedition of Captain William Edward Parry (1790–1885), who was sailing westward along the coast from Prince Regent's Inlet, and the expedition of John Franklin, who was supposed to descend the Mackenzie River and link his earlier explorations in the Coppermine River mouth with the westernmost known points of America by land.

According to the Admiralty instruction, Beechey was to arrive in the Bering Strait no later than July 10, 1826, and there await Parry and Franklin until the end of October or later, weather permitting. In the meantime he was to conduct investigations of America's coasts in the northern and eastern directions by means of a longboat equipped with a sail, for as long as ice conditions would permit. Should the rendezvous not materialize, Beechey was to sail to Kotzebue Sound the following year and wait there until the next fall, Franklin's overland expedition all the while continuing attempts to penetrate farther east and conduct surveys and cartographic work.

In June of 1826 the *Blossom* sailed for Kamchatka out of the Sandwich Islands (Hawaii). In Petropavlovsk the expedition encountered a Russian ship under the command of Baron Wrangell, "an enterprising officer," Beechey wrote in his book, "well known to the world as the commander of a hazardous expedition on sledges over the ice."[162] Here traditional Russian hospitality and letters reporting the return of Captain Parry's expedition were waiting for the travelers. A local priest, following the tradition established by his grandfather to send a calf to each arriving British ship, gave Beechey a heifer and supplied the British daily with fresh milk, butter, eggs, and other products.

On July 5 the *Blossom* left Petropavlovsk Harbor and set her course for Bering Strait. Sailing past a rock 5.5 miles away from Bering Island, Beechey noted: "Kotzebue observes in his narrative that 'this rock has not been laid down in any chart': I presume he alludes to those which are current, as on a reference to the map of Captain Krenitzen's discoveries in 1768, it will be found occupying its proper place."[163] Having noted the Russian priority in the cartography of this rock, Beechey nevertheless found it possible to give it an English name, "Seal Rock." The expedition then continued to St. Lawrence Island, where the expedition members for the first time encountered aboriginal inhabitants of the near-polar region of the Pacific Ocean. Beechey remarked on the closeness of the Natives to the Eskimos of

Greenland and the Canadian Arctic. The inhabitants of St. Lawrence Island visited the British ship, but Beechey writes that they had to leave shortly "in consequence of the fog coming on, which their experience enabled them to foresee sooner than any person on board . . ."[164]

Passing through Bering Strait, members of the expedition found that there, between America and Asia, everything was situated as on Captain Cook's chart but that there were three and not four islands, as Kotzebue had maintained. Beechey notes: "For the sake of convenience I named each of these islands. The eastern one I called Fairway Rock, as it is an excellent guide to the eastern channel, which is the widest and best; the centre one I named after the Russian admiral, Krusenstern, and to the north-western island I transferred the name of Ratmanoff, which had been bestowed upon Kotzebue's supposed discovery."[165]

According to observations by expedition members, the seafloor at the approaches to the Diomede Islands, as well as to St. Lawrence Island, shifted from fine sand to gravel and coarse rock, which could serve, as Beechey noted, as an excellent navigational aid for sailing in those waters.

In the early morning of July 22, 1826, the *Blossom* entered Kotzebue Sound, where the members of the expedition located one more bay which they named Hotham Inlet in honor of one of the lords of the Admiralty. A longboat, commanded by Elson, was dispatched to survey the bay, but could not enter because of shallow water. On July 25 the expedition reached the small Chamisso Island, the designated rendezvous point with Franklin.[166]

On July 30 the *Blossom* once again hoisted anchor and set course to the north. Elson was sent out in a longboat to make a detailed survey of the coast and to put up signals for Franklin (Fig. 47). Elson not only engaged in a hydrographic survey from the longboat but also in a topographic one by means of the theodolite. Moreover, both vessels were engaged in concomitant oceanographic research that included dredging and the measurement of temperature and current velocity at various depths.[167]

To the great regret of all those aboard the *Blossom*, passage along the coast of North America to the Atlantic Ocean was not among their tasks. Nevertheless, Elson was dispatched in this direction in a longboat. With him were First Officer Smith, six sailors, and two marines. On August 27 the *Blossom* was back at Chamisso Island, where members of the expedition found part of a board left by Kotzebue inscribed in Russian: *Ryurik, July 28, 1816. Loyal, 1802.*[168]

In the course of their sojourn in Kotzebue Sound the members of Beechey's expedition met the local inhabitants (Eskimos) numerous times and became convinced of their remarkable geographic and cartographic abilities. Captain Beechey informs us:

> On the first visit to this party, they constructed a chart of the coast upon the sand, of which I took very little notice at the time. Today, however, they

FIGURE 1 *Carte générale de la Siberie et de la Grande Tartaria.* Courtesy of the Newberry Library. Ayer MS Map 30 sheet no. 115

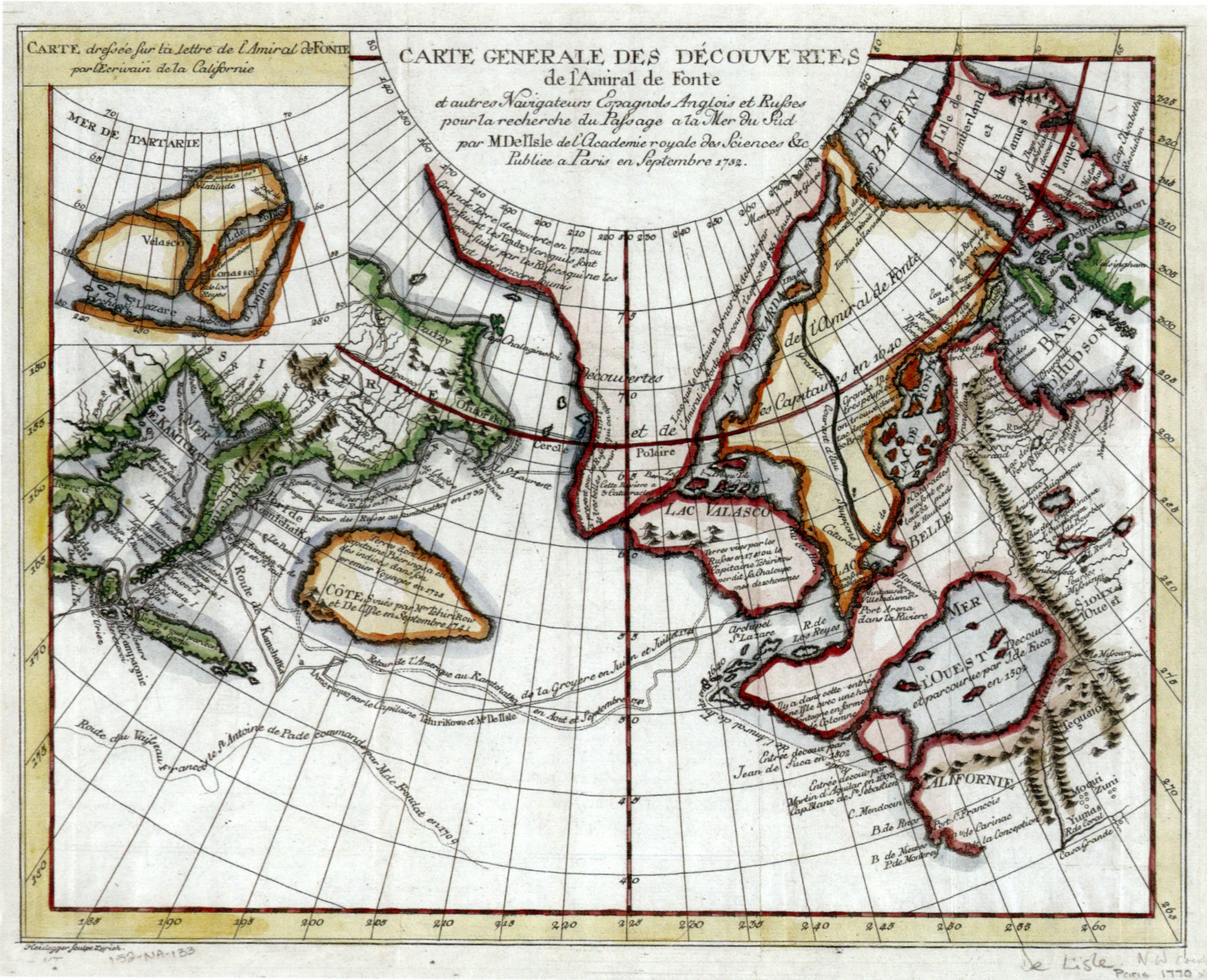

FIGURE 11 Delisle map of the spurious de Fonte geography. UAF Rare Map M0134

FIGURE 12 Müller's Academy of Sciences map, 1754. UAF Rare Map G9236 S12 1754

FIGURE 17 Jacob von Staehlin's Academy of Sciences map. UAF Rare Map G9236 S12 1773 A221

FIGURE 28 Shelikov's Voyage tracks. Courtesy University of Göttingen von Asch Collection 279

FIGURE 66 Portion of Grewingk's mineral map. End papers. UAF Rare Book A0488

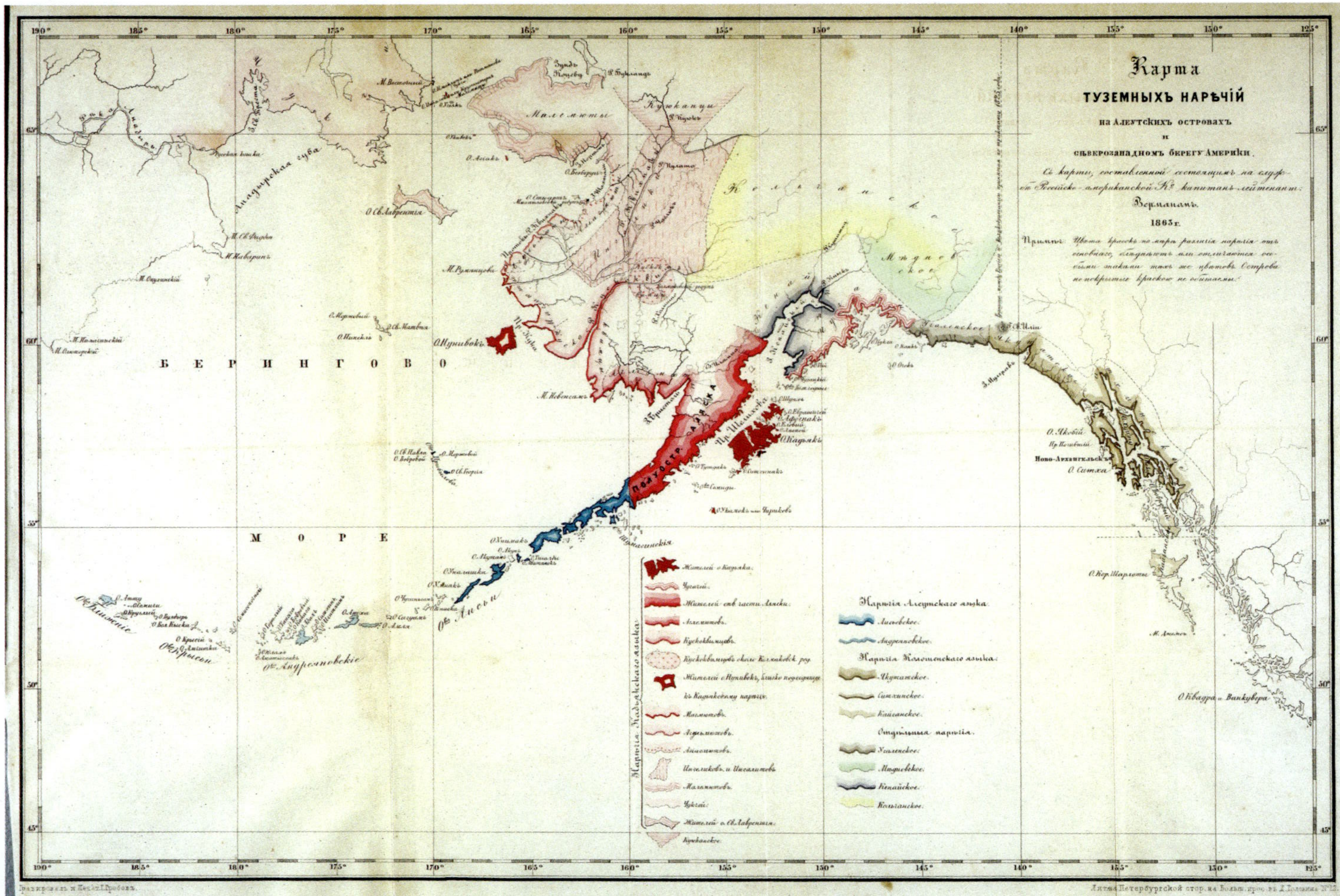

FIGURE 68 Chart of the indigenous dialects of the Aleutain Islands and the northwest coast of America. Prepared by Capt. Lieut. Wehrman, in Tikhmenev, *Istoritchesko obzrienie . . .* 1863. vol. 2 at end UAF Rare Book A0502

FIGURE 76 *North Western America . . . 1867.* UAF Rare Map G4370 1867 U551

FIGURE 47 *Blossom*'s barge on Arctic Coast, oil painting UAF Rare Book vault

renewed their labour, and performed their work upon the sandy beach in a very ingenious and intelligible manner. The coast line was first marked out with a stick, and the distances regulated by the days' journeys. The hills and ranges of mountains were next shown by elevations of sand or stone, and the islands represented by heaps of pebbles, their proportions being duly attended to.

As the work proceeded, some of the bystanders occasionally suggested alterations, and I removed one of the Diomede Islands which was misplaced; this was at first objected to by the hydrographer; but one of the party recollecting that the islands were seen *in one* from Cape Prince of Wales, confirmed its new position, and made the mistake quite evident to the others, who seemed much surprised that we should have any knowledge of such things. When the mountains and islands were erected, the villages and fishing stations were marked by a number of sticks placed upright, in imitation of those which are put up on the coast wherever these people fix their abode.

In time we had a complete topographical plan of the coast from Point Darby to Cape Krusenstern. In this extent of coast line they exhibited a harbour and a large river situated to the southward of Cape Prince of Wales, of neither of which we had any previous knowledge. The harbour communicated with an inner basin, named Imaurook, which was very spacious, and where the water was fresh. The entrance to the outer one was so narrow that two

> baidaras could not paddle abreast of each other. This they explained by means of two pieces of wood, placed together, and motioning with their hands that they were paddling. They then drew them along till they came to the channel, when they were obliged to follow one another, and, when through, they took up their position, as before. The river was between this harbour and the cape, and by their description it wound among lofty mountains, and between high rocky cliffs, and extended further than any of the party had been able to trace in their baidaras. Its name was Youp-nut, and its course must lie between the ranges of mountains at the back of Cape Prince of Wales.
>
> At this last mentioned cape, they placed a village, called Iden-noo; and a little way inland another, named King-a-ghee, which was their own winter residence. Beyond Imau-rook there was a bay, of which we have no knowledge, named I-art-so-rook. A point beyond this, which I took to be the entrance to Norton Sound, was the extent of their geographical knowledge in that direction.
>
> To the Diomede Islands they gave the names of Noo-nar-boak; Ignarlook, and Oo-ghe-eyak; King's Island, Oo-ghe-a-book; and Sledge Island, Ayak. It is singular that this island, which was named Sledge Island by Captain Cook, from the circumstance of one of these implements being found upon it, should be called by a word signifying the same thing in the Esquimaux language.[169] For East Cape they had no name, and they had no knowledge of any other part of the Asiatic coast. Neither Shishmaref Bay nor the inlet in the Bay of Good Hope was delineated by them, though they were not ignorant of the former when it was pointed out to them. It has been supposed that these two inlets communicate, and that the Esquimaux, who intimated to Kotzebue that a boat could proceed nine days up the latter and would then find the sea, alluded to this junction; but our rude hydrographers knew of no such communication.[170]

The description of Eskimo cartographic practices and geographic knowledge presented by Beechey is the first and most exhaustive of its kind. It was utilized in analysis of cartography outside of the European framework by Leo Bagrow and R. A. Skelton in their classic work and has become well known in the history of world cartography.[171]

While Beechey awaited his meeting with Franklin's expedition at Chamisso Island, he conducted scientific observations and surveys (specifically investigating Spafariev Bay, which had not been explored in sufficient detail by Kotzebue). The longboat commanded by Elson was engaged in a description of the coast to the north of Kotzebue Sound. Elson's group reached the northernmost point of the Arctic coast in this area, which they named Cape Barrow in honor of the initiator of polar research, Lord of the Admiralty John Barrow.

Altogether 126 miles of the Arctic coast of North America were added to what was already known during the voyage of 1826. Franklin not having arrived, the expedition set out October 13 for San Francisco and the Hawaiian islands for the winter. Beechey had no way to know that Franklin had already reached the eastern portion of Alaska's Arctic coast, but had turned back in August 1826, arriving in England in September 1827.

In 1827 Beechey's expedition returned to the northern part of the Pacific Ocean. After a short stay at Petropavlovsk-Kamchatka (July 3–18), it continued its work off the Arctic coasts of North America. The travelers were much intrigued by a passage to the east shown by the Eskimos on their map in the area of Cape Prince of Wales. The longboat, commanded by Elson, was dispatched to search for this passage. Elson located the passage, but was unable to investigate it due to foul weather. In this area Elson was visited by "several of the natives while there, one of whom drew him a chart, which corresponded to the one constructed upon the sand in Kotzebue Sound the preceding year."[172]

This harbor was investigated in greater detail by Beechey himself, who returned to the Cape Prince of Wales region after a voyage north to the edge of the polar ice. He entered the passage, called by the Natives Tokshook, and in the interior of the bay found three Eskimo villages. The outer harbor that Beechey investigated he named Port Clarence (in honor of the Duke of Clarence), and the inner Granley Harbour in honor of Lord Granley. At this same time the longboat commanded by Lieutenant Belcher conducted explorations to the northeast of Chamisso Island. In the fall of 1827 the expedition suffered several mishaps, the most important of which was the loss of the longboat at Chamisso Island. Relations with the local inhabitants worsened and on October 6 the *Blossom* turned back for the homeland.[173]

The most important result of the expedition was the detailed surveying and mapping of approximately 600 miles of the coast of North America, one-fifth of which had never been studied before. Navigational maps based on the results of the surveys were drawn during the voyage, and fair copies of them were made under Beechey's leadership at the hydrographic department of the British Admiralty. Despite the rivalry between Britain and Russia in the northern part of Alaska, in 1830 the hydrographers of the British Admiralty Captain (later Sir) Francis Beaufort gave Krusenstern a copy of the *North Polar Chart* with additions made by Beechey.[174]

The most valuable contribution to the creation of a map of Alaska was the expedition's surveys from Cape Rodney to Cape Barrow, as well as those in the region of Kotzebue Sound and Port Clarence. The cartographic results of the voyage are represented in the *Chart of Part of the North West Coast of America, from Point Rodney to Point Barrow by Captain F. W. Beechey R.N.F.R.S. in His Majesty's Ship* Blossom *assisted by Lieu. E. Belcher, Mr. Elson, Master, Mr. Jas. Wolfe, Mate . . . 1826 and 1827.*[175]

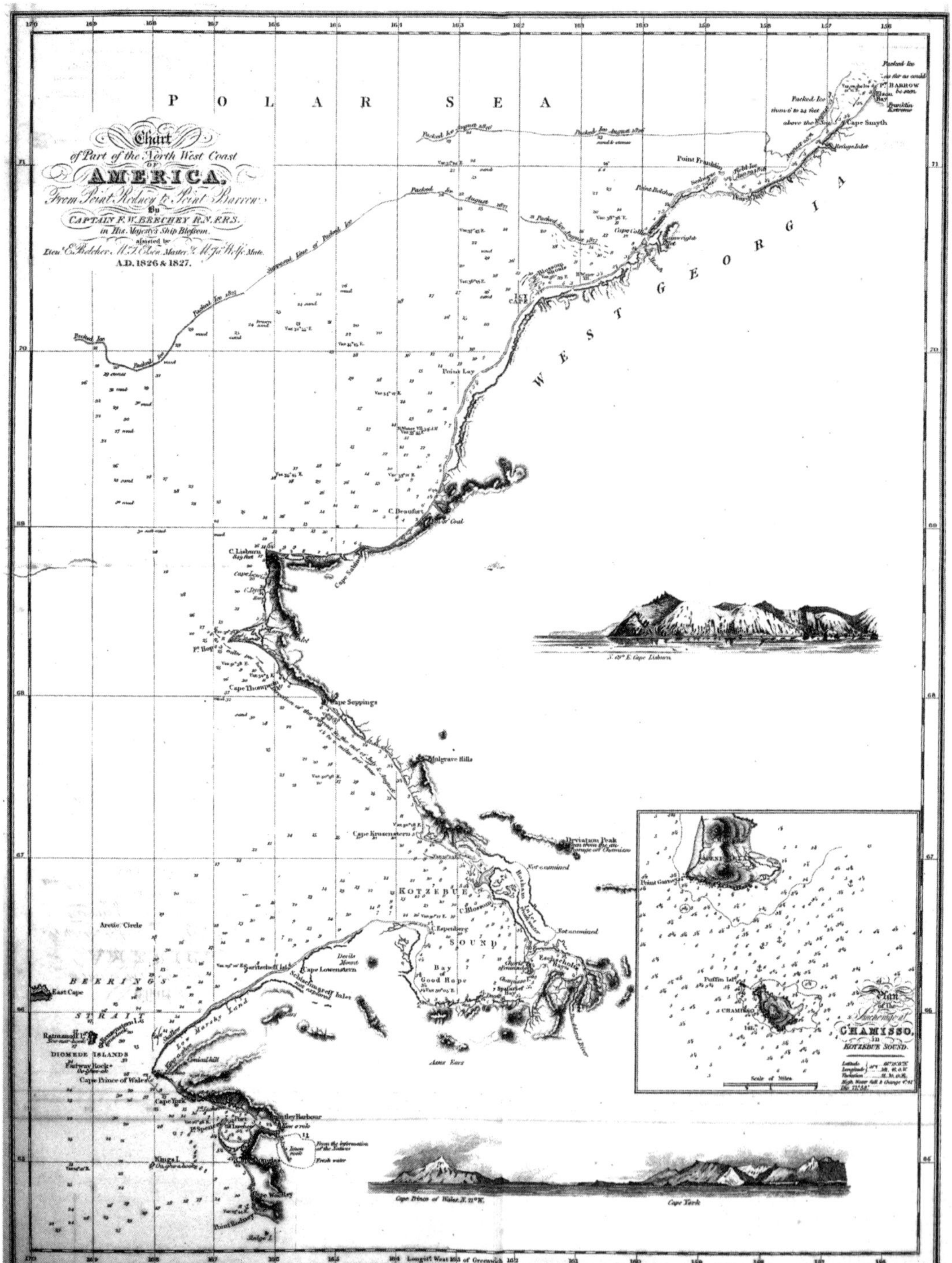

FIGURE 48 F. W. Beechey, *Narrative of a voyage to the Pacific and Beering's Strait . . . 1831.* Vol. 1, endpaper. UAF Rare Book A1133

This is an elegantly executed navigational chart with indications of depth and in some areas indications of sea currents and their velocity. Even though members of the expedition had elicited local geographic names from the Eskimos, there was no record on this chart except in the very southern part. The main toponymy is English, and in part Russian, after Kotzebue's materials.

The Beechey Expedition brought back data on the nature of the Arctic coast of Alaska, summarized in preliminary fashion in Beechey's two-volume work. These materials served as a rich source for British natural scientists. Thus, as early as 1829–1837, Sir John Richardson utilized the expedition's data in his fundamental work on the zoology of the northern part of North America.[176]

In 1839 the results of the fauna collections of the Beechey Expedition and of some other natural science observations were summarized in the collective work by Richardson, Vigors, Lay, Bennet, Owen, Gray, Buckland, and Sowerby.[177] A separate work on the vegetative specimen collection was published in 1841 by a professor of botany from Glasgow, Sir William J. Hooker, and George A. Walker Arnott.[178]

The expedition's work considerably advanced the study of the Eskimos of the Bering Strait region, and its materials continue to be used with success by anthropologists and ethnographers. In particular, Beechey's data were highly praised and widely utilized by anthropologist Dorothy Jean Ray, who emphasized that Beechey was the first European who absolutely identified the Natives he encountered as "Eskimos." His descriptions of the settlement King-a-ghe (modern Wales) and its residents are the first ethnographic notes about the largest Eskimo settlement near the point where Europeans (Gvozdev) sighted the Alaska coast for the first time.[179]

Dease and Simpson Expedition

The search for a passage from the Atlantic to the Pacific Ocean succeeded about ten years later, by a Hudson's Bay Company expedition led by Thomas Simpson and Peter Warren Dease in the years 1836–1839. In 1837 the expedition left Fort Chippeway, descended the Mackenzie River to the Arctic Ocean and proceeded westward by sea, proposing to reach Cape Barrow. On July 23, after overcoming a multitude of ice blockages, the travelers reached Return Reef, beyond which no Europeans had penetrated. Here they began a systematic coastal survey.

During this process the main bays and capes received the names of various functionaries of the Hudson's Bay Company. On July 31, while in the vicinity of Cape Simpson (named not for this expedition but in honor of the governor of the Hudson Bay territories, George Simpson), Simpson and Dease realized that further progress by boat was impossible. A decision was made to dispatch a group of five men headed by Thomas Simpson, on foot and in places by kayak along the shore toward Cape Barrow.

On August 1, 1838, Simpson's party left the last camp reached by boat and named it Boat Extreme. Near Dease Bay, named in honor of the official head of the expedition, the travelers encountered a group of Eskimos who loaned them their umiak to traverse this bay. Simpson questioned the Eskimos about the route his group intended to take and wrote the following:

> procured, from the most intelligent of the women, a sketch of the inlet before us, and of the coast to the westward, as far as her knowledge extended. She represented the inlet as very deep; that they make many encampments in travelling round it; but that it receives no river. She also drew a bay of some size to the westward; and the old man added a long and very narrow projection, covered with tents, which I could not doubt to mean Point Barrow.[180]

The further progress of the expedition proved that the Eskimo map in its general outline fully corresponded to reality. For example, by the next day the travelers found a small bay to the northwest of Dease Bay as indicated by the Eskimos, and at 1:00 a.m. on August 4, 1837, in the rays of the rising sun they sighted Cape Barrow, which proved to be in full accord with the Native map: a long spit with a huge Eskimo burial ground and two settlements of these people, a summer and a winter one, three miles distant from each other.

According to Simpson's observation the Eskimos encountered at Point Barrow had indirect contact with Russians, whom they called *Noonatagmun*. An old Eskimo man even undertook to transmit to the Russians, or some other white people, Simpson's letter confirming the fact that the British had landed at Point Barrow at 71°03′24″ N and 154°26′30″ W.[181]

Thus the search for the Northwest Passage that had occupied European voyagers for over two centuries was concluded. In 1839 the Royal Geographic Society announced the award of the gold medal, and the government announced a pension of 100 pounds sterling per year. These awards could not be delivered to Simpson, as the traveler died under mysterious circumstances in July of 1840 after a quarrel with his companions that ended in a shootout.[182]

The surveys by Deese and Simpson were published on the *Map of the Arctic Coast of America from Return Reef to Point Barrow Explored by Messrs. P. W. Dease & T. Simpson under the Direction of the Honourable Hudson's Bay Company, 1837*, published in the Journal of the Royal Geographic Society of Great Britain.[183]

RUSSIAN COASTAL EXPLORATION 1816–1830

Following the Beechey Expedition, Russian interest in the Northeast Passage into the Atlantic waned considerably. Instead there a was a renewed concentration on the detailed cartography of Russian America's coasts, exploration of the interior regions, and implementation of the old plans of Shelikhov and Rezanov to expand southward, and to establish themselves in California. Three basic "information channels" had been formed, and through them there was an integrated flow of data about the territories transmitted almost continuously to the leadership of the Russian-American Company and to the Admiralty.

The first of these channels were the reports made by naval officers while on the regular circumnavigating voyages of the Russian-American Company. During their voyages oceanographic surveys were constantly conducted and special work was undertaken to survey coastal sectors of Russian America that had not been sufficiently investigated.

The second informational channel, and the most constant and fruitful, were the voyages of the company's commercial vessels stationed in the North Pacific. Besides maintaining the company's commercial activity (by delivering hunting parties, delivering foodstuff, and also exporting the amassed furs and other goods), these vessels were engaged in continuous investigation of Alaska's coasts and the surrounding islands, as well as in constantly updating the existing navigational charts. Special exploratory expeditions were often conducted by these vessels.

The third informational source emerged with the establishment of permanent posts, the redoubts of the Russian-American Company on the mainland. These redoubts emerged as unique centers for collecting information about localities, which was made possible by their widespread contacts with the locals who came to the redoubts for trade. Explorations of new regions were made by company employees during their annual treks into the Interior to collect furs. The trading redoubts and outposts were support points for special exploratory expeditions organized by the company.

Fort Ross

The Russian-American Company settlement in California, Fort Ross, was located sixteen miles north of Bodega Bay (Rumiantsev Bay on Russian charts) in the valley of the Slavyanka River (now the Russian River). Originally ninety-five Russians and eighty Aleuts were settled here, using forty baidarkas for the sea animal hunt. Later the population fluctuated between 200 to 400 people. A small settlement also existed in the Farallon Islands.

By the 1840s Fort Ross had 2,000 head of cattle, 1,000 horses, and 1,000 sheep. In spite of constant protests by the Spanish authorities concerning the Russian presence in California, the Spanish widely utilized the services of Russian craftsmen at Fort Ross, obtaining extremely well-made boats from them. In the period up to 1824 four large vessels were also built there for the Russian-American Company.[184]

In order to achieve safety for the regular sea voyages between Alaska and Fort Ross, accurate charts of America's coast were needed. For this reason, in 1816 the regular circumnavigating voyage of the ships *Kutuzov* (Captain-Lieutenant Ludwig von Hagemeister) and the *Suvorov* (Captain-Lieutenant Zakhar Panafidin) were instructed to visit Fort Ross on their way and to conduct detailed hydrographic surveys of the adjacent regions.

Hagemeister, who was preparing to assume the post of chief manager of Russia's colonies in America, made it a point to acquaint himself with the state of affairs at Fort Ross and the relations between the company employees and local inhabitants. In 1817 he "invited to the settlement the chief *toion* from among those who reside in the vicinity, Ch-chu-oan, and another, Vale-lii-l'ie, together with many of their kinsmen . . . and . . . I discovered that they live in peace with the Russians, and that there are many mixed marriages between the local Natives and the Aleuts."[185]

At that time the first detailed survey of the Russian Californian settlement was accomplished and a large-scale (fifty sazhen to an English inch) plan was made: *Location Plan of Fort Ross. On the Coast of North America in 38°30′24″ N Latitude and Longitude from Greenwich 122°45′ W, Compass Direction 16° East, Compiled by Right Compass in September of the Year 1817*[186] The "Explanation" to this plan is a complete description of Fort Ross, with details such as that the walls there were constructed:

> with thick logs, 3 arshin high and on top, in their entire circumference, were wooden chevaux de fries (a military spiked obstacle). The fort's defenses were strengthened by two two-story octagonal and septagonal blockhouses.

Further along in the "Explanation" we are told that:

> the "house of the manager is built of logs, contains 7 rooms, an entryway, two storerooms, and underneath are warehouses and a powder store, while the employee barracks are built of hewn planks and contain the common and two small separate rooms.

Besides these structures, on the fort's grounds there was one more house of hewn planks with three "separate chambers, 27 Aleut yurts built of planks," two two-story warehouses for foodstuffs and for furs, two "houses of planks, containing a storeroom, common kitchen and the dogs 'quarters,'" a jail, workshop, bakery, blacksmith's shop, steam bath, cattle shed, pigsty, various barns, cemetery, the brig *Rumiantsev* hauled out onto the land, a wheat field, and "vegetable gardens that number up to 50 in the fort's vicinity."[187]

During the voyage of the *Suvorov* and the *Kutuzov* surveys were made south of Fort Ross, and especially to Little Bodega Bay (Rumiantsev Bay). A colorful manuscript chart was compiled: *Chart of Part of the Coast of Northwestern America from Fort Ross to the Cape of Bolshaya Bodega with Depth Indicated in Sazhens at the Ebb. Compiled by Right Compass in September of the Year 1817. By Assistant Navigator 14th Class Kislyakovsky, of the Ship Kutuzov in the Scale of 1 Verst per Inch (1:42000). Copied at the Drafts room of the State Admiralty Department from the Chart Sent by Captain Hagemeister by Student 1st Class Andrei Tokarev. 9th of December in the Year 1819.*[188]

This is a rather elegantly executed chart with an artistically rendered landscape showing coasts, with details such as depth measurements for Rumiantsev Bay. At the bottom of the chart profiles of the shoreline are shown: *Land to N from Bodega toward NNO at a Distance of 8 Miles.*

Homeward bound, the expedition surveyed this area for a second time, extending the survey somewhat more to the south along the California coast, resulting in a *Chart of the Part of the Northwest Coast of America from F. Ross to Monterey. Compiled by Right Compass in the Year 1818 by Navigator Kislakovsky of the Ship of the Russian-American Company, the Kutuzov.*[189]

On this chart Rumiantsev Bay is shown in even greater detail, based on the survey of this harbor summarized in *A Chart of Rumiantsev Bay. On the Coast of NW America with Indication of Depth in Feet at the Ebb. Tide 7 feet. Compass 16° E. Compiled on Right Compass in June of the Year 1818 by Assistant Navigator 14th Class on the Ship Kutuzov, Kislakovsky. Anchorage A. 38°18′48″ N Latitude, 122°31′ W Longitude from Greenwich.*[190]

Besides the Russian place names, the chart also shows Native names, for Rumiantsev Bay—"Chok-liva" and for Cape Rumiantsev—"Tiu-Tiuiia." Depth measures are indicated on this chart not only for the bay itself but also for the River "Avvacha." An inset chart shows the territory between Cape Rumiantsev and the island of "Omuia-lai."

A good supplement to the maps was the description of Fort Ross and its vicinity by the manager (1812–1821), Ivan Kuskov (1765–1823) that contains the following:

> Fort Ross is situated at 38°30′28″ N latitude and 122°41′00″ W longitude from Greenwich, on an elevated and level area. The buildings therein are the manager's house, barracks, warehouses, kitchen, steam bath and barnyard; all is surrounded by a thick palisade in a rectangular shape; at the corners are blockhouses with embrasures; outside the fort is a windmill; inhabitants of both genders, Russians and Aleuts, number up to 80 persons, who are employed in tending vegetable gardens and fields, shipbuilding, and sometimes in sea otter and sea lion hunting.
>
> The climate is most healthy, there is practically no illness whatsoever, the soil is composed of sand and red clay and yields all kinds of vegetables of good size and fine taste.
>
> Winter wheat produces sixty-fold and occasionally even more, but because of the shortage of people cultivated fields are few. Cattle-keeping is in the best condition: bulls, cows, sheep, horses, swine—there is no count of the latter; domestic fowl, chickens and ducks are plentiful.
>
> The small river, the Slavyanka that flows close to the fort provides sturgeon and salmon, one kind of salmon in plenty; the level plain around the fort yields fine grass for the cattle throughout the year.

> Toward the interior, away from the coast, there is forest, oak, laurel and cedar that looks red and is layered like the pine; the first the Company uses to build vessels, the last-named for houses. There are no wildfowl in those woods except for the crested partridge; on the lakes are geese, ducks, and swans. Of animals there are wolves and tigers (cougars), but these are very peaceable, bears that are over bold and ferocious, wild goats and pigs.
>
> There are no very high mountains, but there are low ones. Some are covered with forest, others are bold and sandy. To judge by the character of the ground there should be ores.
>
> The aboriginal inhabitants are Indians, rather tall and stately in appearance, their faces are of somewhat olive color, the hair is straight and black; they pluck from their chins; they have no clothing. There are various tribes but none are numerous. Each tribe has its own language. Their disposition is quiet, not trusting, and cowardly. Their life is nomadic, mostly along the sea coast, on the lakes and rivers. They have no weapons except for bows and arrows made of reeds with an eagle feather and stone blade. They live a peaceful life, they have no boats, they traverse the rivers or small straits on reeds tied together at the ends, and instead of oars they use their palms. Their food consists of wild wheat. When the ears ripen they light the straw for a while and dry out the grains, then they collect them and mill them between two stones. Out of flour they make dough, from which they make pancakes, these they cook over hot stones; but their main food is mollusks, crayfish and fish. They obtain fire by rubbing wood on wood. The Spanish soldiers catch these children of nature with lassos and take them to the presidios to their clergy, who treat them inhumanely and use them to cultivate fields, etc.
>
> A serious drawback as far as the fort is concerned is the fact that the arriving vessels have no anchorage. The coast is straight and the sea is very deep with a rocky bottom. For this reason the vessels anchor at Malaya Bodega (Little Bodega) more than 30 versts distant from the fort, but the roads here are inconvenient as they are open to the ocean, sheltered but by one small island. People come here from the fort by baidarkas or along the shore on horseback.[191]

The inconveniences associated with the location of Fort Ross cited in the above description, its isolation from the base of the Russian colonies in America, and the constant rejection by Spain, and later by Mexico, of the Russian presence in California, forced the leadership of the Russian-American Company to reject any plans for further expansion in the Rumiantsev Bay area and later on to evacuate the Fort Ross population to Alaska. On December 13, 1840, the Russian-American Company concluded a contract with an American, John A. Sutter, according to which the location of Fort Ross, with all the structure thereon, was sold to Sutter for $30,000. The last Russian settlers left the fort in January 1842.[192]

Litke Expedition

The next important scientific expedition was Captain-Lieutenant (later Admiral and founder of the Imperial Geographic Society and president of the Imperial Academy of Sciences) Fyodor Litke (1797–1882) (Fig. 49),[193] in command of the sloop *Senyavin*, and Captain-Lieutenant (and well-known Russian writer of sea stories) Mikhail Stanyukovich, in command of the sloop *Moller*.

FIGURE 49 Portrait of F. Litke. Lithograph 1823. RAJVUSARHIIV F. 2057, op 1, d. 452

This was Litke's second voyage to the northern part of the Pacific Ocean. He was a member of the expedition of Captain Vasily Golovnin in 1817–1819 on the *Kamchatka*. On that voyage Litke made his first surveys in these regions, the results of which are preserved in the Russia's Naval Archive in the form of schematic manuscript charts of Kodiak, the Near Aleutian, and the Commander Islands.[194]

The Litke Expedition left Kronstadt on August 20, 1826, and by June 1827 was off the coasts of Russian America, where Litke, after visiting Sitka, sailed to Unalashka conducting surveys en route and to the Pribilof Islands. He also mapped St. Matthew Island in detail. Stanyukovich, aboard the *Moller*, surveyed the northern coast of the Alaska Peninsula. The expedition spent the winter in the Caroline Islands, where it also engaged in scientific investigations and surveys (Fig. 51). In the summer of 1828 the expedition returned to the northern part of the Pacific Ocean, where hydrographic descriptions of the coasts of Kamchatka and Chukotka were conducted. The *Moller* returned to Kronstadt on August 21, 1829, the *Senyavin* a few days later.[195]

This expedition was of outstanding scientific significance. Litke himself conducted extensive geographic investigations, the results of which were published in his multivolume work that appeared almost simultaneously in Russian and French.[196] Aboard the *Senyavin* were three scientists whose works were important

FIGURE 50 Sitka Harbor by Litke Expedition mineralogist and artist Alexsandr Postels. Plate 3, Litke, *Atlas*. Rare Book C0024 UAF

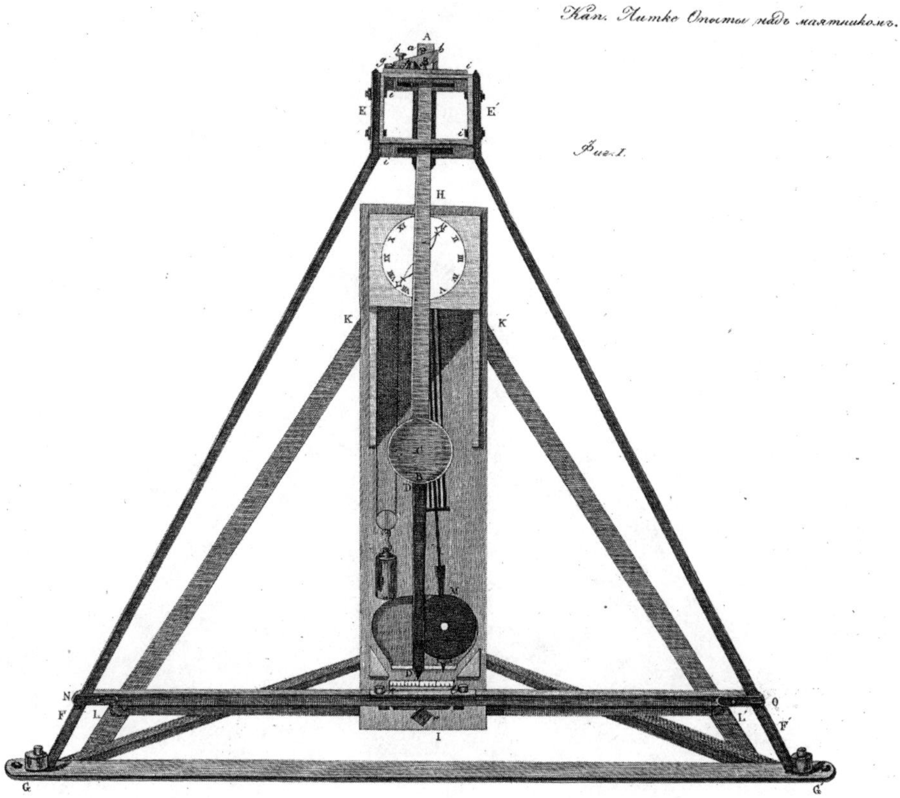

FIGURE 51 One of Litke's instruments to measure gravitational variation. Plate 1, *Opti Nad*. . . Rare Book A1710 UAF

contributions to various branches of knowledge. Thus, the German naturalist Baron Heinrich von Kittlitz (1799–1874) published an account of the expedition and albums of drawings of birds and plants.[197]

The artist and geologist Alexander Postels (1801–1871) amassed a large collection of minerals and botanical specimens for the Imperial Academy of Sciences in St. Petersburg. He also executed a multitude of drawings which were published by Litke, together with Kittlitz's illustrations, as a separate album (Fig. 50–55). The *Senyavin*'s ship's physician, Karl Heinrich Mertens (1796–1830), was engaged during the voyage in studying and drawing mollusks. Regrettably, most of his manuscript materials were not prepared for publication because of his untimely death from a fever in the year 1830.[198] Litke transmitted some of his own scientific observations to Academician E. Lents and to Professor G. Haellstroem, which were then published as a separate series of articles that established Litke's fame as a geographer.[199]

Litke's expedition also brought significant cartographic results, which are reflected in the *Atlas* that accompanies his main work and also in a special navigational atlas containing fifty charts and plans, published by the hydrographic depot

FIGURE 52 The Chief Manager's residence, Sitka. Plate 3, Litke, *Atlas*. UAF Rare Book C0023

FIGURE 53 Log Cabin in Sitka. Plate 7, top. Litke, *Atlas*. UAF Rare Book C0023

FIGURE 54 Sitka showing the church that preceeded Veniaminov's cathedral. Plate 7, bottom. Litke, *Atlas*. UAF Rare Book C0023

FIGURE 55 Vegetation surrounding Sitka. Plate 5. Litke, *Atlas*. UAF Rare Book C0023

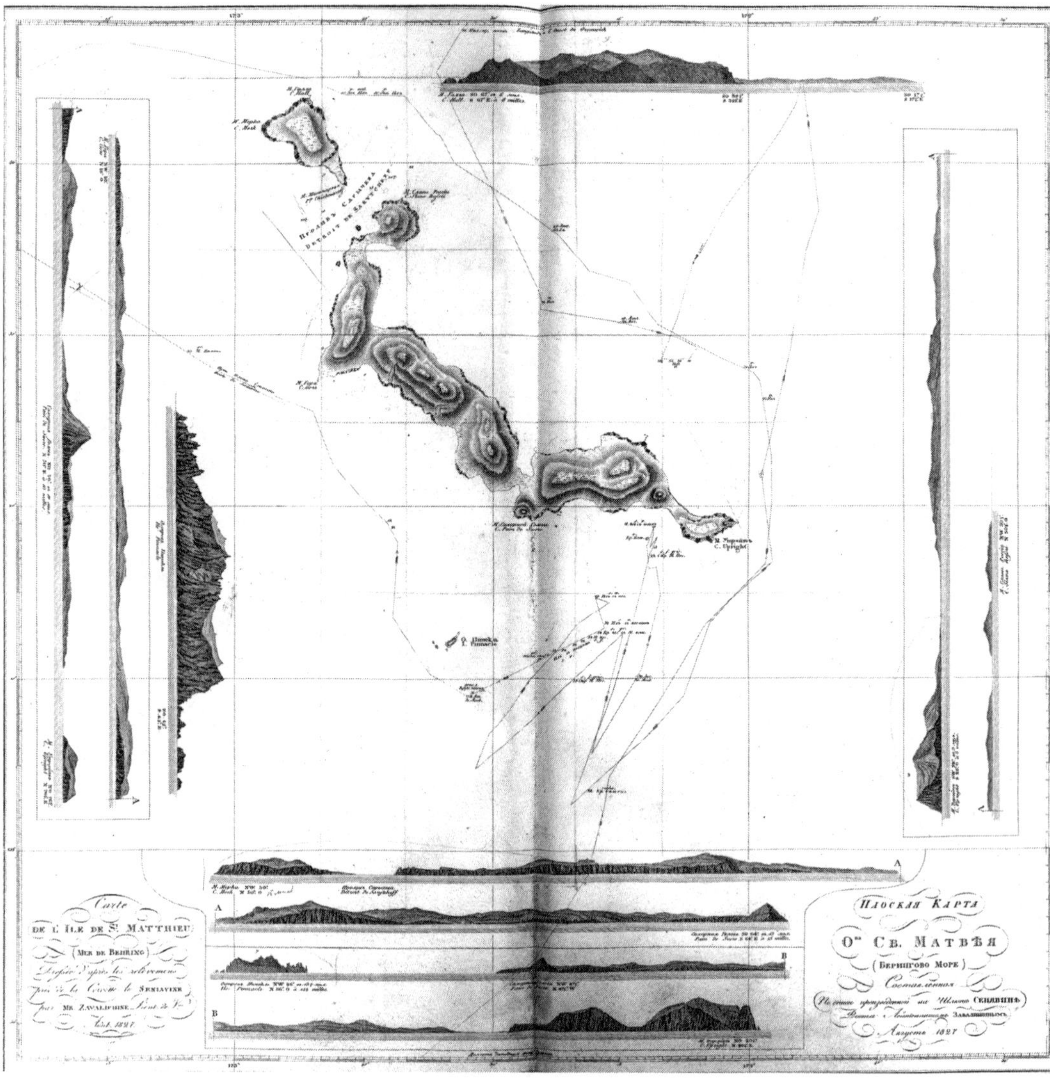

FIGURE 56 Litke's map of St. Mathew Island with coastal views. Litke, *Atlas*. UAF Rare Book C0023

of the Main Navy in St. Petersburg in 1832 (Fig. 56).[200] Besides the navigational charts, this atlas contains very valuable and remarkably accurate data from geomagnetic observations that were conducted in the course of Litke's expedition.

Litke's expedition was a scientific undertaking to satisfy the needs of Russia's Navy; it also had great significance for maintaining the empire's national prestige, as well as the progress of knowledge on a worldwide scale. The explorations undertaken by the Russian-American Company in that period had a more practical application, securing its commercial interests above all, but they also advanced the progress of geographic knowledge and cartography of North America.

After Baranov, the chief managers of Russia's colonies were appointed from among educated naval officers with close ties to the Admiralty Department. They conducted geographic surveys in accordance with the Admiralty programs which guaranteed uniformity in the materials used in cartography although those who carried out geographic investigations and cartography in Russian America did not always possess the same high level of expertise as the navy's professional hydrographers.

A constant stimulus for exploring new regions and refining the maps of territories which were already known was the desire of the Russian-American Company to expand the regions of fur procurement, as by this time the populations of some species had begun to decrease catastrophically due to excessive harvest (Fig. 57). The company's leadership was seriously disturbed by the developing situation in the second decade of the nineteenth century.

Conservation of Fur Seals

In 1819 the Head Office of the Russian-American Company noted, along with the rapacious extermination of fur-bearers in its possessions, the exceptionally irrational utilization of biological resources. Specifically, it was stressed that "in these islands the oil and meat of fur seals is discarded and only the skin taken . . . for this reason

FIGURE 57 Aleut hunters, Aleutian Islands. Plate 10. Litke, *Atlas*. UAF Rare Book C0023

several thousand barrels of oil are lost per year that could probably bring not insignificant profit."[201] It was perfectly apparent that decisive measures were needed to regulate fur procurement and to preserve valuable animal populations.

Recommendations for such measures were formulated, at the request of the Russian-American Company, by a member of the St. Petersburg Academy of Sciences, the German naturalist Langsdorff. He had become personally acquainted with the problem as a member of the expedition under Krusenstern and Lisyansky. Langsdorff stressed that:

> it is necessary not to conduct matters as was done until now, killing these beasts (fur seals) without any reason except that they are many, but to institute a procedure, dispatch a number of people by agreement to the islands of St. Paul, St. George, Matmai [*sic.* St. Matthew?], the Commanders and others and oblige them not to take more than 80,000 to 100,000 fur seals. The skins are to be taken off with care, also cleaned and dried. It is necessary to issue a special instruction that only males are to be killed, not any females, and each year, or every other year, the hunt is to be switched from one island to another.[202]

In accordance with this recommendation, the Russian-American Company began to limit the hunt, opening the way to rehabilitating the fur seal population. Some measures were also taken to preserve the population of white Arctic foxes in the Aleutian Islands, where "the company about twelve years ago prohibited the taking of the same in the Near Islands for about five years and only when an increase was observed was trapping permitted."[203]

New animal populations had been introduced by *promyshlenniki* from Siberia. This was especially true of swine, which adapted well on several islands, multiplied, and in some cases become feral. But some of these, like the local animal species, suffered greatly from natural disasters from time to time, which were characteristic of this zone of active volcanoes.

One such instance is recounted under the date of January 18, 1828, by Pyotr Chistyakov (1790–1862), who served as chief manager of the Russian colonies in America from 1825 to 1830:

> Manager Petrovsky (Rodion Petrovsky of the Unalashka Office) reports to me that on September 12, 1826, the mountain ridge on the SW side of Umnak Island exploded after there was great thunder, an earthquake and a horrifying cracking sound; a mass of ash was emitted so that the nearby islands, but especially Sanak and Chernabura were covered up to half of knee height. According to the baidara sailor Larion Merkuliev, who was on Unimak, he and his people did not see daylight for 8 days; it was not possible to emerge outside into the air, because the winds carried the ash, a kind of dust, and from this all the inhabitants suffered later from a malady of the eyes and a cough. The

> water in the lakes and streams was muddy, somewhat resembling home brew for over two months. At present the mountain emits flame at the spot that suffered the collapse. The swine that were put onto Chernabura Island in 1823 and that have multiplied since then greatly, all perished, either from the strong freezes and deep snows or from the soot that after the earthquake covered the entire island; their carcasses were found in piles near the streams.[204]

Chistyakov strove to regulate the taking of sea animals and improve the safety of the company's vessels and also to intensify the reconnaissance of new fur-rich regions. He demanded strict observance of the fur-hunting regulations developed by scientists and the Russian-American Company, especially with respect to fur seals.

It was necessary to perfect the navigational charts of the islands. For example, Chistyakov, on May 20, 1828, instructed Andrei Ingestrem (Ingestrom), captain of the boat *Sivuch*, to do the following: "Should you visit Copper Island, Amchitka and Kiska, where there are known harbors though no surveys of the same are reflected on the current charts, you have the honor of completing the proposed surveys. However, such surveys should be conducted when occasion presents itself without interference with your assignment."[205]

Vasiliev's 1829 and 1830 Expeditions

It wasn't possible to explore, study, and map new fur-producing regions with routine Russian-American Company hunting parties. It required the organization of special expeditions. In this connection, on September 26, 1826, Chistyakov asked the manager of the Kodiak Office, P. E. Nikiforov, to assemble the information needed to launch an overland expedition "for the establishment of permanent settlements in the interior of America, near the regions where the headwaters of the rivers Nushagak, Kuskokwim, and Kvikhpak (Yukon) come close."[206]

The idea and a general plan had been worked out in the head office, but the implementation was entrusted to Chistyakov, who was instructed to dispatch "a shore expedition upstream of the Nushagak River." Chistyakov took this assignment seriously. In order to prepare a common program and instructions to participants in the expedition, he asked Nikiforov to obtain from Fyodor Kolmakov, at that time the manager of the Novoaleksandrovsky Redoubt (1819–1840?), the following information about the Nushagak River:

"During his tenure as manager there, were any Russians dispatched upstream along the river? If not, is it possible to learn from the inhabitants, through interpreters, how deep it is? That it is for baidarkas or baidaras? Is it linked with the rivers Kuskokwim and Kvikhpak (Yukon)? If they are not linked, how long are the portages?" The chief manager desired that such information be obtained from Kolmakov personally, as "his verbal reports would be more satisfactory than any written ones."[207]

After studying all the information submitted to him, Chistyakov compiled a detailed instruction dated March 4, 1829, for the appointed expedition commander, Sub-Lieutenant of the Navigators' Corps Ivan Yakovlevich Vasiliev (1797–1838?). It is the most detailed and clearest document of its kind, and served as a model for all subsequent instructions for exploring mainland Alaska during the Russian period.

Along with the set of features to be covered obligatorily in the expedition's materials, Chistyakov provides some general information about the natural features and population of parts of interior Alaska that were already known by that time. The instruction, addressed to Vasiliev, a well-educated navigator, but not a professional scholar, is simple and clear. The influence of the St. Petersburg Academy of Sciences and other scientific institutions is reflected in its content.

In March of 1829 the expedition was formed. Navigation student, Creole Pyotr Kolmakov, the son of the manager of the (New) Aleksandrovsky Redoubt Fydor Kolmakov, was appointed as Vasiliev's second-in-command. Two other Creoles became members of the expedition also: Semyon Ivanovich Lukin (he and his children would be among the most active explorers of continental Alaska in the future) and a man named Okhotin. Dispatching the expedition on its way, Chistyakov issued to its participants the following *Orders of the Chief Manager of Russia's Colonies in America*.

> *To the employees of the Russian-American Company setting out on the Northern Overland Expedition.*
>
> I entrust you to the command of Sub-Lieutenant Mr. Vasiliev, of the Corps of Navigators, whom I delegated, on orders from the head office, to describe the lands of our America and the peoples inhabiting the same who have not been previously known to us.
>
> As your commander, I order you to obey Mr. Vasiliev during the voyage and until your return, as is fitting for the true subjects of our Lord Emperor. But I beg of you to earn for yourselves glory and the company's gratitude, through proper behavior, obedience to your immediate commander, fearlessness should there be enemy action, cheerfulness of spirit whenever danger threatens, honest, kind and inoffensive relations with the inhabitants, diligence in labor, patience in need, and harmony among yourselves.
>
> I declare to you that, upon the expedition's end, each one of you may expect a fitting reward upon attestation by Mr. Vasiliev that you carried out your duties; those who should prove insubordinate, cowardly or lazy will receive no reward and may be yet punished.[208]

Vasiliev left Kodiak on board the *Karluk* and sailed to Katmai settlement on the Alaska Peninsula. He then traversed the peninsula, crossing Lake Naknek's ice on skis. On May 1 the expedition arrived at the Novoaleksandrovsky Redoubt in baidarkas. On May 13 or 14, traveling along the Aleknagik River, Vasiliev reached the lake of the same name. On May 31 he traveled upstream along the Nushagak River with four three-hatch and eighteen single-hatch baidarkas. He was accompanied by four Creoles and Russians and ten Eskimos.

On June 2 the travelers reached a settlement of Kiiaten people, numbering seventy persons, and spent a week there. On June 10 they arrived at the settlement of Kakhatuliak. Then, having passed the mouth of the Mulchatna River, the party explored the source of the Ilgaik River. On June 16 they traveled upstream along the river Nuiakuk and reached the lake system at its headwaters (including lake Chauekuktuli). They reached and described the northernmost lake, Lake Nishlik (called "Tuksiu" in Vasiliev's materials). Learning that it was possible to reach the Kuskokwim River by interior streams, Vasiliev tried to persuade his Native companions to go there, but in vain. On July 11 the expedition returned to the Novoaleksandrovsky Redoubt.[209]

Although the main goal set for Vasiliev's expedition was not achieved in the summer of 1829, Chistyakov fully approved the actions of his party, and, after perusing the leader's report of July 26, 1829, wrote to him as follows:

> I discern that you and the expedition entrusted to you reached Lake Chivaktula from Lake Nushagak and that because you were left by your guides, the Kiiaktaigliut and Aglegmiut that you took on at the Aleksandrovsky Redoubt and were unable to persuade the Kuskokwim of the Tuksiu settlement to assign a few to accompany you along the Kuskokwim River, you were forced to return to the redoubt. Having studied your journal and the map you compiled during the journey, I must say, to console you and myself, that though this year the expedition failed to achieve its goal . . . at least your voyage from the redoubt along the Nushagak River, your exploration of the direction of the flow of the Ilgaiak River up to its very source, your return to the Nushagak River and then up to Lake Nushagak where that river finds its source, and then again to Lake Chavaktula, which was previously unknown to us, brings us benefits, makes us familiar with the interior of America, and thus brings us closer to the desired goal . . .
>
> "There is nothing more I can say in conclusion, except to ask you not to despair because of the expedition's misfortunes. The nature of your expedition was such that it was difficult to expect success in the course of a single year. From your report I see that upon your return to the redoubt you set out in three baidarkas to explore a lake to the west of the Aleknashak River. Such zeal on your part for the benefit of geography brings you great honor, and it is a great pleasure for me to note that you do not waste time. I remain convinced that you shall act in the same manner in the future, something I never doubted.[210]

Chistyakov believed that in order to continue the expedition it was absolutely necessary to engage Kuskokwim men as guides and that the latter should leave hostages at the redoubt.

Between August 18 and September 5, 1829, Vasiliev, accompanied by two men (Alexei Baturin and Pyotr Talkviak), set out in four baidarkas once again for Lake Aleknagik. On August 24 they portaged their baidarkas from Lake Ungaktalyk to "a small lake named Kyng, from which flows the river Annanyk, on which they traveled; it flows to the west."[211] From this river's mouth they traveled upstream starting on August 26 in the direction of Togiak River, across the lake of the same name to the river "Anchaguktuli."

In a settlement of the Kuskokwim, Vasiliev observed that one of the elders wore a cross and discovered that many of the Kuskokwim had copper icons; he supposed that they obtained these from the missionary Iuvenalii, who was killed by these people about the year 1796 next to the settlement of Kwinga [Quinhagak—LB].[212] The Natives did not undertake to escort Vasiliev to the Kuskokwim, though according to their testimony the party could descend from the lake by any of the small rivers and reach its mouth in one day of travel. The party went down the Togiak River and by means of an interior stream crossed the peninsula that separates Togiak Bay from Bristol Bay. In effect they traversed in reverse the route Kolmakov and Korsakovsky followed in 1818. On September 5 the travelers were at the Novoaleksandrovsky Redoubt. On November 26, 1829, with only a single companion, Vasiliev set out again on his way, crossed the Alaska Peninsula on foot, and on January 15, 1830, he reached St. Paul Harbor in Kodiak.[213]

Based on the results of the expedition, Vasiliev compiled the *Map Indicating the Flow of Rivers and Lakes, Situated to the NW. + indicates locations where copper plates were buried in the ground. The sea coasts are drawn after the chart by Mr. Khromchenko. Compiled by Sub-Lieutenant of the Corps of Navigators Vasiliev.*[214]

Chistyakov one again gave a high evaluation to Vasiliev's new explorations, writing to him: "On the 8th of the month of April past I received with the brig *Ryurik*, returning from Kodiak to Sitka, your report of the 1st of the same month and the map compiled by you showing the flow of rivers and lakes to the NW of Bristol Bay together with the journal appended to it . . . your second journey in this new direction brought you closer than before to the headwaters of the Kuskokwim River."[215]

Chistyakov sent both maps compiled by Vasiliev and his journal to the head office of the Russian-American Company, and from there to the hydrographic depot. The chief manager, writing to the head office, characterized Vasiliev's successes as follows: "The rivers and lakes he has described were unknown and not shown on any map, except for the mouth of the Aleknagik where he began his journey . . . in spite of the fact that . . . those who saw him upon his return to Kodiak think he has lost much weight, and he himself complains of strong rheumatism, there is no one here who could replace him in case of need."[216]

In the spring of 1830, in spite of feeling poorly, Vasiliev set out on another journey, accompanied by a small party. Chistyakov's official report was dispatched on October 5, 1830, on the very same day the famous Russian scientist, hydrographer, and voyager, Baron Ferdinand Petrovich von Wrangell (1796–1870), assumed the post of chief manager of Russia's colonies. Specifically:

> (Vasiliev) set out from St. Paul harbor (Kodiak) on April 3, and on the 10th of the same month arrived at Katmai. He departed the same on the 17th and having overcome many obstacles arrived at the Aglegmiut settlement of Paugvik. However, huge masses of ice that blocked the mouth of the Nushagak River and the whole of Bristol Bay from Ugashik to Cape Neenham did not permit any kind of communication with the Aleksandrovsky Redoubt. Finally, on May 13, Mr. Vasiliev arrived at the redoubt.
>
> Having left the redoubt on June 19, he reached the Ilgaiak River, ascended the ridges on the 29th of that month and after a torturous crossing of five days came out at the headwaters of the Khulitna (Holitna) River, and then followed it for its entire course. On July 8 he entered the Kuskokwim River in the latitude 62°09′ and 155°43′ longitude.
>
> Enemy action between the Kuskokwim and the Nikaliukhmiut, inhabitants of the headwaters of the Kvikhpak River, and other occurrences noted in his journal did not permit him to proceed further into the interior of America. From the Kuskokwim River, he picked his way with God's help through the numerous crowds of the barbarian Kuskokwim people, numbering up to seven thousand. The time he spent in twenty-two of their and seven of the Kvikhpak settlements was spent in reconciling the warring parties, persuading them to establish trade relations with us that did not exist before, because the Khulitna (Holitna) and the Kuskokwim Rivers are the richest of all rivers that flow through this part of the continent in numbers of amphibious animals. The forests are full of other valuable fur-bearing animals. From the mouth of the Kuskokwim he proceeded along the coast only to Good News Cape, from where by means of portages and lake crossings he came back to the redoubt on July 29.
>
> Mr. Vasiliev, known for his zeal for the Company's benefit, says in his report that he is ready as far as his strength permits to carry out all orders and execute every point of the instruction he received, but his totally ruined health will not allow him to complete the explorations he has begun. He suffers from rheumatism and bleeds from the throat . . . The only thing that remains yet for Mr. Vasiliev to do is to explore the course of the Kvikhpak River, about which he brought valuable information he collected from the peoples, including how to reach the source of that river. He does not refuse, after he recovers, to achieve the goal of the expedition that has been entrusted to him.[217]

But by September 17, 1830, responding to Vasiliev's official report dated August 6 on his expedition's results, Chistyakov fully accepted its results and believed that its program had practically all been carried out. He permitted Vasiliev to return from Novoaleksandrovsky Redoubt to Kodiak to improve the traveler's much-damaged health. On Vasiliev's recommendation, the expedition's chief guide, Creole Semyon Lukin, received an award of 100 rubles and, because of his knowledge of the Kvikhpak language, was ordered to Sitka.[218] In March of 1832 Vasiliev himself was awarded 3,000 rubles, and less than a month later received from the hands of Wrangell the Order of St. Vladimir, 4th Class.[219]

Vasiliev's expedition was a great scientific achievement in the study of the mainland parts of Alaska. The confirmation of the wide distribution of beaver was of great significance for the Russian-American Company. When he forwarded the expedition's journal to St. Petersburg Chistyakov suggested that it should be published: something quite unusual, if one considers the general company policy of secrecy. The journal of Vasiliev's 1830 expedition has not been located so far, but it is possible that it contained more detailed ethnographic information than his materials of 1829. It is possible that Wrangell gleaned his knowledge of the Ingalik, Aglegmiut, Kiatagmiut, and Kuskowagmiut from the 1830 journal.[220]

Based on his explorations, Vasiliev recommended that a redoubt be established in the area of Stewart Island, about 100 miles north of the Yukon delta. The company's leadership agreed to this proposal, and in 1833 the Mikhailovsky Redoubt was founded near this low-lying island in the southern part of Norton Sound. This redoubt opened the way to the Yukon for the Russians and later played a big role in the exploration of this great American river.[221]

Russian-American Company Coastal Exploration 1830–1831

The coasts were not neglected. Russian-American vessels were instructed to search for new islands and to bring navigational charts up to date. For example, in the spring of 1830, in his instruction to Fleet Lieutenant Zarembo, who was sailing on the brig *Okhotsk*, Chistyakov says the following:

1. As you proceed along the Aleutian Islands chain, conduct astronomical observations by means of the chronometers in your possession, circumstances permitting, to determine the situation of these islands; should there be discrepancies with data in the (Admiralty) department atlas, upon your return to Sitka report to me especially on this matter.
2. The Head Office of the Company forwarded to me with the dispatch of March 30 of the past year of 1829 (# 331) a copy of the memorandum by His Excellency Hydrographer-General Gavril Andreevich Sarychev addressed to that Office, dated January 16 of the same year (# 101) concerning the

exploration to the south and north of the chain of the Aleutian Islands for the existence of land nearby. I hereby append for your information a copy of the said memorandum by the hydrographer-general, and, in accordance with the dispatch addressed to me by the Head Office, I recommend that on your voyage to Okhotsk you choose routes not yet traversed by sailors, without, of course, any detriment to your main assignment. In choosing such routes you are to be guided by the charts compiled at the hydrographic depot. . . . Upon arrival in Okhotsk you are to report to the Head Office of the Company any discoveries and observations made by you concerning this matter; you are also to report if wind conditions have prevented you from choosing as yet unexplored routes not indicated on the appended chart. When you return to Sitka you will report all about these matters to me also.

3. In consequence of instructions I received from the Head Office, dated also March 30 of the past year (# 322) concerning information to be forwarded to it about the two rocks discovered by Messrs. Commanders of Company vessels, and which have been seen by Mr. Klochkov, I recommend that should circumstances permit while you are among the islands of the Atkha District, you find the rock lying off the E. Cape of Amlia, 5 Italian miles [1 statute mile = 1.5 italian miles] distant along the compass direction NNE; determine its situation exactly and upon arrival in Okhotsk report also to the Head Office in order to satisfy the hydrographic depot.

As far as the second rock that was seen by Klochkov in Sitka Sound is concerned, I have already relayed the information I obtained from you to the Head Office.[222]

We shall note in passing that the information Chistyakov received from Zarembo was that "the rock or submerged islet noted by Navigator Mr. Klochkov to the N. of Amlia Island has not been observed to this day by anyone due to the thick fogs and darkness that is constant in those regions. As far as the other rock seen by Mr. Klochkov in Sitka Sound near Biorka Island is concerned, it is, indeed, situated at that place, and was seen clearly this past year by Mr. Lieutenant Zarembo. Its situation as shown in the department's atlas, published in the year 1826 [Sarychev's atlas—AP], corresponds exactly to the determination by Mr. Zarembo."[223] On May 4, 1830, Chistyakov officially reported that instructions similar to the one he issued to Lieutenant Zarembo had been circulated to all other captains of company vessels.[224] Moreover, in the summer of the same year, Chistmakov organized a special seagoing expedition to go north. It was to investigate possibilities of commercial activities and conduct hydrographic work. In his instruction dated June 14, 1830, to the expedition commander, Etholen, captain of the brig *Chichagov*, Chistyakov expected him to carry out the following:

1. Acquire furs by trading with the Americans inhabiting Norton Sound and from the latter to Bering Strait on Ukyvok (King) Island. It is known that the fur catch obtained by these peoples is bought from them by the Aziat (Azigmiut?) and delivered for sale to the Chukchi. Therefore, in order to buy the same, it is necessary to visit St. Lawrence Bay, supposing that this is the main point where the goods intended for sale are brought.
2. Acquire from the inhabitants of St. Lawrence Island walrus ivory and collect the same along the coasts of the uninhabited St. Matthew Island, where the Company some time ago had an outpost for this reason.
3. Precisely research what benefits may be accrued from the above-mentioned trade with the wild ones by means of establishing a permanent settlement in Norton Sound, Golovnin Bay or on an island, or on Stewart Island, in the localities already known to us, or in another locality where the Americans from the surrounding regions could bring for sale to us their catch; such a settlement ought to have a decked vessel that could be used more efficiently for the execution of the points set out in 1. and 2. than a larger vessel . . .

   If time permits, do not neglect to make astronomical observations to determine the coordinates of the localities you will visit throughout your voyage, nor surveys of the coasts, so that the expedition brings profit not only to the Company but will also benefit hydrography. If you should note discrepancies with the situation of the same as represented on the chart by the Admiralty Department in your possession, report on this directly to me upon your return.[225]

It appears that these results opened a new stage in the hydrographic and navigational exploration. This was initiated by Admiral Sarychev, who to the end of his life retained an active interest in the northern parts of the Pacific Ocean, which he investigated first as a young captain-lieutenant with the Joseph Billings Expedition. Due to Sarychev's initiative, hydrographic surveys off Alaska coasts became obligatory from 1830 on for all Russian-American Company vessels, while the hydrographic depot began to regularly supply the ships' pilots with the newest navigational charts. This guaranteed constant additions to them, more precise definition, and their renewal.

Ferdinand von Wrangell

This process was energetically followed by Wrangell, who was the chief manager of the colonies from 1830 to 1835 and who, like Sarychev, retained a scholarly interest in Alaska to the end of his long life. During the years of his tenure as chief manager, navigational and hydrographic investigations acquired primary importance in the work of the company's expeditions. The starting points of these investigations were well prepared, as is attested to by a report dispatched by Chistyakov to the head office of the Russian-American Company on the transfer of his office to Wrangell:

262, October 5, Report To the Head Office

Concerning the discoveries in hydrography and new coastal surveys conducted by commanders of Company vessels in the course of my five-year tenure as manager of Russia's colonies in America, I have the honor to report to the Head Office of the Company that the charts specified on the appended list compiled at various times by the vessels' commanders were transferred by me to Baron Ferdinand Petrovich (Wrangell) so that he would be able, when occasion affords itself, to order that these descriptions be checked and present them to the Head Office at the proper time. As the Head Office knows, the maps compiled by the commander of the Northern Overland Expedition, Mr. Lieutenant Vasiliev, and the charts of Port Chichagov on Attu Island compiled by Midshipman Etholen, were transmitted in the original, together with their journals, to the Head Office.

Charts of coastal surveys compiled by Mssrs. Commanders of the Company vessels between 1825 and the year 1830.

1. Chart of Port Chichagov on Attu Island, compiled by Midshipman Etholen in the year 1827 aboard the brig Chichagov.

2. Chart of the harbor on Bering Island compiled by Conductor (now Sub-Lieutenant) Mr. Chernov, in the year 1827 aboard the brig Chichagov.

3. Two charts of the harbor on Amchitka Island and harbor on Copper Island compiled by Sailor Mr. Ingenstrem in the year 1828 aboard the boat *Sivuch.*

4. Chart of Kiska Island, compiled by him also.

5. Chart of the harbors on Unga island, compiled by Sailor Mr. Stepanov aboard the boat Unalashka.

6. Chart of St. Quentin Bay in California, compiled by Mr. Etholen in the year 1828 aboard the sloop Baikal during the loading of salt.

7. Map showing the directions of the flow of the rivers Nushagak and Chivaktul and 8. showing the course of the rivers and lakes situated to the NW of Bristol Bay, compiled in the year 1829 by Mr. Sub-Lieutenant (now Lieutenant) Vasiliev, commander of the Northern Overland Expedition to explore the interior of America.

9. Special chart of Kenai Inlet (Cook Inlet) with Nikolaevsky Redoubt and showing the mouth of the River Kaknu and adjacent shores thereof. And 10. Plan of the Konstantinovskaya harbor and part of Nuchek Bay (Hinchenbrook Island), compiled by Mr. Sub-Lieutenant Chernov aboard the boat *Bobr.*

> (Chart of the entrance to Columbia River, copied by Mr. Lieutenant Tebenkov from the chart by the commander of the English vessel Codboro, lieutenant of the Royal British navy, Mr. Simpson.)[226]

Of the maps by Vasiliev on the above list, only one has been found. This report does not contain any mention of the results of Etholen's voyage on the brig *Chichagov* in 1830. But this expedition had significant consequences for the expansion of the Russian-American Company's possessions to the north. In full accordance with the instruction, the *Chichagov* sailed into Norton Sound and from there past the islands of St. Lawrence, Aiak (Sledge Island), and Ukivok [King Island—LB]. Upon his return to Sitka, Etholen recommended that an outpost be established on Stewart Island.

By 1831 Wrangell desired to reaffirm Etholen's success, so he gave the assignment to the commander of the brig *Chichagov*, Lieutenant Tebenkov, "to carry out continuance of the activities and relationships begun in the past year by Mr. Etholen with the inhabitants of the islands and coasts to the north of our settlements in America and also in Asia."

Tebenkov was supposed to sail to Unalaska and Pribilof Islands, then explore Stewart Island and its vicinity, and obtain the following information:

> About means of subsistence, at what seasons and in what quantity do fish species occur here, are there migratory birds and in what numbers, do they molt here and bring up their chicks, or do they visit the shores only for a brief time; do roots that are used for food grow there, berries and edible grasses, such as wild onions, garlic, wild sorrel etc.? At what seasons of the year and from what direction do the strongest winds blow, does winter come early, how long does it last, how severe is it, does the sea freeze, is there drift ice and when does the sea become free of ice, is the summer weather clear or is it rainy? All of this information you will be able to obtain in part from the Natives, and in part acquire through your own investigations. Beyond the above, note the tides, currents, depths, bottom off the easterly and southerly coasts, if a small harbor may be found or a small bay suitable for anchorage; describe the latter, find a location most suitable for settlement, on an elevation, near a source of fresh water and of driftwood. Do not commence any of these investigations without first offering civilities to the inhabitants in the form of gifts and through barter and telling them that we plan to settle here next year for greater convenience of trade with them; that a vessel will be arriving each year. Note how they react to this announcement. Should they be pleased, then you may begin to question them, if not, attempt to attain your goal in a way they will not notice, declaring that as they are not willing to have us as neighbors we shall settle at Tachik, where we are being invited by the Natives. Gather from the local inhabitants information about the River Kvikhpak, in what season

> is it possible to engage with the inhabitants of its mouth, is it deep, how far upstream is it possible to travel by baidarka, is the population thereabouts numerous, are there many beavers and other fur-bearers, do they communicate with the Kvikhpak always by water or over the mountains?[227]

Tebenkov carried out this program only two years later when he entered Norton Sound aboard the sloop *Urup* and founded Mikhailovsky Redoubt, which became a base for further explorations into the continent's interior.

Wrangell's instruction to Tebenkov demonstrated that enhanced attention to hydrographic surveys on the part of the chief manager of the colonies did not mean reduced interest in scientific aspects of the exploration of the American possessions.

Wrangell brought with him the "Appeal of the Conference of the Imperial Academy of Sciences" dated January 21, 1830. The employees of the Russian-American Company were requested to organize collections of natural history specimens for further study by scholars and for the academy's collections. On the basis of this appeal, the chief manager sent an instruction to Fyodor Kolmakov, the manager of the Novoaleksandrovsky Redoubt, demanding regular collections of specimens of flora, fauna, ores, and minerals. Similar instructions were circulated to all other offices, and to specific islands and redoubts, including Fort Ross.[228]

However, hydrography remained a top priority for Wrangell at this time. This is apparent in his refusal to continue the explorations of the Northern Expedition, even though its commander, Vasiliev, had recovered somewhat from his illnesses in the course of the winter of 1831–1832 and was eager to return to the taiga. Instead, on February 17, 1831, Wrangell ordered him "should your health permit, to take up the command of the baidarka expedition for the survey of the coasts from Cape Douglas to the southeast along the Alaska (Peninsula) coast." Nothing was left to Vasiliev but to forget the thought of continuing continental explorations and to begin preparations for a new expedition in accordance with Baron Wrangell's instructions.

These instructions represent a further development of Sarychev's instructions concerning naval hydrographic surveys. More advanced navigational instruments are referenced in this work. The central goal of his instruction was to guarantee the geometric precision of navigational charts and to ensure the safety of navigation (basic features of orientation, shoals, banks, subaerial and submerged rocks, and so on).

The purely hydrographic aim of the expedition is clearly evident, in contrast to the instructions by Chistyakov. There is no suggestion here that the survey is a sideline to the commercial goals of the Russian-American Company. The goal is the survey itself.

Vasiliev once again conscientiously carried out his assigned work. He conducted the survey in the summer of 1831 and compiled nine large-scale charts of the southern and eastern coasts of the Alaska Peninsula. The charts themselves have not been located, but we may be sure that they were subsequently used in the compilation of navigational charts by the Admiralty Department. In 1832 Vasiliev was engaged in charting the islands and straits of the Alexander Archipelago. The summary chart showing the results of these hydrographic surveys was published in 1835 by Litke.[229] Fedorova found two summary charts compiled on the basis of the nine field charts by Vasiliev in the National Historical Archive of Estonia, in the collection of maps assembled by Krusenstern. These are:

1. *Chart of a Part of the Eastern Coast of the Alaska Peninsula and Shelikof Straits. Compiled from the Survey Conducted by Lieutenant Vasiliev of the Corps of Navigators on the Orders of the Russian-American Company. 1831.* It shows the southeastern coast of the Alaska Peninsula between Cape Douglas and Cape Kubugakli, that is, the shore across from Shuyak Island, and Afognak and Kodiak.
2. *Chart of a Part of the Eastern Coast of the Alaska Peninsula and Shelikof Strait. Compiled from the Survey Conducted by Lieutenant Vasiliev of the Corps of Navigators on the Orders of the Russian-American Company, 1832.* This chart contains the southwest coast of the Alaska Peninsula between Cape Kubugakli and Cape Kumliun.[230]

It is possible that the results of Vasiliev's survey in the navigational season of 1831 are reflected to a degree in two manuscript charts from that time period, author's name not stated, which are preserved in the Russia State Archive of the Navy. One of these represents the eastern part of the Aleutian Islands (Unalaska region, Umnak, and the western part of the Alaska Peninsula).[231]

This chart, besides oceanographic information that even includes specifics about currents between the islands, presents population data for the year 1831. Individual settlements are marked by numbers, and their names and population (number of persons by gender) are given in table form. This chart also shows volcanoes. The legend contains their names and explanations, as, for instance, "Makushin Peak, smoking. Shishaldin Peak, emitting flame. Nesanakhski [Isanotski—LB] Ridge, smoking. Pavlovskaya Peak, Sevidovskaya Peak, Tulikovskaya Peak."

In the chart's eastern part, geographic names are transcribed not only in Russian but also in French in red. In this time period the detailed cartography work is also represented by the *Flat Chart of the Kenai Gulf. Compiled by Right Compass, Deviation of which is 30° E, in the Year 1830.*[232] It shows the localities in the area somewhat to the south of Kasilov Cape and the Kasilov River to the territory to the north of Nikolaevsky Redoubt, including *Part of the Island of Kulgiakh.* Moreover, a large-scale (150 sazhen to the inch) inset represents *Plan of the Entrance into the Kaknu River with the Adjacent Shores. Depth, in Feet, at Ebb Tide, Elevation 3 Sazhen.* There are two examples of this chart in the archival file, one of which is probably the original survey field chart, as on it headings and fixes are drawn in pencil.

The results of the hydrographic surveys conducted by Vasiliev in 1831 and 1832 are on an engraved chart located by James VanStone. It was published in Russian and French: *Chart of a Part of the Eastern Coast of the Alaska Peninsula, Described by Sub-lieutenant (correctly Lieutenant) of the Corps of Navigators Vasiliev in the Years 1831 and 1832* (*Carte d'une Partie de la Côte E de la Presquill d'Alaska Reconnue par M. Wassilieff Officer du Corps des Pilotes en 1831 et 1832*). It shows the southeastern

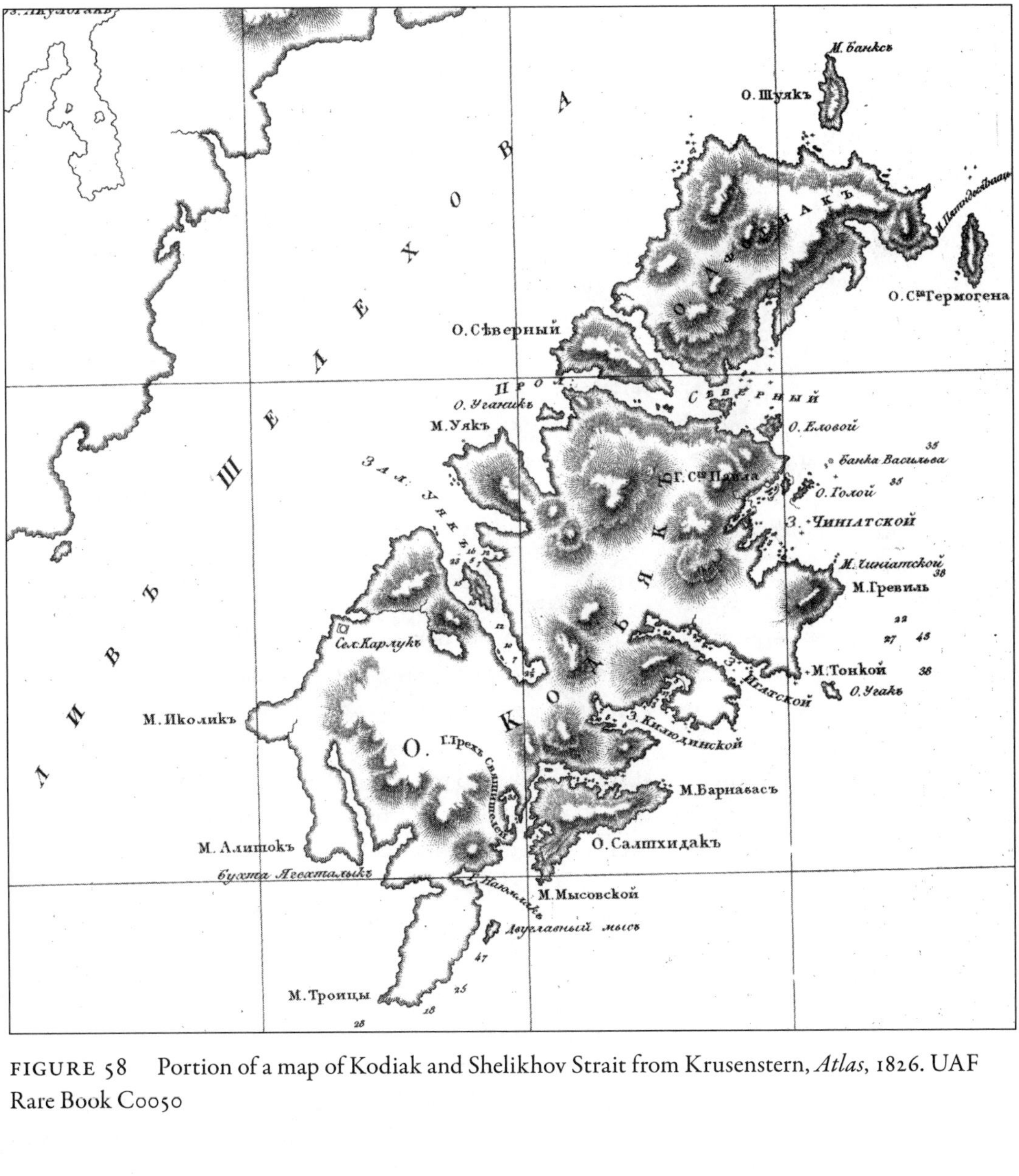

FIGURE 58 Portion of a map of Kodiak and Shelikhov Strait from Krusenstern, *Atlas*, 1826. UAF Rare Book C0050

coast of the Alaska Peninsula and also Lake Iliamna, Bristol Bay, and the middle course of the Nushagak River.[233]

An example of how Vasiliev's new data appeared on subsequent general charts is a comparison of two Krusenstern maps, one published in Russian in 1826 (Fig. 58) and one published in French in 1835 (Fig. 59). Both of these charts depict the same part of the Alaska Peninsula, Shelikhov Strait, and Kodiak Island.[234]

On the whole, Wrangell was pleased with the surveys conducted in the summer of 1831, and he wrote Vasiliev as follows: "Having perused your reports about last year's survey, I did not find the astronomical journal. The survey itself has been carried out in sufficient detail and the charts compiled by you are excellent, though the described extent is very small. Foul weather might have been an obstacle, in part, for passages, but the fact that your family accompanied you on your voyage would have been a hindrance. In general, it is not smart to take along a wife on such a journey, as she could have become ill and thus stopped the entire expedition. Therefore, this year you shall leave your wife with her parents and you shall strive to extend the survey for a much greater distance than was the case last year."[235]

It should be noted that such tactful reprimands from Baron Wrangell about the participation in the expedition by Vasiliev's wife were apparently called forth not so much by an insufficient volume of survey work in the summer of 1831 but by a scandal that shook Kodiak after the expedition returned. Vasiliev and his wife had a disagreement and she left him, and he started drinking and engaged in a drunken rampage.[236] Vasiliev exhibited such willfulness not only in his private life, but he also annoyed Wrangell by his clear preference for overland continental explorations.

This is well illustrated by an excerpt from the assignment given to Vasiliev by the chief manager for the field season of 1832:

> The survey is to be conducted in accordance with the instruction issued by me last year, chronometer observations excepted . . . I expect from you in this employment the same zeal you have demonstrated in your overland journey to the interior of America, having no doubts about the essential benefit of your assignment. But I counsel you not to reason about benefits, but to carry out precisely the directives issued to you. Then your work will not be forgotten, but will bring you honor. Negligence brings consequences, same in the Colonies as in Government service.[237]

Midshipman N. Y. Rozenberg was entrusted in the summer of 1832 with the task of conducting a hydrographic survey of the straits linking Sitka Sound with the ocean and Chatham Strait. Wrangell directed to Rozenberg an instruction similar to the one Vasiliev had received earlier.[238]

In accordance with Baron Wrangell's assignments for the navigational seasons of 1832–1833, the navigators of the Russian-American Company, but especially Vasiliev, continued to conduct hydrographic surveys and cartography of the islands

FIGURE 59 Portion of a map of Kodiak Island and Shelikhov Strait from Krusenstern, 1835. UAF Rare Book C0056

and straits of the Alexander Archipelago. The two manuscript charts located by Fedorova in the Wrangell Collection in the National Archive of Estonia (in Tartu) are but one result of this work.

These are:

1. *Chart for the Entrance into Novo-Arkhangelsk Port by the Neva and Olga. And Cross Strait with the SW Part of the Peril Strait Fort Passage to the Chatham Strait. From the Survey Performed by Lieutenant Vasiliev in the Year 1833.*

On this chart are shown Baranov, Chichagov, and Kruse islands and the passage to the Novo-Arkhangelsk port from the northwest through the Klokachev Strait (modern Salisbury) and then by Cross Strait, and also the initial segments for circumnavigation of Baranov Island along its northern coast by Peril Strait;

2. *Chart of the Biorko Island Group Situated at the Entry into Sitka Sound and Rakovoi Bay with Islands Within it. Compiled from the Survey Performed in the Year 1833 by Lieutenant of the Corps of Navigators Vasiliev.*

   This chart shows groups of islands to the southwest of the entry into Sitka Sound, the most significant of which are Biorka and Elagin islands.[239] The summary chart showing the results of these hydrographic surveys was published in 1835 by Litke.[240]

Precisely in this period the head office decided to reestablish the northern continental expeditions under Vasiliev's leadership and, in connection with the founding of a settlement in the southern part of Norton Sound, asked the chief manager (in a letter dated March 24, 1832) to dispatch such an expedition, postponing the hydrographic survey of Alaska.[241] This directive was motivated to a certain degree by the leadership's intention to begin a search for gold in Alaska.

Wrangell was specifically informed as follows: "Of course, the time is not ripe yet for us to begin gold extraction, and perhaps it is not even to be found in our colonies. At present it is not a case of immediately commencing gold mining, but gradually exploring for lodes. You have already achieved the beginning, as can be seen from the instruction you issued to Mr. Vasiliev to split off small pieces of rock from the coastal cliffs and to note which sample had been taken where."[242]

Wrangell did not carry out the head office's directive to reestablish the Northern Expedition and organize a baidarka voyage along the Yukon, thus clearly incurring the displeasure of the leadership in St. Petersburg.[243] However, he did not completely neglect the northern parts of the Russian-American possessions. Immediately following Etholen's expedition Wrangell worked out and transmitted to Tebenkov a program for the exploration of areas in the vicinity of Stewart Island.

EXPLORING THE YUKON AND DESCRIBING THE COLONY

In 1832 Wrangell again tried to use another hunting voyage of Lieutenant Tebenkov to organize contacts with the Indians of the basin of the River Kvikhpak. In his instruction to Tebenkov dated June 1832 Wrangell ordered him to go through to Tebenkov Bay (St. Michael Bay) and leave there the *promyshlennik* Yukhovsky, who had volunteered for this. Yukohovsky was to stay for the winter with two Aleuts and a baidara to "travel to the interior of the land and especially on the River Kvikhpak, to familiarize yourself with the inhabitants, encourage them to hunt river beavers, and incline them towards coming to receive our trade goods at Tebenkov Bay."

Baron Wrangell then wrote: "It is extremely desirable for Yukhovsky to be on the River Kvikhpak now and to return to Tebenkov Bay soon enough that on the return voyage from Michilmen Bay, on entering Tebenkov Bay for the second time, he could obtain trade goods and gather information about the River Kvikhpak and the disposition of its inhabitants—in order that we can finally make a decision about the establishment of a Company settlement in that country."[244] Besides the latest attempt to obtain information about the River Kvikhpak and the inhabitants of its valley, Lieutenant Tebenkov was instructed: "If time and circumstances permit, then from the Pribilof Islands direct the voyage of the brig to the West and try your luck at finding the unknown island about which there are contradictory stories. For this enterprise the two attached maps may be useful for you, indicating the routes of the *Golovnin*, the *Baranov*, the *Unalashka* and the *Okhotsk*. For new information you will have the right to the Company's gratitude, and you may promise half a quart of rum and 100 rubles in money as a reward to the sailor who first sees the island."[245]

The excerpts from Wrangell's instruction to Tebenkov provide evidence of his desire to use investigations during the regular hunting activity of small parties from the company consisting mainly of Creoles and Aleuts to obtain information about the mainland areas of Alaska. Wrangell was also counting on expanding fur hunting to new islands with marine mammal breeding grounds. In accordance with a directive of the head office of the Russian-American Company dated March 31, 1832 (No. 397), he had been instructed to expand the hydrographic surveys so that "systematic sailing directions" for Russian America could be drawn up at the department of the hydrographer-general of the Russia Navy.

The head office required him to "make an instruction for the Commanders of vessels on gathering this information by topic, whenever possible, and to deliver it to this Office to be presented to the Department of Mr. Hydrographer-General."[246] Besides that, in a dispatch dated March 28, 1832 (No. 275), the head office of the Russian-American Company again confirmed the need to search for new land in the waters adjacent to the company's possessions:

> In a special dispatch copies of two statements from Mr. Hydrographer-General have been sent to you, in which His High Excellency writes to affirm that based on indisputable signs unknown lands are located in the vicinity of the Aleutian Islands on the south and north sides. Besides this, according to the Head Office's information there are signs of the existence of land opposite the island of Kodiak on the south side, which is also mentioned by Khlebnikov in the second part of his notes on America. According to old legends there are cat dens on that land or island.
>
> There is also some basis to assume the existence of an island between the northern islands St. Paul and St. Matthew. All these suppositions may only be based on guesses, but without the appropriate investigation they cannot be

> refuted with certainty, and so the Head Office, being obliged to concern itself with the distribution of local knowledge about its colonies, and at the same time bearing in mind the benefits for the Company which may follow from the discovery of land anywhere in the above-mentioned places, commissions you and humbly begs you, as soon as circumstances permit, to equip a special vessel under the leadership of an experienced commander, especially for the purpose of searching for islands that are unknown until now, whose existence is assumed in the above-mentioned places, around which the searches should take place.
>
> In accordance with that aim be so good as to create such an expedition based on your own plan and send it at a time that is convenient for you, supplying its Commander with an appropriate instruction by which he will be guided in his actions. For its part, in the event of the discovery of an unknown land, and outside the boundaries of the Hydrographic position on newly-discovered land and its description, the Head Office will not hesitate to make the proper reward to the leader of the expedition and the crew of the vessel, in proportion to the importance of the discovery and the extent of the benefit that may be accrued in the newly-discovered place for the Company's business. The Head Office will await detailed notification from you about your instructions relating to sending this expedition and about its successes.[247]

The navigation of 1832, however, did not bring either new information about the mainland areas or discoveries of previously unknown islands. Tebenkov did not manage to establish trade relations with the Indians who populated the Kuskokwim and Yukon basins this time either. In connection with this, Wrangell decided that "one method remains for us to establish ourselves in this country, and it is to settle on the River Kvikhpak itself."[248]

To implement this plan Wrangell thought that "we need to find out:

> 1) Is the mouth of the River Kvikhpak navigable?—And this can be learned with a sailing vessel and two baidarkas.
>
> 2) Are beavers found in the vicinity of the river?—This we have already found out circumstantially from two directions. Consequently as far as is permissible I will be confirmed in my opinion about the uselessness of a baidar expedition on the River Kvikhpak."

Defending his opinion, Wrangell then writes: "In its dispatch dated March 24, No. 302, the Head Office, discussing this subject, notes that with insufficient methods for undertaking a baidar expedition on the River Kvikhpak an end should be put to the coastal surveying of Alaska. From this remark I conclude that the idea has taken root at the Head Office that I, having destroyed the expedition to the interior of America and appointing another one on the coast of Alaska, have lost sight of the benefit to the Company and sacrificed it for a generally useful enterprise. To this I will have the boldness to reply that such an opinion does not correspond either with my actions, or with the nature of my ideas, and he or they who brought forth these suspicions are severely mistaken about me."[249]

Fyodor Kolmakov

Wrangell's actions in this period confirm the truth of his statement: In the summer of 1832 he began to actively encourage the head of the Aleksandrovsky Redoubt, Fyodor Kolmakov, to study the interior areas of Alaska. Kolmakov was instructed (dispatch No. 319, dated July 10) to go to the Kuskokwim to trade and to look for locations for outposts on that river. The plan was to overwinter on the Kuskokwim.

In his assignment to Kolmakov Wrangell wrote: "You have been able to get to know the peoples here, they are attached to you and your name is known throughout the entire country and along the rivers Kuskokwim and Kvikhpak. And so for the benefit of the Company I am choosing you to be the one to carry out a new attempt to penetrate into the interior of the land and to acquire as many more beavers for trade as possible."[250]

Kolmakov's trading expedition was carried out in the winter of 1832–1833. The party got through from the Aleksandrovsky Redoubt to the Kuskokwim River in the vicinity of the mouth of the Holitna River "on foot and dogsleds."[251]

In the summer of 1833 Wrangell again dispatched an expedition by sea under the overall command of Lieutenant Tebenkov to search for an entrance to the mouth of the Kvikhpak (Yukon) River and to establish an outpost of the Russian-American Company in Norton Sound. The Creole Andrei Glazunov, who had been educated in the colonies as a ship's pilot and was able to determine astronomical coordinates and reconnaissance surveys, was assigned to the expedition. The immediate exploration of Norton Sound and the mouth of the Kvikhpak were entrusted to Lieutenant Rozenberg commanding the schooner *Kvikhpak*.

The instruction addressed to him by Baron Wrangell, dated May 9, 1833, # 179, directed him as follows:

> After taking on ballast and water proceed to St. George Island, where you shall deliver cargo. Then proceed to Stewart Island and drop anchor in Tebenkov Bay; should the *Urup* not be there, await her arrival. The main goal you shall strive to achieve is an entrance into the Kvikhpak River. Make every effort to achieve this goal as long as the safety of your vessel and crew permits. The necessary instruments and instructions will be given to you by Lieutenant Tebenkov. Being acquainted with the Natives and with the

localities themselves, he will take into account circumstances and will inform you about anything you need to know in this matter.

Lieutenant Tebenkov will try to persuade one or two Natives who have been to the Kvikhpak to accompany you to show you the fairway, and for this same purpose assign to you a baidara and paddlers. After passing through the straits that separate Steward Island from the mainland, it seems best to keep near the mainland, not approaching it at less than 2 sazhen of depth, when the wind is full, to be able to stand off shoals if necessary. It may occur that the Natives, knowing the extent of the shallows, will suggest sailing farther out to sea; you, of course, will take such advice into consideration when setting the course. Currents, rip tides, driftwood, the color of the water—all of this shall not escape your attention and should serve to guide you. I intend to make here but a few remarks.

The vessel's speed is always to be measured by lead and near the shoals, while keeping to good depths, the lead should drift for not less than 7 sazhen. Assuming that the tide rises not more than one half sazhen, the Kvikhpak should not have less than 2 sazhen of water, thus to cross a bar it shall be possible to risk it even at ebb tide and at lesser depth. Should the swells be running, or the seas be high, the depth under the vessel is to be increased. Should the depth drop to less than two sazhen, drop the stream anchor, or anchor and dispatch baidarkas to sound the depth. At low speed it is advisable to have in front of the vessel baidarkas taking turns, and sounding lead.

Should you encounter difficulties in finding the fairway from the sea, you may, depending on circumstances, dispatch a baidara with your assistant to the mouth of the river by a route known to the Natives. When he enters the river, order him to sail along its course while you take bearings on the baidara from the vessel, and the baidara should take bearings on the vessel. The assistant should record the depth soundings and compass bearings, the time of rowing and the course in order that an approximate chart may be compiled. Should the weather become inclement while the baidara is away, sail the vessel to Stewart for a pre-arranged rendezvous and there await the baidara.

The baidara is to be ordered not to sail in foul weather but to make for shore.... Prepare convenient signals for communication with paddled vessels that are engaged in taking soundings. You are not to be satisfied with one or two unsuccessful attempts, but repeat the effort until you are truly convinced that it is impossible to enter the *Kvikhpak* river.

Having entered the river, you should strive to penetrate upstream, sounding depths and putting the bends of its shores on the chart even approximately, noting the population and industries of the Natives. It is desirable that you reach a place where there is a change in language and where the forests begin and there are fur-bearing animals in the forest. Halt here and try to

> become more closely acquainted with the inhabitants, showing them kindness through small gifts, friendly intercourse and barter in goods.
>
> Collect detailed information about the upper course of the river, about the relationship of the river inhabitants among themselves and with the Kuskokwim River and the sea coast near Stewart, and what that country yields, etc. . . . propose to them to conduct trade with us permanently and persuade them to collect beaver, river otter and silver foxes for the next year, promising that a vessel bringing goods will arrive then. In between, investigate a locality where it would be convenient, in time, to establish a redoubt.
>
> When you collect information about that land, I ask you not to be satisfied with answers to your questions by a single person; repeat these questions at various times and get to the truth by considering answers that will be in all probability contradictory. The more you learn about topography, industries, and trade relations in that country, the better.[252]

Wrangell's instruction shows how thoroughly he grasped the study and cartography of Alaska. The directive generalizes the company's customary mode of using Native information and a method of approximate hydrographic surveying from aboard a large vessel paired with a baidarka. Although the expedition was headed by naval officers in Russian-American Company service, Tebenkov and Rozenberg, the role that brought the most significant results was played by ordinary civilian employees, especially the Creole Glazunov.

Tebenkov's expedition was unable to enter the Yukon mouth from the sea, but its other task was completed successfully: "in 1833, with the Natives' consent in latitude 63°28′ N and longitude 161°53′ W from Greenwich, a redoubt named Mikhailovsky[253] (modern St. Michael) was founded at the expense of the Russian-American Company" in Norton Sound near Cape Stephens (modern Cape Stevens). The redoubt was named for St. Michael, the guardian angel of the expedition's leader, Mikhail Dmitrievich Tebenkov.[254]

Andrei Glazunov

In this new settlement, Redoubt Mikhailovsky, "in the same year of 1833 an expedition was outfitted to cross mountains and rivers and reach the Gulf of Kenai (Cook Inlet). It was commanded by assistant pilot Andrei Glazunov, a Creole educated in the colonies, and a fluent speaker of the Kadiak language, a common tongue in the region that the expedition would have to traverse initially. Four Russians volunteered to accompany him."[255]

Lavrenty Zagoskin provided a striking portrait of the expedition's leader: "Glazunov is kind and cordial in his daily dealings with the Natives; and clever in trade, he knew the Native language perfectly and the tender sex served him as his police."[256] One of the goals of the expedition was to reach the Kvikhpak River mouth in the area of Pastol Bay. However, because of rumors of hostile Natives on the Pastol (Pastoliak) River no guide was willing to take this route to the Yukon. But three Natives "agreed to guide our detachment to the locality where the language changes" in the northeastern direction.

On December 30 Glazunov's party set out on their way with two sleds of five dogs each. After traveling to the northeast for eight hours on the ice of Norton Sound, they reached a Native settlement, where they were hospitably received and supplied with fish and three additional dogs. The travelers halted here for a day and "Glazunov learned that Inkalit [plural of Ingalik, Yup'ik name for the Lower Yukon Athabascans—LB] from the Kvikhpak River came each year to this settlement of Kikhtaguk (modern Klikitarik)[257] bringing beaver skins for sale to Aiakmiut who halt here on their way to Pastol."[258]

On January 1, 1834, the party continued on its way and crossed into the Yukon basin in the upper reaches of the Anvik River. The travelers continued downstream along the frozen Anvik in the direction of the Kvikhpak. The route along the Anvik was not easy because of a thaw and "due to the weight of fallen snow there was water covering the ice." One of the *promyshlenniki* fell through the ice and suffered frostbite on one foot. The party was forced to halt for three days, waiting for a hard freeze, to repair snowshoes, and for other necessities. Because of the difficulties of the trail to the Kvikhpak, Glazunov sent three of his men back.

On January 17, having left the sled, the party traveled lightly, with only two dogs, toward the nearest settlement, and finally came to a winter camp. The Native who gave them shelter told them the settlement was located at the place where the Anvik enters the Kvikhpak "and that the inhabitants of this settlement had assembled in full strength, according to the words of this wild one, expecting an attack by the people of the Unalakleet River."[259]

On January 26 Glazunov's party arrived at this settlement and was cordially received by the Natives. Glazunov spoke to the Natives about the Russians and invited them to come to trade at the Mikhailovsky Redoubt. Strong winds from the southwest kept the travelers in this settlement at the mouth of Anvik River until January 30. Glazunov used this enforced halt to question the Natives in detail about their country. He managed specifically to learn about portages from the Kvikhpak to the Kuskokwim valley, knowledge that the Natives initially concealed from him. In his daily journal Glazunov provides a description of the mores and customs of the inhabitants of the settlement at the mouth of the Anvik River, who had had practically no contact with Europeans; he also provides descriptions of the local natural environment.

He notes that "the Kvikhpak River is teeming with various fish, which are taken by fish traps made of willow twigs." The Natives told the travelers about "savage" people living in the interior of the country, who they named "Inkalikhmiut or

Kylchane (Goltsane of the Kenai or Kolchane of the Copper River people)." After paying their hosts for the dried fish that the party had taken from their caches while traveling, the travelers went on their way on January 30, 1834, following the Kvikhpak (Yukon) downstream.[260]

On February 3 the party came to the settlement of Magimiut, where they stayed for two days. "In the meantime, two men from the River Chagiliuk arrived at the settlement. From them Glazunov learned that the above-mentioned river flows between the Kvikhpak and the Kuskokwim and enters the first-named. Because there are all kinds of animals, its shores are densely settled. They counted for him 40 large winter settlements where two different languages are spoken. The distant people are the Kolchane or Ingalikhliuaty, who trade with those who live closer."[261] From the description of the river Chagiliuk it is clear that the River Innoko, a major tributary of the Yukon, is meant. The Shageluk settlement survives on it to this day.

After changing their two Native guides in the settlement of Magimiut, the travelers continued downstream along the Kvikhpak and on February 8 came to the settlement of "Anenulykh Tykhlak," situated on the right bank of the river (near the modern population center of Holy Cross). Glazunov ascertained that people came to this settlement from the rivers "Anvik," "Pshanukshak," and "Chagiliuk" and that Anenulykh Tykhlak was the boundary of the Ingalit language. Further downstream, along the course of the Kvikhpak, other peoples lived who spoke the "coastal" language. Eight large and several small settlements were located there.

The travelers also learned that "according to those who had visited the sea coast, the River Kvikhpak empties into the sea through five arms which are named, from north to south in order, Ankhun, Kvikhpak, Kvikhliuak, Kiinaiak, and Kizhunak."[262] After obtaining guides who knew the way to the Kuskokwim, Andrei Glazunov and his companions set out on February 14 and ascended the watershed between the two great rivers of Alaska, from where they glimpsed the Kuskokwim valley.

The next day, having come out to the river itself, the travelers encountered seven Kvikhpak men who were returning from Kolmakov. The travelers gave to Semyon Lukin, the interpreter who was accompanying them, all the furs they obtained by barter. On February 20 Glazunov's party went upstream along the Kuskokwim (or, as the local inhabitants called it here, the "Kushkukhvak"). The next day, once again, in the settlement of "Paina-imiut" they encountered Lukin, who brought the travelers to the outpost that was established by Kolmakov in the year 1830.

Two guides were willing to show Glazunov the way to the mouth of the river Tkhalkhuk. "These people told Glazunov that there cannot be any portage from the headwaters of the Kuskokwim River to the east, to the Gulf of Kenai (Cook Inlet) because of the high impassable mountains in that direction. But the Kolchane who live at the headwaters of the rivers Kuskokwim and Chagiliuk (Innoko) exchange beavers for tobacco, which they obtain from the inhabitants of the River Tkhalkuk, who obtain tobacco from the Kenai. However, because of the great distances involved, this trade is insignificant."[263]

The detachment traveled through inhabited territory until the 28th; in every settlement their reception was hospitable. Many Indians were already acquainted with the Russians, having been baptized by Kolmakov. On March 2 Glazunov passed the mouth of a small river, Khytynalo. At the headwaters of it, according to the guides, there was a pass to the Chigiliuk River. They passed the river mouths of Iunilnuk and Takhvilkhanakot, which were rich in beaver, according to the Natives. On March 7 at the approach to the mouth of the Tkhalkuk River, the travelers sighted ahead along their route a huge mountain, Tenada, at a distance estimated by Glazunov of seventy to eighty versts.

"According to the guide the source of the River Tkhalkuk lies on that mountain and the Kenai people live there. The guides left Glazunov here, refusing, no matter what recompense was offered, to continue on the way, where inevitable death awaited them. The party went on without guides, going upstream along the Tkhalkuk River . . . On the 14th they were apparently at a considerable elevation; the mountains were treeless, and the snow was not deep, but they observed no birds or animals and not the slightest indication of habitations."[264]

The travelers were totally exhausted, and the food situation was desperate to the point that for their supper they "boiled a piece of leather bag that they had carried along for repairing the soles of boots." Glazunov was forced to take the party back. On March 17 they reached the Kuskokwim. Having eaten one of the dogs, they set out downstream. The sorry state of the party, including its leader, is vividly described by Glazunov in the journal entry he made the next day: "My people boiled the rest of the dog meat. I was feeling terrible pain in my eyes, which have become bloodshot, my feet were swollen and I had a pain in my left side. I was frightened. I took off one of my skin boots, cooked it in the fire, and ate it."[265]

On March 19, about six versts before they could reach the summer camp, Glazunov was so unwell that he had to lie down on the sled that his companions then pulled. The party rested at Ushkugalik until March 23. On the 24th they encountered the interpreter Lukin once again, who was going with three large sleds to the Holitna River with "*the catch; he carried with him two baidaras in case the rivers flooded*."

In the end Glazunov reached the Yukon and from there, over Pastol Bay, "the expedition returned to the Mikhailovsky Redoubt on April 13, where the people continued to suffer for a long time with eye pain and swollen legs. The journey lasted 104 days, of which 43 days were spent advancing and 22 coming back, 39 were spent resting; Glazunov calculated that they had covered 2,080 versts."[266]

In the course of the entire expedition Glazunov conducted a route survey; on the basis of it he compiled a map which was highly evaluated by Wrangell who published it, with some additions, in his fundamental work on Russian America.[267]

Glazunov's original map has not been found, but it was widely used by other travelers and explorers, specifically by Tebenkov and Lavrenty Zagoskin.

The delay in establishing an outpost in Norton Sound was met with disapproval by the company's management, who demanded a renewal of the continental expedition to be led by Vasiliev.[268] Wrangell, however, continued his tactic of exploring and taking control of American continental regions where fur-bearing animals were located. Here, in his opinion, the main role was to be played by experienced *promyshlenniki* such as Kolmakov, who had lived side by side with the Natives and enjoyed their trust.

Explaining his position in a report to the head office (#56, dated April 10, 1834), Wrangell wrote: "I do not consider the dispatch of Vasiliev's expedition into America's interior wise, so as not to damage our relations with the Natives . . . as . . . the Kuskokwim inhabitants (next to the Kvikhpak) have known Kolmakov for a long time and he knows how to treat them, while Mr. Vasiliev, who went there only once (his other trips were in the areas to the south of the Kuskokwim in the vicinity of the Aleksandrovsky Redoubt) managed, in a short time, to enrage the Natives and his guides and left in those regions a poor opinion of himself; and all the time a party trading with the Natives is in less danger than one who passes through the settlements of the savages with military threats.

"Kolmakov, whom I sent last year to the Kuskokwim and the Holitna, returned safely to the redoubt and brought 1,100 bartered beavers; again, on my orders, he went to these places, while Glazunov was dispatched, also with the goal of trade, to the Kvikhpak and beyond from the new redoubt on Stewart Island. Is all of this less valuable than Vasiliev's expedition that did not acquire anything [Wrangell means fur catch—AP] but cost the Company plenty?"[269]

At the same time (April 10, 1834) Wrangell sent an extract from Kolmakov's journal to the head office "preserving therein Kolmakov's own words and expressions, at all points where these are notable for interest and originality and demonstrate the intellect, experience, bravery, diligence and sharpness of this old serviceman of the Company, a native of Tobolsk." Wrangell adds to this that "Kolmakov, his interpreter Lukin and his entire party (seven men) are worthy of special praise and reward to encourage them in the future."[270]

Immediately following the establishment of the Mikhailovsky Redoubt, Wrangell began to use it actively as a base for exploring adjacent areas, especially the Kvikhpak and Kuskokwim basins, by company employees who were sent there to obtain fur. He envisioned such explorations as the settlers' everyday work, paralleling their commercial fur procurement.

Instructive in this respect is the admonition sent by the chief manager on May 19, 1833, # 234, to the "manager of the new [Mikhailovsky—AP] Redoubt in Tebenkov Bay, Company Employee Ivan Kuzmin."[271] The instruction specifically mandated that an all-out effort be made to clarify if it were possible to deliver catches from the Pastol and Kvikhpak rivers.

Toward this end "one or two Russians accompanied by an interpreter are to be sent together with the locals to the River Kvikhpak across the mountain to barter for furs. If this is not entirely convenient, then send an emissary, even without trade goods, to that river to declare to the local inhabitants that we are friends to them and are asking them to come to the redoubt for trade."[272]

This assignment was fully carried out in the course of Glazunov's journey. Wrangell's general instruction to the manager of the Mikhailovsky Redoubt devotes significant content to relations with the Natives. "Should various tribes war between themselves and should one of them ask for your assistance, under no circumstances are you to take sides but you are always to impress them that you want to live in peace with everybody; do not permit any one to challenge you but do not interfere in somebody else's quarrels; by all possible means try to make peace between the enemies, making it clear to them that from conflicts both sides lose and that it is better to hunt and trade than to kill each other. They probably have heard of Kolmakov and know that he made peace between the Aglegmiut and the Kuskokwim and that nowadays both are grateful to Kolmakov for it."[273]

Wrangell also ordered Kuzmin to compile a journal noting the activities of the employees and events at the redoubt: "You are to . . . enter daily observations on weather, wind, how the people were employed, events, when the sea froze, when did species of fish, birds, or animals appear, what kind of contact you had with the inhabitants and what were able to learn about the country. Such *daily entries* are to be forwarded to me each year."[274]

The first meteorological observations were conducted at Novoarkhangelsk in the summer and fall of 1819 by Semyon Yanovsky, and these materials are preserved in the Archive of the Russian Geographical Society in St. Petersburg.[275]

Unification of Sea and Navigational Charts

In 1833–1834 Wrangell continued to utilize every occasion for hydrographic exploring and compiling navigation charts of local seas. Intensive methodological scientific work was being conducted in Russia aimed at creating new instructions for compiling maps and using symbols on large-scale charts. Great attention was paid to unifying the content and form of large-scale military-topographical land maps and navigational charts.

Between 1827 and 1837 the military topographical and the hydrographic depots of the general staff of the navy were under the unified command of the remarkable Russian cartographer and geodesist Fyodor Fyodorovich Shubert (1789–1865).[276] The hydrographic depot decided that the creation of sea surveys and of navigational charts in Russian America were also to be unified.

To ensure compliance, the head office notified Wrangell that " all conductors of the (navigational) Corps must have this (Sarychev manual) in their possession."[277]

Also published at that time were unified topographical maps, using symbols from the military, *Naval Symbols to be Used on Hydrographic Charts and Plans, Compiled at the Hydrographic Depot of His Imperial Majesty's General Staff of the Navy (St. Petersburg), 1833*. One year later this work, too, was sent to Russian America.

Wrangell, who himself was an outstanding hydrographer, expended much effort to perfect the navigational charts of Russia's American possessions. Almost every skipper of company ships received assignments to search for new lands off the Alaska coasts and to improve the existing charts. Thus, in 1833 Rozenberg, captain of the schooner *Kvikhpak*, received an instruction along with a voyage assignment to Norton Sound to search for "the islet that according to the tales of the Umnak Aleut is situated to the south of Samalga Island or off the southern end of Umnak."[278]

He was also instructed to "pass to the point where the islet of Transfiguration is shown on old charts, Lt. Sindt's lost discovery of 1764, that is from St. Paul roughly WNW on the right compass for 240 Italian miles."[279] Wrangell's instruction in this case demonstrates how persistent Sindt's cartographic fantasies were.

DIONISY ZAREMBO

Lieutenant Dionisy Zarembo, captain of the brig *Chichagov*, received orders (under # 241, dated May 31, 1833) pointing out the necessity of carefully exploring the ocean between 40°–46° N latitude and 144°–155° W longitude from Greenwich, where "various sailors at different times have noted signs of the proximity of land and where more than once fur seals have appeared, thus giving reason to think that thereabouts is an unknown islet."

Wrangell gave Zarembo detailed recommendations on methods to determine the proximity of land in the ocean and to compile search charts, advising him particularly as follows:

> The flight of birds in the evening, but especially the unexpected appearance of kelp and marine mammals, will indicate to you the most likely direction where land might be found, and the lead will give you assurance of its nearness. . . . Make a chart for yourself at such a scale that each mile can be shown; mark the vessel's route by dotted lines and the approximate distance of the horizon observed from the vessel, according to the clarity of the air; on the same charts mark which signs of land have appeared and especially note the points determined astronomically. In this way your chart will serve you, in future attempts, too, as a rather convenient guide in choosing a course.[280]

In the hydrographic work in this period the reliance on Native guides and pilots acquired even more importance. The captain of the boat *Beaver*, Alexander

Kashevarov, who was sailing on a routine voyage from Novoarkhangelsk to Kodiak, was instructed on May 10, 1833,

> at every opportunity to engage in observations and surveys, checking previous point determinations wherever possible, keeping a detailed journal. Should you note anywhere an error in soundings in a harbor or bay, engage in correcting the same, if at all possible, but do not expend too much time on this.... Should you deem it necessary, request from the Kadiak Office a Native who knows the area. After the conclusion of your voyage, return him to the office.[281]

Wrangell sent a dispatch to the "commander of the vessel sailing to Nushagak" in which he asked him to use the voyage to bring the charts of this area to greater accuracy. Wrangell's assignment reads in part:

> append here a *Nushagak Map* and the register of observations I have made. I ask you, if possible, to check this chart and expand it. An Agligmiut from the settlement on Cape Ekuk has a certificate. He will approach you when you raise the flag and fire your cannon. This Agligmiut knows the fairway in the river; though having no idea how to steer a vessel and work the sails, he can, at the very least, tell you if you are approaching shallows and in which direction the depth continues.[282]

The Aleuts and other local inhabitants of the Russian colonies played a great role in Vasiliev's land and sea explorations which continued in the navigational season of 1833, when he was given the assignment of charting the southeastern coasts of Russian America in detail. This work was conducted in accordance with the following instruction issued by Wrangell to Vasiliev:

> The work you have accomplished testifies to your experience in this matter, and therefore I shall not speak of methods of survey. I am only calling to your attention to soundings, bottom, investigation of currents, high and low water and marking of anchorages, in the straits themselves and also in the entrances and exits of the same [underlined in the original—AP], compiling in this way reliable sailing directions and charts for the benefit of vessels that must sail from the Ocean through Peril Strait to Chatham Strait or from Sitka Sound through the Neva or Olga Strait to the same destination, or from Khutsnovsky [Hootsnowoo, Angoon—LB] Strait though the Peril and Olga Strait to the Port of Novoarkhangelsk, and the same in respect to Biorka islands and the rocks near them.

The excerpt from the instruction cited here makes it clear that Wrangell had in mind the necessity of compilation of a detailed navigational chart of the areas adjacent to Baranov at the highest scientific and technological level. This was assigned to Vasiliev, whose crew included the Tlingit Pavel.

It should be noted that Pavel had, along with Vasiliev himself, great experience in expeditionary explorations and enjoyed an excellent reputation in the colonies. On the eve of Vasiliev's expedition Wrangell wrote as follows: "The Tlingit Pavel, who was aboard the *Chichagov* during its last voyage, has proved himself to be, according to the testimony of Captain-Lieutenant Etholen, an excellent pilot for the straits; he behaved diligently and earned encouragement. Therefore I propose that the Novoarkhangelsk office assigns him a salary of 180 rubles per year as of June 1 of this year; the Office will be kind enough not to issue it in money but in vouchers for receipt from the store of items he needs."[283]

The hydrographic work that was actively pursued on Wrangell's assignments in the years 1830–1833 brought substantial results, as we have already seen. One of these was evidence of the non existence in the northern part of the Pacific Ocean of the legendary islands that had been sought by many sailors. Wrangell reported this to the head office in his dispatch # 75, dated April 10, 1834, *On Zarembo's Voyage with the Brig Chichagov on a Search for Islands and Failure to Find Such*. Along with Zarembo's chart, Wrangell appended to his dispatch the *Chart Indicating the Routes of Vessels that have Searched for Land in the Northern Sea to the West of the Pribilof Islands in the Years 1831, 32, and 33*.

This chart, as well as all the testimonies by navigators of the Russian-American Company, demonstrated, in Wrangell's opinion, the total uselessness of future searches for the legendary islands. He wrote the following:

> As far as renewing the search for new lands throughout the entire expanse of the Northern part of the Great Ocean to the south of the Aleutian Chain is concerned, guided by *possibilities* and *likelihood*, I find any such future enterprise superfluous. Should it be the will of the Head Office to publish each point in the ocean with *mathematical precision*, this would demand centuries of uninterrupted searches.
>
> I am even of the opinion that should an island and fur seal rookery be discovered to the south of the Aleutian Chain, the Company would suffer a clear loss should these fur seals be harvested. The fur seals, in all probability, are in the south just before their pups are to be born. Consequently, in order not to lose the increase of the young, there should be no hunting south of the Chain under any circumstances; in the north they are taken when the young are strong enough to swim away to the south with the onset of winter.[284]

Wrangell believed that the perfection of cartographic knowledge about the coasts and seas of Russian America had much greater significance. Thus he reported to the head office the discovery by Lieutenant Zarembo of a direct route between "the Stikine, by means of which . . . the distance between Stikine and Novoarkhangelsk is so diminished that the brig *Chichagov* accomplished the return voyage, with halts and waiting for the currents and winds, in 7 days."[285]

Wrangell asked all captains of company vessels to describe the seas known to them, and he put at their disposal all the old journals and charts which were then present in the Novoarkhangelsk archive. Mikhail Dimitrievich Tebenkov took the most active part in carrying out this assignment. It was at this time that he began his remarkable cartographic activity, the results of which were collected in 1852 as *Atlas of the Northwest Coasts of America*, which we shall analyze in an appropriate section.

In 1834 Wrangell wrote:

> Lieutenant Tebenkov presented me with 9 charts of all our possessions in America, compiled by him from surveys already known to the public: of Captains Beechey, Litke and Stanyukovich and those not yet known: by Messrs. Vasiliev (a navigator who is presently departing), Ingstrem, Chernov, Zarembo, Khramchenko, and Etholen, with analysis of the same and remarks about a voyage from Novoarkhangelsk to Unalaska Island and to the North, to the Bering Strait. I am forwarding these 9 charts, the analysis and the comments compiled by lieutenant Tebenkov to the Head Office for transmittal to the Office of the Hydrographer-General, as a very useful and not unimportant work of Mr. Tebenkov. As I had not time to copy the analysis and comments, I ask the Head Office to send me such copies so that they can be expanded as materials in this matter are constantly collected here.

Altogether, in addition to the two summary charts by Vasiliev showing the area from Cape Douglas along the southern coast of the Alaska Peninsula, which Wrangell sent off on May 1, 1833, he dispatched to the head office of the Russian-American Company on April 28, 1834, with memo #199, twenty-three charts and one plan described in the *Register of Charts and Plans* as follows:

> *Under 1st cover:*
>
> *9 charts of Russian-American Company possessions compiled by Lieutenant Tebenkov (with appended packet addressed to F. Litke).*
>
> *Under 2nd cover:*
>
> *1 chart of the voyage of the Brig Chichagov under the command of Lieutenant Zarembo in the year 1833 in search of islands to the south of the Aleutian Chain.*
>
> *1 chart of the voyages of vessels to the west of Pribilof Islands for purposes of search in the years 1831, 32 and 33.*
>
> *1 chart of the neighboring Tlingit Straits within the possession of the Russian-American Company by Captain-Lieutenant Etholen.*
>
> *2 detail charts of Etholen Harbor and part of Prince Frederick Sound compiled by Lieutenant Zarembo.*

9 detail charts from the survey by Lieutenant Vasiliev of the south coast of the Alaska Peninsula, compiled by Mr. Vasiliev.

1 plan of Ozerskoi Redoubt compiled by Sub-Lieutenant Chernov.[286]

This list of charts dispatched by Wrangell to the head office of the Russian-American Company demonstrates how much attention he devoted to questions of the geographic exploration and cartography of the Russian colonies. Wrangell came to count more and more on regular commercial trips by parties of company employees, on information collected at redoubts and outposts and on Native informants.

Clear evidence of the fact of such an approach is the instructional letter to the commander of the schooner *Kvikhpak*, "Sub-Lieutenant of the Corps of Navigators Kashevarov," dated May 9, 1834:

> Last year nothing satisfactory was accomplished by the schooner *Kvikhpak* in search for the mouth of the Kvikhpak River and exploring its entry; however, the ship's mate, Glazunov, managed to explore with baidarkas the communication between the Kvikhpak River and the redoubt (Mikhailovsky) by means of sea channels and river sloughs and compiled an approximate chart of the same, a copy of which I append here for your information [underlined by the author—AP].
>
> The extent of the sea shoals and the low coasts indicate the torturous course of the river and its division into many small arms, and therefore finding a passage suitable for vessels from the sea will demand not only great effort but would also cost extended time, which I cannot grant you, except in a limited sense.
>
> Besides, for the benefit of commerce at the new redoubt you ought to establish acquaintance with the inhabitants of the River Kvikhpak and learn the most convenient route of communication with them so that a decision can be reached on whether it is worthwhile to establish an outpost on that river where the catch can be collected and then delivered by oar-propelled vessels to the redoubt.
>
> Another matter is making the Pastol trade its own, a matter that the *baidarshchik* has probably already taken in hand. It follows that these two matters should have, for the benefit of commerce, priority over the search for the river mouth from the sea side. To them, then, you shall devote your time and effort.
>
> The communication routes established by the Natives with the land rich in beaver and foxes are, no doubt, the best, and you ought to follow these routes should circumstances and time permit. It is most desirable, and I am sure you will not let an opportunity pass, that you should penetrate inland as far as your safety and other circumstances permit (by means of baidarka, of which, together with those at the redoubt, you will have 6, not counting

> the longboat that you will probably leave behind in a convenient place when shallow water and a rapid current will obstruct progress).
>
> During such explorations try to compose an approximate map (there will be no time to construct a precise one) and do note in which location it would be most convenient, safe and profitable to establish an outpost into which catches can flow, if it proves unsafe to have one at the mouth of the Pastol. From the Pastol inhabitants . . . collect as much information as you can about the country's interior in general, about the lakes and rivers, the peoples, and the abundance of subsistence resources on the Pastol and Kvikhpak Rivers. Try to find out the disposition of the Natives in regard to visitors from the redoubt (the Mikhailovsky) and trade with them.
>
> Should Glazunov be at the redoubt, after handing him my instructions, take him along in your travel to Pastol and beyond, so that by utilizing his practical information about the land and its inhabitants you might succeed in penetrating further.
>
> Receive (at the Aleksandrovsky Redoubt) from Kolmakov his report to me and obtain from him information about how he proposes to reestablish the fur-procuring expedition to the Kuskokwim from where by that time interpreter Lukin should have returned to the redoubt. Having finished all your assignments start immediately on the return voyage to Novoarkhangelsk, choosing to sail through the Unimak Straits (Pass) as the shortest route, course.
>
> It is superfluous for me to remark that wherever circumstances permit you should try to make astronomical and marine observations, determination of coordinates and surveys, and that upon your return you shall deliver to me your report with appended journals, charts, etc. Order that the Aleuts should receive the same rations as the crew and use them in work and on watch equally with the others.[287]

Sub-Lieutenant Kashevarov had no chance to complete the large program sketched out for him by Wrangell, as his voyage on the *Kvikhpak* extended only to the islands of Unga, Unalaska, St. George, Stewart, and the Aleksandrovsky Redoubt on the Nushagak River. A Russian outpost on the Kvikhpak was established in 1839 by the Creole Pyotr Malakhov about half a verst from the Native village of Nulato.

In the years 1838–1840 Malakhov made two journeys to the Kvikhpak. In 1841 the structures at the outpost—a dwelling, storehouse, and steam bath—were destroyed in a fire. The Russian post was reestablished and an *artel* (work camp) of Russian *promyshlenniki* was based in the Nulato settlement, on the middle course of the Yukon. It became the northernmost Russian outpost in Alaska's interior.[288]

Earlier, in 1832, as a result of Vasiliev's expedition, the first Russian trading post was established on the Kuskokwim, at the mouth of its tributary, the Holitna.

Called the Khulitnakskaya outpost, it was commanded by Semyon Lukin. Somewhat later the Holitna outpost was transferred downstream, to the other shore of the Kuskokwim, across from the River Kvygym. In 1841 it was transformed into a redoubt called Kolmakovsky in memory of its founder, Fyodor Kolmakov, who died in 1840.[289]

The Russian outposts on the Yukon and the Kuskokwim became bases for organizing commercial and exploratory expeditions and active centers for collecting information about interior Alaska. Almost immediately after establishment of these bases, the journey of Assistant Navigator Pyotr Fyodorovich Kolmakov (son of Fyodor Kolmakov) began. In 1838 and 1839 he and the Creole Markel Matrosov reached the River Tlegon (the upper course of the Innoko River) and the watershed between the Yukon and the Kuskokwim and compiled a map of this region.[290] The presence of these Russian settlements in Alaska's interior contributed to the success of Zagoskin's expedition, which shall be discussed below.

Wrangell continued to devote attention to exploring those regions that were immediately adjacent to the metropolitan area of Russian America—Baranov Island. In this work he widely employed the Tlingits as informants and guides. When dispatching "Sub-Lieutenant of the Corps of Navigators Kuznetsov" with the schooner *Chilkat* to "sail by Olga or Peril Strait across Christian Sound into Chatham Strait and . . . to the north of this strait into the Lynn Canal (Chilkat)," Wrangell firmly recommends that the captain "find a Tlingit for the identification of localities and . . . information about the country." The chief manager had already obtained beforehand an agreement in this matter with the "Tlingit elder of the Kukhantan Anakhuts" who resided on the shores of the Lynn Canal. He had promised to send a fine pilot to meet the schooner. Kuznetsov was directed, while in the Anakhunts' settlement:

> To collect detailed information relating to the following questions:
>
> 1. By what routes and in what seasons of the year does the Chilkat Tlingit communicate with the inhabitants of the interior, to the north and east, and what kind of fur catches they obtain from there?
> 2. Are there any significant rivers entering the Lynn Canal from the east? Is the country there mountainous or a plain, is it covered by lakes? What are the most abundant animal species? If there should be a significant river, how far inland it extends and if the Chilkat Tlingit navigate it or do the inhabitants (of the interior) travel down it to them (the Chilkat)?
> 3. Is there on the Lynn Canal a gathering place where the Chilkat Tlingit congregate for fishing or for other reasons and in what season do they gather there? Do they at such a time have with them their catch and would they like for our vessel to call on them at such a time for trade? Try to learn all possible details about this.

4. It appears that somewhere a large river rich in beaver and land otter flows down from the continent, probably into the Stephens Passage. Collect most detailed data about the mouth of that river, and also about the inhabitants that live on it and to what clan or settlement the catch from that river belongs and at what time they gather this catch.

> You are to proceed from the Lynn Canal to Stephens Passage. According to Vancouver, it cannot be entered by a sailing vessel except through Chatham Strait and Prince Frederick Sound. On this leg of your voyage you probably will be able to learn from the residents of that Sound, the Kake (Tlingit, Kekutsy), information about that river and to obtain a guide. Approach its mouth as near as vessel's safety will permit and collect from the primary source, so to speak, as much information about this river as will be possible.

From there Kuznetsov was to proceed to the mouth of the Stikine River and be at the disposal of Lieutenant Zarembo in order to protect Russian-American Company interests from encroachments by the British. Interpreter Gedeon was assigned to the expedition to facilitate communication with the Tlingit.[291] The essence of the dispute with the Hudson's Bay Company stemmed from the fact that in 1833, on the brig *Dryad*, the British had attempted to establish a permanent post on the Stikine River. The Russian sailors of the brig *Chichagov* and schooner *Chilkat* forced the British ship to depart and then built the Redoubt of St. Dionysius on a narrow peninsula in the mouth of the Stikine.

The conflict was finally resolved peacefully at a conference in Germany in 1839, when Wrangell, as the official representative of the Russian-American Company, concluded an official agreement with Sir George Simpson, the representative of the Hudson's Bay Company. According to this treaty the Hudson's Bay Company agreed to supply Russia's colony in America with foodstuffs for moderate prices. In return, certain territories in Russian America were to be leased to the Hudson's Bay Company for ten years.

Wrangell's lively interest in these regions is apparent in the map (Fig. 60) appended to his work on Russian America.[292] American scholars James Kari and James Fall believe that all subsequent nineteenth-century charts (by Tebenkov, 1848; Dall, 1870; and Petroff, 1884) represent the Susitna River basin in much lesser detail than is shown on Wrangell's map, which was compiled, they think, from cartographic sketches and toponymy by the Dena'ina Indian tribes inhabiting these regions.[293]

Chief Manager Ivan Kupriyanov

On October 29, 1835, Captain of the First Rank Ivan Antonovich Kupriyanov (ca. 1799–1857) took the post of chief manager of the Russian colonies in America,

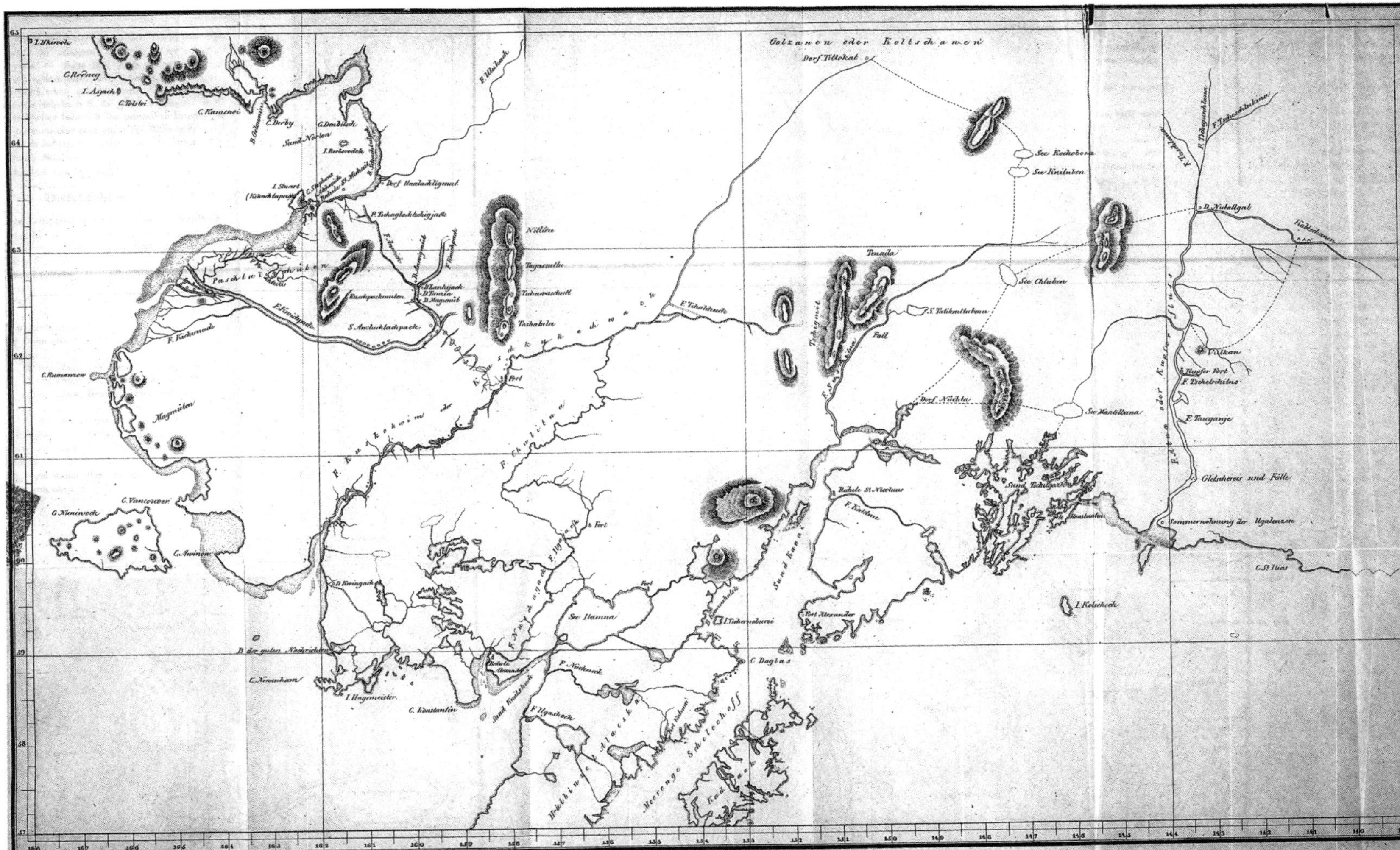

FIGURE 60 Wrangell's map. End paper, *Stadtische und Ethnographische Nachrichten* . . . 1839. UAF Rare Book A0503

vacated by Wrangell who had departed for Russia. Immediately following Kupriyanov's appointment, the head office decisively moved against any new expeditions in search of legendary islands, declaring "that two expeditions dispatched for discoveries have disproved the legends . . . and there will be no future dispatch of vessels for purposes of their discovery; nevertheless, all vessels sent out to sail south of the Aleutian Chain will be alert and note in the journals any phenomena that might be taken for signs of land's proximity."[294]

Kupriyanov took an active part in exploring and charting Alaska's coasts and its islands. In 1836 he dispatched Lieutenant V. K. Voronkovsky (died 1837) to conduct a survey, ordering him "to continue and finish the one conducted by Vasiliev, formerly in company service, of the east coast of the Alaska Peninsula, beginning at the spot where he concluded it and left a marker."

The main goal of the survey, according to Kupriyanov's instruction, was "to make this coast well known for the vessels sailing along it; consequently, one must locate descriptive features and tie them to all the islands; attention is to be given only to those bays that constitute a good harbor; it is useless to spend time on the others; one must only indicate precisely the headlands in order to tie together the coasts generally; one has to compose a chart of this area with measurements, with the degree of precision that would satisfy you yourself for a sail with a vessel entrusted to you. Append to it the necessary description of significant coastal features and short sailing directions."[295]

The same means and instruments that Vasiliev used were put at Voronkovsky's disposal. For the most part, the same Aleuts who participated in Vasiliev's survey in 1831–1832 were on this expedition also. Voronkovsky sailed out of Novoarkhangelsk on March 6, 1836, and came to Cape Mumliuk, where Vasiliev had terminated his surveys. Then, according to the official report by the head office, "Voronkovsky completed the survey to Cape Khitkuk, for the extent of 300 Italian miles, and tied this to the survey conducted by Captain Stanyukovich, the commander of the naval sloop *Moller*. En route he surveyed Unga Island, the Shumagin island group, and Unga Harbor, where he made very accurate soundings which we did not possess before and which were absolutely necessary for our vessels that visit yearly the dangerous Unga Harbor. Having finished the survey, Mr. Voronkovsky returned to port at Novoarkhangelsk and presented, together with his report, a chart which Mr. Kupriyanov delivered in the original."[296]

A calamity befell the expedition: on May 13, 1836, during measurement of the velocity of the current in the straits between Cape Kumliuk and a group of rocks called Liagat, two of the expedition's baidarkas capsized and three Aleuts died of hypothermia. In spite of this occurrence, the results of Voronkovsky's voyage were judged highly and the expedition's commander received the rank of staff-captain. In 1837, when Voronkovsky was commanding the schooner *Chilkat*, he suffered a shipwreck near Baranov Island and died with the rest of his crew.[297]

The Kashevarov Expedition

When Wrangell arrived in St. Petersburg, the former chief manager began to argue for a renewal of Russian exploration of the Arctic coast of Alaska, which had remained unexplored since the expeditions of Beechey and Franklin, and thus to gain priority over England in the discovery of the Northeast Passage from the Pacific to the Atlantic Ocean. Captain Beechey's book, which was delivered to the head office, was carefully studied, and "many useful pointers for guidance" were found in it, while "on the basis of his and Franklin's charts a special chart, in large scale, was created here" [in the head office—AP], which was forwarded to Kupriyanov "via the Okhotsk office with the mail."[298]

National pride demanded Russia's priority in geographic explorations in regions immediately adjacent to Russian America. This desire was "driven on" by the fixed interests on the part of the British government in the exploration of the Arctic coast of America. In this respect, Krusenstern wrote:

> As this (territory) near the Bering Strait is closer to our possessions than to the English, it should have been apparent that this honor belongs to Russian sailors; it is even more desirable in view of the fact that the Englishmen, not wishing that another nation should find the final solution to this question, which is more than 300 years old, are very much afraid of this.
>
> This is obvious from the remarks by the president of the geographical society, which I cite in the original: 'The idea of the question of the North-West passage being taken up by some other nation is far from being chimerical. There are two naval powers, either of whom would be ready to take it—were they persuaded that we had abandoned it—it has been the subject of discussion of both; and one of them happens just now to be placed under so many favorable circumstances for attempting it, that I consider it will be matter for surprise if she (Russia) should not avail herself of them. With improving settlement on the North-West coast of America, close to Behring's strait, with two corvettes always on the station and above all, with a both intelligent and enterprising governor (Wrangell), who passed 58 days on ice of the Arctic sea to the North of Siberia and whose mind is turned towards geography. There is every reason to believe that the consent of his government only is wanting to induce him to try his fortune on an adventure, the success of which would confer on his name immortal honor.
>
> Here I must mention one circumstance that ought not to be lost from attention. The Russians were the first to renew the search for the Northwest Passage in 1815 by dispatching the ship *Ryurik*. And barely three years later the English began to dispatch expeditions for the solution of this question. Should our government decide to begin explorations of this small portion of the coast, this should not be delayed, or the English will be before us. This

> can be easily accomplished with the aid of the American Company; all that would be required of it are several baidaras and Aleuts who can be sent to the Bering Strait aboard one of the American Company's vessels that are in the colonies and are used for various voyages; this vessel can await the baidaras' return either in Kotzebue Sound or in another location nearby.
>
> There is no doubt that such an expedition will be successful as it is known that the Aleuts [underlined in the original—AP] are most gifted in respect to such enterprises; however, the command over them should be entrusted to a naval officer, a very able one, who should 1) determine by means of astronomical observations the latitude and longitude of all distinctive features, specifically of the headlands where the survey begins and ends; 2) describe this coast with the utmost precision and accurate soundings and 3) compose a chart in accordance with all the surveys; compose a precise and satisfactory report about all he observes. It is not easy, of course, to select such an officer, as he must have, in addition to the above-indicated skills, a strong constitution, be strong in spirit and decisive, and he must have a passionate desire to undertake this expedition without regard to any personal aggrandizement, having nothing else as a goal than the enhancement of the glory of Russia's naval officers.[299]

Krusenstern's proposals were examined and further developed by Wrangell, who then worked out a program for an Arctic expedition and presented it to the Board of Directors of the Russian-American Company. The program proposed action "utilizing the time before the English with renewed efforts have penetrated further." There should be an expedition "using an Aleut crew, commanded by a Creole, with baidaras and baidarkas. A Company vessel is to be used to transport them so that the vessel is not forced to winter in those regions, and the activity of the baidara and baidarkas should be completed in a single summer."[300]

Developing his proposals, Wrangell wrote that "the expedition is to be composed of one six-oar bench baidara, that is of 12 oars, and five three-hatch baidarkas." He considered it necessary that the "crew be composed half and half, of Russians or Creoles and Kadiak men, volunteers who are healthy, keen, and are used to local foods of all kinds. Among the numbers composing the baidara crew one may include 2 or 3 Kuskokwim or Agligmiut whose language has a similarity to the language spoken by the Eskimos on the shores of the Arctic Ocean."

Wrangell underscored the absolute equality of Russian and Native members in the expedition in terms of supplies with foodstuffs and hunting equipment, declaring, in particular, that "the entire crew, Russians and Kadiaks, and others, must be clothed from head to foot warmly and comfortably, adding to the warm clothing esophagus or gut kamleias and torbasa."[301]

According to Wrangell's project the expedition was to travel from Novoarkhangelsk to the Mikhailovsky Redoubt aboard a large vessel (a brig), where it was to be augmented with men of experience, those who were with Glazunov on his journeys, and Native volunteers . . . From the redoubt the expedition is to proceed to Cape Lisburne, or as far as the sea, being free of ice, will permit. The expedition is to be disembarked there, so that they can set out on their route not later than July 1. The expedition's commander is to be instructed that he is to make every effort to return, without fail, in the first days of September. The brig is to await him until that time, or even to the middle of September, in Margat Inlet or in Kotzebue Sound, wherever it will be found to be more convenient and safe.

Should the expedition not return at the appointed time (about September 20–25), the brig is to leave with a reliable Native all possible support in the form of foodstuffs for the expedition. An agreement should be obtained with the Native to give hostages as a guarantee of the safety of the goods being left, under the condition that they will be returned next year, and then the brig is to proceed to Novoarkhangelsk."[302]

The expedition's landing party, traveling in the baidara and baidarkas, was to proceed:

along the coast to Point Barrow, the last point determined by Captain Beechey at latitude 71°23 and ½′ and 156°21′12″ W longitude from Greenwich. On this journey they are not to stop to determine locations which were accurately placed on Captain Beechey's chart, but they are to check their own observations against them. The voyage of Captain Franklin from the East terminates at Beechey Point, located at 70°30′ latitude and longitude 149°45′; it follows that the unexplored area extends for 6°½, or about 146 Italian miles almost along the latitude.

The goal of the expedition is to explore this expanse, and to determine by astronomical observation, as far as possible, the location of islands, capes, mountains and other significant features. Should circumstances and time permit, do not omit survey of bays, especially those that are suitable, with depth soundings, current observations, tide elevation etc.

The expedition is constantly to stay near the coast and put out to sea only when solid ice prevents a forward advance along the shore. When the ice is smooth there is the possibility of putting the baidara and baidarkas on runners (prepared beforehand for the purpose from whale ribs) and to haul them for as long as possible. Should the baidara be damaged so as to be unusable, or demanding a long delay, it should be left with extra men in a locality that is easily noticeable and the expedition is to proceed with baidarkas . . .

When the sea is clear, the expedition is to speed ahead without spending valuable time on observations; unavoidable stops will allow them to make observations. This suggestion is supported by the fact that the return voyage

> will be more sure than the forward journey, because by the middle of August the sea is most often free of ice. Thus, proceeding toward the E, one should not lose a single good minute.
>
> Here one can recall the remarks by a famous traveler that one should not embark upon too detailed a description of a coast that no one in all likelihood will ever see again, it is sufficient to determine their general outline. Landing in convenient places along the coast, the expedition's commander, or his deputy, should ascend the elevations personally and observe where there are areas free of ice.
>
> It is understood that the expedition's commander must keep a journal, entering observations daily as well as remarks about the localities and inhabitants, should these be encountered.[303]

It was proposed that after the expedition's successful return each member should receive an award equal to his annual salary and that "the expedition's preparation and its goals be kept secret [underlined in the original—AP], especially from the foreigners who frequent Novoarkhangelsk to pursue trade relations."[304]

This program for an expedition to the Arctic, worked out by Krusenstern and Wrangell, was presented to the chief of staff of the navy, Prince Alexander Sergeevich Menshikov. He, in turn, "had the happiness to bring it to the attention of Lord Emperor" and received "the royal blessing" from Nicholas I for its organization.[305]

Orders were sent to Kupriyanov in the spring of 1837. As we saw earlier, this was the navigational season when employees of the Hudson's Bay Company, Dease and Simpson, traveled westward from the mouth of the Mackenzie River to Elson Lagoon, completing the exploration of the passage from the Atlantic to the Pacific Ocean along the mainland coast of the Beaufort Sea.

Not knowing about this expedition, the Russian-American Company actively worked to prepare the Arctic voyage that Krusenstern and Wrangell had planned. Finally, in the summer of 1838, a party of eighteen men, Aleuts and Creoles, headed by a Creole, a graduate of the Corps of Navigators, Sub-Lieutenant Alexander Kashevarov (1809–1866), sailed with the brig *Polyphemus* out of Novoarkhangelsk for Kotzebue Sound. From July 5, beginning at Cape Lisburne, the exploration was carried out, in baidarkas, under the most difficult conditions. They traveled thirty miles to the east of Point Barrow and located bays which received the names of important men in the Russian-American Company: Prokofiev and Kupriyanov.[306]

The baidarka party was taken on board the *Polyphemus* on September 5, 1838, near Chamisso Island in Kotzebue Sound. It is difficult to accurately assess Kashevarov's contribution to the advancement of geographical knowledge about Alaska, as no chart has been published in concert with his journal. However, the journal contains several positive comments about the accuracy of Captain Beechey's chart. Therefore it is possible that Kashevarov considered it superfluous to publish his own chart as it would have added little to the published materials of the British voyager. [Kashevarov incorporated his data later in his atlas; the chart of the north coast of Alaska shows several Inupiat settlements not shown by Beechey.—LB]

At the same time, Kashevarov's journal contains much more detailed geographic information than that which appears in the description from Captain Beechey's longboat expedition or in the materials by the Franklin search expeditions.[307] Consequently, there is a more vivid picture about the land and the specifics of its ecology than in any other contemporary source. Although Kashevarov's addenda to the geographic nomenclature for this region did not survive, Zagoskin's map shows a considerable number of Eskimo settlements located by Kashevarov.

Kashevarov's contribution to ethnography is considerable. In contrast to Elson, who visited these regions with a longboat as part of Beechey's expedition, the Russian explorer closely questioned the Eskimos he chanced to meet and described their settlements and camps. This was the first expedition that had an Iñupiaq interpreter on board. This permitted him to obtain precise information about the land, the settlements' names, the rivers, and the tribal groupings.

In particular, the Eskimo interpreter, whose name was Utuktak, drew a map of the coastline between Kotzebue Sound and Point Barrow. Kashevarov made it more precise and it was later used by Zagoskin.[308] Kashevarov's detailed data on the character of the settlements are most useful. With a considerable degree of certitude, seventeen permanent and seasonal settlements were identified. This is a much greater number than mentioned by other expeditions of that period. Most of them were put on subsequent maps as a result of Kashevarov's exploration.

With other information about settlements, the number of dwellings is mentioned in the journal, and this permits us to estimate the population. Kashevarov's journal also contains the first published account of the dwellings and summer tents of the northwest Alaskan Eskimos. It is possible that Kashevarov's most important ethnographic contribution was determining the names of the various tribal groupings and gathering information on inter-tribal relations.

These data are important historical testimonies for modern research to determine the territorial distribution of traditional societies and tribes in northwest Alaska in the mid-nineteenth century. Kashevarov's travel journal also contains information on flora and fauna, as well as means of subsistence employed by the Eskimo, and their trade connections, specifically with northeastern Siberia.

Kashevarov's observations on the relationship of man to nature sound quite modern. For example, he draws a comparison between the natural environment of the Arctic seas and that of more southerly regions, and notes sharp contrasts in temperature and changes of seasons in the north, which determined in part certain seasonal cycles of human activity. Kashevarov expressed great concern about the changes that he observed in the traditional life of the Eskimos of northwest Alaska and the unhappy fate that awaited them should they be alienated from their habitual lifestyle by European values.[309]

The leadership of the Russian-American Company expressed its appreciation in a letter (# 134 dated March 31, 1839) to Kupriyanov. It stated that "although the English Hudson's Bay Company preceded the Russian-American Company in the part of the American coast that remained unexplored, Russian observations are more accurate and the notes are more complete than those made by Messrs. Dease and Simpson, which gives Mr. Kashevarov a right to acknowledgement and recompense."[310]

In the same year of 1839 the head office of the Russian-American Company reported the following about Kashevarov's expedition at the general meeting of company shareholders:

> In 1836, upon His Excellency Baron Ferdinand Petrovich Wrangell's return from America, the Head Office proposed to outfit an expedition for a final survey of the small section of American mainland that lies within our boundaries, between Point Barrow and Cape Beechey, which remained unexplored. This proposal was brought to the attention of His Imperial Majesty through the Chief of Staff of the Navy.
>
> Subsequently, precisely two years later, it became apparent that the English Hudson's Bay Company had developed at the same exact time an analogous intention and as one of the Directors, Mr. Simpson, was at the time in the northern possessions of the (Hudson's Bay) Company they put the plan into immediate execution. Thus in the next year of 1837 the English expedition, on two oar-propelled vessels, departed from the mouth of the Mackenzie River westward, in the footsteps of Franklin, and in five weeks completed the voyage and the survey to Point Barrow and back, thus gaining for England the honor of the final survey of the northern coasts of America, which the English Captains Beechey and Franklin had sought in vain.
>
> The annual mail to the Colony brought the instruction of the Head Office to Sitka not before the autumn of 1837, precisely at the time when the English had accomplished the enterprise, though the Chief Manager of the Colony, Captain of the 1st Rank Kupriyanov commenced to prepare for it with the burning zeal of a patriot who wished to reserve the honor for Russia. With the opening of the summer navigational season of 1838, the expedition set out from Novoarkhangelsk under the command of the Creole, a fosterling of the Company, Sub-Lieutenant of the Corps of Navigators Kashevarov.
>
> He sailed to sea with one 6-oar baidara and 4 three-hatch baidarkas, with 18 men, Aleut and Creole, aboard the Company brig *Polyphemus*, commanded by the Sub-Lieutenant of the Corps Chernov. According to the prepared plan, the brig *Polyphemus* was to penetrate beyond the Bering Strait as far as ice would permit and then approach the American shore and launch the baidara and baidarka, and Kashevarov was to continue along the coast sailing between the ice; after completing the survey to Cape Beechey, he was to

return to Kotzebue Sound, where the brig *Polyphemus* was to take his expedition on board for the return trip to Sitka.

This plan, in its essence, is completely identical to the one which was to be followed in 1821 and 1822 by Captain Vasiliev with the navy sloop *Discovery* and in 1826 and '27 the English Captain Beechey with the navy sloop *Blossom*, extending here the hydrographic work of the famous Cook.

Thus, Mr. Kashevarov had to enter into competition with Cook, Vasiliev, and Beechey, who, as is known, extended the coastal survey with a decked oar-propelled vessel to Point Barrow, 120 miles beyond the point reached by Captain Vasiliev; the latter, for his part, sailed 35 miles beyond Cook. Cook, Vasiliev, Beechey, and Kashevarov all had a single goal in mind: to extend the survey of the northern coast of America from west to east for the greatest possible distance, while in the meantime the last Hudson Bay's Company expedition under the command of Simpson and Dease proceeded from the east to the west, in the same direction as Mackenzie and Franklin.

The complete success of Simpson and Dease deprived Russia of the honor of completing the survey of the northern American coast, but it did not deprive Mr. Kashevarov of honor in victory in a noble contest always linked to self-denial. Overcoming the greatest obstacles and danger from ice and inimical Natives, Kashevarov, who was launched into the sea in the vicinity of Cape Lisburne, passed the Icy Cape. Here he had to part with the baidara and part of his crew and continue only with baidarkas, reaching Point Barrow, the final point reached by Captain Beechey; having passed beyond it and surveyed the coast for 30 miles he found it necessary to find shelter among ice from pursuing Natives; Seeing no possibility of saving his crew from the numerous savages enraged against whites by a smallpox infection that had recently appeared here, Kashevarov, with heavy heart, decided to turn back. After covering 850 Italian miles, the expedition reached safely Chamisso Island in Kotzebue Sound and was reunited there with the *Polyphemus*. They returned aboard it to Novoarkhangelsk with the entire crew well. The Head Office hopes to receive with the autumn mail detailed journals of this remarkable expedition.[311]

Kashevarov's results were first published in his articles in the years 1840–1846, while the full text of the journal was not published until 1879.[312]

Kashevarov's expedition remained virtually unknown outside of Russia and only one researcher, Lavrenty Zagoskin, appreciated his achievement. Zagoskin apparently used the journal, then not yet published, and certainly perused Kashevarov's article printed in 1840. Also, as we saw above, Zagoskin utilized Kashevarov's survey materials to compile his map, including the sketch by interpreter Utuktak. Kashevarov is an outstanding example of what great success could be achieved by a gifted Creole [His father, Filip Kashevarov, was a serf.—LB]. Born in Russian America, he steadily progressed upward on the service ladder of the navy.

Kashevarov ultimately achieved the status of hereditary nobleman and the rank of brigadier general.[313]

After Kashevarov's expedition the management of the Russian-American Company practically stopped exploration of the northern border regions of its possessions. The head office once again turned its attention to the interior regions of Alaska and to territories in the immediate vicinity of the colonial metropolis region (Baranov Island and Novoarkhangelsk). It should be noted that by this time, chiefly due to Wrangell and his wife Elisabeth, Novoarkhangelsk had been transformed into a flourishing center of European civilization and culture on the northwest coast of North America.

Sitka

A vivid description of the Russian-American capital of this period is given in the book by Edward Belcher, captain of the corvette *Sulphur*, who was conducting hydrographic surveys in 1836–1842 in the Pacific. *Sulphur* was the first foreign naval vessel to visit Novoarkhangelsk.[314] The *Sulphur* approached Baranov Island on September 12, 1837, and was met by Kupriyanov's secretary and a pilot. Then the deputy chief manager, N. Rozenberg (1809–1857), guided her into the harbor with several longboats. Kupriyanov received Belcher very hospitably and put at his disposal "all the facilities of the port and the arsenal."

Belcher noted that the chief manager of Russia's colonies was fluent in English and that "his civilities were overpowering." With assistance from the Russians, the British erected a field observatory on the island, directly across from the arsenal and were able to "secure our meridian before midnight." Belcher highly appreciated Sitka's craftsmen of the time and noted that he witnessed "comparative cleanliness and comfort, and much to admire, particularly in the school and hospital." He was enchanted by Father Veniaminov, whom he describes as a powerful man, of athletic stature, "quite Herculean, and very clever."

Belcher saw in the house of this remarkable Russian priest an organ, barometer, and other items that he had made himself. Veniaminov repaired barometers aboard the British vessel that were incorrect. In spite of the language barrier, the Englishman and the Russian cleric developed a firm friendship. On September 25 Kupriyanov gave a ball in honor of the British expedition. At this event the local ladies amazed the British with their refinement, beauty, and ability to dance.[315] A beautiful watercolor, "drawn by the Creole Alexander Olgin on July 30, 1837" preserved in the State Naval Archive of Russia, provides a good idea of the capital of Russian America of the time.[316]

The Russians' presence in America began to show observable scientific results. This was demonstrated before the entire world by Wrangell's book on the geography and ethnography of Russian America published in German three years after his return from the colony.[317] It appeared as the first volume in the new series of *Beitraege zur Kenntniss des Russischen Reiches.*

This work, which has not been translated into Russian to this day, not only summarizes data from many of Wrangell's predecessors and contemporaries, but also includes his first hand observations during lengthy inspection trips through the colonies. He studied all this information deeply and analyzed it through comparisons of materials from geographic explorations in other polar regions of the globe, which were well known to him not only through literature but by his own experience in the Arctic Ocean and on the Siberian coasts. This comparative analysis added much to the understanding of the natural processes and facilitated the development of physical geography as a branch of science. The book is also acknowledged as a fundamental work in ethnography.

The first four chapters of the monograph are Wrangell's presentation about Russian America based on statistical information. He describes methods of fur hunting in detail, discusses the problem of contact between the aboriginal inhabitants of North America and those of Siberia, and focuses on various ethnic groups of Natives of Russian America, including those residing in the vicinity of Fort Ross. Describing the latter, Wrangell uses his own observations and those of the commandant of Fort Ross in 1830–1836, Pyotr Stepanovich Kostromitinov.

These materials served later as an important source for subsequent works. The fifth chapter, added by the editor, naturalist, and member of the St. Petersburg Academy of Sciences Karl Ernest (in Russia, Karl Maximovich) Baer (1792–1876), is an excerpt from the journal of Glazunov's expedition. An article by Veniaminov about the Aleuts is also included (chapter eight), as is a vocabulary of Native languages (chapter nine). Baer himself presented a summary of the pioneering work on the languages of the aboriginal inhabitants of North America by American scholars (chapter ten), and also a first of its kind essay on the Alaska climate (chapter eleven).[318]

In later years Academician Baer returned many times to materials from Russian America in his geographical studies, though he never visited Alaska. The works of Wrangell and his contemporaries were not only used by Baer in his theoretical geographical works, but also permitted him to advance several practical recommendations for economic development and agriculture in Russian America. Baer expressed his thoughts on this matter in a letter to the Russian-American Company (# 145, dated April 7, 1839), in which, among other matters, he says the following:

> The weather observations conducted in the course of three years on the Island of Sitka by Vice Admiral Wrangell serve as a sure measure of the local climate at various seasons of the year. Comparing summer temperatures at Sitka with the degree of warmth in other agricultural lands, I reached the following conclusions:

1. Rye cannot be successfully grown at Sitka. It can be hoped, though, that in time, as forests are felled on the mountainsides, valleys may be found that are reached by sunshine sufficiently to permit the experimental cultivating of rye
2. On the other hand, it is feasible to suppose that barley will ripen almost every year, especially if it is planted at the lower southern slopes of the mountains.
3. I give preference to raising in our American Colonies the bread-providing plant Quinoa[319] which has become known but most recently and which is cultivated in great quantities in the mountains of Peru and Chile at such elevations where rye and barley cannot survive. I have the honor to include here a small brochure on this topic that I have published; I think that it would be superfluous to add to it anything more about the merits of this plant. Its leaves provide greens that resemble goosefoot, which is used as food in all parts of Russia. These greens can be used with benefit even if the seeds do not ripen due to some circumstance.

I also enclose a small quantity of seeds of Quinoa, as I consider it to be most useful to learn from experiment if this plant can grow in Sitka and produce seeds. If these seeds will not produce in Sitka, they will in all probability produce in Kamchatka.[320]

We do not know if the initiative of Academician Baer to introduce quinoa found any application in Russian America.

Ilya Voznesensky

Following the publication of the work of Wrangell and Baer, the St. Petersburg Academy of Sciences made a decision to organize the regular collection of scientific data and specimens from the possessions of the Russian-American Company. In 1839 Ilya Gavrilovich Voznesensky (1816–1871), a laboratory specialist at the Zoological Museum of the Academy, was dispatched for this purpose to Russian America. He spent almost ten years engaged in scientific observations of extreme importance to zoology, botany, ethnography, and geography. He spent time in California, on the islands of Kodiak, Baranov, Unga, Unalaska, Pribilof, Attu, and St. Lawrence, at the Gulf of Kenai (Cook Inlet), at the Mikhailovsky Redoubt, and in Kotzebue Sound. A set of his drawings and the ethnographic collection he assembled are in the Museum of Anthropology and Ethnography in St. Petersburg.

When recommending Voznesensky for an award, the Academy of Sciences gave the following assessment to the results of his assignment:

The results of this remarkable expedition have exceeded the Academy's expectations in all aspects, in their richness, variety and importance. The items he

> collected that pertain to the three natural kingdoms and to ethnography were in 150 boxes; their contents provided invaluable material for our naturalist scholars. Many new species, animal and plant, are already described, and their number will reach, even exceed, 400. The mineral specimens that came to the Mineralogical Museum made it possible for the conservator of the Academy, K. I. Grewingk, to publish in 1848 an orthographic and geognostic outline of the northwest coast of America and of the Aleutian Islands.[321]
>
> Besides these material benefits that accrued to the Academy through Voznesensky, his voluminous notes and large set of accurate drawings deserve special attention. Moreover, he trained many persons resident in those lands in the preparation of specimens and they continue to collect and prepare natural science specimens for the Academy according to his instructions.[322]

Although Voznesensky himself did not engage in cartography, his materials became the basis for the geological and geomorphological maps in Grenwingk's work, and that will be discussed later.

One of Voznesensky's capable students was Lavrenty Zagoskin,[323] whose expedition in continental Alaska we will now proceed to examine.

LAVRENTY ZAGOSKIN'S EXPEDITION OF 1842–1844

The Russian-American Company never abandoned the idea of a new large-scale expedition into the interior of continental Alaska. The ground for such an expedition had been solidly prepared by the company's fur trade and by the preceding journeys by Vasiliev, Kolmakov, Glazunov, Malakhov, and Lukin. The Russians were by now acquainted in general terms with the geography and peoples inhabiting the lower courses of the Kvikhpak and Kuskokwim. The tribes inhabiting these regions, for their part, were informed, at least indirectly, about Russian *promyshlenniki* through the process of the fur-barter trade, and by this time did not avoid contacts with them.

In June of 1840, Adolf Etholen replaced Kupriyanov as chief manager of Russia's colonies in America. Etholen, as we saw earlier, possessed substantial experience working in Russian America and knew it well. In his person, the management of the Russian-American Company acquired a scholar of natural history and the peoples of Alaska as enthusiastic as Wrangell himself. Lieutenant Lavrenty Alekseevich Zagoskin (1808–1890) became one of the most outstanding executors of Etholen's plans.

Like the majority of Russian-American Company navigators, Zagoskin obtained his naval education at the Naval Cadet Corps in Kronstadt. On October 6, 1839, Zagoskin arrived at Novoarkhangelsk in command of the company brig *Okhotsk*.

He spent the next two years on commercial voyages off the American coasts. In particular, in 1841 he sailed aboard the corvette *Elena* to Fort Ross. On the return trip he was joined on board by Voznesensky. Voznesensky gave Zagoskin many tips about natural science and the collection and conservation of flora and fauna specimens.[324] In the spring of 1842 Etholen proposed that Zagoskin undertake an expedition into America's interior.

Zagoskin outlined his plan in 1840 in a letter sent to Wrangell in St. Petersburg. One of the goals of the expedition was to reach the sources of the Kvikhpak and Kuskokwim as well as to find the best route between the basins of these two rivers.[325]

In July 1842 Zagoskin was brought to the Mikhailovsky Redoubt, where, aside from preparing for the expedition, he immediately began to record weather conditions and phenological changes. He collected geographical information from experienced *promyshlenniki*, interpreters, and local inhabitants. Zagoskin paid close attention to the details of the preceding journeys and the traditional migration routes of the Natives. In particular, he determined that "the closest route from the sea coast to the Kvikhpak and the Nulato River, used by the Natives, is the one from the Unalakleet River."[326] In the *Journal of the Secret Overland Expedition under the Command of Lieutenant Zagoskin*, we find detailed information he obtained from Assistant Navigator Glazunov and the names of "many concentrations of mountain ranges" reported by Glazunov "in the Native language."

When speaking about the Mikhailovsky Redoubt he notes the following: "At the founding of the redoubt one Native woman from the neighboring settlement advised against settling on the present locality, saying that in her lifetime twice the area became flooded; her words were discounted as fable, but the truth about the past is evidenced by large half-rotted driftwood logs found on the elevated parts of the island, more than a mile from shore."[327]

One of the most important assignments given to Zagoskin was to determine the routes and tribes through which Alaska furs were reaching Chukotka, bypassing the employees of the Russian-American Company. The middlemen in this trade with the peoples of northeast Asia were known then under the collective name of Malemiut. According to the opinion of Dorothy Ray, in the 1830s and 1840s this term was applied to any Eskimo trader from the north.[328]

"Information about the local inhabitants, the Mallemiut [*sic*] that was needed in order to check the flow of the furs from the American shore to the Asiatic one" was collected by Zagoskin from the "former interpreter for the northern expedition of Mr. Kashevarov and checked several times against information from other Natives of that region."

Zagoskin entered the data he obtained onto a map, along the coast of Stewart Island, and in his journal he expressed the following thoughts, rather aptly characterizing the importance to cartography of even such approximate knowledge: "It is very possible that I placed the location of the settlements on the basis of information from the Natives not very accurately, but 5 to 6 miles have no importance in this matter. I know from experience that it is necessary for a traveler on his first journey to know after how many days or hours of marching one shall reach a settlement, its population, and where the summer and winter camps are; on such a basis estimates are made for needed food supplies and for caution."[329]

Besides showing settlements on his map, Zagoskin compiled a *List of Settlements along the Unalaklik [sic] River to the North to Point Barrow.*

Zagoskin also critically examined materials from the earlier journeys. In particular, he referred to Malakhov's map and Glazunov's travel journal. Zagoskin learned from Malakhov's map that "from Unalaklik [*sic*] to the emergence on the Kvikhpak along the straight line . . . is 100 miles or 175 versts, and from there to Nulato 30 miles or 52 versts, that is, 227 versts in total."

He noted that on Malakhov's map the locations' longitudes were farther to the east of their correct situation, as Malakov did not take into account time spent on stopovers when calculating longitude; the latitudes were off by two minutes due to the instrument's inaccuracy.[330] In perusing these materials Zagoskin paid particular attention to relations between the travelers and the Natives.

In this respect he noted that:

> when in accordance with the orders of the colonial authorities instructions were given to investigate and then settle these localities, the clever (Native) traders, understanding Russian strength and anticipating their losses, resorted to the politics of Machiavelli. Offering their services as guides, they conducted Glazunov and Malakhov to the Kvikhpak for four winters, always concealing the easiest and shortest route; finally, when through a young Native Malakhov managed in 1838 to find the real portage, they resorted to cutting loose bags with provisions in transport, robbed the dried fish at Unalaklik [*sic*, Unalakleet], terrorized gullible Natives of the Kvikhpak with the name of the Russians and pushed their tribesmen, the Takaiksan, to get rid of us with knives; in order to remain outside it all and avoid vengeance, they themselves sent warnings of such events.

Zagoskin underscores that it was said about these traders that "it is seldom that one of the Ulugak-miut is not a shaman." Such was the great fame of these skillful and cunning travelers.[331]

Zagoskin remained at Mikhailovsky Redoubt until the end of December 1842. Finally, when "two side suns (sun dogs), or as it is dignified in Holy Russia, ears, promised steady frost," the expedition party set out upstream along the Unalakleet River.

Zagoskin wrote: "I counted us to be 10, excluding the guides, or, to be more precise, foot stompers, as the route of the portage was known to many Russians." The Creole Vertoprakhov was replaced by "Tungus (Evenk) Nikitin, a former hunter with Kozmin's expedition to survey the Shanta Islands." For this"I had to thank

God; I confess that without this man's intelligence and all possible skills, his experience in overland travel and in shooting caribou in the course of a year and half, the expedition would have been compelled to expend much of its goods in purchasing foodstuffs."[332]

On January 15, 1843, the travelers arrived at Nulato without special complications. if one does not count that Zagoskin, "not feeling very well . . . did not record the bends of the river."[333] At that time the Nulato outpost was managed by the *baidarshchik* Deriabin, a participant in all the journeys undertaken by Glazunov and Malakhov. He, therefore, was in a position to inform Zagoskin about certain details of these expeditions. Having acquired at Nulato three small, light sleds of local manufacture and hired a guide, "a local Native," Tatlik, "nicknamed Volosatik," Lieutenant Zagoskin went on a reconnaissance upstream along the Yukon that was "called in the language of the sea coast Kvikhpak, in the dialect of the Lower Ingalik—Iuk-khana and the Upper ones—Iun-a," which meant "in translation the same: the Big River."[334]

Movement through the snow-covered taiga and among the ice hummocks on the river was difficult. Zagoskin remarked in his journals: "Natives go toward their selected goal through the forest with assurance; we had to travel by the stick, that is, seeking to find the snow-covered trail by feel." Zagoskin did not manage to go sufficiently far from Nulato by such means, and therefore had to satisfy himself through the winter season by questioning Natives and making geographic observations. From detailed questionings he managed to clarify "the important information . . . about the portage to Kotzebue Sound and communication between the local Natives," but Zagoskin reported in his materials "only such information that was later verified by my own experience or similar testimonies by the Natives with whom subsequently . . . he had ". . . numerous encounters."[335]

In the winter and spring Zagoskin made phenological observations, the first for the Yukon area. He records, for example, that "on April 18 the first merganser appeared on the (river) Nulato, and on the 20th the first goose. It was killed. A bad sign, adds the traveler, to kill the scout goose, there won't be many birds in the coming year, but what was there to do? The goose went for a 24-hour ration for two men; shooting began on the 22nd and for the rest of the month no one went hungry."

With the onset of summer weather, Zagoskin's party went up the Yukon. He wrote the following about their goal and the results: "All of us hoped that we shall reach the mountains that divide the British from our possessions; I proposed to test Mackenzie's hypothesis (that the Yukon and Kvikhpak are the same river) regarding the great river that flows to the west of Stony Belt; such a trek by water was not foreign to me, because of my two years' travels from Astrakhan to Tsaritsyn (Volgograd). But I must confess, at the beginning, that our efforts were almost totally in vain because exploring the Yukon for 300 or 350 versts upstream from Nulato is not sufficient for solving the problem."[336]

His explorations left little doubt that the Kvikhpak had its source in a ridge (more correctly a plateau) separating the Hudson's Bay Company possessions from those of the Russian-American Company, and that the Kvikhpak was the very same Yukon on the upper reaches of which the British had established themselves.

The general geographic considerations, based on the orographic and hydrographic structure of the land clarified by Zagoskin and his predecessors, pointed to this conclusion, as did the data collected by Zagoskin from the Natives about the river's names. The concordance of the two forms (Iun-a and Iuk-khana) and "Yukon" could hardly be ascribed to coincidence. However, Zagoskin did not assert categorically that the two rivers were one and the same.

This fact was demonstrated with finality only in 1863 when Ivan Lukin, the son of the commander of the Kolmakovsky Redoubt, went up the Yukon to the vicinity of the modern Dawson in Canada. Thus this mighty river had been explored by the Russians almost over its entire course,[337] and it was demonstrated that this was, indeed, Alexander Mackenzie's "Great River."

One of Zagoskin's main assignments was to conduct surveys and mapping en route. In his field journal there is a description of such work as he proceeds up the Kvikhpak: "I record, by compass, just the main directions; at each night camp, having noted the estimated distance covered, I sketch the shores. As far as the overland survey permits, astronomical observations make it possible for me to correct errors that are inescapable under such a mode of reckoning."[338]

Zagoskin paid special attention to local geographic names, establishing the versions which were used by the representatives of various tribes. He also attempted to learn the meaning of these names, such as, for example:

> The outermost mountain cliff on the right shore that the Natives call Khamynchikhten, 'Of Blocks', in accordance with the layers of its mineral strata. In this spot the Kvikhpak is 1½ miles wide. On the south shore, across from the cliff, one may see the winter dwellings of the settlement Tlialilkakhat, 'the one on the edge', 'the border one'—from the word Tliat—stop, do not come any further. According to Tuesday (a guide), the Natives of this settlement are the last ones who come down to Nulato to trade. Further upstream, Tuesday told us, there is a different tribe, speaking a different dialect, who have their winter settlements in the upper reaches of various rivers and come to the Kvikhpak only to fish.[339]

It should be noted that it was necessary to overcome considerable linguistic difficulties in order to obtain such details. Zagoskin vividly describes this when talking about information obtained by questioning Natives about the locality:

> This is the essence of the tale by the savages about their country. However, to avoid criticisms in the future, I deem it my duty to represent here that all

> the information I collected from the Natives of the tribe Tlegon-khotana, as well as from those whom I encountered subsequently, reached me in the following way: the Natives responded to my questions by answering the second interpreter; the latter passed the answers on to Tatlek, he to the interpreter, a Spanish-Creole-Aleut [Pavel Aglayuk—LB], and the last-named to me. In this way even correctly-stated testimony may be distorted by verbal transmission though interpreters who barely understand each other. Beyond those mountains, so said the savages pointing to the south, is the river Tlegon, where our tribe or kinsmen live, which in their understanding means the same; we do not know about its source, because we do not travel there . . . I asked them to indicate the direction toward the portage, and they pointed to ESE by the right compass.[340]

Earlier we mentioned that in 1838–1839 the Creoles Pyotr Kolmakov and Markel Matrosov visited the Tlegon River. Throughout the entire course of the expedition, Zagoskin used the map of this region compiled by Kolmakov in 1839. However, checking this map against information supplied by the Natives, Zagoskin, at the outset, "could not at all figure out that he (Kolmakov) went along the entire length of the Tlegon [underlined in the original—AP], which he called the Legon, from its source to its mouth. Kolmakov's journal would have easily resolved the problem, but his journal is unavailable in Novoarkhangelsk."

Later, when Zagoskin sought information about the routes from the Kvikhpak River basin to the Kuskokwim valley, he:

> was greatly aided by a Native from the lower course of the Ittege, called Kantelnuk, a constant trader at Kolmakov's redoubt, who visited Ikogmiut (Russian Mission) on September 8, 1843. Kantelnuk explained to Zagoskin "that Kolmakov's Legon River [more correctly, Tlegon—AP], is the headwater of the river Tiltononto, or the Inno-ka. The traveler had heard in the summer about this river, which we know in the language of the Ttynan and Kan-iuglet people, under the name Itteke and Chageliuk, from the peoples inhabiting the Upper Kvikhpak."[341]

By 1844 Zagoskin's questions and doubts about Kolmakov's map were resolved with finality, because of testimonies by one of Kolmakov's paddlers who was under Zagoskin's command, the manager of the Kolmakovsky Redoubt, Semyon Lukin, and by the assistant to Pyotr Kolmakov, Markel "Matrozov" (Matrosov).[342]

In order to reconcile the information obtained from the Natives and his own surveys and observations made during the journey to the upper reaches of the Kvikhpak, Zagoskin gives the following "essay on the work" of Pyotr Kolmakov, which has its own significance.

On the orders of the Chief Manager Kupriyanov, Pyotr Kolmakov has been dispatched out of the Aleksandrovsky Redoubt with 5 baidarkas to explore the country around the headwaters of the Kuskokwim. Kolmakov, having received assistance from the manager of the Kuskokwim outpost, Lukin, that is, interpreters and dried fish, ascended to the River Tochotno, or as it is also called, the Tokhonotut, and learned about the possibility of portaging to another river that according to statements by the Natives empties into the Kvikhpak. He set out on June 12 and describes the portage as not difficult for baidarkas, and estimates it to be 36 versts—but in all probability even shorter.

The headwaters of the river that Kolmakov reached, the Natives call Tlegon, or Legon, and on his map the entire river is shown under this name. Kolmakov was received by the Natives everywhere in a friendly manner. According to the Creole Matrozov, who participated in the expedition, these do not have large settlements and are not numerous. Kolmakov suffered need in food provisions, as the sea fish run has not started yet; nonetheless he collected a sufficient quantity of beaver skins, over 800, and on the 25th of the same month arrived at the settlement of Ttality.

Having in sight the mountains on the shores of the Kvikhpak, he intended to cross to Ikogmiut, but the inhabitants counseled him to turn back. He listened and they showed him the shortest portage to the Kuskokwim, from the headwaters of the river Ialchikatna or Tachaichagat. Kolmakov designates this portage, to the headwaters of the Tochotno, as being very difficult in the summer time.

All the tales by the Natives about the direction of the river Shiltonotno and the names of various localities and tributary streams agree with those indicated by Kolmakov, but it must be explained that when compiling his map he did not take into account compass deviation, nor the determination of the four compass directions, nor a constant scale for designating distances traversed. The Natives added that the inhabitants of these headwaters connect with the River Iun-a, the Kvikhpak, by means of a short portage from the River Ttachegno to the Iulutno or Iukutno.

This past summer when we traveled along the Kvikhpak we passed the localities where the tribesmen of the Tlego-Khotana emerge, but not having a good interpreter I was not able to elicit details about the portage and the designation of the portage river; its mouth in all probability ought to be in the Notliageliata slough, between the observations of July 16 and 18. Malakhov, on his exploration along the Kvikhpak, added the River Iukunak, similar to the Iukutno, according to the tales of a Native who lives near the Bankatan cliff.

The locations of settlements are shown on Kolmakov's map under the general name Kalchane—or more correctly Goltsane—tribe of the Beautiful Dress—the name given by the Ingalik to the Kaniulit people. Here are the

> names of their main settlements transmitted to me by the Natives, beginning with the river's headwaters: Tlegon, Tlegokokhkakat, Tlekaet, and Kholikakat. The three last-named are individual hunting or fishing camps.[343]

According to Zagoskin's observations, information supplied by the local inhabitants, no matter how accurate, demanded careful analysis and good knowledge about Native concepts of topography, which were conditioned by their mode of subsistence, and even about the narrative style pertaining to the routes of their travels. He underscores:

> The savage, talking about his route, will not omit anything: here did he smoke his pipe, here he drank water; saw an animal etc., etc. Recounting each of these events, he crooks a finger, keeping count of rests and stops; one has to have special skill to understand, but I myself, counting fingers, cannot assume the authority to maintain positively that 10 Native stopping camps are equal to 10 days of march. But there are many entertaining journeys that rest on such conclusions about the distant and little known lands, and these are not in the Russian language.[344]

On July 1, 1843, having failed to negotiate an extensive shoal, Zagoskin's party was forced to turn back and proceed down the Yukon.

Upon his return to Nulato, taking advantage of the short stay there, Zagoskin continued to collect information about the locality and started a description of the Nulato district. Here he noted that communication from Nulato "with the St. Michael Redoubt was maintained by two routes: by water upstream on the Kvikhpak River and overland by the portage to the Unalakleet River. The first route has been simplified by the transport from Novoarkhangelsk of a wooden baidara; this provides the means to supply Nulato in a single trip with an annual ration for a 10-man crew, as well as with all the items needed for the most extensive trade this place is capable of conducting."[345]

Zagoskin also determined that even before the establishment of a permanent outpost at Nulato, "the *baidarshchiks*, traversing the entire Kvikhpak, began to trade in the settlements." Zagoskin, who was a hereditary nobleman, was very favorably impressed with the ordinary men serving at the Nulato outpost, especially the *baidarshchik* Deriabin.

He wrote the following: "The outpost management is the best that can be wished. The *baidarshchik* engages in the most difficult tasks in fishing, he alone carries the burden of provisioning the crew, and issuing the supplies and trade goods that are sent from the redoubt; he does not distance himself from the laborers; he is illiterate . . . A year passed that I spent in the society of low-class persons," so continued Lieutenant Zagoskin, "and daily among the savages; but the first-mentioned do care for the benefit of the fatherland and conversations with them are never boring, they often were instructive for me, and the second ones in their mores and customs aroused in me curiosity."[346]

On August 5 the expedition resumed their travel down the Yukon. They were joined by "Ulakagmiut Tumachugnak, former guide to Malakhov, who was making merry on the Kvikhpak." The expedition's work along this route is reflected in the following note in Zagoskin's travel journal:

> Not having the opportunity [because of weather—AP] to make observations, I note by compass the direction of the course of the river from promontory to promontory and at each night's camp record on the map distances sailed, indicating some islands, sharp bends, the river's width, and the extent and distance of mountain ridges on both shores. The calculations agreed fairly well with the determinations made by Malakhov in the year 1838; therefore, I admit to my map the latitude he indicated for the mouth of the River Kakokhgotna, that is 63°59′41″, longitude by calculation 156°41′ from Greenwich. For geognosy (then a branch of geology) I make the effort to collect several stones that compose the main rock strata of the adjacent ridges.[347]

Zagoskin made this entry at the winter habitation of Tuttago, known to the Russians "under the name Igudovskoe, from the name of the Ulukagmiut Igudok who lived here in winter in order to trade with the Natives of Chageliuk."

The travelers began to note signs of approaching autumn. Already at Nulato Zagoskin recorded in the night of July 30–31 a temperature drop to −1.5° (−1.5 Reaumer = 28.6° Fahrenheit), and in the following week "during the gloomy and rainy weather the willows began to turn yellow and the leaves to drop."

The fall weather became the travelers' constant companion, and on August 16 Zagoskin recorded the following observation:

> Half a month it rains, at times it pours, at times it is a drizzle; old-timer *promyshlenniki* tell us that it has been noticed that August is the rainiest month of the year. I have no right not to believe them—instead of checking the hydrometer the working folk check on their own persons the degree of moisture in the air, and many, judging by their chest pains or joint aches, predict the weather as well as the barometer does.[348]

On November 30 Zagoskin's party reached the Kuskokwim River and on December 3 the exhausted travelers came to the Kolmakovsky Redoubt, where they were hospitably received by the redoubt's manager, the Creole Semyon Lukin, a man who was enjoying well-deserved fame in the Russian colonies. Zagoskin was deeply impressed by the redoubt and by the manager. He compares the Mikhailovsky Redoubt with the Kolmakovsky Redoubt and writes: "In the Mikhailovsky District the Russians clothed themselves in Native dress; here, in contrast, the Natives wear our broad cloth even in winter. There the manager is Mister, here he is Tata (father), and the first among laborers."

The manager of the Kolmakovsky District, "Semyon Lukin was brought up by the never-to-be-forgotten Alexander Baranov, and since the year 1819, that is since the establishment of the Aleksandrovsky Redoubt, never left this region." According to Zagoskin's observations, the absence of any kind of defensive fortification at the redoubt was to be explained by the fact that Lukin, and all those close to him, had completely "joined" their natural and ethnic environment. An entry in the daily travel journal attests to this: "Should Lukin himself, through marriage with a Native from the settlement of Ugavik, and his entire command, through kinship ties either by descent or through their wives linked to the Natives of the five settlements I listed, fear death from their kinsmen?[349] Also, most of the Natives from the adjacent region had been converted to Orthodoxy by that time by Lukin and Kolmakov.

Zagoskin was greatly struck by Lukin's knowledge of the geography and peoples of the Kuskokwim and Kvikhpak river basins. Besides having been an interpreter for Wrangell and a participant in Vasiliev's expedition, Lukin had been dispatched in 1832 to establish an outpost on the Kuskokwim, and by 1833 had "descended to the settlements of the lower course (of the Kuskokwim) to trade, and also established relations with the Natives living along the Tkhalkuk and Tychannanik, the name of the Kuskokwim used above the entry into it of the Hulitnak (Holitna)."[350]

One of the important practical results of Lukin's many journeys was to establish that "the natural boundaries of the mountain ridge show that the entire Kuskokwim system falls into the province of the Kolmakov Redoubt."

As far as the Kuskokwim River is concerned, Zagoskin noted that "Experts in the language of the Kaniulit, Glazunov and Lukin, maintain that they call this river the Kuskokwi*k*, and that the ending *im* indicates the instrumental case, but I," Zagoskin continues, "retain the name under which this river became known from Vasiliev's expedition and in this way offer the sole tribute to his efforts, because all the rest of his determinations are wrong and this will be shown at an appropriate place. In local dialects Kuskokwim is heard as Kushkokvan. Not a single local etymologist was able to explain to me the meaning of these names."[351] Moreover, Zagoskin learned that "the Kuskokwim is called by the Ingalik settlers Ttykani; there exists also the form Ttychannanik."[352]

Lukin's children from his first marriage, Ivan, Konstantin, Anna, and Matrona, took an active part in many labor-intensive tasks at the redoubt. Zagoskin noted especially that Ivan Lukin worked without pay, "is used on all sorts of errands and works alongside the crew."[353]

By the time Zagoskin's expedition arrived at the Kolmakovsky Redoubt, the commercial and fur-extracting activity had become routine. Of course, each of the regular trips, without any doubt, brought new information about the geography and local peoples of Alaska's interior. In spite of the fact that such information was seldom fixed in surveys and maps, just the same it was included in what we can call in modern language the "traditional geoinformational system" that was used by the Natives and by the *promyshlenniki* of the Russian-American Company.

Forty to sixty guides from among the local population were hired for the regular winter and summer trips out of the Kolmakovsky Redoubt. In the summer, usually "two different trips were accomplished, one outward bound upstream, the other on the return, in the course of which two or three three-hatch baidarkas and from 10 to 15 single-hatch ones were sent out from the Kolmakovsky Redoubt up the Kuskokwim between June 15 and 20. They then ascended the Holitna and Taku, portaging across the hilly tundra for about 7 miles to the river Uksiuakhchulik, a tributary of the Chichitnak; then, following the course of the Chichitnak they descend into Ilgaiak, from where they sail to the Nushagak River; they return by portaging from the Chichitnak into the Chiniuakhtakhtuli, a tributary of the Aim, and then by the Holitna and the Kuskokwim. Over two months are spent on the journey."[354]

As was the case in the past season at Nulato, while staying at the Kolmakovsky Redoubt and during his trips along the Kuskokwim in the winter of 1844, Zagoskin conducted regular meteorological and phenological observations. Thus he noted in the expedition's journal on April 9, with a temperature of 15 degrees Réaumur (66° Fahrenheit): "During the daytime the heat was unbearable, insects appeared on the snow, butterflies were flying, *cold is coming, the promyshlennik naturalists decided* [emphasis ours—AP]." The forecast by the experienced company employees proved true—the next day there were two degrees of frost.[355]

On May 19, 1844, Zagoskin's party set out up the Kuskokwim in two three-hatch expedition baidarkas. Lukin, the manager of the Kolmakovsky Redoubt, traveled along with the explorers, also with a three-hatch baidarka and a small baidara carrying trade goods. Once again the daily routine of topographic surveys conducted en route set in. The work was complicated because the Kuskokwim meandered through a rather broad valley.

Zagoskin notes in his travel journal: "It is impossible to avoid errors when determining passage through reaches and bends, as due to the irregular current velocity and the speed of the baidarkas the log is totally unsuitable and the estimated distance, or the distance measured by the watch, is only an approximate one. I tried to reconcile the compass heading of the bends and their extent with the points determined by observation."

Lieutenant Zagoskin notes the following: "The determination of points made by me does not agree with the survey of this river which was conducted by Mr. Vasiliev, who, as I was told by the redoubt's manager who served as Vasiliev's interpreter, had no occasion to make a single astronomical observation; this moved me to show all the reaches on the Kuskokwim and their estimated extent, checked by daily observations."[356]

In particular, Lieutenant Zagoskin "determined the location of the mouth of the Holitna River by observation at latitude 61°41′55″ and longitude 157°01′01″ W from Greenwich. The difference from Mr. Vasiliev's determination along the latitude to the south is 20′35″ and in longitude to the west 1°05′28″."[357]

On the Kuskokwim Zagoskin became a witness to the tragic consequences for the animal world of the taiga, not only due to the predatory hunt for fur-bearers, but also through irrational actions by the Natives. He notes that "at the present time the best beaver lodges in the waters of the Kuskokwim are destroyed by the Native, who, wishing to take a bear cub that has climbed a spruce tree, set it on fire."[358]

We saw earlier that the Russian explorers of continental Alaska paid particular attention to investigating the possibility of crossing from the Kuskokwim River basin to the Gulf of Kenai (Cook Inlet). It was thought that one of these passes was at the headwaters of the River Tkhalkuk (Stony River). Zagoskin made an effort to collect as much information about this as he could, and summarized it in his journal in the following manner:

> The River Tkhalkuk became known to us in 1834 through the expedition of the asst. pilot Glazunov; but a year earlier the present manager of the Kolmakov Redoubt [Semyon Lukin—AP] visited the Tkhalkuk inhabitants in the settlements Glazunov did not reach. The inhabitants of that river belong to the nation of Ttnai, of the same tribe as the Kenai, who are known to the Company. At its mouth the Tkhalkuk flows from the E, but, according to the Redoubt's manager, has its source between Northeast and East; he could not go further by his baidarkas than the settlement of Tkhalkuk Ttynai because of rapids, but the Natives say that it flows from a lake located in the midst of a mountain range [it is possible that they meant Lake Telakuana, located at the headwaters of Stony River—AP] known by the name of Chigmiut, the Stony.
>
> Several peaks of this range, that are the closest, can be seen from a hill on the right shore of the Ttychannanika across from the lower mouth of the Tkhalkuk, from a distance of about 50 miles. The Chigmiut Range divides the waters of the Kuskokwim basin from those that flow into the Gulf of Kenai and perhaps is a branch of the real stone belt seen by Mackenzie during his journey toward the Arctic Ocean.[359]

On June 1, 1844, Zagoskin obtained from three Natives the following description of the upper course of the Kuskokwim (the Ttychannanika) and its tributaries:

> This river is formed by many mountain streams, but its most important tributary is the river Togtygchagno, entering it (the Kuskokwim) from the Gulf of Kenai side. They do not know the names of the uppermost sources of the Ttychannanika, but below, or from the Tochotno, here are the names of the most remarkable:
>
> 1. Zlagyzatno, its source they indicate as flowing from a lake from the river Mantashtano or Tkhalkuk.

2. Shakhchagno.
3. Tochinnago.
4. Chatamchagat.
5. Khogchekhnikho. And
6. Togtygchagno; all of them are rich in beaver, and all flow, except for the Chatamchagat, from the Kenai Gulf direction or from the south-east.

> The Kenai cross on winter trails to the people of the upper Ttychannanika, who assemble for major trade events at the settlement of Itstsyno located near the mouth of the Togtygchagnno. There are no large settlements among the local Natives. They live in small family groups at the headwaters of rivers, hunting caribou, moose and beaver . . . The source of the river Ttychannanika has been pointed out to me as lying to the east, the source of the Togtygchagno to the south-east; the distance between the mouth of the Tochotno and the settlement of Itstsynno was indicated as seven days of travel, or the same as the distance from this place to the Holitna mouth.[360]

The description cited above is typical of the information from Natives by Zagoskin and which was used to compile the expedition's map in the parts for which he lacked data. The investigator approached such questioning with great care, noting in this context: "In general, I was reluctant to include information about lands unknown to us that I was not able to verify by personal observation or testimonies of several Natives."

Apparently because it was impossible to obtain verifications through independent sources and personal observation, Zagoskin did not mark information on his map "about the existence of a large lake or beautiful Mediterranean sea, located in all probability between 150 and 154 degrees West longitude from Greenwich and between 63 and 65 degrees latitude. Access to this lake," according to the traveler's opinion, "should be sought through the river Noggoia, which enters the Kvikhpak" where Lieutenant Zagoskin "obtained the first news about the great lake Mynkkhatokh, rich in fish, about the numerous population inhabiting its shores, and about the multitude of beavers found about its tributaries."[361]

We will probably never be able to determine with certainty which lake the Natives had in mind when talking to Zagoskin, but it is possible that information about the northern Great Lakes or Hudson Bay was reaching the inhabitants of the Yukon and Kuskokwim valleys.

Although the travelers wished very much to reach the source of the Kuskokwim, Zagoskin calculated that this would demand no fewer than two additional weeks and "detaining the Redoubt's crew would violate the Manager's orders about the timely dispatch of the fur catch." On June 4 the party turned back.

The travelers then moved to the Kvikhpak once again. On the way Zagoskin made interesting observations at the settlement of Khkhalkas-miut that once again confirmed the system of environmental traditions and prohibitions that had been established among the local population. In this settlement the members of the expedition witnessed the beginning of the fish run and learned that "no noise near the river ought to disturb the fish and that all the work was done away from the settlement and its shores. In Kkhkhalkag-miut they strictly observe this superstition." However, in order to supply the travelers with foodstuffs an exception was made to this strict rule.[362]

On June 10, having crossed the watershed between the Kuskokwim and Kvikhpak basins, Zagoskin's party came out to the Kvikhpak a mile-and-a-half from the settlement of Ikogmiut. In this settlement they were hospitably met by the *baidarshchik* Glazunov. During the expedition's rest, which lasted until June 13, Glazunov told Zagoskin some particulars about his trip to the Kvikhpak.

On his way down the Kvikhpak Zagoskin used the Angun (Ankhun) Slough [*sic*, Apkhun or today Apoon Pass—LB], which had been found by Glazunov, and traveling by it came to "a camp of the Pastol called Aimgagiag-miut, where one could see the remains of something resembling a fortress, consisting of a trench and earth wall." This fortress, according to Glazunov's information, which Zagoskin confirmed, played a role in the history of the local tribes, specifically, "as the Pastol tell it, in former years the Chnag-miut were at war with the Magg-Miut; after suffering a crushing defeat they retreated to the fortress, which later was destroyed by the enemy."[363]

On August 5, 1844 Zagoskin's expedition was taken on board by the brig *Okhotsk* and on September 26 they arrived safely in Novoarkhangelsk.[364] Chief Manager Etholen, reporting the conclusion of the expedition to St. Petersburg, characterized its results primarily with respect to the benefits to the Russian-American Company's interests thus:

> I have the honor to transmit to the Head Office the detailed journal kept by Mr. Zagoskin during his expedition's activity and also the map of the routes the expedition traversed in the interior of our northern part of America. I must say that though Zagoskin was not able to live up to my expectations and could not personally extend his investigation and explorations to Kotzebue Sound (the main goal of his expedition), and traveled but a very short distance upstream along the Kvikhpak beyond the areas already known to us, the results of his labors are not small . . .
>
> These notes [the journal—AP] indicate the easiest ways to deliver trade goods and various other supplies to the redoubts and outposts by water along the rivers Kvikhpak and Kuskokwim and their tributaries, and also, in general, provide information about the Native industries and trade in those lands; they have also discovered various shortcomings and abuses on the part of local managements and indicate the most efficient means to obtain the desired end:

that the fur catch does not bypass the Company's hands when delivered from the American to the Asiatic coast.[365]

Zagoskin's expedition was an outstanding achievement with respect to the study of the environment and population of Alaska's interior, a fact acknowledged by his contemporaries and modern scholars. Zagoskin's first article about his expedition's results was published in the *Notes of the Hydrographic Department* immediately after his return to St. Petersburg.[366]

In the winter of 1846 Zagoskin was given leave and went home. On January 8, 1847, at the general meeting of the Imperial Geographical Society, a report about his journey was read, in his absence, by S. I. Zelyony; the text of this report with a map was published in the *Notes* of the society.[367] Immediately afterwards, Zagoskin's main work, *An Overland Survey of Part of the Russian Possessions in America . . .* was published in two volumes.[368] Accounts of the journey were also published outside Russia.[369]

The scholarly circles in Russia showed great appreciation for Zagoskin's works, and in 1849 he was elected an active member of the Imperial Geographic Society. The St. Petersburg Academy of Sciences awarded him a prize.

Zagoskin's contemporaries noted that even though he was not a professional geographer, his studies were outstanding in their scrupulous accuracy and the directness and clarity of his geographic observations and descriptions, based solely on what the traveler saw himself or carefully ascertained himself. Precisely because of this, even now, 150 years later, his observations remain a valuable source for scholars. The Russian naturalist and traveler, Academician Alexander Middendorf (1815–1894), gave a high assessment to the cartographic work of Zagoskin's expedition, stressing that it became the basis for the cartography of this vast region.

The cartographic results of Zagoskin's expedition were indeed remarkable, not only because he personally conducted precise surveys en route and made astronomical determinations, but also because he summarized the work of his Russian and foreign predecessors, carefully checking their data against information provided by the local inhabitants and experienced company employees. The *Mercator Chart of Part of the Northwest Coast of America, Compiled by Lieutenant Zagoskin during the Expedition of the Years 1842, 43 and 44* (Fig. 61) that reflects the immediate results of the expedition is preserved in the Russian State Archive of the Navy in St. Petersburg.[370]

Judging by the official report of Etholen, the chief manager of the colonies, this chart was sent as part of a set together with other surveys of the Russian-American Company's possessions, attached to dispatch # 382 dated May 15, 1845.[371] Zagoskin determined the general orographic and hydrographic schema of the lower and middle courses of the Yukon and Kuskokwim rivers, and he represented the relative size of the basins of these great Alaska rivers precisely, due to three portages

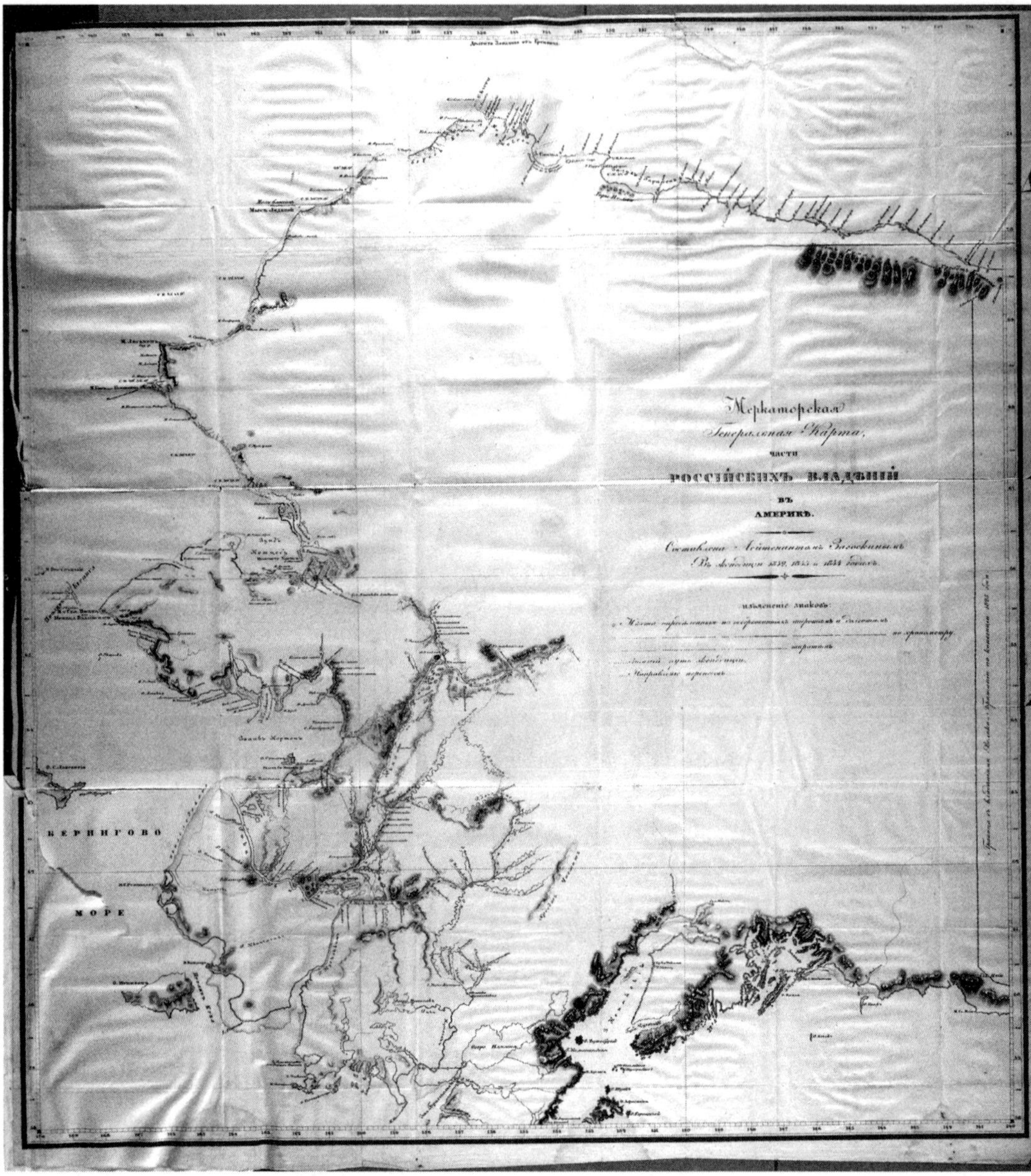

FIGURE 61 A portion of Zagoskin's map. End papers, *Peshekhodnaia͡ opis' chasti russkikh vladie͡niĭ v Amerikie͡,* vol. 2. UAF Rare Book A0504

used by the local tribes since times immemorial which the expedition traversed and surveyed on their journey.

The lower course of the Kvikhpak is geographically and cartographically "tied" to the coast of Norton Sound, because from Khogoltlinde the route along the Yukon is completed through the Unalaklik River (in its mouth Zagoskin completed observations on a full astronomical point, determining the coordinates of latitude and longitude) to the ocean, then along it by sea through the full astronomical point at the Mikhailovsky Redoubt to the mouth of Angun (Apoon Pass) Slough, and then to the astronomical point "Ikogmiut camp." This closed route is augmented by two more points on the Kuskokwim ("Kyvygym palma" and "Khunannilinde") where astronomical coordinates of latitude and longitude were established.

This system of full astronomical points tied to latitude and longitude by chronometer, and Zagoskin's routes linking these points by reconnaissance surveys, created a sufficiently firm "inner structure" for the composition of a summary map that showed for the first time the true basic orographic and hydrographic features of the region explored by Zagoskin.

In Zagoskin's book the geographic boundaries of the represented territory were considerably enlarged through the inclusion of information, maps, reports, and accounts by his predecessors: Russian, Natives, and Creoles. Russian and British charts also served as materials for compiling the map which was published in Zagoskin's book, the *Mercator General Map of a Part of Russia's Possessions in America, Compiled by Lieutenant Zagoskin in the Course of the Expedition in the Years 1842, 1843 and 1844*.[372]

Thus the coastline of Alaska is reproduced from the original charts by Franklin (1825–1826), Dease and Simpson (1837), Beechey (1827), the survey by Cook, the chart and survey by Khramchenko (1822), from the original chart of navigator Ustyugov (1819), Litke's chart (1828), and others. To represent the interior territories, besides his own survey Zagoskin used journals and surveys by Ivan Vasiliev and maps by Kolmakov and Lukin.[373]

Zagoskin's investigation has been highly valued by twentieth-century scholars and the information he obtained about the population of Alaska's interior is still a reliable source on the ethnography of these peoples for the period of their initial contact with Europeans. One American scholar of the Athabascan Indian culture, Cornelius Osgood, noted that "Zagoskin's report is the outstanding source for the history, geography, and science of western Alaska while under the Russian flag." And: "No one probably contributed more ethnography on the Ingalik during the nineteenth century than Zagoskin."[374]

Wendell H. Oswalt, who devoted many years to the study of the Eskimos who inhabit the Kuskokwim River valley, maintained that Zagoskin's description of the peoples of the Kuskokwim remains to this day the main ethnographic source for the region around the middle course of that river.[375]

The significance ascribed by anthropologists and historians to the ethnographic data in Zagoskin's statements depends, in part, on the fact that he observed the peoples of the Kuskokwim and the Yukon at a time when their way of life had not been changed at all through contact with the Russians. Also, Southwest Alaska became a focus of close ethnographic attention only relatively recently. Modern ethnographic investigations on the Yukon began with the work of Osgood and de Laguna in the 1930s and 1940s, and Oswalt on the Kuskokwim in the 1950s.

James W. VanStone draws attention to Zagoskin's contribution in solving one of the most difficult ethnographic problems in Southwest Alaska, to wit: the complex sub cultural divisions in the region of the Athabascan and Eskimo community along the lower course of the Yukon River and its tributaries. Zagoskin's observations in regard to this problem were accurate and were confirmed by de Laguna during her investigations in the Yukon valley in 1935. Both Osgood and de Laguna rise to Zagoskin's "defense" against an unjust accusation by Dall, who maintained that the Russian traveler didn't receive information about the Ingalik through direct contacts but, allegedly, through rumor.[376]

Of great value for the modern scholars is the fact that whenever possible Zagoskin always recorded the name of each settlement that he visited or was informed about in the course of his lengthy journey. He noted population numbers and wrote down his personal observations about the Native way of life and mores, which are of great interest for modern specialists. For example, in 1843, when he halted at the settlement of Tulukagnachmiut (Raven Village) on the middle course of the Kuskokwim somewhat below Aniak, he noted that there were ninety Natives in five dwellings and that twenty of them were Christians. Two of the Christian families were baptized at the Aleksandrovsky Redoubt at the mouth of the Nushagak River and had moved to Raven Village only recently.

At the time that Zagoskin visited, all the inhabitants, with the exception of one man and three women, were taking part in a memorial feast at Ukkhagmiut lower down the course of the Kuskokwim. One of the old women gave him fish for his dogs, while the man stole an axe. Information of this kind, given sometimes in greater detail for other settlements, has exceptional value for historians as well as for the ethnographers who are attempting to reconstruct the character of the distribution of and the specifics of Native settlement in the nineteenth century.[377]

Zagoskin's field diaries are filled with brief but useful notes on the region's ethnography that describe the subsistence life cycle of the Eskimos of the Kuskokwim River, trade relations between various Eskimo groups and also between the Eskimo and the Indians of the Lower Yukon, dwelling construction arrangements, burial of the dead on the Yukon, and much more. Ethnographers also use information supplied by Zagoskin about the regions adjacent to the territories he investigated.[378]

Zagoskin's recommendations to improve fur procurement, taking measures to ensure the preservation of the fur riches, helped Etholen formulate new instructions to local management. In part these were aimed at continuing the exploration of interior Alaska. Specifically, Semyon Lukin, the manager of the Kolmakovsky Redoubt, was ordered in the following *Particular Instructions*:

> A party is to be dispatched from the Kolmakovsky Redoubt not later than May 15 for trade with the inhabitants of the headwaters of the Kuskokwim; it is to consist of 5 three-hatch baidarkas and 11 men. In the first year the

> manager of the redoubt has to go himself in order to learn about the region. After trading at the trade assembly point near the hill Vinisali (Vinisale Mountain), the party is to proceed to the inhabitants at the source, that is, to those with whom the Kenai maintain relationships.
>
> Having made their acquaintance, and not losing sight of questioning about other mainland inhabitants, water communications, and the quantities of fur-bearing animals, the manager, or the party's leader, is to turn to the mouth of Tochno-Tukotno; here he is to construct a raft, load it with the catch and send it off to the redoubt escorted by two baidarkas; he himself is to take a Native guide from among those who live on the Tochno, and portage, with three baidarkas, to the headwaters of the Tlegon and descend to the Kvikhpak to his temporary camp Chinik.
>
> This operation has to be completed, as the current manager knows, by July 1. Counting from May 15, an estimated 8-day journey to the assembly point at Vinisali, two from there to Tochtno, from there an 8-day journey to settlements at the source of the Kuskokwim; 5 days for the return downstream to Tochtno and raft construction, and 23 days sailing along the Tlegon to Chinik camp.
>
> Also counting from May 15, 10 men from the remaining 19-man crew are to be dispatched in two small Native baidaras to transport the catch collected during the winter to the Kvikhpak camp . . .

The manager was required to sail down to the Mikhailovsky Redoubt by July 10, where by that time the vessel from Sitka was expected to arrive. In his reports the manager had to present "his own notes about the number of the inhabitants, the localities, and the region's potential fur catch."[379]

Semyon Lukin was extremely useful to Zagoskin's expedition. Gratitude was conveyed to Lukin in a letter from Etholen dated May 14, 1845 (# 272): "In accordance with your wish conveyed to me by Lieutenant Zagoskin, your son Konstantin is admitted into company service with a salary of 250 rubles per year. Your son Ivan is appointed as your assistant, and the Creole Alexei Matrozov as traveling headman, with salaries as specified in the budget, effective January 1 of this year of 1845. On representations by Mr. Lieutenant Zagoskin, I award to the Severnovsk *toion* 'Kantelnuk' for his loyalty to the Russians and the aid rendered to Mr. Zagoskin's expedition, the silver medal 'Allies of Russia.'"[380]

As Zagoskin recommended, a special role in further investigations of part of interior Alaska was assigned to the Nulato outpost. The entire basin of the Kvikhpak, from the Tutago-Igudovskoe settlement upstream to "its very source," was included in its sphere of fur-procurement and exploration activity. In the *Instruction for Management and Production of the Nulato Outpost* worked out by Etholen in 1845, much attention was devoted to regulating the hunting of fur animals and measures to preserve fur-bearers from complete extinction.

The following limits were set: "Shooting beavers and otters with firearms is forbidden and the manager will be held strictly accountable; he is also to take care that the destructive habit of taking beavers by breaking up their lodges, especially in the spring when only pregnant females remain there, is stopped; explain to the Natives that in taking them this way they only deprive themselves of future income."[381]

The chief manager demanded that the implementation of his instructions was to be achieved only through friendly persuasion and under no circumstances was force to be used. "Gentleness is necessary in dealing with the Natives," Etholen wrote, "and as their chief shortcomings are laziness and lack of care, it is the manager's responsibility, through example, to demonstrate the benefit of labor and thrift." In contrast, it was demanded that the strictest discipline be observed among the Russians and Creoles who were official employees of the Russian-American Company. To maintain it, especially in their dealings with the Natives, even corporal punishment was used. Thus, "one caught by the manager stealing from the Natives is to be punished corporally, no more than 20 strokes, and the punishment entered into the [penalty—AP] book."[382]

The general order of the commercial and fur-procurement and search operations for the field season at the Nulato outpost was set out by Etholen in an instruction as follows:

Each year, "when the river becomes free of ice, and not waiting for debris to be carried off completely, the manager is, with 3 baidarkas of Kodiak construction and a crew of 7 men, to set out upstream along the Kvikhpak to buy up the fur catch, making all possible effort to notify beforehand the Native traders living up the river and on the Kuyukak (Koyukon). It is not possible to determine the ascent along the Kvikhpak. According to verbal information from Natives, Zagoskin's expedition failed to reach the mouth of the River Nogchoya by two-day travel in a baidara, where their large meeting place is located. Subsequent annual trips will permit the extension of exploration further, but even in the area looked at by the expedition, there is in summer such a number of Natives in residence that it is possible to conclude that the buy will be rich. At the River Minkhotletna the party can have a rest stop and put up fresh fish that enters this river in large numbers. A raft will guarantee the floating of the bought-up catch down the river." The party was expected to return to Nulato between July 1 and 10.[383]

For further exploration and mastering of the Lower Yukon the same instruction directed development of "the new outpost which is being established called the Andreevskaya. A special directive has entrusted to it the entire country of the Lower Kvikhpak from the Chinik mark to the sea coast, including all the arms of the river." Moreover, "settlement of the Andreevskaya outpost is to be placed on the right shore of the Kvikhpak, where the River Nypyklaik (Nychiglik) enters into it, as this is the most convenient place for travel on all the branches of the Kvikhpak, and especially for access to the Agulmiut."

Because of the imperfect knowledge of the lowest reaches of the Yukon [delta—LB], Etholen's directive instructed that "the manager must himself undertake the first trips into the Kvikhpak arms which are poorly known to us, leaving his assistant in charge of the outpost. The trip journals are to be presented to the Chief Manager of the Colonies. In them should be noted the names of settlements, the number of inhabitants of both genders, the appearance of the country, its natural resources, the direction of water flow, and in general anything that is worthy of mention with regard to trade in that region. Sketches (maps) are to be appended to them."[384]

In the very same year of 1845 the Kvikhpak Mission (modern Russian Mission) was established in the settlement of Ikogmiut, about "200 versts" distant from the Mikhailovsky Redoubt upstream along the Yukon. It was headed by Priest Yakov Netsvetov, a Creole. The mission was to bring the word of God to the inhabitants of the Kvikhpak and Kuskokwim basins and their tributaries.[385] The Andreevskaya outpost was destroyed by the Natives in 1855.

Vozensensky sent 150 huge boxes at various times to the Academy of Sciences. Etholen, writing to the Imperial Academy of Sciences on May 15, 1845 (#314), says that on leaving the colonies he sent thirty-four boxes belonging to Voznesensky through London and "8 boxes that are actually mine, marked AK.E containing items of a similar nature, addressed to the Zoological Museum of the Academy. I humbly beg the Academy," Etholen continues, "when the boxes arrive, keep them until my return. At this time, when Mr. Voznesensky departs the Russian-American Colonies, it is my duty, and justice demands, to testify before the Imperial Academy to his ceaseless labors, the honest and excellently fine carrying-out of his duties, never letting an opportunity pass to enrich the museum. He has earned, through his modesty, nobility of character and exemplary behavior general respect in the Colonies and gained the willingness and zeal of everyone to help him by word and deed. That Mr. Voznesensky was able to amass such a huge collection of various items for the Academy while his monetary means were limited can be ascribed to this fact."[386]

The materials of Etholen[387] and Voznesensky, together with the reports about the remarkable expedition by Zagoskin, attracted the close attention of the leadership of the Geographical Society. The latter decided on a study of Russia's natural environment and economy, by means of a questionnaire the academy devised, beginning with the territory of Russian America.[388]

Immediately following the presentation at the society's meeting about Zagoskin's expedition, the vice president of the Russian Geographical Society, Admiral Litke, sent a letter to Innocent, the bishop of Kamchatka (Ioann Veniaminov), asking that the collecting of geographical data about Russian America by missionaries of the Russian Orthodox Church be initiated in Alaska and the Aleutian Islands. The clergyman scholar responded to the hydrographer, whom he had known since the latter's days in Russian America, in a letter preserved in the Archive of the Russian Geographical Society:

To His Excellency, Vice-President of the Russian Geographical Society Fyodor Petrovich Litke.

Your Excellency, My dear Sir:

I had the honor to receive Your Excellency's memorandum of July 21 of the past year of 1846, # 351, in which Your Excellency requests on behalf of the Russian Geographical Society my assistance in furthering its work in September of the last year of 1847 in America via Kamchatka.

First of all, I deem it my inescapable duty to express my sincere gratitude to the Society for such a flattering assignment and then to declare that although I personally am not able, to my deep regret, to collect information about this country, as I shall not be visiting any unexplored regions, I shall through the Missionaries and other personnel under my jurisdiction obtain, inasmuch as this will depend on me, such information, and deliver the same to the Society.

Looking at the time I received the abovementioned memorandum by Your Excellency, you shall discern that I received it after various other documents from the Society. Of course, that has no impact on my communications with my subordinates, but it is of great significance for me personally; had I received it in a timely fashion, that is, when I was in Okhotsk in March 1847, then, traveling almost as the first one along the new Ayan winter route, I would not have omitted, in spite of many difficulties, to collect the information that might have been accessible to me; I would have done the same when traveling from Ayan to Udskoe by baidara and baidarka.

To my misfortune, until I returned to Udsk I did not even know of the existence of the Society, as I was traveling in such places, as for example Gizhiga, which even the most important communications do not reach in less than a year. For this reason I paid but very superficial attention to the things I encountered, and the information I obtained in these travels I do not deem worthy of presentation to the Society today.

With absolute respect and devotion, I have the honor to remain Your Excellency's humble servant, Innokenty, Bishop of Kamchatka.

10th day of May, 1848
Novoarkhangelsk[389]

Earlier we saw that the environment and aborigines of Russian America were at the center of Father Ioann Veniaminov's attention, alongside his pastoral activities. Veniaminov's letters to Litke, preserved in the Manuscript Division of the Russian National Library, in the personal collection of Innokenty, Metropolitan of Moscow, testify to the fact that the learned missionary regularly transmitted to St. Petersburg his observations in natural sciences made in Russian America from 1823 to 1833.

For example, in his first letter to Litke, dated July 3, 1828, he writes: "With this I have the honor to present to Your Excellency, as promised, the meteorological observations made by me at Unalaska starting on October 7, 1827, that is, beginning with the time when circumstances permitted me to conduct them regularly."[390] Moreover, Veniaminov included in his reports materials collected by other persons: thus, on his visit to the Pribilof Islands he discovered the results of meteorological observations that had been made on St. George Island from 1824 to 1829, and on St. Paul Island from 1823 to 1829.

In these letters Veniaminov pays much attention to environmental data obtained from the Natives. Thus, Veniaminov reports that "many of the aboriginal local inhabitants have noted that in spring and summer the sea water flow decreases, that is, from the northwestern into the southeastern sea, while in the fall and winter the opposite is the case; for this reason the tides are less in the first case and occasionally for the entire day; in the second case the opposite occurs."[391]

In Veniaminov's letter from August 19, 1831, there is a detailed description of the eruption of the Shishaldin volcano in November–December of 1830 and in March–April of 1831.[392] Strange as it seems, Litke did not respond in any way to Veniaminov's letters. This aroused a fully justifiable feeling of hurt on the part of the priest, who, sending his meteorological observations for the last time on July 20, 1833, noted: "And thus, as I do not have the honor of receiving from you even a line about them [the notes—AP], I think you have no use for them, perhaps because in much, if not in totality, you find them unsatisfactory and I, for my part, am not able to do better out of my own ignorance and lack of means . . . I confess that I transmit these copy books with sorrow, as with their termination one ought, following good manners, to stop writing to you also."[393]

This time Litke responded with a warm message from April 19, 1834, expressing his deep gratitude to the priest and acknowledging that his observations were of great use.[394] From September 1 Litke officially entrusted to Veniaminov the keeping of meteorological observations on Baranov Island (Sitka).[395] A year-and-a-half later (May 3, 1837) Kupriyanov sent to the Imperial Academy of Sciences a packet with "the notes kept in Novoarkhangelsk port by the Priest Ioann Veniaminov, beginning with first of April of the past year of 1836 to the first of April of this year of 1837, and observations on variation of the magnetic pointer also made at the Novoarkhangelsk port by Mr. Captain-Lieutenant Etholen, Lieutenant Zarembo and the mechanic, a fosterling of the Russian-American Company, the Creole Terentiev, from April 22 of the past year of 1836 to April 24 of this year of 1837."[396]

After Zagoskin's expedition the Russian Geographic Society began a systematic investigation, by means of special questionnaires, "to obtain ethnographic, climatic and other data." These questionnaires were sent out for the first time by the society to all the regional governments of Russia in March of 1848. Among those who received this questionnaire in several copies were the leaders of the regional nobility, directors of schools, high-ranking administrators, bishops, monastery abbots and selected priests of the Orthodox Church, foresters (who received the questionnaires through the Regional Government Office of the Ministry of State Properties), county presidents of the assemblies of the gentry, and selected landowners.[397]

The request addressed to Innokenty, Bishop of Kamchatka, to collect geographical information for the Russian Geographical Society was sent to him earlier, back in 1846. Judging by the correspondence preserved in the Archive of the Russian Geographical Society, Innokenty also received the 1848 questionnaire together with the "Additional Guidelines for Collecting Information that Serves to Determine Climate." These clarified particular sections of the 1848 questionnaire and were sent to all Russian cities and boroughs on March 20, 1849.[398]

In spite of his overwhelming workload dealing with the largest diocese in all of Russia, Innokenty once again took an active part in assisting the Geographical Society in the completion of its program. In 1857 he sent from Yakutsk to St. Petersburg the results of observations which had been made in Russian America in accordance with the program of the Russian Geographical Society. The following letter, preserved in the Archive of the society in the original, accompanied the above materials:

> #750, from 24 December 1856, Yakutsk City. To His Excellency, the secretary of the Russian Geographical Society, Evgeny Ivanovich Lamansky.
>
> Your Excellency, My Dear Sir!
>
> I have the honor, My Dear Sir, to inform you that I have received the Annual Report for the year 1855 of the Imperial Geographical Society which you sent me.
>
> I also have the honor to transmit to Your Excellency several answers to certain questions regarding the climate of Russia (incomplete) and meteorological observations kept in the deep North of our America.
>
> Your Excellency will no doubt be surprised by the lateness of the response; you will be even more surprised when I tell you that from some persons I have received no answers to this day; but this is due precisely to the circumstances wherein I find myself and because of the war our region was subjected to [apparently he has in mind Great Britain's attempt to launch intervention in Kamchatka during the Crimean War of 1853–1856—AP]. Besides, the questions themselves I received only at the beginning of the year 1851 on my way from Kamchatka, through Gizhiga, to Okhotsk. I was able to send them out only at the end of the year. Then I had to sort out misunderstandings by those who wished to reply to the questions. And all of that took dozens of months.

I have the honor to append to this one hundred and fifty rubles for the benefit of the Society, as well as thirty rubles with my humble request to send me, for this sum, the Society's Messenger for all the past years and if possible bound (in a simple cover).

With utmost respect and devotion, I have the honor to remain Your Excellency's Obedient Servant,

Innokenty, Archbishop of Kamchatka.[399]

A vivid impression of the conditions under which the missionaries of the Russian Orthodox Church fulfilled their Christian duties and obtained geographical information in Alaska is to be gained from the following account by Bishop Innokenty in the official statement *On the Condition of the Kamchatka Diocese*, directed to the Synod in 1850 and preserved in the Archive of the Foreign Policy of the Russian Empire:

In the last year of 1849, as I have reported, I dispatched to the River Kvikhpak Hieromonk Filaret to assist the local missionary, having also in mind to open a new mission at the first opportunity. However, information on this matter can be obtained no earlier than in September of this year of 1850.

The Kenai missionary, Hieromonk Nikolai, continues to serve with his customary activity and the same zeal. From March 28 to June 10, 1849 he traveled *by baidarka* to Nuchek and Copper River, and in winter, from December 7 to 25 and from January 19 to 25, *on foot, by snow shoe* to the settlements of the Kenai who live in the mainland interior and whom it is absolutely impossible to visit in summer. He covered on the last two journeys no fewer than 400 versts, wilderness, where there are no roads. He undertakes such journeys not for the first time.

As soon as a possibility opened up, the missionary undertook a journey from Nuchek to *Copper River.* His journey was the first one since the time of Russian settlement there. They traveled two days by baidarka from Nuchek to the mouth of the Copper River and ascended it for another day, reaching the first settlements of the Ugalentsy (Eyak). . . . The missionary did not ascend any further because the disturbances have not quieted down (after the murder of the Creole topographer sent out by the Company in the year 1847 and who was killed by the inhabitants of the mountains).

In a note, Innokenty stresses that nothing was taken from the topographer by the Natives and that even his instruments were safe and not damaged.

The activity of the Kvikhpak mission in the past year was limited to the missionary's one trip to the Kuskokwim River, to the Kolmakovsky Redoubt."

The head of the mission, a priest who was a Creole, Yakov Netsvetov, was unable to visit any other locality that year because, Veniaminov says, "a) after meeting with me at the Mikhailovsky Redoubt in 1848, he returned to his mission only on August 24, and therefore it was too late in the season for him to travel elsewhere, and besides, b) that summer there was very little fish there, so that in the settlements along the Lower Kuskokwim (where the priest had not visited yet) in the spring many died of starvation and for this reason, c) by fall of 1848 almost all of the Natives were traveling; even the residents of the settlement where the mission is located were away from home through the entire winter looking for food; they returned, but not all, for the Lenten communion time, and soon thereafter left again.

Remaining in residence, the missionary with his clerical assistants and one more Aleut laborer was busy organizing the Mission. Before severe frosts set in they managed to build store rooms and a steam bath and began to prepare the spot where the church was to be constructed. While on the Kuskokwim River from February 4 to 28, among other things he consecrated the new chapel built by the efforts of local manager Lukin and his subordinate Creoles and Natives; besides the usual services for various occasions and regular church services, whenever a convenient opportunity arose he assembled the children and taught them the Law of God; he newly joined to the church 22 persons from among the pagans.

After the River Kvikhpak became free of ice, the missionary and his subordinates departed on June 5 as usual down the river for the Mikhailovsky Redoubt, where they arrived on the 15th. Residing here while waiting for the arrival of the vessel, he performed services in preparation for communion, both for the redoubt's residents and for some of the Natives. Passing the settlements along the river, he saw very few people there; because of a shortage of food all had departed for the mountains to hunt (this happened in 1849).

The missionary's successes in converting the Natives were limited from August 1848 to July 1849 solely to the 22 abovementioned persons who joined, and 16 more during his travel downriver in two settlements, Takchagmiut and Kanigmiut . . .

The Creole Ivan Zhukov, although he was a good interpreter of the Tlingit language, could not be used by the mission because of his poor behavior . . .

"8th day of May, 1850, Novoarkhangelsk.[400]

The excerpts from Bishop Innokenty's report cited here, while creating a vivid picture of the activity of the Russian Orthodox Church on American soil, also demonstrate how deeply they were "submerged" in the natural environment and the ethnic milieu, and this, in turn, testifies to the accuracy and precision of the information they supplied.

The materials sent by Innokenty were received at the Russian Geographical Society with great approval once again, a fact that is confirmed by a letter from Litke to the archbishop of Kamchatka:

> # 495, of 29th March 1857; To His Grace, Innokenty, Archbishop of Kamchatka.
>
> Your Grace,
>
> My dear Sir and Archpastor,
>
> On behalf of the Council of the Imperial Russian Geographical Society I have the honor to let Your Grace know our sincere appreciation for your donation for the benefit of the Society of 150 R., that was delivered here together with your most esteemed letter dated December 24, 1856, and which has been added to the capital that is the means by which the Society executes its scientific undertakings—and also for your enlightened assistance to the Society's work. At this same time, wishing to secure for the Society your sympathy for the benefit of our fatherland's science and your activity in support of the Society's endeavors, the Council deems it its duty to inquire of Your Grace if you would not benevolently accept for yourself the status of Active Member of the Society. The Bylaws of the Society are appended.
>
> Entrusting my own self to your prayers,
>
> I humbly beg to accept my assurance in my absolute respect and devotion,
>
> F. Litke
>
> P.S., your admirer of thirty years' standing is joyful that God brought him to a meeting with Your Grace in the field of science. He is no less joyful in the hope that soon he shall meet you here.[401]

THE CREATION OF MIKHAIL TEBENKOV'S *ATLAS OF THE NORTHWEST COASTS OF AMERICA* OF 1852

In January 1845 Captain Mikhail Dmitrievich Tebenkov (1802–1872)[402] was appointed chief manager of Russia's colonies in America. In April of that same year, in Okhotsk, he accepted the office from Captain Etholen. On September 1 the family of the new chief manager arrived in Sitka.[403] We have encountered

Tebenkov's name more than once. He was in active service in the colonies from 1825 to 1831, and commanded the company's vessels, the brigs *Golovnin*, *Ryurik*, and *Chichagov*, and the sloop *Urup*. He began his fruitful activity as a cartographer in 1829, specifically conducting the surveys of Norton Sound and the Alexander Archipelago.[404]

Previous chief managers had put matters of geographical exploration and cartography on a firm scientific, technical, and organizational basis. This guaranteed the uninterrupted arrival of new navigational charts and surveys at the colonial management and the Hydrographic Department of the Russian Admiralty. Such materials were dispatched to St. Petersburg by Etholen prior to his departure from the colonies. Etholen's "Report" cited below demonstrates that by this time the compilation of charts in Novoarkhangelsk was well organized.

382, May 15 (1845). To the Head Office.

REPORT

I had the honor to inform the Head Office in my report of May 8 of the past year under # 274 about the assignment I gave to Sub-Lieutenant Gavrilov in respect to the drawing of charts, of which one set (as the Head Office knows from Report # 274) was given by me to Lieutenant Kashevarov, who left the colonies last year. Mr. Gavrilov, having spent the winter engaged diligently in the same work, on my instruction, made several copies of these charts, in which all the additions and corrections made by our vessels are included. I have the honor now to present to the Head Office out of these sets eleven charts, listed in the appended register. With them I also present the general map of North-Western America from Mr. Lieutenant Zagoskin's expedition (that follows a separate report by me) and a plan of the Port of Novoarkhangelsk, in its latest form, as I ordered it to be compiled this spring.

Register of charts.

1. Aleutian Chain, charts with corrected latitudes and longitudes, compiled from the charts of Lieutenant Tebenkov.
2. Kurile Chain, 1 part, compiled by Sub-Lieutenant Gavrilov.
3. Bering Sea chart, compiled from the charts by Tebenkov and from the department's atlases.
4. Chart of the Coast of Kamchatka, from the charts of Krusenstern, Litke and Beechey.
5. Chart of Avacha Bay.
6. Chart of Tlingit Straits, two charts, corrected by vessel Commanders who sailed the Straits.

7. Chart of the Okhotsk and Bering Seas.
8. Chart of Ayan Bay, corrected by Sub-Lieutenant Gavrilov.[405]

The materials sent from the colonies were received in St. Petersburg with approval, but in a letter of March 8, 1846 (#493), the head office instructed Tebenkov, in accordance with the request by the Hydrographic Department, to send appropriate journals and surveys with the charts, on the basis of which the corrections were made.[406] It was also confirmed that Tebenkov was to continue keeping geomagnetic and meteorological records at the observatory in Novoarkhangelsk established by Etholen.[407]

Altogether Tebenkov had spent over twenty years in the colonies, and throughout this entire time he paid close attention to the cartography of the islands and coasts of Russian America. The charts listed in the register cited above testify to that, as many of them were compiled by him personally. Having accepted the office of chief manager, this remarkable hydrographer received a chance to complete his work and to prepare for publication the *Atlas of the Northwest Coasts of America*

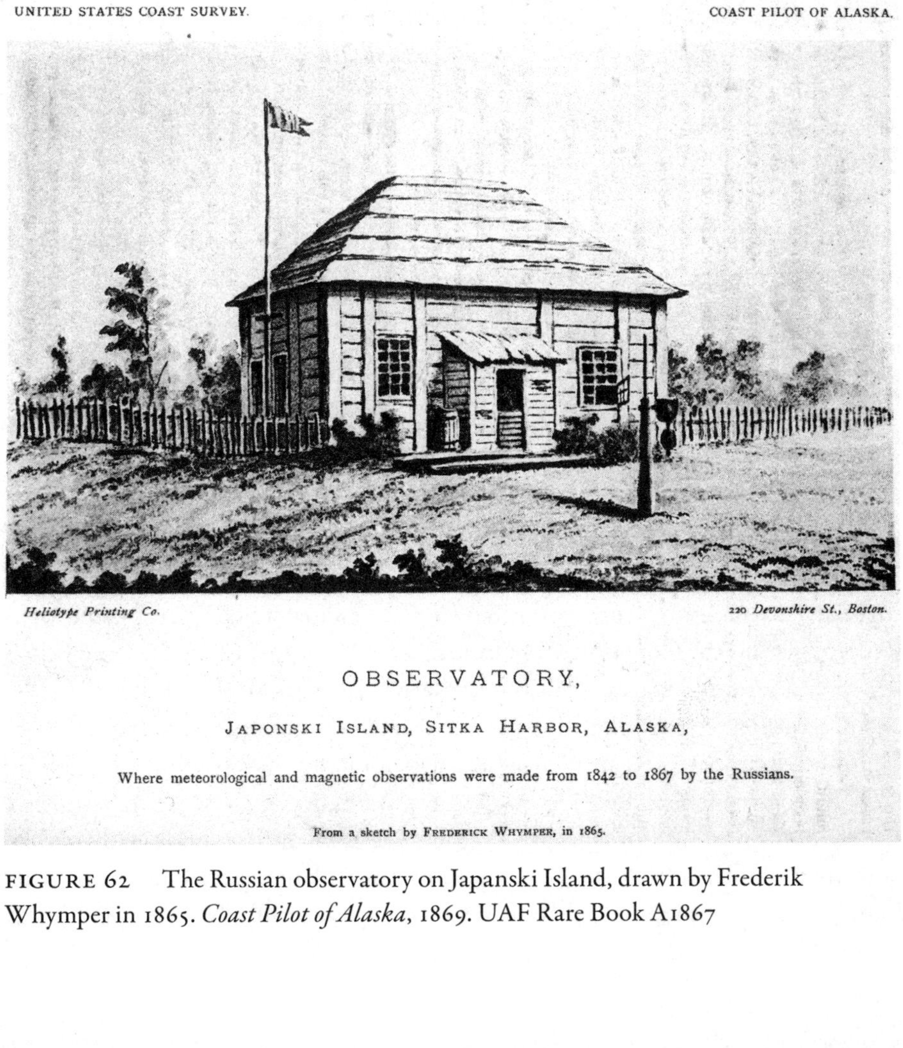

FIGURE 62 The Russian observatory on Japanski Island, drawn by Frederik Whymper in 1865. *Coast Pilot of Alaska*, 1869. UAF Rare Book A1867

from the Bering Strait to Cape Corrientes and the Aleutian Islands with Several Segments of the Northeast Coast of Asia.

Tebenkov utilized every trip through the possessions of the Russian-American Company to collect cartographic and descriptive materials and also to correct charts based on his own observations. In the process of this daily work he became persuaded of the necessity of conducting special hydrographic surveys to fill in the gaps which were present at the time. The volume of such work was so large that Tebenkov decided to dispatch specially equipped baidarka expeditions for surveys and descriptions of the coasts.

On May 10, 1848, Tebenkov informed the head office of the Russian-American Company about the organization of two expeditions to the islands of Unga, Unalaska, and Atka. These initiatives on Tebenkov's part found full support.

Some supplementary data about the northernmost borders of the Russian possessions in America came at this time because of the involvement by the Russian-American Company in the search for the lost Franklin Expedition. Franklin had been dispatched with the vessels *Terror* and *Erebus* in search of the Northwest Passage in 1845. It was last seen in Baffin Bay in July of that year. A series of searches were launched from England. One of the major searches, this one from the west, was that of the ships *Plover* and *Herald* in 1848 with substantial assistance from the Russian-American Company.[408] The company provided interpreters and baidarkas with experienced crews.[409] The cartographic results obtained by the search parties were reflected in official British publications.[410]

Of course, the basic materials for Tebenkov's atlas were surveys and descriptions by the sailors of the Russian-American Company. At the time of its preparation such hydrographic surveys were being carried out not only in poorly known regions, but also in areas that had been surveyed many times in the past in order to update the charts. In particular, at that time new surveys were made of Kodiak Island, Sitka, and the "Chugach shore."[411]

Not limiting himself to hydrographic work, Tebenkov investiged the mainland interior. In 1846 he dispatched an expedition out of Nulato up the Kvikhpak. In 1847 Serebrennikov's party was sent up the Copper River in order to reach the Kvikhpak or one of its tributaries. The party consisting of three baidarkas set out from the mouth of the Copper River on August 14. It wintered at the mouth of the Chistochina, in the spring went around Lake Ploveshnoe (Tazlin), and then went farther up the Copper River. On June 25, 1848, Indians attacked and killed Serebrennikov and his crew. One of the Natives brought Serebrennikov's journal and descriptions to the company post at Nuchek. Thus the substantial geographical information obtained on this journey was not lost.[412]

The continental explorations carried out on Tebenkov's initiative permitted the verification of information obtained by Zagoskin and, and especially important, they confirmed that the Kvikhpak and the Yukon were a single river. This found expression, along with other new data, on a summary map engraved at

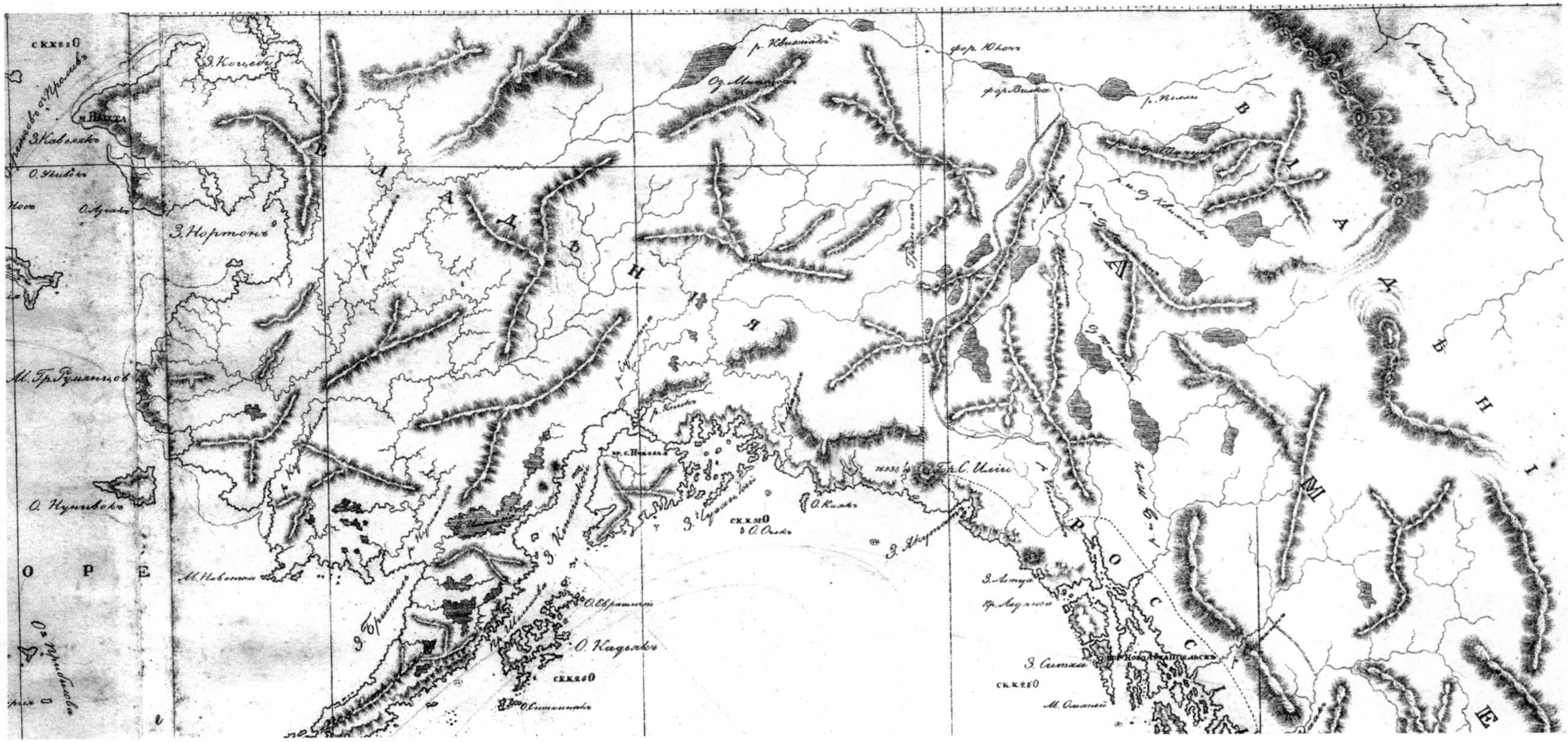

FIGURE 63 Portion of the summary map of the North Pacific from Tebenkov, *Atlas* showing Interior Alaska. UAF Rare Book C0035

Novoarkhangelsk in 1849, the *Chart of the Northern Part of the Pacific Ocean*. There the forts Yukon and Vilka are represented on the upper course of the Kvikhpak, shown by a dotted line (Fig. 63).

The compilation of the atlas was accomplished by Tebenkov in Novoarkhangelsk.[413] According to him the maps were engraved on copper by the Creole Grigory Klimovich Terentiev (born 1818). However, George Davidson of the US Coast Survey states that the maps were engraved by another Creole, Captain Mikhail Kadin (ca. 1830–1868),[414] who was one of the best draftsmen in the service of the Russian-American Company, producing many valuable maps that were later used by the US Coast Survey in Washington.

American scholars Clarence Andrews and Richard Pierce with good reason assume that Mikhail Kadin drafted the charts, while Terentiev engraved them.[415] Be that as it may, the main author of the atlas is Tebenkov, and he accomplished it at the highest scientific and methodological levels of his time. This is evident from the

FIGURE 64 Plate VIII showing Southeast Alaska, from the Tebenkov *Atlas*. This map is heavily annotated by George Davidson in preparation for the publication of the 1869 *Coast Pilot*. UAF Rare Book C0035

charts themselves and from his "Hydrographic Notes to the Atlas of the Northwest Coasts of America and the Aleutian Islands and Several other Localities in the North Pacific Ocean." In the latter, besides detailed descriptions of the environment and sailing conditions, he included analytical information about the sources for the charts and the methodology used by the compiler.

Let us consider in greater detail the main scientific and methodological premises of Tebenkov's atlas.

In analyzing the available navigational materials for the northern part of the Pacific Ocean, Tebenkov noted that the data most precisely established by astronomical and geodesic means were those in the atlas by George Vancouver. However, Tebenkov stresses that "Vancouver's atlas has become extremely rare and moreover new observations have shown that the coasts for which he determined the longitude are not in those locations."

Tebenkov continued: "Since Vancouver's time (1791) many details have been learned and I decided to extend the coverage along the northwest coasts of America to latitude 20° N, within the limits of Vancouver's atlas."

Tebenkov states that after examining all of the journals kept at Sitka, all of the reports of voyages he had sent out, his own personal observations, and all available publications, he still knew that there could be errors.[416] After returning to St. Petersburg, Tebenkov studied materials there and included them with the materials he had brought from Russian America.

In this way Tebenkov located and used in his compilation of the charts of the northern part of the Pacific Ocean practically all the available sources, some of which have survived to our time. For this reason the charts and descriptions in the atlas contain specific facts and items absent or poorly elucidated in other documents. Professor Pierce notes that "Indian or Rezanov Bay" (modern Humboldt Bay) is shown on the inset to chart #13, while the explanatory text states that Rezanov visited this bay, which was discovered by "United States citizens" and visited in 1806 by an Aleut hunting party commanded by the *promyshlennik* Sysoi Slobodchikov, who composed a chart of the mouth of the Columbia River.[417] In another instance Tebenkov notes that Russians had landed on Kodiak even before Glotov, whose visit to that island in 1763 is usually considered the first. Perhaps further research will confirm this.[418]

Of great value for the history of cartography are the many mentions by Tebenkov of the sketches and verbal information supplied by the Inuit and Aleut. These have aided Europeans in charting their settlements and particularly ensured the preservation of many aboriginal geographic names.

Tebenkov's survey of Russian America in the *Hydrographic Notes . . .* is given from north to south, with a brief sketch about the history of explorations of specific coasts and an analysis of the trustworthiness and accuracy of materials. Thus, the author confirms the high degree of accuracy of the surveys by the Beechey Expedition, which were, on his orders, verified by the "Russian skipper Pavlov,"

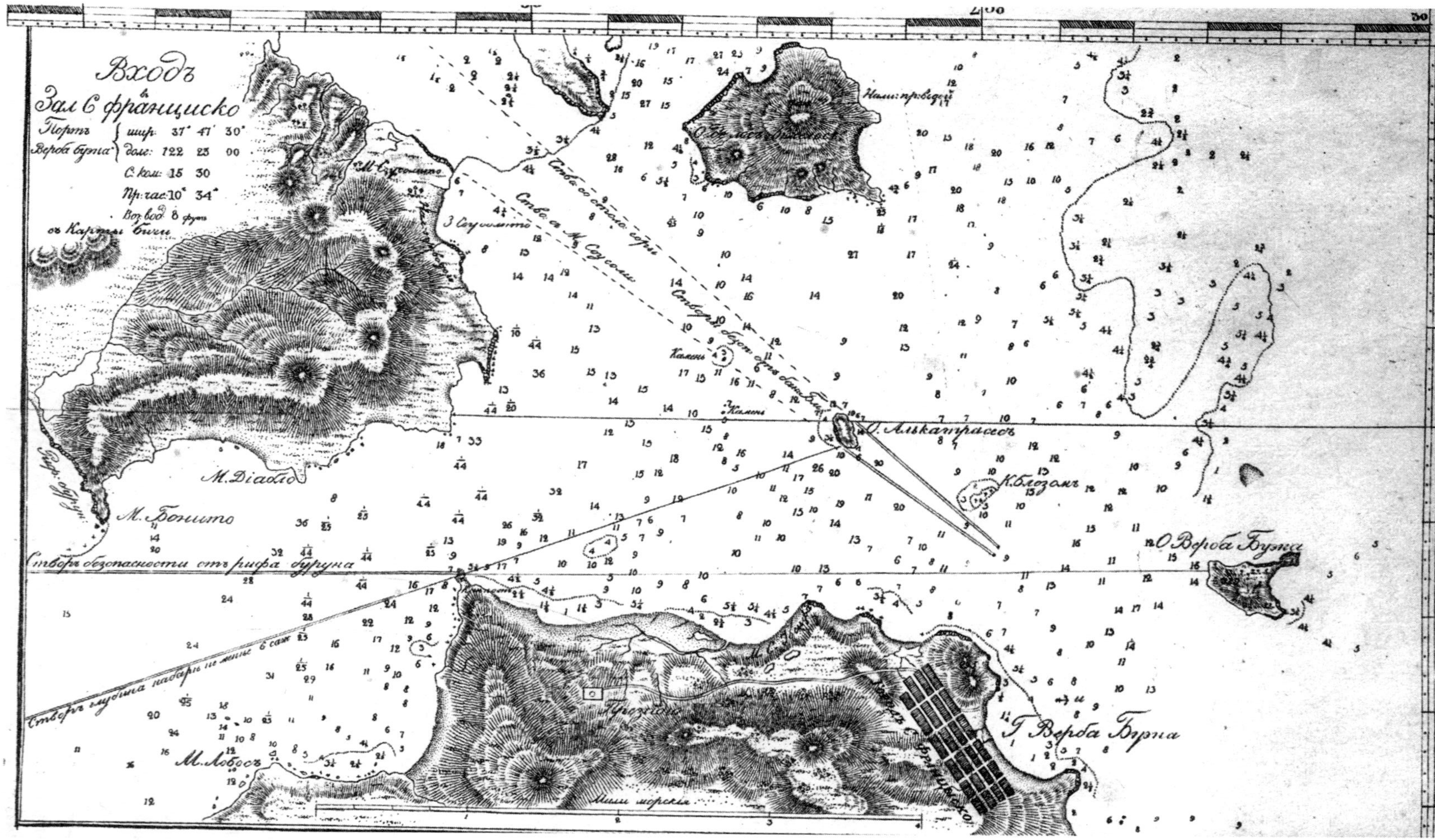

FIGURE 65 Inset map on Plate XIV, Tebenkov *Atlas*, depicting San Francisco on the eve of the California Gold Rush. UAF Rare Book C0035

who found the location of the Gvozdev (Diomede) islands as shown "by Beechey to be rather accurate."[419]

According to Tebenkov's description, the coast to the southwest of Cape Aziachag "is low and shallow for a great distance; for this reasons no sailor sighted it before." For the same reason the cartographer showed it "according to the sketch by the Natives. This extended area contains, almost to 60½° latitude, the sloughs of the great but shallow Kvikhpak River, whose source lies at about 130° longitude and 66° or 67° latitude. The coast along mainland America from the Kvikhpak River to the Kuskokwim River is also shown from the sketch by the Natives."[420]

Further to the south information provided by Native inhabitants is used by Tebenkov to indicate the "trend of the coast to Good News Bay, called by the Natives *Imagpichuak*."[421] Tebenkov was highly appreciative of Vancouver's survey in Chugach Bay (Prince William Sound). Using this survey as the basis for his own chart, he noted that "from the sloughs of the Copper River to the SE, the east coast of Chugach Bay is shown in accordance with testimonies by the Natives, the Chugach and Koloshi (Tlingit)."[422] He also notes that the Koloshi call themselves "*klinkit*."

In his *Hydrographic Notes* Tebenkov confirms many times the veracity of Native cartographic drawings and descriptions of localities and that these were used in compiling the atlas. He also stresses the accuracy of Native observations pertaining to natural processes. We may cite as an example the account of high/low tide ice abrasion and transfer of sediment materials in the Gulf of Kenai (Cook Inlet): "High tide does not leave the ice to itself, often it breaks it and lifts within it huge boulders that are carried about the gulf in the spring . . . According to Native testimony, a reef was formed of such rocks near Chernobury Island (St. Augustine), and in the not-too-distant past."[423]

Upon his return to St. Petersburg, after he had surrendered the management of the colonies on October 14, 1850, to Captain of the Second Rank Nikolai Yakovlevich Rozenberg (1809–1857), Tebenkov supervised the completion and publication of his atlas. It received a prize and earned recognition among his contemporaries and later generations. Wrangell maintained that Tebenkov's work filled important gaps in the hydrography of these regions and was much more perfect then any preceding works.

The charts from this atlas were used for many years and became one of the most important sources for the first navigational guide to Alaska, George Davidson's *Coast Pilot*.[424] To this day the charts from this atlas are used as a source for Alaska's Native toponyms. They were utilized for this in the compilation by Donald Orth of the *Dictionary of Alaska Place Names*.[425]

Beyond the compilation of general geographical and navigational charts there were thematic maps. Geological cartography was most widespread, testimony to the growth in interest in the natural resources of Russian America. It must be noted that the earliest geological investigations were conducted at the very end of the eighteenth century by the mining engineer Dmitry Tarkhanov. Toward the end of 1795 he was on a modest expedition to the Copper River in search of copper ore, which was to be found, according to the Indians, at the headwaters of this river.

The explorer reached the Indian settlement of Takekat on the upper course of the river. Tarkhanov left us his notes, which contain the first data on the geography, geology, and ethnography of the Copper River basin, and which were introduced to science for the first time by A. V. Grinev.[426] Geological explorations were conducted, as we have seen, by Zagoskin, Voznesensky, and other investigators, but interest in the sub surface riches of Russian America received stimulus from the discovery of rich gold fields in Upper California in 1848, in the region that up until 1842 belonged to Fort Ross.

Precisely at that time (1848–1850), mining engineer Pyotr Petrovich Doroshin was active in the colonies. He found sufficiently thick anthracite coal strata in English Bay on the Kenai Peninsula and began to work them. In 1848, having learned about the California gold strike, Tebenkov dispatched Doroshin, accompanied by a team of four Russians and six Tlingit, to explore for gold supplies in the Sacramento River basin.

On February 26, 1849, the exploration party came out to the Sacramento tributary, the Yuba, but stopped all further work by April 16 because of rising flood waters. They had panned more than eleven pounds of gold. Company searches in California found no further gold. On March 28 the head office issued the following directive:

> On geognostic investigations in the Colonies.
>
> In regard to gold, although the Head Office does not expect significant results from working this metal, the presence of which apparently is not subject to any doubt, it finds it necessary to ask the colonial management to send the Mining Engineer, with a party of reliable employees, in case Your Excellency has not done it already, to the headwaters of the River Kaknu and its tributaries to investigate the gold-bearing capacity of these rivers and the feasibility of organizing there a gold panning operation, provided it will be found possible and proven through experiment. The mouths, and if necessary the headwaters, of other rivers flowing through the Colonies are to be also investigated with this aim in view.[427]

Somewhat ahead of this order, in the field seasons of 1850–1851 Doroshin conducted a reconnaissance for gold in the areas surrounding the Gulf of Kenai (Cook Inlet). He found signs of gold in the valleys of the rivers Chunu-Ktny and Tazlikh-Ktnu and their tributaries. Even earlier he noticed gold shining at the mouth of the Kenai River and at the Nikolaevsky Redoubt.

"No matter how meager are the results of my two year-long reconnaissance in the Kenai Mountains," Doroshin wrote, "they provide the most positive grounds for further efforts and hopes that gold fields of respectable content of the noble metal will be found."[428] The leadership of the Russian-American Company did not share Doroshin's optimism, and in an official letter of August 19, 1852, addressed to Rozenberg they ordered the following:

"Finding that the results of the two summer expeditions by Mr. Doroshin did not bring firm hopes of finding gold in sufficient quantities, the Head Office in response to Your Excellency's dispatch of November 29, 1851 under # 837 instructs you to set aside the further search for gold and engage the means of the Colonies in the timber, ice, and fishing industries, though continuing a search by small parties for anthracite coal deposits."

Thus the search for gold in Russian America was not developed. This was not only the reluctance of the leadership of the Russian-American Company to take manpower away from its other activities, but also the fear of attracting the attention of the neighboring United States. Alaska had already become an object of easy pickings by foreign fur traders and whalers, whose vessels traversed Russian America's waters and visited her hospitable ports.

The results of the geological and geomorphological investigations conducted in Russian America in the first half of the nineteenth century were summarized by the keeper of the mineralogy office at the St. Petersburg Academy of Sciences, Konstantin Ivanovich Grewingk (1819–1887) in his works of the years 1848–1850. This summary was based on the collections kept in the office.[429] These works encompass three schemata that in essence are the first thematic maps devoted to the relief and geology of Alaska.

The first of these is *Gebirgsferbreitung in der Westlichen Haelfte Nord-Amerikas* (*The Extent of Mountains in the Western Half of North America*). It is an orographic schema in which the main mountain ranges are indicated by solid lines with letter signs that refer to the legend. The *Karte zur Abhandlung ueber die Geogn. U. orogr. Beschaffenheit der N.W.—Kueste Amerika's n. der anliegend Inseln mit Zugrundelegung der Karten des Hydr. Dep. Des See-Ministeriums zu St. Petersburg. Von C. Grewingk 1848* (*Map Appended to the Paper about the Geognostic and Orographic Diagnostic Department of the Marine Ministry in St. Petersburg*) is the first geological map of Alaska and its islands. With the help of hand painting in various colors it shows rock and sediment formations of various origin and age: trachytes, basalts, dolerites, granite-gneiss, metamorphic, deluvial and alluvial rocks, and so on, (Fig. 66).

The third schema, *Karte der westlichen Haelfte der Halbinsel Tschugatsk nach Wasnessensky* (*Map of the Western Half of the Chugach Peninsula after Voznesensky*) is a successful attempt to represent in a large scale, after Voznesensky's materials, sections of the Cook Inlet coast and the mountain lake "Tustamena See," with characteristic forms of the relief and the glaciers and with a geological cross-section (Fig. 67).

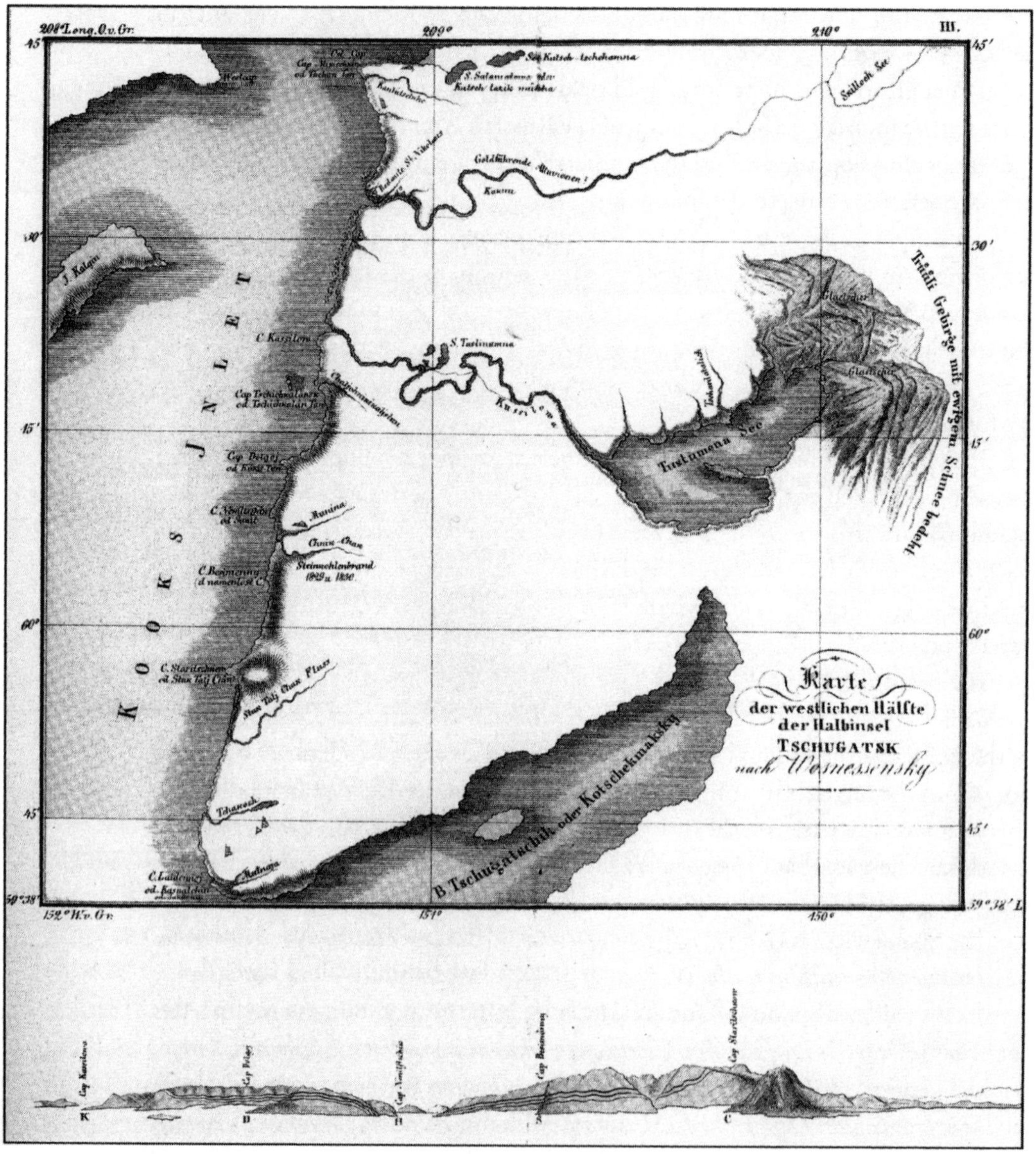

FIGURE 67 Grewingk's map of the Kenai Peninsula. UAF Rare Book A0488

WHALING

Whaling in the North Pacific was increasing rapidly. Whaling maps published in America were sent to the head office of the Russian-American Company by the Russian ambassador in Washington, A. A. Bodisco, on January 8, 1852. These maps showed, "in the ambassador's opinion with precision and clarity the main routes followed by large and small whales, as well as the localities where they are found in larger or smaller quantities."[430]

On August 19, 1852, the head office of the Russian-American Company forwarded two "whaling charts" (*Maury's Sailing Directions, November 1851*) obtained by the Russian ambassador in Washington for use by colonial whalers.[431] It should be noted that in this period foreign whalers began not only to cause much disquiet for the Russian-American Company, but also inflicted perceptible harm upon the ecosystems of the North Pacific. For example, in his letter to the Ministry of Finance of October 8, 1847, under #141, Tebenkov characterized the condition of the natural environment as a result of the unbridled excesses by the whalers:

"In the Kurile District entire shores are covered with whale skeletons or the carcasses of whale dead from wounds. The whalers come ashore at night and make fires to render whale oil. The large fires at night, the smoke and the whale oil floating near shore drive away not only the sea otters, but also the sea lions and seals that constitutes the main food for the Natives. According to the testimonies by local managers, as well as by the Natives, the sea lions and seals no longer come because the grass on shore is burned out by the fires made."[432]

The considerable profits brought in by the whalers, and the large quantities of whale oil used in military production, attracted the attention of the military in the United States of America.[433]

The first whaler to penetrate into the Arctic Ocean through the Bering Strait was Captain Roys in 1848, with the vessel *Superio*. In the course of a few weeks he filled his holds to capacity with whale oil. In 1849 Roys returned, leading a fleet of 154 whaling vessels; in the course of a single navigational season they obtained 206,850 barrels of oil and 2,481,600 lbs. (1,126,646.4 kg) of baleen. In 1850 144 American vessels were hunting in the Arctic waters, taking 243,680 barrels of oil and 3,65,400 lbs. of baleen.

In 1851, 145 vessels entered the hunt, but the costs of equipping this fleet were not recovered due to more than seven vessels perishing in severe storms and the poor quality of the American navigational charts. The total profit from whaling for these four years, according to the paper's author, amounted to 17.5 million dollars. The United States Department of Navy was presented with the need to perfect the hydrographic knowledge of the Bering Strait and the nearby seas of the Arctic Ocean by means of organizing a special naval expedition with this goal in mind.

All this information, with a request to cooperate in the undertaking of such an expedition, was sent to the Russian Ministry of Foreign Affairs, where a special commission was formed with the participation of the leadership of the Russian-American Company to discuss the possibility and expediency of providing help. In the Russian-American Company the prospect of the appearance of a large hydrographic expedition by an American naval fleet did not provoke enthusiasm, and it was considered preferable to put the newest Russian charts and surveys at the disposal of the Americans.

An announcement was made at a meeting of the Commission of the Russian Ministry of Foreign Affairs on June 17, 1852, that:

> The originals of the charts made by the Company and journals are always sent to the Hydrographic Department of the Naval Ministry and are included in the atlases that are published by it, which generally and particularly for the Aleutian and Kuril island chains may be considered fully satisfactory and complete. It is suggested that to avoid an expedition America be presented with copies of these charts and atlases. From this we can conclude that the delivery of copies of our charts to the American government would eliminate at least part of the plausible reason for visiting our possessions with this aim.[434]

At the Commission of the Russian Ministry of Foreign Affairs the desire to strengthen the neighborly relations that had been developing by that time with the United States prevailed.[435] On February 6, 1853, the head office of the Russian-American Company was ordered to provide all possible assistance to the American expedition.[436] Carrying out this order, in its dispatch No. 317 on March 20, 1853, the leadership of the company instructed the chief manager of the colonies to help the American government expedition that had been tasked to study the Bering Strait and the Arctic Ocean, "in the form of investigating those seas and various coasts there."

At the same time it was clarified that: "in the event of its appearance near the coasts of the Colonies the appropriate attention to its needs and all possible cordiality should be provided, within the framework, however, of the necessary prudence and caution." Subsequently in the instructions it was directed that "on a plausible pretext and without showing the slightest indication of such, when possible try to reduce the length of the American expedition's sojourn in our waters, as a visit to the Colonies by foreign vessels, as is well known to the head office from the information received by it in regard to the expeditions which searched for Captain Franklin, for the most part causes harm to the company.

"And information about the whereabouts in the Colonies of fossil coal and other statistical data about the products of our Colonies should be hidden from the Americans on a plausible pretext. Ask the members of the expedition not to engage in the fur trade or to sell alcohol." To aid the expedition it was instructed that its leadership should be offered "interpreters from among those who work for the Company," for which it was necessary to issue a special "Instruction on Observing the Activities of the Expedition in Our Waters."

Captains of Russian-American Company vessels that sailed close to the areas where the expedition was active were ordered "when convenient to approach the places where the Americans are doing their work, and absolutely, if not on the same sail then on the next one, to visit the places where the American expedition has come on shore, and to watch them carefully, observing at the same time the influence that the Americans may have on the Natives."[437]

The American expedition took place in 1854–1856 under the leadership of Captain John Rogers on the flagship *Vincennes*, the steamer *John Hancock*, under the command of Lieutenant Henry Stevens, and the schooner *Fenimore Cooper*, under the command of Lieutenant William Gibson. Detailed surveys close to the shores were carried out by Lieutenant John M. Brooke on a whaleboat from the flagship *Vincennes*. During this large-scale expedition the *John Hancock* actively engaged in surveys of the western coasts of Kamchatka and the coasts of the Sea of Okhotsk up to the Tartar Strait. The *Fenimore Cooper* went to Novoarkhangelsk and made its way along all the Aleutian and Kuril islands. The expedition's flagship carried out a hydrographic survey of the coasts of Chukotka and then headed north.

The expedition caused no serious problems for the Russian colonies. The charts produced from the results of its work were: *Track Chart of the U.S. North Pacific Surveying Expedition. John Rogers, U.S. Navy Commanding, 1854–56. Drawn by Louis Waldecker/United States Navy Department, Bureau of Navigation, Hydrographic Office (Washington, 1856)* and *Behring's Sea and Arctic Ocean, from Surveys of U.S. North Pacific Surveying Expedition in 1855, Commander John Rogers, Commanding and from Russian and English Authorities. Compiled by E. R. Knorr. Drawn by Louis Waldecker.*

These charts, in comparison with the Russian manuscript and published materials of the same period, did not advance the hydrographic study of the northern part of the Pacific Ocean significantly. George Davidson, who became head of the US Coast Survey on the west coast, stated that Tebenkov's atlas and hydrographic descriptions of the waters of Russian America became available to the Coast Survey the year after Lieutenant John Rogers's expedition.[438]

At the same time foreign whalers and hunters of fur-bearers continued to hold sway in the waters of the northern part of the Pacific Ocean, and their activity was having an increasingly destructive effect on the environment. The Russian-American Company and the Russo-Finland Whaling Company also suffered serious material losses from this activity. The Russian leadership decided to study the problem and to conclude special international agreements to regulate the harvest of marine mammals in the northern part of the Pacific Ocean. The first unilateral regulations of this type were approved by Grand Duke Konstantin on December 11, 1853, in the form of a "Project for Foreign Ships off the Coasts of Siberia and Russian America."[439]

According to this document foreigners were prohibited from entering the gulfs and bays north of the parallel 54°40′ and from coming within a distance of less than three miles of the coast and islands of the Kodiak archipelago, the Aleutian chain, the Pribilofs, the Komandors, or others situated in the Sea of Okhotsk north of the parallel 45°36′ N. Besides that, putting to shore was not permitted without the permission of the local Russian authorities. An exception was made only in the case of an accident happening to the vessel. However, whaling in the Sea of Okhotsk was permitted, even though it, in the opinion of the commission that had examined the issue, was unacceptable, as it was taking place "geographically in an internal Russian sea."

The unilateral regulations introduced by the Russian authorities were hardly followed at all. However, in 1860 the Russo-Finland Whaling Company officially demanded that the whales be protected. In the company's official petition addressed to Grand Duke Konstantin:

> All these foreigners know that the catch is taking place within the borders of Russia, in waters where it is specifically prohibited, and so, expecting to be banned, thinking that this will be the last summer for them, they act completely imprudently and chaotically in the Sea of Okhotsk; they shoot whales with explosive projectiles, from which many die without any benefit for the one or the other; solely out of jealousy they frighten away each other's whales, and thus the Russo-Finnish whalers, who are the farthest from their homeland, suffer most of all; and, of course, if no restrictions are placed on this mischief and willfulness in time, then within a few years the whales in the Sea of Okhotsk will be completely exterminated.[440]

To end these outrages the Russian whalers suggested completely prohibiting foreigners from hunting in Russian waters, and bringing two Russian naval vessels to permanently guard those waters to enforce the ban.

The commission appointed to investigate this matter in 1860 established that a hunt took place annually by up to 300 vessels under American, British, and French flags in the Sea of Okhotsk and the other waters of the Russian-American Company. The number of Russo-Finnish whaling ships at that same time did not exceed four.[441]

It was decided to draw up clarified regulations for protecting the hunt of the mammal in the northern part of the Pacific Ocean by means of a continuous patrol by naval vessels to study the situation on the spot. In the winter of 1861 an expedition was sent on the screw-propelled corvette *Kalevala*, under the command of Captain-Lieutenant Davydov the 5th, "to gather information about the whaling industry in the northern part of the Pacific Ocean."[442]

The expedition's report, which was presented to the government commission, confirmed the information it had received earlier about the foreigners' abuses in the possessions of the Russian-American Company, where they sometimes amounted to all-out robbery. In 1860 the Americans robbed a warehouse of walrus tusks worth a total of around $15,000. At the same time the head of the expedition established that "for the whole of their long existence our trade has suffered from the interference of American *promyshlenniki*."[443]

Presenting to the commission, naval Counter-Admiral Kozakevich, relying on the international regulations for use in neutral waters, did not consider it possible to completely prohibit the whale hunt in the Sea of Okhotsk. On examination of all the materials that had been presented, the commission formulated some general recommendations. All these issues required a special treaty with the United States and Great Britain, which would envisage "free trade and whaling industry along the entire coast with (Russian) naval ships to observe them and an international commission to settle all difficulties."[444]

It was acknowledged that there was a need to significantly improve the hunting methods used by the Russian-American Company itself. In reply to this, the head office of the Russian-American Company officially announced the following:

> The Company has always been convinced of the need to establish the proper authority over the hunt, and by this means, at present, has succeeded in protecting all species of creatures, and has taken one of them (the stingray) to extraordinary abundance, thus establishing a constant source of gain for the Natives, and along with that also ensuring its own profit from that sector and industry.
>
> These results have been achieved with care and caution, specifically by strictly observing that the catch takes place at no time of year other than the stipulated one, is undertaken by a defined number of *promyshlenniki*, is overseen by experienced foremen, and most importantly takes place in locations appointed by turn, leaving the rest in peace for a specific period of time for the runs . . . Among the Russians who are still living who have sailed on the Company's vessels and also visited the colonies and had occasion to see the Company's navigation, and who will testify to the same, we can name here the admirals: Litke, Baron Wrangell, Mashin, the 1st and 2nd Voevodskys, Etholen, Zavoiko and others."[445]

THE ERA OF ROZENBERG, RUDAKOV, AND FURUHJELM

Nikolai Rozenberg (chief manager, 1850–1853) continued the land and sea investigations. Some of them were associated with searches for the lost expedition of John Franklin. In particular, a British search party arrived at Novoarkhangelsk in October 1850 on the ship *Enterprise*, under the command of Captain Richard Collinson, who was working in polar waters in 1850–1854. Captain Collinson, describing his stay on Baranov Island, talks about the exceptional hospitality of the Russians, who supplied the British with everything they needed (including running rigging), taking no money for the food and goods that were produced in the colonies.

Mikhail Tebenkov showed the British captain all the Russian charts of the northern part of the Pacific Ocean, and also gave him a summary chart of the Russian possessions in Asia and America, which had been engraved in Sitka. Nikolai Rozenberg and his officers provided all possible assistance to the British search party and placed twenty baidarkas and eighteen Aleuts at their disposal for work in the northern waters. Several officers from the *Enterprise* were left in the Russian colonies to investigate the interior regions, one of whom, J. J. Barnard, was killed in February 1851 along with the Russian *baidarshchik* Deriabin (Vasilli) on the Kvikhpak River in an attack on Nulato by Natives. The Natives were offended by Barnard's disrespect toward their leader. Later the fort at Nulato was rebuilt and its garrison was reinforced.[446]

Despite the concerns associated with the Crimean War (1854–1856), skirmishes with the Tlingits (1855),[447] and a typhoid epidemic brought to the colonies by soldiers, Alexander Rudakov (1817–1870), who was manager of the Russian colonies for a year (June 1853–April 1854), and Stepan Voevodsky (1805–1884), who replaced him, leading the colonies in 1854–1859, continued investigations in Russian America. They also started to take control of the coal reserves that had been found by Doroshin.

To organize coal mining, engineer Hjalmar Furuhjelm (1823–1886) came to the colonies with Voevodsky and embarked on a detailed study of and the exploitation of the deposits in the vicinity of English Bay on the Kenai Peninsula.[448] Over the course of several years a mine was built here, a steam pump was installed, and a population formed rivaling the number of residents of Novoarkhangelsk. The quality of the coal turned out to be poor, and although it was used on the steamers of the Russian-American Company, there was no export demand for it in California, due the development of local deposits there and imports of coal from other parts of America.[449]

On June 22, 1859, the new chief manager of the colonies, Captain First Rank Johan Hampus Furuhjelm (1821–1909) arrived in Novoarkhangelsk; he ran the colonies until 1864. The brother of the abovementioned mining engineer Hjalmar Furuhjelm, J. H. Furuhjelm was one of the brilliant representatives of the Finnish diaspora in Russian America. Under his management hydrographic surveys continued *en route* during hunting trips, and special expeditions were also organized to investigate specific areas and to correct inaccuracies and errors that appeared on maps.

Journals of hydrographic surveys have been preserved that were kept on baidarkas by Russian skipper N. Arkhimandritov from July 1 to August 8, 1860. The journals describe "Pine Island, Igrut or Cholitskaya Bay on Afognak Island, Karluk Lake and River and Uyak Bay on the NW side of the island of Kodiak."[450] In 1864 he started on "the surveying and astronomical determinations of the Pribilof Islands, Tebenkof Bay, Norton Sound, and part of the coast up to Cape Stefan and the Main part of Stewart Island."[451]

An examination of these survey journals shows that Arkhimandritov's expedition was very well equipped and carried out detailed hydrographic surveys at an adequate scientific and technical level. For the surveys, "two compasses, a patent log, a magnet, lead and line, a sextant, an artificial horizon and a half chronometer" were used.[452] The journals were presented in the table format which had been finalized by that time, and very detailed descriptions of the work process and the geographical peculiarities of the mapped territory were given for each survey.

In 1861 news of the discovery of gold at the headwaters of the Stikine River reached Novoarkhangelsk. The leadership of the Russian-American Company and Russian diplomats demanded detailed information about those regions. The Russian ambassador to Great Britain, Baron Filipp Brunnov, asked the Russian Minister of Foreign Affairs Alexander Gorchakov to urgently send a "detailed copy of a chart of the localities adjacent to the Stakhin (Stikine) River and our boundaries with the English possessions in North America."[453]

Furuhjelm sent a prospecting party to the Stikine River in 1862. In 1863 the corvette *Rynda* arrived in the colonies under the command of Captain Barsagin to protect Russian interests. On board this vessel the chief manager sent a new expedition to the mouth of the Stikine with a professor of geology from California, William P. Blake. The expedition was divided into three brigades, one for detailed measuring and surveying of the mouth of the river, another was sent for detailed mapping of the valley and channel of the river for thirty miles along its course, and the third was directed to prospect as high up on the upper reaches of the river as was possible in the two weeks allowed for the expedition.

Although the search for gold in 1862–1863 brought only news of rather poor layers of gold-bearing sand in the areas that were investigated, Professor Blake, who had done a successful study of the geology of the southern part of Alaska, concluded that good-sized deposits of this precious metal would inevitably be found in its interior regions.[454] During the expedition by the Russian hydrographers a detailed survey was conducted and a map was made of the Stikine River.[455]

Johan Hampus Furuhjelm did a great deal to improve relations between the Tlingit and the Russians, which had considerably worsened under his predecessor. The chief manager's strong interest in the local tribes, their languages, and their culture, found its reflection in the first thematic ethno-linguistic map known to science. It was created toward the end of his tenure, using different colors for the various languages of the aborigines of Alaska. This *Map of the Native Dialects in the Aleutian Islands and the Northwest Coast of America. From the Map Made by Captain-Lieutenant Verman while in the Service of the Russian-American Co. 1863* was first published by Pyotr Tikhmenev (Fig. 68).[456]

Furuhjelm wanted to increase the level of education and culture of company employees. So, with his encouragement, the head office announced on October 5, 1860, that "desiring by all available methods to provide more resources for those who are serving in the colonies to satisfy their inquisitiveness, we have acquired for this purpose a library of several thousand volumes of various works by the best national and foreign authors, and have sent it on the next Company ship to the colonies, the *Kamchatka*."[457]

The colonial school in Novoarkhangelsk by this time had practically become the only educational institution that was accessible to the ordinary people who worked for the company, especially the Creoles. In 1853 the head office decided to stop sending Creoles to St. Petersburg for training and education.[458] However, the colonial school in Sitka was as effective as the maritime educational institutions of St. Petersburg, as evidenced by the good theoretical and practical knowledge of its graduates, who right from their childhood also received the unique professional skills of "navigator-kayakers."

The *Register of Books for the General Colonial School* gives us an impression of the level of theoretical preparation at the Sitka school. Besides general educational textbooks, there were numerous manuals "for the study of the English language" and dictionaries; special textbooks on mathematics, geography, astronomy, navigation, and maritime practice for the Marine Corps; instructions on sea surveys; various geographical atlases and maps; and also books such as *The Theory and Practice of Ship-Building* and *On the Draining of Bogs*.[459]

On May 26, 1864, the last chief manager of the Russian colonies in America Prince Dmitri Petrovich Maksutov (1832–1889) arrived in Novoarkhangelsk. His unhappy rule was distinguished by intensive geographical investigations within the framework of the search for a route for and the construction of a telegraph link between North America and Asia. The planning and implementation of this expedition was closely connected with the issue of the sale of the Russian transoceanic possessions to the United States.

NOTES

1. N. N. Bolkhovitinov, *Rossiya otkryvaet Ameriku* . . . 1991:186.
2. For more on the role of Shelikhov's widow, see *Natalia Shelikhova: Russian Oligarch of Alaska Commerce*. Edited and translated by Dawn Lea Black and Alexander Yu. Petrov. Fairbanks, 2010.
3. M. E. Wheeler, "The Russian-American Company and the Imperial Government: Early Phase," in *Russia's American Colony*, 1987:43–64, 374–378.
4. *Polny sobranie zakonov Rossiiskoi imperii* (*Complete Collection of Laws of the Russian Empire*) (PSZRI), vol. 25:699–700.
5. PSZRI, vol. 25:700–704; Bolkhovitinov, ibid., 1991:188–189.
6. AVPRI, fond 339, RAK, opis 888, no. 251 (January 18, 1818–January 24, 1819), *Donesenie glavnago pravleniya Rossiiskoi Amerikanskoi kompanii v Ministerstvo vnutrennikh del ob uchastii kreolov na ostrove Kadiake v raznykh rabotakh, proizvodimykh dlya zashchity portov, krepostei . . . pri uchenii k voennoi sluzhbe, k morekhodstvu, torgovle i proch . . .* (*Report of the Chief Management of the Russian-American Company to the Ministry of Internal Affairs on the Participation of Creoles on the Island of Kodiak in Various Jobs Undertaken for the Defense of Ports and Fortresses . . . with Training in Military Service, Navigation, Trade and Others . . .*) Folio 1 and verso. J. Lincoln Starr, *Education in Russian Alaska*, Alaska Historical Library, 1972:2.

7. AVPRI, fond 339, RAK, opis 888, no. 251, folio 2.
8. Richard A. Pierce, ed., *The Russian Orthodox Religious Mission in America, 1794–1837 with Materials Concerning the Life and Works of the Monk German and Ethnographic Notes by Hieromonk Gedeon*. Translated by Colin Bearne. Kingston, 1978:26.
9. Ivanov Bolotov. *Kratkoe opisanie ob Amerikanskom ostrove Kadiake, sobrannoe iz dostovernykh zapisok i raspolozhennoe na topograficheskoe, klimaticheskoe, statisticheskoe i esteticheskoe otdeleniya* (*Short Description of the American Island of Kodiak, Collected from Reliable Notes and Arranged in Topographical, Climatic, Statistical and Aesthetic Sections*). Moscow, 1805.
10. AVPRI, fond Sankt-Peterburgsky Glavny Arkhiv, ii–33, opis 74, 1803, no. 2 *Opisanie ostrova Kadiaka, soobshchennye Novgorodskim mitropolitom Amvrosiem Grafu Nikolaiu Petrovichu Rumiantsevu* (*Description of the Island of Kodiak, Reported by the Novgorod Metropolitan Amvrosii for Count Nikolai Petrovich Rumiantsev*). Folio 1.
11. On this, in greater detail, see Postnikov 1989:46–54.
12. AVPRI, fond St. P., Glavny arkhiv, ii–33, opis 74, 1803, no. 2, verso 2, verso 3.
13. Ibid., verso 4, verso 5.
14. Ibid., verso 8, verso 9.
15. Ibid., verso 9–13.
16. Ibid., verso 14–17.
17. Ibid., folio 4, folio 14.
18. Ibid., folio 3.
19. Ibid., folio 6, folio 7.
20. Ibid., folio 9, folio 14.
21. Ibid., folio 14, folio 15.
22. Ibid., folio 9, folio 13, folio 15, folio 17.
23. Ibid., folio 11, folios 15–18.
24. Ibid., folio 18.
25. Materials tied to the activity of Hieromonk Gedeon have been published in the collection of essays *K stoletnemu iubileyu Pravoslaviya v Amerike (1794–1894). Ocherk iz istorii Amerikanskoi Pravoslavnoi dukhovnoi missii. S prilozheniem 1) Portreta amrikanskogo missionera monakha Germana i ego faksimilie. 2) Vida na ostrove Kadiake. 3) Vida ostrova Elovogo. 4) Rukopisi ieromonakha Gedeona i perepiski—materialov dlya istorii Kadiakskoi missii, ranee ne byvshikh v pechati*. Published by the Valaam Monastery, St. Petersburg, 1894. English language edition: Richard A. Pierce, ed., *The Russian Orthodox Religious Mission in America, 1794–1837 with Materials Concerning the Life and Works of the Monk German and Ethnographic Notes by Hieromonk Gideon*. Translated by Colin Bearn. Kingston, 1978. In the present work we are using the original file *Po predlozheniyam ieromonakha Gedeona . . .* preserved in the Russian State Historical Archive in St. Petersburg. RGIA, fond 796, Kantseliariya Sviateishego Sinoda (Clerical Office of the Holy Synod), opis 90, no. 273 (years 1803–1809).
26. RGIA, fond 796, opis 90, no. 273, 52 verso and folio 53.
27. RGIA, fond 796, opis 90, no. 273, folios 53 and 47.
28. Ibid., folio 51.
29. RGIA, fond 796, opis 90, no. 273, folio 47 and verso.
30. Ibid., folio 48.

31. Ibid., folio 2.
32. RGIA, fond 796, opis 90, no. 273.
33. Ibid., folio 186.
34. A. Prisadsky, *K istorii pravoslavnoi tserkvi v Amerike. 200-letie otkrytiya Alyaski* (*Towards a History of the Orthodox Church in America. The 200th Anniversary of the Discovery of Alaska*). Izdanie Russkogo istoricheskogo obshchestva v Amerike pod predsedatelstvom M. D. Sedykh (Publication of the Russian Historical Society in America under the Chairmanship of M.D. Sedykh). San Francisco, 1943:77.
35. His diaries have been published in English: *Journals of the Priest Ioann Veniaminov in Alaska, 1823 to 1836*. Translated by Jerome Kisslinger, introduction and commentary by S. A. Mousalimas. Fairbanks, 1993.
36. Published in English: Richard A. Pierce, ed., *Notes on the Islands of the Unalaska District*. Translated by Lydia Black and R. H. Geoghegan. Kingston, 1984.
37. For Veniaminov, his works, and the history of the Russian Orthodox Church in Alaska, see Paul D. Garrett, *St. Innocent and the Mission on the Nushagak River. Orthodox Alaska* VII, no. 2, 1979:13–21; Afonsky (Gregory), *A History of the Orthodox Church in Alaska (1794–1917)*, St. Herman's Theological Seminary, Kodiak, Alaska 1977. Lydia Black, "Ivan Pan'kov, an Architect of Aleut Literacy," *Arctic Anthropology*, XIV, no. 1, 1977:94–107. Robert S. Rathburn, "Indian Education and Acculturation in Russian America," *Orthodox Alaska*, VII, nos. 3 and 4, 1979:79–101. Vsevolod Rochcan, *The Origins of the Orthodox Church in Alaska, 1820–1840, Orthodox Alaska* III, no. 1, 1971:1–23 and III, no. 2, 1–15. James W. VanStone, *Eskimos of the Nushagak River: An Ethnographic History*, Seattle and London, 1967. V. Basanoff, "Archives of the Russian Orthodox Church in Alaska in the Library of Congress," *Pacific Historical Review*, II, no. 1, 1933:72–84. John T. Dorosh, "The Alaskan Russian Church Archives," *Quarterly Journal of the Library of Congress*, XVIII, no. 4, 1961:193–203. Library of Congress. *The Alaskan Orthodox Church Archives: Records of the Russian Orthodox Greek Catholic Church of North America—Diocese of Alaska*. Washington Manuscript Division, Library of Congress, 1984.
38. *Tvoreniya Innokentiya Mitropolita Moskovskago* (*The Works of Innokenty, Metropolitan of Moscow*), book three. Collected by Ivan Barsukov. Moscow, 1888:VII–VIII.
39. Ibid., 74, 76.
40. Ibid., 79.
41. Ibid., 95–212.
42. Ibid., 212–213.
43. *Tvoreniya . . .* 1888:279.
44. Ibid., 278–279.
45. Ibid., 281.
46. Ibid., 443–444.
47. Ibid., 445–446.
48. Ibid., 455–456.
49. Ibid., 574.
50. AVPRI, fond RAK, opis 888, no. 997 (microfilm roll 35), January 5–October 12, 1833. *Iskhodyashchie dela Glavnago pravitelya Rossiiskikh kolonii, Flota Kapitana 1ranga*

Barona Vrangelya (*The Outgoing Affairs of the Chief Management of the Russian Colonies, Fleet Captain 1st Rank Baron Wrangell*), 1., 110 verso.

51. Lydia Black, ed. and trans., *The Journals of Iakov Netsvetov. The Atkha Years, 1828–1844*. Kingston, Ontario, 1980:XXVI.
52. G. I. Davydov, *Dvukhkratnoe puteshestvie v Ameriku Morskikh ofitserov Khvostova i Davydov. Pisannoe sim poslednim* (*The Two Journeys to America of Naval Officers Khvostov and Davydov. Written by the Latter*). Part 2, St. Petersburg, 1812:140–144. Translated into English as *Two Voyages to Russian America, 1802–1807*. Translated by Colin Bearne. Kingston, 1977.
53. V. N. Berkh, *Khronologicheskaya istoriya vsekh puteshestvii v Severnuyu polyarnuyu stranu* (*A Chronological History of All the Journeys to the Northern Polar Country*). Part I, St. Petersburg, 1823:129–130 ; M. M. Chernenko and G. A. Agranat, *Lavrenty Alekseevich Zagoskin. Puteshestviya Zagoskina v Russkoi Amerike v 1842–1844 gg.* (*Lavrenty Alekseevich Zagoskin. Zagoskin's Journey in Russian America in 1842–1844*). Moscow, 1956:15, 16; *Pokhod V. Ivanova v glub Alyaski* (*V. Ivanov's Journey to the Interior of Alaska*), *Izvestiya ANSSSR*, Seriya geograficheskaya, no. 3, 1963:97–102. M. B. Chernenko, *Lavrentiy Alekseevich Zagoskin: An Account of his Life and Works. Lieutenant Zagoskin's Travels in Russian America, 1842–1844*. Edited by H. N. Michael, 3–34. Toronto, 1967:9–10, 29–30. S. G. Fedorova, *New Data on Russian Geographic and Ethnographic Investigations in Alaska (First Half of the Nineteenth Century)*. Paper read at the 9th International Congress of Anthropological and Ethnological Sciences, 1973. S. G. Fedorova, *The Russian Population in Alaska and California. Late 18th Century–1867*. Translated and edited by R. A. Pierce and A. S. Donnelly. Kingston, 1973:121; G. I. Davydov, *Two Voyages* 1977:200–201. Pierce, ibid., 1980:117.
54. Davydov, ibid., 1812:25.
55. Davydov, Ibid., 24.
56. Davydov, ibid., 1812:107.
57. N. V. Dumitrashko, "Y. F. Lisyansky and Russian Circumnavigating Voyages." In *Puteshestvie vokrug sveta na korable "Neva" v 1803–1806 godakh* (*A Voyage Around the World on the Ship "Neva" in 1803–1806*), Moscow, 1947:8, 11, 15–22.
58. For a well-researched account of the battle of Sitka from a Native point of view, see Nora Dauenhauer, *Anoooshi lingit ka = Russians in Tlingit America: the battles of Sitka, 1802 and 1804*. Seattle, 2009.
59. Y. F. Lisyansky, *Puteshestvie vokrug sveta na korable Neva v 1803–1806 godakh*. Moscow 1947:158–159.
60. Ibid., 161–162.
61. Lisyansky, ibid., 167.
62. Ibid., 167.
63. Ibid., 173–174.
64. Ibid., 188.
65. Iibid., 177, 192.
66. Ibid., 196.
67. Iibid., 217.
68. AVPRI, fond Morskie karty, opis 504 M, no. 14.

69. Y. F. Lisyansky, *Puteshestvie vokrug sveta v 1803, 4, 5 i 1806 godakh, po poveleniyu Ego Imperatorskogo Velichestva Aleksandra Pervago, na korable Neve, pod nachalstvom Flota Kapitan-Leitenanta, nyne kapitana 1–go ranga i kavalera Yuriya Lisyanskogo* (*A Voyage Around the World in 1803, 4, 5 and 1806, by the Command of His Imperial Majesty Alexander I, on the Ship Neva, under the leadership of Fleet Captain-Lieutenant, now Captain 1st Rank and Decorated Officer Yuri Lisyansky*). Parts 1 and 2, St. Petersburg, 1812.
70. Y. F. Lisyansky, *A Voyage Round the World 1803 to 1806.* Weybridge, 1814.
71. G. H. von Langsdorff, *Bemerkungen einer Reise um die Welt in den Jahren 1803 bis 1807*, II, Frankfurt am Main, 1812.
72. G. H. Langsdorff, *Voyages and Travels in Various Parts of the World during the Years 1803–1807*, 2 vols., London 1814.
73. Langsdorff, *Voyages and Travels*, p. 72; Fedorova, *Russian Population*, 233.
74. P. A. Tikhmenev, *Istoricheskoe obozrenie obrazovaniya Rossisko-Amerikanskoi kompanii i deistvii ee do nastoyashchogo vremeni* (*Historical Review of the Formation of the Russian-American Company and its Activities up to this Day*), Part II, St. Petersburg, 1863. An English edition appeared in two separately published volumes: *The History of the Russian-American Company.* Translated by Richard Pierce and Alton Donnelly. Seattle, 1978; and Richard Pierce, ed., *A History of the Russian-American Company. Vol. 2: Documents.* Translated by Demitri Krenov. Kingston, 1979. N. P. Rezanov, *Pervoe puteshestvie rossiyan okolo sveta* (*The First Russian Voyage Around the World*), *Otechestvennye zapiski* (*Notes from the Fatherland*), X, XI, XII, 1823; XX, 1824.
75. Khlebnikov's Notes. *O Amerike* (*On America*). *Materialy dlya istorii russkikh zaselenii po beregam Vostochnogo okeana*, III, St. Petersbur g, 1861:137. In English as *Colonial Russian America: Kyrill T. Khlebnikov's Reports, 1817–1832*. Translated and edited by Basil Dmytryshyn and E. A. P. Crownhart-Vaughan. Portland, 1976.
76. On the details of the investigations of Bodega Bay and its name changes, see Clinton R. Edwards, "Wandering Toponyms: El Puerto de la Bodega and Bodega Bay," *Pacific Historical Review*, vol. 33, no. 3 (Aug. 1964):253–272.
77. Fedorova, ibid., 1973:134–135.
78. AVPRI, fond 339, RAK, opis 888, no. 251. 18 January 1818–24 January 1819. *Donoshenie Glavnogo Pravleniya Rossiiskoi Amerikanskoi kompanii v MVD ob uchastii kreolov na ostrove Kadiake v raznykh rabotakh, proizvodimykh dlya zashchity portov, krepostei . . . pri uchenii, k voennoi sluzhbe, k morkhodstvu, torgovle i proh.* (*Report by the Chief Management of the Russian-American Company to the Ministry of Internal Affairs on the Participation of Creoles on the Island of Kodiak in Various Work, Undertaken for the Defense of Ports, Fortresses . . . in Military Service, in Navigation, Trade and Others*). Folio 4. 29.
79. AVPRI, fond 339, RAK, opis 888, folio 2.
80. Ibid., folio 2.
81. Ibid., folios 5–16, *Spisok Kreolam Rossisko-Amerikanskoi Kompanii* (*List of Creoles of the Russian-American Company*).
82. Ibid., folio 31, verso 31.
83. Fedorova, ibid., 1973:245–246.

84. A. I. Andreev, *Russkie otkrytiya v Tikhom okeane i Severnoi Amerike v XVIII v.*, Moscow,1948:346.
85. Fedorova, ibid., 1973:247.
86. AVPRI, fond 339, RAK, opis 888, no. 174, folio 2–verso 2.
87. *Puteshestvie na shlyupe Diana iz Kronshtadta na Kamchatku, sovershennoe pod nachalstvom flota leitenanta Golovnina v 1807–1811 gg.* (*Voyage on the Sloop Diana from Kronstadt to Kamchatka, Completed under the Leadership of Fleet Lieutenant Golovnin in 1807–1811*), Moscow 1960:329. Fedorova, ibid., 1973:253. Fedorova published two plans and drawings by Vasiliev from the materials of Vasiliev's work (see Fedorova, ibid., 1973, illustrations 7–10).
88. Fedorova, ibid., 1973:347.
89. RGAVMF, fond 1331, opis 4, no. 132. And fond 1331, opis 1, no. 211.
90. RGAVMF, fond 131, opis 4, no. 133.
91. About Japanese who chanced to come to the Aleutian Islands, see V. M. Konstantinov, *Orosiyakoku Suimudan (Sny o Rossiim)*. Pamyatniki literatury narodov Vostoka (*Literary Documents of the People of the East*). Teksty. Malaya Seriya. XI, Moscow:1961. V. M. Konstantinov, "Svedeniya ob ekipazhe korablya 'Sinse-Maru'" ("Information about the Crew of the 'Sinse-Maru'"). *Problemy vostokovedeniya* (*Problems of Oriental Studies*), no. 3, 1959. S. Nakamura, *Yapontsy i russkie. Iz istorii kontaktov* (*The Japanese and the Russians. From the History of Contacts*). Moscow, 1983.
92. Richard A. Pierce, ed., *Documents on the History of the Russian-American Company*. Translated by Marina Ramsey, Kingston, 1976:192–197.
93. AVPRI, fond 339, RAK, opis 888, no. 210, January 20, 1811. *Predstavlenie Admiralteiskogo departamenta na imya Traverse o poruchenii Golovinu opisaniya severozapadnogo berega Ameriki* (*Declaration by the Admiralty Department in the Name of Travers on the Assignment of Golovin to the Description of the Northwest Coast of America*). Folios 1–4.
94. Ibid., folios 2–3.
95. Otto von Kotzebue, *A Voyage of Discovery into the South Sea and Bering's Straits, for the Purpose of Exploring a North-East Passage, Undertaken in the Years 1815–1818 at the Expense of His Highness the Chancellor of the Empire, Count Rumantzoff, in the ship "Riurik," under the Command of the Lieutenant of the Russian Imperial Navy, Otto von Kotzebue*, vol. 1, London, 1821:193–270.
96. Kotzebue, ibid., vol. 2, 1821:174–177.
97. Pierce, ibid., 1980:120.
98. Otto von Kotzebue, *Puteshestvie v Yuzhny okean i v Beringov proliv dlya otyskaniya severo-vostochnogo morskogo prokhoda, predprinyatoe v 1815, 1816, 1817 i 1818 godakh izhdiveniem ego siyatelstva grafa Nikolaya Petrovicha Rumiantseva . . .* , 3 vols., St. Petersburg, 1821–1823. Otto von Kotzebue, *Entdeckung-Reise in die Sued-See und nach Bering-Strasse zur Erforschung einer nordoestlichen Durchfahrt. Untenommen in den jahren 1815, 1816, 1817 und 1818 auf Kosten Sr. Erlaucht des herrn Reichs-Kanzlers Grafen Rumanzoff auf dem Schiffe Rurich.* 3 vols., Weimar 1821. English translation Kotzebue, ibid., 1–3 vols. London, 1821. Reprint, Amsterdam and New York, 1967. Besides the above Kotzbue work, scientific information and drawings were

published by A. Chamisso and L. Choris. Adalbert von Chamisso, *Reise um die Welt mit der Romanzoffischen Entdeckungs-Expedition in den jahren 1815–18 auf der brig Rurik, Kapitain Otto von Kotzebue*, Leipzig, 1836. English translation: Adalbert von Chamisso, *A Voyage Around the World with Romanzov's Exploring Expedition in the Years 1815–1818 in the Brig "Rurik," Captain von Kotzebue*. Translated and edited by Henry Kratz. Honolulu, 1986. Louis Choris, *Voyage Pittoresque Autour du Monde . . .*, Paris, 1820–1822.

99. Summary charts of the results of Otto von Kotzebue's expedition accompany all foreign-language editions of his work, and three large-scale plans based on the expedition's surveys appeared in Paris in *Novye obzory pouteshestvie* (*New Surveys of Voyages*), as *Carte du Droit de Behring. Plan du Group des Iles Rumanzoff. Cartes des Iles Radack et Ralik*. See *Nouvelle Annales des Voyages 1822, ler serie 13–14*, Paris, 1822. Marvin Falk, *Alaskan Maps: A Cartobibliography of Alaska to 1900*, New York, London, 1983:86.

100. P. A. Tikhmenev, *A History of the Russian-American Company*. Translated and edited by R. A. Pierce and A. S. Donnelly. Seattle 1978:158–159.

101. M. B. Chernenko, "Lavrentiy Alekseevich Zagoskin: An Account of his Life and Works," *Lieutenant Zagoskin's Travels in Russian America in 1842–1844*. Edited by H. N. Michael, Toronto, 1967:3–34. The full text of Korsakovsky's journal has been published in English: James W. VanStone, ed., *Russian Exploration in Southwest Alaska: the Travel Journals of Petr Korsakovskiy (1818) and Ivan Ya. Vasiliev (1829)*. Translated by David H. Kraus. Fairbanks, 1988.

102. AVPRI, fond 404, RAK, no. 963, roll 1. File Microcopies of records in the National Archives, no.11, roll 26, *Records of the Russian-American Company 1802–1867*. Vol. 1, January 11, 1818–April 28, 1819, folio 7.

103. This was the Russian-language edition of the accounts by the English travelers Samuel Hearne (1745–1792) and Alexander Mackenzie (1763–1820), who investigated northwest America and Canada's Arctic coast. Due to these explorations, the possessions of the British Hudson Bay Company extended directly to the borders of Russia's colonies in America. The activities of the Hudson Bay Company were followed with great attention in Russia. The Russian-language publication of the works of Hearne and Mackenzie in 1808 attests, in part, to this. See *Puteshestvie po Severnoi Amerike k Ledovitomu moryu i Tikhomu okeanu, sovershennoe gospodinom Khernom i Makenziem* (*Journey across North America to the Icy Sea and the Pacific Ocean, Completed by Mssrs. Hearne and Mackenzie*), translated and with an introduction by V. N. Berkh, parts 1 and 2, St. Petersburg, 1808. For English publication of these works, see Samuel Hearne, *A Journey from Prince of Wales' Fort in Hudson Bay to the Northern Ocean Undertaken by Order of the Hudson Bay Company, for the Discovery of Copper Mines, North West Passage, in the Years 1769, 1770, 1771 & 1772*, London 1795; Alexander Mackenzie, *Voyages from Montreal, on the River St. Lawrence, through the Continent of North America, to the Frozen and Pacific Oceans, in the Years 1789 and 1793. With a Preliminary Account of the Rise, Progress, and Present State of the Fur Trade of that Country*, London, 1801.

104. AVPRI, fond 404, RAK, no. 963, roll 1, folios 1–7.

105. AVPRI, fond 404, RAK, no. 963, roll 1, folio 8 and verso 8.

106. AVPRI, fond 404, RAK, no. 963, folio 68 and verso 68.

107. State Library of Russia (Rossiiskaya Gosudarstvennaya biblioteka) [further—RGB], Manuscript Division, fond 256, Rumiantsevs, no. 487.
108. Fedorova, ibid., 1973:308; V. N. Berkh, *Khronologicheskaya istoriya vsekh puteshestvii v severnye polyarnye strany*. Part 2, St. Petersburg, 1823.
109. RGB, Manuscript Division, fond 256, Rumiantsevs, no. 487, verso 5.
110. Ibid., folio 10–11 verso.
111. RGB, Manuscript Division, fond 256, Rumiantsevs, folio 12–15 verso.
112. Ibid., 24 verso.
113. Ibid., folio 30, Fedorova, ibid., 1973:309.
114. RGB, Manuscript Division, fond 256, Rumiantsevs, no. 487, verso 39.
115. Ibid., folio 42–verso 43.
116. James W. VanStone, Introduction in *Russian Explorations in Southwest Alaska: the Travel Journals of Petr Korsakovskiy (1818) and Ivan Ya. Vasiliev (1829)*. Edited with an introduction by James W. VanStone, translated by David H. Kraus, Fairbanks, 1988:7
117. RGB, Manuscript Division, fond 256, Rumiantsevs, no. 487, folios 46–verso 46.
118. Ibid., folio 47, Fedorova, ibid., 1973:310.
119. AVPRI, fond 404, RAK, no. 963, roll 1, December 4, 1818. Outgoing correspondence number 286. Proposal to Kadiak Office, verso 131.
120. AVPRI, fond 404, RAK, no. 963, roll 1, verso 131–verso 132.
121. AVPRI, fond 404, RAK, no. 963, roll 1, verso 132–folio 134.
122. AVPRI, fond 404, RAK, opis 888, no. 963, roll 1, folios 134–136 verso.
123. Berkh, ibid., 1823:22–23.
124. S. G. Federova, *New Data on Russian Geographic and Ethnographic Investigations in Alaska (First Half of the Nineteenth Century)*. Paper read at the 9th International Congress of Anthropological and Ethnological Sciences 1973:8. S. G. Fedorova, *The Russian Population of Alaska and California. Late Eighteenth Century–1867*. Translated and edited by R. A. Pierce and A. S. Donnelly. Kingston, Ontario, 1973:68–69.
125. AVPRI, fond RAK, opis 888, no. 964, roll 2. Papers of the Main Office of the Russian-American Company, 1820–1821, verso 175.
126. V. N. Berkh, *Khronologicheskaya istoriya vsekh puteshestvii* . . . II, 1823:22–23. S. G. Fedorova, *The Russian Population* . . . 1973:69.
127. I. Veniaminov, *Zapiski ob ostrovakh Unalashkinskogo otdela* (*Notes on the Islands of the Unalashka Section*), vol. 2, St. Petersburg, 1840:16.
128. Kyrill T. Klebnikov, *Russkaya Amerika v neopublikovannykh zametkakh K. T. Khlebnikova* (*Russian America in the Unpublished Notes of K. T. Khlebnikov*). Compiled with an introduction and commentaries by R. G. Lyapunov and S. G. Fedorova. Leningrad, 1979:221, 276.
129. RGAVMF, fond 1331, opis 4, no. 156.
130. S. G. Fedorova found a manuscript chart of San Francisco Bay in the St. Petersburg department of the Academy of Sciences that is apparently part of Mikhail Rudalev's work. Archive of the Academy of Sciences, St. Petersburg, fond 53, opis 1, no. 34, folio 4. S. G. Fedorova, *The Russian Population* . . . 1973:258, 349.
131. V. A. Esakov, A. F. Plakhotnik and A. I. Alekseev, *Russkie okeanicheskie i morskie issledovaniya. XIX—nachalo XX veka* (*Russian Ocean and Sea Explorations. Nineteenth–early Twentieth Centuries*). Moscow, 1964:69–72.

132. Pierce, ibid., 1980:120–121.
133. *Atlas*... 1964, no. 189:121. RGAVMF, fond 1331, opis 4, no. 45.
134. *Atlas*... 1964, no. 188:121, RGAVMF, fond 1331, opis 4, no. 47.
135. RGAVMF, fond 1331, opis 1, no. 175.
136. AVPRI, fond RAK, opis 888, no. 964, roll 2, papers of the Main Office, 1820–1821. Letter from the Main Office of the RAC to Muraviev dated March 15, 1821, under no. 179. Folios 191–191 verso.
137. AVPRI, fond RAK, opis 888, no. 964, roll 2, 278 verso–279. Letter from the Main Office of the RAC to Muraviev, dated October 6, 1821, under no. 592.
138. In 1822 the dispute about the discovery of Nunivak Island was decided by the Admiralty Department in favor of Vasiliev. The island received the name Otkrytie (Discovery) in honor of Vasiliev's vessel. See A. P. Lazarev, *Zapiski o plavanii voennogo shlyupa Blagonamerennyi*... (*Notes on the Voyage of the Naval Sloop Loyal*). Moscow, 1950:412–420.
139. Fedorova, *The Russian Population*, 1973:257.
140. James W. VanStone, ed., *V.S. Khromchenko's Coastal Explorations in Southwest Alaska, 1822. Fieldiana Anthropology. A Continuation of the Anthropological Series of the Field Museum of Natural History*. Vol. 64. Edited with an Introduction by James VanStone. Translated by David H. Kraus. Chicago, 1973:23–24.
141. (## 1–13/14–18). Besides the above-mentioned publications of materials from Khromchenko's and Etolin's expedition and research on them by VanStone and S. G. Fedorova, excerpts from Khromchenko's journal were utilized by A. F. Burykin. See A. F. Burykin, "Four Voyages of Captain Khromchenko," *Priroda* (*Nature*), no. 3, 1957:77–80.
142. AVPRI, fond RAK, opis 888, no. 966, roll 4, 33 verso, folios 108, 114.
143. Ibid., folio 223.
144. RGVIA, fond VUA, opis 16, no. 23424. RGAVMF, fond 1331, opis 4, no. 176. S. G. Fedorova, who found the copy in Russia's State Archive of the Navy (RGAVMF), was the first to note the significance of this chart. See Fedorova, *The Russian Population* ... 1973:258, 349.
145. RGVIA, fond VUA, no. 23509, part 1.
146. Fedorova, *The Russian Population*..., 1973:258–259, 349. RGAVMF, fond 402, opis 1, no. 2.2.
147. RGAVMF, fond 1331, opis 4, no. 188. A copy published in *Atlas*... 1964, no. 190:121.
148. Marvin W. Falk, *Alaskan Maps. A Cartobibliography of Alaska to 1900*. New York and London, 1983:87.
149. AVPRI, fond 404, RAK, opis 888, no. 966, roll 4, folios 165–166.
150. Ibid., folio 166.
151. AVPRI, fond 404, RAK, opis 888, no. 966, roll 4, folio 213.
152. Ibid., folio 395. Letter no. 47, dated March 17, 1825, from the Ministry of the Interior of Russia—*O sodeistvii k oboigashcheniyu Imperatorskogo Botanicheskogo sada rasteniyami i prochim* (On assistance in the enrichment of the Imperial Botanical gardens with plants and others).
153. AVPRI, fond 404, RAK, opis 888, no. 966, roll 4, folio 121.
154. Ibid., roll 5, folio 208–verso 208 (March 31, 1827).

155. Cit. by L. P. Kirwan, *A History of Polar Exploration*, New York, 1960:77.
156. Cited after Wendell H. Oswalt, *Eskimos and Explorers*, Novato, California 1979:174.
157. Wendell H. Oswalt, *Eskimos and Explorers*, Novato, California, 1979:169–175. In greater detail, see William E. Parry, *Journal of a Voyage for the Discovery of the North-West Passage*, London, 1821; William E. Parry, *Journal of a Second Voyage for the Discovery of the North-West Passage*, London, 1824. William E. Parry, *Journals of the First, Second and Third Voyages for the North-West Passage*, 5 Volumes. London, 1828.
158. In Parry, ibid., 1824, and G. F. Lyon, *The Private Journal of Captain G. F. Lyon of H.M.S. Hecla*. . . London, 1824. The chart is reproduced in Coolie Verner, *Explorers' Maps of the Canadian Arctic 1818–1860*. Cartographica, Toronto, 1972, no. 23.
159. Barry M. Gough, editor and author of the Introduction, *To the Pacific and Arctic with Beechey: The Journal of Lieutenant George Peard of H.M.S.* Blossom *1825–1828*. Cambridge, 1973:9–12.
160. Gough, ibid., 1973:22.
161. Gough, ibid., 1973:47.
162. F. W. Beechey, *Narrative of a Voyage to the Pacific and Bering's Strait to Co-operate with the Polar Expeditions; Performed in His Majesty's Ship* Blossom, *under the Command of F. W. Beechey, R.N. F.r.s. & c., in the Years 1825, 26, 27, 28. Published by Authority of the Lords Commissioners of the Admiralty*. Vol. 1, London, 1831:326.
163. Beechey, ibid., 1831:329.
164. Beechey, ibid., 1831:332–333.
165. Beechey, ibid., 1831:337–338.
166. Beechey, ibid., 1831:342–351.
167. Beechey, ibid., 1831:356–369.
168. Beechey, ibid., 1831:377–392.
169. This observation by Beechey is incorrect, as the Eskimo island's name Aiak means "broken off and cast away from the shore," i.e., a separate rock, a base or a column, Ray, ibid., 1975:82.
170. Beechey, 1831: 399–401.
171. Leo Bagrow, *History of Cartography*. Revised and enlarged by R. A. Skelton, second edition, Chicago, 1985:27.
172. Frederick W. Beechey, *Narrative of a Voyage to the Pacific and Bering's Strait*. . . vol. 2, London, 1831:255.
173. Beechey, ibid., vol. 2, 1831:256–290.
174. Berry M.Gough *To the Pacific and Arctic*. . . p. 47.
175. Beechey, ibid., vol. 1, 1831, map between pages 354–355. This map was first published separately in 1830 by the British Admiralty as *Chart of Part of the North West Coast of America, from Point Rodney to Point Barrow. By Captain F. W. Beechey, Admiralty*. London, 1830. Falk, Cartobibliography., 1983:97.
176. Sir John Richardson, *Fauna Boreali-Americana, or Zoology of the Northern Parts of British America*. 4 volumes. London, 1829–1837.
177. Sir John Richardson, et al. *The Zoology of Captain Beechey's Voyage; compiled from the collections and notes made by Captain Beechey, the officers and naturalist of the expedition* . . . . London: Lords Commissioners of the Admiralty, 1839.

178. Sir William Jackson Hooker and George A. Walker Arnott, *The Botany of Captain Beechey's Voyage*. 1841. Reprint, London, 1965.
179. Ray, ibid., 1975:77–86.
180. Thomas Simpson, *Narrative of the Discoveries on the North Coast of America; Effected by the Officers of the Hudson's Bay Company during the Years 1836–1839. By Thomas Simpson, Esq.*, London, 1843:149.
181. Simpson, ibid., 1843:127–167.
182. Ibid. Introduction by Alexander Simpson: xix, xvi.
183. J. H. Pelly. *An Account* . . . 1837: 213–225. Facsimile published in V. Coolie, *Explorers Maps* . . . no. 60.
184. R. A. Thompson, *Russian Settlement in California, Fort Ross, Sonoma County*. Santa Rosa, CA 1896:3–8.
185. AVPRI, fond 339, RAK, opis 888, no. 236, 1816–1830. *Po doneseniyu Glavnogo Pravleniya Rossiisko-Ameikanskoi kompanii ob otpravlenii eskadry v Rossiisko-Amerikanskie kolonii, sostoiashchei iz dvukh korablei Kutuzov i Suvorov i o vozvrashchenii korablya Suvorov k 1818 godu (In Accordance with the Dispatch of the Head Office of the Russian-American Company on Sending Fleets to the Russian-American Colonies Consisting of the Two Ships Kutuzov and Suvorov and on the Return of the Ship Suvorov by 1818*). Folios 2, 42.
186. RGAVMF, fond 1331, opis 1, no. 221. Published in Fedorova, ibid., 1971, and Fedorova, ibid., 1973: illustr. 13.
187. RGAVMF, fond 1331, opis 1, no. 221, *Obyasnenie* (*Explanation*).
188. RGAVMF, fond 1331, opis 4, no. 154. There exists yet another version of this chart, executed in black-and-white with orientation to the south "from the original, taken from the American Company . . . at the Draftsroom of the State Admiralty Department in the Year 1818, December 11," RGAVMF, fond 1331, opis 4, no. 153.
189. RGAVMF, fond 1331, opis 4, no. 155.
190. RGAVMF, fond 1331, opis 4, no. 151.
191. Russia's State Historical Archive, St. Petersburg RGIA, fond 994 (Mordvinovy), opis 2, no. 834. *Zapiska Glavnogo Pravleniya Rossiiskoi Amerikanskoi kompanii v Sovet, uchrezhdenny pri Kompanii o zaselenii Kolonii Ross osedlymi khlebopashtsami i opredelenii granits etoi Kolonii. Zapiska o pravakh Kompanii na vladenie koloniey Ross. 1819 i 1825 gody (Note from the Head Office of the Russian-American Company to the Council Established within the Company on the Settlement of the Ross Colony by Non-Nomadic Farmers and the Determining of the Borders of that Colony. Note on the Company's Rules for Managing the Ross Colony. 1819 and 1825)*. Folios 17–19.
192. Thompson, ibid., 1896:13.
193. For a biography of Litke, see A. I. Alekseev, *Fedor Petrovich Litke*. Translated from the Russian edition of 1970 by Serge LeComte and edited by Katherine Arndt. Fairbanks, 1996.
194. RGAVMF, fond 1331, opis 4, no. 150, folios 1–3.
195. Frederic Litke, *A Voyage Around the World, 1826–1829, by Frederic Litke*. Vol. 1, "To Russian America and Siberia," translated from the French edition by Renee Marshall, supplemented with a parallel account by E. H. Baron von Kittlitz, translated from the

German with an Introduction by Joan Moessner. Edited by Richard A. Pierce. Kingston, Ontario, 1987: i–ii.

196. F. P. Litke, *Puteshestvie vokrug sveta, sovershennoe po poveleniyu Imperatora Nikolaya I, na voennom shlyupe "Senyavin" v 1826–1829 gg. Flota kapitanom Fyodorom Litke. Otdelenie Istoricheskoe* (*A Voyage around the World, Completed by Command of Emperor Nicholas I, on the Naval Sloop Senyavin in 1826–1829. By Fleet Captain Fyodor Litke. Historical Department*). 3 volumes with an atlas. St. Petersburg, 1834, 1835. F. P. Litke, *Puteshestvie vokrug sveta sovershennoe po poveleniyu Imperatora Nikolaya I, na voennom shlyupe "Senyavin" v 1826, 1827, 1828 i 1829 godakh . . . Otdelenie morekhodnoe* (*A Voyage Around the World, Completed by Command of Emperor Nicholas I, on the Naval Sloop Senyavin in 1826, 1827, 1828 and 1829 . . . Nautical Department.*) With an atlas.

197. F. H. von Kittlitz, *Vierundzwanzig Vegetations-Ansichten* . . . Friedrichsche Verlagsbuchhandlung, 2 vols., Siegen und Wiesbaden, 1844. Friedrich Heinrich von Kittlitz, *Denkwuerdigkeiten einer Reise nach dem russischen Amerika*. Vol. 1, Gotha, 1858. Friedrich Heinrich von Kittlitz, *Twenty-Four Views of the Vegetation of the Coast and Islands of the Pacific, with Explanatory Descriptions taken during the Exploring Voyage of the Russian Corvette "Senjawin" under the Command of Captain Luetke, 1827–1829*, London, 1861.

198. Pierce, ibid., 1980:122.

199. Litke, ed. By R. A. Pierce, ibid., 1987:ii.

200. Fredric Litke, *Atlas du voyage autour du monde de la corvette Seniavine fait en 1826, 1827, 1828 et 1829 sons les ordres de Frédéric Luetke du membre du Comite Scientifique de l'Etat Major de la Marine, de l'Academie Imperiale des Science, de la Société Geographique de Londres* . . . Hydrographic Depot, Naval General Staff H.I.M., St. Petersburg, 1832.

201. AVPRI, fond RAK, opis 888, no. 125 (1797–1839), *Bumagi naidennye posle smerti grafa Speranskogo, otnosyashchiesya do torgovli . . . Obozrenie sostoianiya i deistvii Rossiskoi Amerikanskoi kompanii s 1797 po 1819 god* (*Papers Found After the Death of Count Speransky, Relating to Trade . . . Review of the Condition and Operations of the Russian-American Company from 1797 to 1819*). Folio 162.

202. AVPRI, fond RAK, opis 888, no. 125. *Primechaniya o Kamchatskoi torgovli, uchinennye G. Langsdorfom* (*Notes on Kamchatka Trade, Made by G. Langsdorf*), folio 161–161 verso.

203. AVPRI, fond RAK, opis 888, no. 125 (V. M. Golovnin) *Obozrenie sostoianiya i deistvii Rossiskoi Amerikanskoi Kompanii* . . . Year 1819, folio 226.

204. AVPRI, fond 404, RAK, opis 888, no. 993, roll 31, Outgoing correspondence, January 2, 1828–December 12, 1828. Folios 8–9.

205. Ibid., folio 111, 118.

206. Ibid., folio 194.

207. Ibid., 194–verso 194.

208. AVPRI, fond 404, RAK, opis 888, no. 993, roll 31, folios 288–289.

209. Fedorova, ibid., 1973:254.

210. AVPRI, fond 404, RAK, opis 888, no. 993, roll 31, folios 478–481.

211. Cited after Fedorova, ibid., 1973:254.

212. Ibid., 254–256.
213. Ibid., 256.
214. RGAVMF, fond 1331, opis 4, no. 243.
215. AVPRI, fond RAK, opis 888, no. 994, roll 32 (January 7–December 30, 1830), verso 137.
216. Ibid., folio 172, 174 verso.
217. Ibid., folios 268–271.
218. Ibid., folios 245–246 verso.
219. James W. VanStone, ed., *Russian Exploration in Southwest Alaska. The Travel Journal of Petr Korsakovskiy (1818) and Ivan Ya. Vasiliev (1829)*. Edited with an Introduction by James W. VanStone. Translated by Davis H. Kraus, Fairbanks, 1988; S. G. Fedorova, "Shturmany Ivany Vasilievy i ikh rol v izuchenii Alyaski (pervaya polovina XIX v.)" ("Ivan Vasiliev's Navigators and their Role in the Study of Alaska [First Half of the Nineteenth Century]"), Part 1, *Letopis Severa* no. 10, 1982:195–198.
220. F. P. von Wrangell, *The Inhabitants of the Northwest Coast of America*. Translated and edited by J. W. VanStone, *Arctic Anthropology* 6(2):13–19, 1970; VanStone, ibid., 1988:13.
221. Fedorova, ibid., 1973:256. About Vasiliev's expedition, see also S. G. Fedorova, ibid., 1979:167–211, part 2, *Letopis Severa*, no. 10, 1982: 141–160; B. P. Vishnevsky, "Iz istorii izucheniya rek i ozer Severo-Zapadnoi Ameriki" ("From the History of the Study of the Rivers and Lakes of Northwest America"), *Izvestiya Vsesoyuznogo Geograficheskogo obshchestva*, vol. 94, no. 6, 1962:507–510.
222. AVPRI, fond RAK, opis 888, no. 994, roll 32, 30 April 1830, folios 17–19.
223. Ibid., folio 67.
224. Ibid., 67 verso.
225. Ibid., folios 230–234.
226. AVPRI, fond RAK, opis 888, no. 994, roll 32, folios 277–279 verso.
227. Ibid., no. 995, roll 33, May 26, 1831, folios 247, 250–251 verso.
228. Ibid., 321 verso.
229. F. P. Litke, *Puteshestvie vokrug sveta, sovershennoe po poveleniyu imperatora Nikolaya I, na voennom shlyupe Senyavin v 1826, 1827, 1828 i 1829 godakh* . . . Otdelenie morekhodnoe s atlasom, St. Petersburg, 1835.
230. National Historical Archive of Estonia, Tartu, fond 1414, Krusenstern Family Papers, opis 2, no. 43, folios 97, 100. Fedorova, ibid., 1979:195–198.
231. RGAVMF, fond 1331, opis 4, no. 245. [My research shows that this chart was drawn by Veniaminov and used by Litke in his atlas. Veniaminov sent it to Litke, they corresponded.—LB]
232. Ibid., no. 244.
233. Fedorova, ibid., 1979:198.
234. The two atlases by Krusenstern are *Atlas iuzhnago moria, sochinennyĭ flota kapitan" komandorom" Kruzenshternom*. 2 vols. Saint Petersburg, 1824–1826; and *Atlas de l'ocean Pacifique dressé par M. de Krusenstern*. 2 vols. in 1. St. Petersburg, 1827 [i.e., 1835?].
235. AVPRI, fond 404, RAK, opis 888, no. 996, roll 34, folio 62, letter no. 89, dated April 22, 1832.

236. Ibid., folios 59–60 verso.
237. Ibid., folios 19–19 verso.
238. Ibid., folios 251–253 verso.
239. National Archive of Estonia, Tartu. Fond 2057 (Wrangell Family Papers), opis 1, no. 527, 533. Fedorova, ibid., 1979:199–201.
240. F. P. Litke, *Puteshestvie vokrug sveta, sovershennoe po poveleniyu Imperatora Nikolaya I, na voennom shlyupe "Senyavin," v 1826, 1827, 1828 i 1829 godakh* . . . Otdelenie morekhodnoe s atlasom, St. Petersburg, 1835.
241. AVPRI, fond RAK, opis 888, no. 970, roll 8, folios 30–31.
242. Ibid., folios 34–34 verso.
243. Ibid., folios 312–314.
244. Ibid., roll 34, folios 243–243 verso.
245. Ibid., no. 996, roll 34, folios 247–247 verso.
246. Ibid., no. 970, roll 8, folio 216.
247. Ibid., folios 178–179.
248. Ibid., no. 996, roll 34, folio 381. Letter to the head office dated October 31, no. 471.
249. Ibid., folio 381 verso–382 verso.
250. Ibid., folio 479 verso, 484, 486–486 verso, 482.
251. ARGO, fond Category 99, opis 1, no. 61. F. P. von Wrangell, *Extract from the Travel Journal of Assistant Navigator Andrei Glazunov (Creole by heritage, taught in the colonies) from December 30, 1833 to April 13, 1834*. Folio 2.
252. AVPRI, fond 404, RAK, opis 888, no.s 997, roll 35, folios 95–98.
253. ARGO, fond razryad 99, opis 1, no. 61. Wrangell, Extract. Folio 2. For the English version, see A. Glazunov, "Extract of a Diary of Ship's Mate Andrei Glazunov, Written During a Journey into the Northwest American Interior," in *Russian America Statistical and Ethnography Information By Rear Admiral Ferdinand Petrovich Wrangell* . . . , 69–89; and A. Glazunov, "Russian Exploration in Interior Alaska. An Extract From the Journal of Andrei Glazunov." Edited by James W. VanStone. *Pacific Northwest Quarterly* 50, no. 2 (1959):37–47.
254. ARGO, fond razryad 99, opis 1, no. 61, folios 2–2 verso.
255. Ibid., folio 2, verso.
256. AVPRI, fond 404, RAK, opis 888, no. 1039, roll 77. Journal of the Activities of the Northern Overland Expedition under the Command of Lieutenant Zagoskin. Folio 152.
257. Ray, ibid., 1975:124.
258. ARGO, fond razryad 99, opis 1, no. 61, folio 3, verso 4.
259. Ibid., Folio 9 verso.
260. Ibid., Folio 12, verso 14.
261. Ibid., Folios 15–16.
262. Ibid., folios 18–20.
263. Ibid., folios 21, verso 25.
264. Ibid., folio 26, verso 27.
265. Ibid., folios 28, verso 28.
266. Ibid., folios 29, verso 29.
267. *Karta k puteshestviyu Andreya Glazunova vo vnutrennost Ameriki v 1833–1834 gg.* (*Andrei Glazunov's Map of his Journey into the Interior of America, 1833–1834*). F. P. von

Wrangell, *Statistische unde Ethnographische Nachrichten ueber die Russischen besitzungen an der Nordwestkueste von Amerika. Beitraege zur kenntniss des Russischen Reiches*, vol. 1, St. Petersburg, 1838. In English translation as *Statistical and Ethnographic Information*. Translated by Mary Sadouski and edited by Richard Pierce. Kingston, 1980.

268. AVPRI, fond 404, RAK, opis 888, no. 970, roll 8, *Letter from the Main Office*, dated March 31, 1833, no. 260, folios 312–314.
269. Ibid., roll 36, Communications Sent, Baron Wrangell, January 7–August 31, 1834. Folios 39–42.
270. Ibid., folios 91, versos 92, 94.
271. Ibid., no. 997, roll 35, folio 118, verso 126.
272. Ibid., folio 125.
273. Ibid., folio 125 verso.
274. Ibid., folio 126.
275. ARGO, fond razryad 99, opis 1, no. 14. *Zhurnal pravitelya Novo-Arkhangelskoi kontory*. 1818–1823 gg. (*Journal of the Manager of the Novo-Arkhangel'sk Office. 1818–1823*). Folio 31, verso 43.
276. For more detailed treatment, see A. V. Postnikov, *Razvitie krupnomasshtabnoi kartografii v Rossii* (*The Development of Large-Scale Cartography in Russia*), Moscow, 1989:118–143.
277. AVPRI, fond 404, RAK, opis 888, no. 970, roll 8, folios 432–433.
278. Ibid., no. 997, roll 35, folios 115, verso 115.
279. Ibid., folio 132.
280. Ibid., folios 140–141.
281. Ibid., folios 106–106 verso, 110 verso.
282. Ibid., folios 102, verso 103.
283. Ibid., folio 139.
284. Ibid., roll 36, *Iskhodyashchie dela barona Vrangelya, 7 Yanvarya–31 Avgusta 1834 g.* (*Outgoing Business of Baron Wrangell, January 7–August 31, 1834*). Folios 98–99.
285. Ibid., folio 174.
286. Ibid., folios 187–189.
287. Ibid., folios 238–243 verso.
288. Fedorova, ibid., 1971:130.
289. Fedorova, ibid., 1971:129.
290. Fedorova, ibid., 1979:205.
291. AVPRI, fond RAK, opis 888, no. 998, roll 36, folios 294–297.
292. Wrangell, *Russian America*, 1839.
293. James Kari and James A. Fall, eds. and comps., "The Russian Presence in Upper Cook Inlet." *Shem Pete's Alaska: The Territory of Upper Cook Inlet Dena'ina*. Fairbanks, 1987:17.
294. AVPRI, fond RAK, opis 888, no. 972, roll 10, folio 97 verso (from a dispatch by the head office of the RAC to Kupriyanov dated March 12, 1836).
295. Ibid., no. 1000, roll 38, March 3, 1836, folio 10, verso 11.
296. Ibid., no. 359, *O sostavlenii karty na Alyasku poruchikom Voronkovskim* (*On the Creation of a Map of Alaska by Lieutenant Voronkovsky*). Folios 2–3. Photocopy. Original in RGAVMF, fond Hydrographic Department, no. 870 (1837), folios 1–6.

297. Khlebnikov, Kyrill T. *Russkaya Amerika v Zapiskakh Kirila Khlebnikova*. Compiled, with an introduction and commentaries by S. G. Fedorova. Executive editor, V. A. Aleksandrov. Moscow, 1985:286.
298. Ibid., no. 972, roll 10, folio 395 (from a secret letter from the head office of the RAC to Kupriyanov dated March 19, 1837).
299. AVPRI, fond RAK, opis 888, no. 972, roll 10, folios 396, verso 399, *O polyarnoi ekspeditsii* (*On the Polar Expedition*), letter from the head office of the RAC dated March 19, 1837.
300. Ibid., folios 400, verso 400.
301. Ibid., folios 401–402.
302. Ibid., folio 402, verso 403.
303. Ibid., folios 403, verso 404.
304. Ibid., folio 405.
305. Ibid., folios 455, verso 455 (letter from the head office to Kupriyanov, no. 563, dated April 2, 1837).
306. Fedorova, ibid., 1971:235.
307. James W. VanStone, Introduction, *A. F. Kashevarov's Coastal Explorations in Northwest Alaska, 1838*, Edited with Introduction by James W. VanStone, translated by David H. Kraus. *Fieldiana Anthropology*, Field Museum of Natural History, vol. 69, Chicago, 1977:13.
308. L. A. Zagoskin: *Lieutenant Zagoskin's Travels in Russian America, 1842–1844. The First Ethnographic and Geographic Investigations in the Yukon and Kuskokwim Valleys of Alaska*. Edited by Henry N. Michael and translated by Penelope Rainey. Toronto, 1967:124. Translated from *Puteshestviya i issledovaniya leitenanta Zagoskina v Russkoi Amerike 1842–1844 gg*. Moscow, 1956.
309. VanStone, ibid., 1977:14–16; on A. F. Kashevarov and his contribution to the study of Northwest Alaska, see also B. A. Lipshits, "Kashevarov as an Alaskan Explorer" ("Kashevarov kak issledovatel Alyaski"), *Sovetskaya etnografiya*, no. 1, 1952:175–178; B. A. Lipshits, "On Collections in the Museum of Anthropology and Ethnography Assembled by Russian Travelers and Explorers in Alaska and California" ("O kollektsiyakh Muzeia antropologii i etnografii, sobrannykh russkimi puteshestvennikami i issledovatelyami Alyaski i Kalifornii), *Sbornik Muzeia antropologii i etnografii*. Anthology of the Museum of Anthropology and Ethnography, 16, 1953:358–369; S. G. Fedorova, ibid., 1971:193–194; Fedorova, ibid., 1973:338–339; Ray, ibid., 1975:131–134.
310. AVPRI, fond RAK, opis 888, no. 974, roll 12. Incoming correspondence March 31–August 18, 1839, folio 163 verso. Three volumes of the journal *Syn Otechestva* (*Son of the Fatherland*) for the months of August 1838, January and February 1839, containing the description of the voyage of Dease and Simpson, were appended to the letter.
311. Ibid., folios 344–346 verso.
312. A. F. Kashevarov, "Obozrenie beregov Severnoi Ameriki i mysa barrova sovershennoe Rossiiskoi ekspeditsiei v 1838 godu" ("Overview of the Coasts of North America and Cape Barrow Completed by a Russian Expedition in 1838"). *Syn otechestva* no. 1, 1840:127–144; A. F. Kashevarov, "Otryvki iz dnevnika, vedennago pri obozrenii poliarnogo berega Rossiiskoi Ameriki po porucheniyu Rossiisko-Amerikanskoi kompanii v 1838 godu" ("Excerpts from a Diary Kept during the Overview of the Polar

Coast of Russian America by Order of the Russian-American Company in 1838"). *Sankt Peterburgskie vedomosti*, no. 190–192, 195; A. F. Kashevarov, "Zametki ob eskimosakh v Russkoi Amerike" ("Notes on the Eskimos in Russian America"). *Severnaya pchela* (*The Northern Bee*), no. 227, 1846:907–908; no. 228, 1846:911–912; A. F. Kashevarov, "Zhurnal, vedennoi pri baidarochnoi ekspeditsii, naznachennoi dlya opisi severnogo berega Ameriki, 1838 goda . . ." ("Journal Kept during a Baidar Expedition Sent to Describe the Northern Coast of America, 1838 . . .") Imperatorskoe geograficheskoe obshchestvo (IRGO) (Imperial Geographical Society), *Zapiski po obshchei geografii* (*Notes on General Geography*), 8, St. Petersburg, 1879:275–361.

313. *Obshchii Morskoi Sbornik* (*General Maritime Anthology*), vol. 12, appendix, 1900:496.
314. Edward Belcher, *Narrative of a Voyage Round the World, Performed in Her Majesty's Ship "Sulphur," during the Years 1836–1842* . . . 2 vols., London: Henry Colburn, 1843.
315. Richard A. Pierce, *The Russian Governors: Builders of Alaska, 1818–1867*, Kingston, Ontario, 1986:18–22.
316. RGAVMF, fond 1331, opis 3, no. 61, folio 20. First published in Fedorova, ibid., 1971, fig. 14.
317. Wrangell, ibid., 1839. In English translation: Richard Pierce, ed., *Russian America. Statistical and Ethnographic Information.* Translated by Mary Sadouski. Kingston, 1980. See Yu. V. Davydova, *Ferdinand Vrangel* (*Ferdinand Wrangell*), Moscow, 1959 for further details on Wrangell's life and work.
318. Wrangell, ibid., 1980. Introduction by R. A. Pierce, v–vi.
319. Also quinoa or rice goosefoot (Chenopodium quinoa). An annual from the genus Chenopodiaceae, it is cultivated high in the mountains (4,000 m) of Peru, beyond the area of cultivation of bread grains. The small floury fruits are used in gruel and also for the production of flour. The young leaves are used like spinach.
320. AVPRI, fond RAK, opis 888, no. 974, roll 12, folios 188–189.
321. C. Grewingk, *Beitrag zue Kenntniss der orographischen und geognostischen Beschaffenheit der Nord-West Kueste Amerikas mit den anliegenden Inseln.* Von Dr. C. Grewingks (Aus verhandlungen der Mineralogischen Gesellschaft zu St. Petersburg, für die Jahre 1848–49 besonders abgedruckt.). St. Petersburg, 1850. Translated into English and published as Marvin Falk, ed., *Grewingk's Geology of Alaska and the Northwest Coast of America, with the Adjacent Islands.* Translated by Fritz Jaensch. Fairbanks, 2003.
322. On Voznesensky and his collections, see also M. V. Stepanova, "I. G. Voznesensky i etnograficheskoe izuchenie Severo-Zapada Ameriki" ("I. G. Voznesensky and the Ethnographic study of Northwest America"), VGO, vol. 76, number 5, 1944; B. A. Lipshits, "Etnograficheskie materialy po Severo-Zapadnoi Amerike v arkhive Voznesenskogo" ("Ethnographic materials on Northwest America in Voznesensky's Archive"), *Izvestiya VGO*, vol. 82, no. 4, 1950; E. E. Blomkvist, "Risunki Voznesneskogo (ekspeditsii 1839–1849 gg.)" ("Voznesensky's drawings [expedition of 1839–1849]"), *Sbornik Muzeia Antropologii i Etnografii*, vol. 13, Leningrad 1951:22, 23, 83, 184, 186; R. G. Liapunova, "Ekspeditsiya I. G. Voznesenskogo i ego znachenie dlya etnografii Russkoi Ameriki" ("I. G. Voznesensky's Expedition and its Significance for the Ethnography of Russian America"), *Sbornik Muzeia Antropologii i Etnografii*, vol. 24, Leningrad, 1967; A. I. Alekseev, *Ilya Gavrilovich Voznesensky*, Moscow 1977; A. I. Alekseev, *Russkie geograficheskie issledovaniya na Dal'nem Vostoke i v Severnoi Amerike, 19–nachala 20*

v. (*Russian Geographical Research in the Far East and in North America, Nineteenth–Early Twentieth Centuries*), Moscow: Nauka, 1976:17–19; A. I. Alekseev, *The Odyssey of a Russian Scientist: I. G. Voznesenskii in Alaska, California and Siberia 1839–1849*. Edited by Richard A. Pierce and translated by Wilma C. Follette, Kingston, 1987.

323. Blomkvist, "Risunki . . ." 1951:247.
324. Blomkvist, ibid., 1951:247.
325. Lawrentii A. Zagoskin, *Lieutenant Zagoskin's Travels in Russian America 1842–1844. The First Ethnographic and Geographic Investigations in the Yukon and Kuskokwim Valleys of Alaska*. Edited by H. N. Michael. Introduction by James W. VanStone, Toronto, 1967:14–15; on Zagoskin and his expedition, see also M. B. Chernenko, and G. A. Agranat), "Ocherk zhizni i deyatelnosti" ("Essay on Life and Works") in *Puteshestvie Leitenanta Lavrentiya Zagoskina v Russkoi Amerike v 1842–1844 gg.* (*The Travels of Lieutenant Lavrenty Zagoskin in Russian America in the Years 1842–1844*), Moscow, 1956:16–17; B. A. Lipshits, "Zagoskin kak issledovatel etnografii Alyaski" ("Zagoskin as an Ethnographic Explorer of Alaska"), in *Puteshestvie i issledovaniya leitenanta Lavrentiya Zagoskina v Russkoi Amerike v 1842–1844 gg.*, Moscow, 1956.
326. AVPRI, fond RAK, opis 888, no. 1039, roll 77: *Zhurnal deistvii severnoi sukhoputnoi Ekspeditsii pod Nachalstvom leitenanta Zagoskina. V 1842, 1843 I 1844 gg* (*Journal of the Activities of the Northern Overland Expedition under the Command of Lieutenant Zagoskin. In 1842, 1843 and 1844*). Folio 3.
327. Ibid., folios 10–10 verso.
328. Ray, ibid., 1975:130.
329. AVPRI, fond RAK, opis 888, no. 1039, roll 77, folio 16 verso.
330. Ibid., folios 26–27.
331. Ibid., folio 27 verso.
332. Ibid., folios 19, 29 verso, 18 verso.
333. Ibid., folio 35.
334. Ibid., folios 38–40.
335. Ibid., folio 42 verso–43 verso.
336. Ibid., folio 53.
337. Michael, ed., ibid., 1967:33, note 77.
338. AVPRI, fond RAK, opis 888, no. 1039, roll 77, folio 56.
339. Ibid., folio 58 verso.
340. Ibid., folio 60 verso.
341. Ibid., folio 91.
342. Ibid., folio 60.
343. Ibid., folios 116–117.
344. Ibid., folios 60 verso–61.
345. Ibid., folios 73–73 verso.
346. Ibid., folios 74, 79 verso.
347. Ibid., folio 80 verso.
348. Ibid., folio 84.
349. Ibid., folio 99.
350. Ibid., folios 97–97 verso.
351. Ibid., folios 108–108 verso.

352. Ibid., folios 140 verso, 142 verso.
353. Ibid., folios 101, verso 102.
354. Ibid., folio 103 verso.
355. Ibid., folios 125, verso 126.
356. Ibid., folios 138–139 verso.
357. Ibid., folio 141.
358. Ibid., folio 142.
359. Ibid., folios 142–143.
360. Ibid., folios 147–147 verso.
361. Ibid., folios 147 verso–148.
362. Ibid., folios 148 verso–149.
363. Ibid., folios 157–157 verso.
364. Ibid., folio 159.
365. Ibid., no. 1012, roll 77, folios 61–61 verso.
366. L. A. Zagoskin, "Redut Sv. Mikhaila na yuzhnom beregu zaliva Norton" ("St. Michael's Redoubt on the South Coast of Norton Sound"), *Zapiski Gidrograficheskago departamenta*, vol. 4, St. Petersburg, 1846:86–101.
367. S. I. Zelenoi, "Izvlechenie iz dnevnika leitenanta Zagoskina, napisannogo v ekspeditsii" ("Excerpt from the Diary of Lieutenant Zagoskin Kept during the Expedition"), *Zapiski Imperatorskogo Geograficheskogo obshchestva*, Book 2, St. Petersburg, 1847: 135–202.
368. L. A. Zagoskin, *Peshekhodnaya opis chasti russkikh vladenii v Amerike. Proizvedennaya leitenantom L. Zagoskinym v 1842, 1843 i 1844 godakh. S merkatorskoyu kartoyu, gravirovannoi na medi . . .* (*Foot Survey of Part of the Russian Possessions in America. Produced by Lieutenant L. Zagoskin in 1842, 1843 and 1844. With a Mercator Chart, Engraved on Copper . . .*) 2 volumes. St. Petrsburg, Karl Krai, 1847. The first part of this work was originally published in the journal *Biblioteka dlya chteniya . . .* in July and August 1847. Translated and published in English as *Lieutenant Zagoskin's Travels in Russian America, 1843–1844*. Edited by Henry M. Michael. Toronto, 1967.
369. Erman, "Ueber die Reise und Entdeckungen des lieutenant L. Zagoskin im Russischen Amerika," *Archiv für wissenschaftliche kunde von Russland*, Berlin: G. Reimer, 1848, vol. VI:499–552; vol. VII:429–512; L. A. Zagoskin, Expedition auf dem festlande von Nord-Amerika, *Denkschriften der russischen geographischen gesellschaft zu St. Petersburg*, vol. 1, Weimar 1849. See also Peterman., (A.,) *Geographischen Mittheilungen*, 1857, heft IV und V:211–312.
370. RGAVMF, fond 1331, opis 4, no. 255.
371. AVPRI, fond RAK, opis 888, no. 1012, roll 50, folio 100.
372. *Peshekhodna͡ia opis chasti russkikh vlad͡ienii v Amerik͡ie : proizvedenna͡ia Leĭtenantom L. Zagoskinym v 1842, 1843 i 1844 godakh Zagoskin*, Moscow, 1847–1848. 2 v. This map has been reproduced in *Atlas . . .*, Efimov, Moscow, 1964, no. 192:123.
373. L. A. Zagoskin, *Puteshestviya i issledovaniya leitenanta Lavrentiya Zagoskina v Russkoi Amerike v 1842–1844 gg.* (*The Journey and Research of Lieutenant Lavrenty Zagoskin in Russian America in 1842–1844*). Moscow, 1956:322 ; Fedorova, ibid., 1971:235–236.
374. C. Osgood, *Ingalik Material Culture*. Yale University Publications in Anthropology no. 22, 1940:39, 45.

375. W. H. Oswalt, *Mission of Change in Alaska. Eskimos and Moravians on the Kuskokwim.* San Marino, 1963:13.
376. Osgood, ibid., 1940:480; F. de Laguna, *The Prehistory of North America as Seen from the Yukon*, Society for American Archeology, Memoir no. 3, 1947:31; James W. VanStone, "Introduction," in L. Zagoskin, *Lieutenant Zagoskin's Travels*, 1967:xii–xiii.
377. See, for example, de Laguna, ibid., 1947; Oswalt, ibid., 1962; Wendell H. Oswalt, "Population in Western Alaska and Migration Theory." *Anthropological Papers of the University of Alaska*, vol. 11, no. 1, 1962:1–14.
378. VanStone, ibid., 1967:xiii.
379. AVPRI, fond RAK, opis 888, no. 1012, roll 50, folios 68–70 verso.
380. Ibid., no. 1011, roll 49, folios 317 verso–318.
381. Ibid., folio 73.
382. Ibid., folios 74 verso–75.
383. Ibid., folios 78 verso–79.
384. Ibid., folios 80 verso–81, 88.
385. P. A. Tikhmenev, *Istoricheskoe obozrenie . . .*, part 2:256–257 ; S. G. Fedorova, *Russkoe naselenie*, 1971:130–131.
386. AVPRI, fond RAK, opis 888, no. 1011, roll 49, folios 370 verso–371 verso.
387. Pirjo Varjola (with contributions by Julia P. Averkieva and Roza G. Liapunova), *The Etholen Collection . . .* Finland National Board of Antiquities, Helsinki, 1990.
388. For the initial stages of this work, see ARGO, fond 1–1846, opis 1, no. 19: O sostavlenii statisticheskoi programmy . . . (On the Devising of the Statistical Program).
389. ARGO, fond 1–1848, opis 1, no. 2. Various content. Original, signature of Bishop Innokenty, autograph. Document published for the first time.
390. Manuscript Division of the Russian National Library, fond 314, opis 1, no. 9, folio 1.
391. Ibid., fond 314, opis 1, no. 9, folios 3–3 verso.
392. Ibid., folio 6, verso, 7 verso.
393. Ibid., folio 8, verso 9.
394. Ibid., folio 10.
395. Ibid., folio 15.
396. AVPRI, fond RAK, opis 888, no. 1001, roll 39, folios 52 verso–53.
397. Ibid., ARGO, fond I–1849, opis 1, no. 9. Delo Rosskogo geograficheskogo obshchestva o dostavlenii etnograficheskikh, klimatologicheskikh i drugikh svedenii (File of the Russian Geographical Society on the Delivery of Ethnographic, Climatological, and other Data). Folios 11, 16.
398. Ibid.
399. Ibid., no. 10. O dostav lenii Innokentiem Arkhiepiskopom Kamchatskim 150 r. v polzu obshchestva po vysylke emu Vestnika, a takzhe proglashenie prinyat na sebya zvanie Deistivitelnogo chlena Obshchestva February 23–March 29, 1857 (File on Receipt from Innokenty, Archbishop of Kamchatka of 150 r. for the Benefit of the Society, Dispatch to him of the *Vestnik*, and also an Invitation to Accept the Status of Active Member of the Society), folios 1–1 verso. Published here for the first time.
400. AVPRI, fond Innokenty, opis 874, no. 1, folios 19 verso–26 verso, 32.
401. ARGO, fond 1–1857, opis 1, no. 10, folios 2–2 verso.
402. AVPRI, fond RAK, opis 888, no. 1012, roll 50, folio 242 verso.

403. Pierce, *Builders of Alaska . . .* :128.
404. Ibid., 27.
405. AVPRI, fond RAK, opis 888, no. 1012, roll 50, folios 99 100.
406. Ibid., no. 978, roll 16, folios 116–16 verso.
407. Ibid., folios 171–172.
408. For a description of these voyages, see John Bocstoce, "The Search for Sir John Franklin in Alaska," in *The Franklin Era in Canadian Arctic History, 1845–1859*. Edited by Patricia Sutherland, 93–113. Mercury Series, Ottawa: National Museum of Canada, 1985; and John Bocstoce, ed., *The Journal of Rochfort Maguire, 1852–1854: Two Years at Point Barrow, Alaska, Aboard HMS* Plover *in the Search for Sir John Franklin.* London, 1989. [MF]
409. Ibid., no. 879, roll 17, folios 3, 4.
410. See, for example, *Chart of the Arctic Shores of the American Continent. From Baffin Bay to Behring Strait. From Admiralty Surves* [*sic*] *Corrected to Feb. 1850* in John J. Shillingshaw, *A Narrative of the Arctic Discovery*. London, 1850; *Chart of the North Polar Sea*. Standidge & Co. Lith. Henry Hansard Printer. Corrected to 1849. 1850 in *Arctic Expedition. Return of an Order . . .* in continuation of Parlimentary papers no. 264 & 386 of session 1848 and no. 188 & 387 of session 1849. (G.B. Parliament, *Sessional Papers, Accounts & Papers*,1850, v. 35, no. 107); *Map Arctic America: Discoveries of the Searching Expeditions under the Command of Captn. H. T. Austin and Captn. Penny, 1851* (Poultry, 1851) *Arctic Expeditions, Report of the Committee . . . Sessional Papers, Accounts and Papers*, 1852, vol. 50, no. 1435.
411. RGAVMF, fond 1331, opis 4, no. 261: *Merkatorskaya karta ostrova Kadiaka. S opisi Rossiiskogo shkipera Arkhimandritova. Uchinennaya v 1848 godu. S originalom verno poruchik Moshtaev 2* (*Mercator Chart of Kodiak Island from the Survey by the Russian Skipper Arkhimandritov. Conducted in 1848. True to Original, Lieutenant Moshaev*; also no. 256: *Merkatorskaya karta ostrova Sitkha ot Mysa Omani do porta Marii. Sostavlena po opisi, proizvedennoi v baidarochnoi ekspeditsii Korpusa flotskikh shturmanov Nikolaem Benzemanom. 1847 goda* (*Mercator Chart of Sitka Island, from Cape Omaney to Port Maria. Compiled on the Basis of the Survey Conducted in the Course of the Baidarka Expedition by Sub-Lieutenant of the Corps of Navigators of the Fleet Nikolai Benzeman. 1847*); also, no. 263: *Merkatorskaya karta Chugatskogo Berega. Chast sostavelnnaya s opisi Rossiiskogo shkipera Arkhimandritova. Uchinennaya v 1850 godu . . .* (*Mercator Chart of the Chugach Shore. This Part Compiled from the Survey by the Russian Skipper Arkhimandritov. Done in the Year 1850 . . .*).
412. Pierce, *Builders of Alaska . . .* :30.
413. The atlas itself was published in St. Petersburg in 1852 from the plates engraved in Sitka. However, the individual maps were completed in 1848, 1849 and 1850. A number of them exist in Alaska as single sheets, most likely proof sheets rendered in Sitka. A facsimile atlas with a translation of the hydrographic notes was published by Richard A. Pierce in 1981. [MF]
414. George Davidson, *Coast Pilot of Alaska*, 1869:53.
415. Clarence L. Andrews, *Sitka*:72; Tebenkov, *Atlas of the Northwest Coast . . .* :xii.
416. M. D. Tebenkov, *Gidrograficheskie zamechaniya k Atlasu Severovostochnykh beregov Ameriki, ostrovov Aleutskikh i nekotorykh drugikh mest Severnogo Tikhogo Okeana*

Kapitana I-go Ranga Tebenkova (*Hydrographic Notes to the Atlas of the Northeastern Coasts of America, the Aleutian Islands and some other Places in the North Pacific by Captain of the First Rank Tebenkov*), St. Petersburg: v Tipografii Morskogo kadetskogo Korpusa, 1852:2, 3.

417. Fedorova. Ibid., 1972:284; Pierce, ibid., 1981:xiv; Kiril Khlebnikov, *Russkaya Amerika v "Zapiskakh" Kirila Khlebnikova* (*Russian America in the "Notes" by Kiril Khlebnikov*). Moscow: Nauka 1985:46, 47, 294, 295.
418. Pierce, ibid., 1981:xiv.
419. Tebenkov, *Gidrograficheskie zamechaniya . . .* 1852:2.
420. Ibid., 1852:4.
421. Ibid., 1852:6.
422. Ibid., 1852:25.
423. Ibid., 1852:20.
424. Pierce, ibid., 1981:xiii.
425. Donald J. Orth, *Dictionary of Alaska Place Names*, Washington, 1967.
426. A. V. Grinev, "Zabytaya ekspeditsiya Dmitriya Tarkhanova na Mednuyu reku"("The Forgotten Voyage of Dmitry Tarkhanov to the Copper River"). *Sovetskaya Etnografiya* 4, 1987:88–100. For geological work in Russian America, see E. A. Kirillov, "Geologicheskie issledovaniya v Russkoi Amerike, (konets 18 v.—1867)" ("Geological Investigations in Russian America, end of the Eighteenth Century—1867"), *Tikhookeanskaya geologiya* (*Pacific Ocean Geology*) no. 2, 1993:153, 159; E. A. Kirillov, "Geologiya Russkoi Ameriki" ("The Geology of Russian America"), *Dalnevostochny ucheny* (*Far Eastern Studies*), November 13–19, 1993: 153–159.
427. AVPRI, fond 341, RAK, opis 888, no. 980, roll 18, folios 287–288.
428. P. Doroshin, "Neskolko podrobnostei o rasprostranenii zolota v russkikh Severo-Amerikanskikh vladeniyakh" ("Some Details about the Extent of Gold in the Russian Possessions in North America"), *Gorny Zhurnal*, 2, 1866:277–282. See also P. Doroshin, "Zoloto v Verkhnei Kalifornii" ("Gold in Upper California"), *Gorny Zhurnal* 2, 1850:133–162; P. Doroshin, "Iz zapisok vedennykh v Russkoi Amerike" ("From Notes Kept in Russian America"), *Gorny Zhurnal* no. 3, 1866:365–400; P. Doroshin, "O nekotorykh vulkanakh, ikh izvezheniyakh i zemletryaseniyakh v byvshikh amerikanskikh vladeniyakh" ("On Volanoes, their Eruptions and Earthquakes in the Former American Possessions"), *Zapiski Sankt Peterburgskogo mineralogicheskogo obshchestva* (*Notes of the St. Petersburg Mineralogical Society*), 2 series, part 6, 1870:25–44.
429. Published in English translation as *Marvin Falk, ed., Grewingk's Geology of Alaska and the Northwest Coast of America*. Translated by Fritz Jaensch. Fairbanks, 2003. C. J. Grewingk, *Die orographische und geognostische Beschaffenheit der N. W. Kueste Amerikas mit den angrenzenden Inseln*, St. Petersburg, 1848; Dr. C. J. Grewingk, *Beitrag zur Kenntniss der orographischen und geognostischen Beschaffenheit der Nord-West kueste Amerikas mit den anliegenden Inseln*. Von Dr. C. Grewingk (Aus den Verhandlungen der Mineralogischen Gesellschaft zu St. Petersburg für die Jahre 1848–49 besonders abgedruckt. St. Petersburg. Gedruckt bey Karl Krey, 1850) ; Grewingk, *Orograficheskу i geognostichesky ocherk severo–zapadnogo berega Ameriki i sosednikh ostrovov* (*An Orographic and Geognostic Study of the Northwest Coast of America and the Neighboring Islands*), St. Petersburg, 1850.

430. AVPRI, fond Glavny arkhiv, 1–9, 1819, no. 8 (1819–1899), Dokumenty raznye k voprosu o Rossisko-Ameikanskoi kompanii . . . (Various documents on the issue of the Russian-American Company . . .), folios 352–352 verso.
431. AVPRI, fond RAK, opis 888, no. 982, roll 20, folio 467.
432. AVPRI, fond Glavny arkhiv, 1–9, 1819, no. 8 (1819–1899), folios 306–306 verso.
433. For a history of whaling, see John R. Bocstoce, *Whales, Ice and Men: The History of Whaling in the Western Arctic*. Seattle, 1986. The first whaler arrived in 1848 and the last commercial kill was made in 1921. In 1890 there were as many as 220 vessels in the fleet. [MF].
434. Ibid., folio 23 verso–24 verso.
435. N. N. Bolkhovitinov, *Russko-amerikanskie otnosheniya . . .* 1996:71–88.
436. AVPRI, fond Glavny Arkhiv, 1–9, 1852–1853. Opis 8, no. 15, folio 45.
437. AVPRI, fond RAK, opis 888, no. 982, roll 20, folios 741–745.
438. National Archives, U.S. Coast Survey, Superintendent's Correspondence for 1856, vol. XVI:179: Davidson to A. D. Bache (then Superintendent of the U.S. Coast Survey), San Francisco, Feb. 3, 1856, no. 80. Information about this fact and also copies of the corresponding archival materials were provided to the author by historical researcher of the early period of the development of the U.S. Coast Survey Mr. James E. Ratcliff from San Francisco. Tebenkov's materials and their use by American cartographers are also examined in William Forrest King's biography of Davidson, a 1973 Ph.D. dissertation at Clermont Graduate School, which is available from University Microfilms International of Ann Arbor, Michigan.
439. AVPRI, fond 161: St. Petersburg. Glavny Arkhiv, 1–9, opis 8, no. 15, 1858: *Ob ograzhdenii kitolovnogo promysla v Rossiiskikh vodakh ot sovmestnichestva inostrannikh kitoboev. Ob ustanovlenii tochnikh pravil otnositelno poryadka proizvodstva kitoboinogo, ribnogo i drugikh promislov v nashikh territorialnikh vodakh Tikhogo okeana i Okhotskogo morya (On the Protection of the Whaling Industry in Russian Waters from the Interference of Foreign Whalers. On the Establishment of Precise Regulations Regarding the Procedure for Whaling, Fishing and other Industries in Our Territorial Waters of the Pacific Ocean and the Sea of Okhotsk*). Folio 3–3 verso.
440. Ibid., folios 11–12.
441. Ibid., folio 130 verso.
442. AVPRI, fond 161: St. Petersburg. Glavny Arkhiv, 1–9, opis 8, no. 15, 1858, folios 47–49: *Otchet komandira vintovogo korveta "Kalevala" Kapitan Leitenanta Davidova 5–ogo. Po vozlozhennomu na nego porucheniyu v zimu 1861 g. o sobranii svedenii o kitolovnom promisle v severnoi chasti Tikhogo okeana (Report of the Commander of the Screw-Propelled Corvette "Kalevala" Captain-Lieutenant Davydov the Fifth. On the Assignment Entrusted to Him in the Winter of 1861 to Gather Information about the Whaling Industry in the Northern Part of the Pacific Ocean*).
443. Ibid., folios 52 verso–53.
444. Ibid., folios 31, 71, 79.
445. AVPRI, fond 339, RAK, opis 888, no. 181: Sobranie bumag i dokumentov, otnosyashchikhsya k istorii RAK. 1807, 1824–1888 (Collection of Papers and Documents Relating to the History of the RAC. 1807, 1824–1888). Folios 24–24 verso, 31 verso.

446. Richard Pierce, *Builder of Alaska . . .*:32–36; Richard Collinson, *Account . . .*:194–206. A map is appended to the article: *Sketch of the Shores of Arctic America to Illustrate the Search for Sir John Franklin by Capt. Collinson of H.M.S. Enterprise, 1850–1–2–3–4. John Arrowsmith.* Published by J. Murray. London, 1855. The results of the work of Richard Collinson's expedition were reflected, along with data from other search parties, on the map: *From Chukotsk Peninsula Eastward to Greenland Showing Tracks of Ships of Parry 1819–20, McClure in the Investigator 1850–54, Collinson in the Enterprise 1850–54, and Dr. John Ray 1854; with Notes on the Service, and on the Expeditions of McClure, Maguire (Plover 1852–53), and Collinson*; *Report of the Select Committee on Arctic Expedition . . .* GB Parliament. Sessional Papers: Reports from Committees, 1854–55, vol. 7, no. 409. London, 1856.
447. For more detail on the history of the conflicts between the Russians and the Natives, see Edward S. Petersen, *The Russian-Indian Wars*:7–8, 14); Andrei V. Grinev, *The Tlingit Indians in Russian America 1741–1867.* Translated by Richard L. Bland and Katerina G. Solovjova. Lincoln and London, 2005.
448. For more detail on Hjalmar Furuhjelm and his work, see *Pirjo Varjola* (with contribution by J. P. Averkieva and R. G. Liapunova). *Etholén Collection* 1990:47–51.
449. Pierce, *Builders of Alaska . . .*:38–39.
450. AVPRI, fond RAK, opis 888, no. 1039, roll 77, folios 1–34.
451. Ibid., folios 35–68.
452. Ibid., folio 2.
453. AVPRI, fond St. Petersburg. Glavny Arkhiv, 1–9, opis 8, no. 9: *Po povodu otkritiya zolotikh priiskov po Oregonu v Britanskoi Kolumbii i po khodataistvu Rossiisko–Amerikanskoi kompanii o komandirovanii nashego voennogo kreisera k. r. Stakhinu dlya sodeistviya kolonnalnomu nachalstvu v okhrane polosi nashikh vladenii v Amerike, gde takzhe otkriti zolotie priiski, ot proizvolnikh deistvii inostrannikh prisheltsev. 1862–1863* (*On the Discovery of Gold Fields in Oregon in British Columbia and on the Petition of the Russian-American Company for the Dispatching of Our Naval Cruiser to the Stakhin River to Assist the Colonial Leadership in Protecting the Regions of Our Possessions in America, where Gold Fields have also been Discovered, from the Arbitrary Actions of Foreign Newcomers*). Folio 38 verso.
454. William P. Blake, *Geographical Notes upon Russian America and the Stickeen River: being a Report Addressed to the Hon. W. H. Seward, Secretary of State.* Washington: Government Printing Office, 1868.
455. *Plan reki Stakhin s opisi 1863 goda, proizvedennoi s Korveta "Rinda" Prap. K. Fl. Sht. Butirkinim. Pechatan s kamnya v Gidrograficheskom Departamente v 1867 godu* (*Plan of the Stakhin River from the Survey of 1863 Conducted from the Corvette "Rynda" by Warrant Officer Butirkin. Printed from Stone at the Hydrographic Department in 1867*). Inset map: *Plan ustya reki Stakhin s opisi 1863 goda, proizvedennoi s korveta "Rinda" Prap. K. Fl. Sht. Durkinim i volnim shturmanom Kadinim* (*Plan of the Mouth of the Stakhin River from the Survey of 1863 Conducted from the Corvette "Rynda" by Warrant Officer Durkin and Free Navigator Kadin*). An example of the map has been preserved, in particular, in the Archive of the Hydrographic Office of the Royal Navy of Great Britain (in Taunton). Number: A. 1501.

456. P. Tikhmenev, *Istoricheskoe obozrenie* obrazoraniya Rossiikse Amerikanskoi knmpanii. Vol. 2. Saint Petersburg: E. Weimar, 1863, at end. Also published separately, see ARGO, Razryad 3, opis 1, no. 232. Edited by A. V. Efimov:194.
457. AVPRI, fond RAK, opis 888, no. 985, roll 23, folio 178.
458. Ibid., no. 982, roll 20, folios 861–863. In particular, on the basis of this decision one of the senior employees of the company, Skipper Ivan Kashevarov, was not allowed to have his two sons educated at the merchant marine school.
459. Ibid., no. 985, roll 23, folios 176–177.

CHAPTER V

The Sale of Alaska and the International Expedition to Effect a Telegraph Link between North America and Europe via Siberia

THE SALE OF ALASKA

The sale of Alaska to the United States of America resulted from an involved set of economic and geopolitical causes. Analyzing these causes is far beyond the scope of this study, especially since this problem has been examined from all angles by foreign and Russian scholars in recent years. Foremost among these examinations are the fundamental works of Academician Nikolai Bolkhovitinov.[1]

The majority of scholars who have examined this question agree that the treaty regarding the sale of Alaska concluded on March 18 (30), 1867, between Russia and the United States was the mutually beneficial result of American geopolitical ambitions and was based on a sober assessment by Russia to focus her efforts on the Amur and other regions joined to the empire in 1860. The geographical explorations that played a significant role in achieving this result are the subject of this section.

The idea of an "inevitable" spread of North Americans throughout the continent came into being as early as the seventeenth century, when the first English colonies were founded in the New World. The very name of the new state that emerged in 1776, the United States of America, contained the idea of the spread of the young republic throughout the entire American continent.[2] In 1819 the USA's secretary of state, John Quincy Adams, declared that all other states throughout the world must concede the idea that the entire continent of North America was the lawful possession of the United States: "From the time we became an independent people, it was as much a law of nature that this should become our pretension as that the Mississippi should flow to the sea. Spain had pretensions on our southern, Great Britain on our northern borders. It was impossible that centuries should elapse without finding them annexed to the United States."[3]

As far as the Russian possessions in Alaska were concerned, Adams held exactly the same view and believed that the best policy was to exercise patience and bide time. In his letter to James Lloyd from Massachusetts dated July 15, 1823, he denied that Russia had any rights in North America: "What kind of rights does Russia have on the *continent* of North America? Does she have any that we must recognize? And is not it time for American *nations* to inform the European rulers that American continents are no longer open to European colonization."[4]

This was, in fact, the initial formulation of the Monroe Doctrine: "America for Americans." In respect to Russia, however, the United States at that time had to postpone Adams's expansionist plans and to acknowledge, by a treaty of April 17, 1824, the southern border of Russia's colonial possessions in the New World at the parallel 54°40′ N latitude.[5]

In the 1840s United States expansionism in North America assumed the character of clearly expressed state policy. Texas was annexed in 1845, and then California was joined, and the border was determined to run along the Rio Grande as a result of the war against Mexico (1846–1848). It is precisely in this period that the editor of both the journal *Democratic Review* and the New York *Morning News*, John L. O'Sullivan, coined the phrase "Manifest destiny," which for the apologists of expansionism meant that Providence itself had predestined the United States to reign over the entire North American continent.[6]

The most consistent advocate of this policy was William H. Seward, governor of the state of New York from 1838 to 1842, senator (1849–1861), and United States secretary of state (1861–1869). He considered himself a pupil of the late John Quincy Adams who had died in 1848. In that year Seward firmly declared: "Our population is destined to roll its resistless waves to the icy barriers of the north, and to encounter oriental civilization on the shores of the Pacific."[7]

According to Seward in 1860, the United States was to spread over the entire North American continent, the islands in the Pacific and Atlantic Oceans, and in time possibly to include South America down to Cape Horn. The United States, embracing the North American continent, was to consist of sixty states, with the capital in St. Paul, Minnesota. This scheme for expansion was to be realized by the year 1960. To make this grandiose geopolitical project a reality, Seward planned to use, not conquest, the fruits of which are colonies ("the source of weakness of empires"), but "the extension of republican institutions, whose fruits were equal states ('a source of strength')."[8]

California Senator William McHenry Gwin became Seward's ally in this expansionist policy, and even the diametrically opposed personal views of the two political actors were no hindrance. Seward was a leader of the Republican Party who actively opposed slavery; Gwin was a slave holder, closely associated with the Democratic Party. Parting ways with respect to party passions, both constantly collaborated in "securing governmental support for the Pacific railway, the China trade, and Pacific Coast whaling and trading projects," and other commercial enterprises "so vital to the prosperity . . . of the whole union"[9] and the spread of its influence to the neighboring territories.

As noted by American scholar Hallie May McPherson, "Gwin had conceived the idea of a world empire with the United States at the center. His plan embraced the linking of the Atlantic and the Pacific across Panama," constructing a transcontinental railway, telegraph, and so on. Like Seward, Gwin called for attention to be given to "hundreds of millions of people who inhabit China and Japan" and also to Russian possessions along the Amur as a market for American goods that would give the United States "control of the commerce and the exchanges of the world."[10]

Senators Seward and Gwin well knew the importance of geographical studies of the regions that, according to their views, held potential geopolitical interest for the United States. Witness, for example, the fact that it was they who initiated the bill for organizing the first serious hydrographic expedition to the northern parts of the Pacific Ocean under the command of Lieutenant John Rogers that we examined earlier. Upon his return in 1855, Rogers stressed huge potential for the Amur River, which in his opinion was "the great highway of nature's making from the shores of the Pacific to the centre of Asia" and that in the future "a vast commerce" would be conducted along its banks and a city built at its mouth that would become a Russian San Francisco.[11]

American geopolitical aspirations in the 1850s and 1860s were well known in Russian governmental circles, not least because of the steady drawing together of Russia and the United States during the Crimean War in the face of a common rival, Great Britain.[12] The tsarist government realized especially clearly, in connection with the Crimean War, the necessity of strengthening the position of the Russian empire in the Far East, as well as the necessity of expanding relations with the United States.

Count Nikolai Muraviev-Amursky (1809–1881), the governor-general of Eastern Siberia, initiated such action. In the spring of 1853, upon his arrival in St. Petersburg, he presented a working paper to emperor Nicholas I in which he gave his opinion that it was necessary to strengthen the position of the Russian empire in the Far East. He expressed concerns about the future of the fate of the Russian possessions in North America and the importance of close relations with the United States.

Muraviev recalled that a quarter of a century earlier, "the Russian-American Company approached the government petitioning for occupation of California, then free and almost not possessed by anybody, but stating that this region would soon become the prey of the United States . . . Even then it should have been apparent that these States, once firmly established on the Eastern Ocean, would soon become foremost among all the maritime powers and would stand in need of the entire northwest coast of America. The dominion of the United States throughout the whole of North America is so natural that we should not regret that twenty-five years ago we did not establish ourselves in California—we should have had, by necessity, sooner or later, to yield it. Yielding it peacefully, however, we could have acquired other benefits from the Americans in exchange.

"But nowadays, with the invention and development of railways, we should be convinced more than ever that the United States will spread, without fail, over the whole of North America. We cannot escape the realization that sooner or later we shall have to give up our North American possessions. On the other hand, the realization of another factor is also inescapable: it is absolutely natural that Russia, even if not having dominion over the entire eastern Asia, is to be dominant over the entire Asian coast of the Eastern Ocean. Due to circumstances, we permitted the English to invade this part of Asia . . . but this matter can yet be remedied through our close links with the United States."[13]

The St. Petersburg authorities received Muraviev's memorandum with approval. The proposal of the governor-general of Eastern Siberia to fortify the imperial position in the Amur region and in Sakhalin was studied in detail, with the lord high admiral, Grand Duke Konstantin, and members of the management of the Russian-American Company participating. One of the direct results of this study was the emperor's "command" of April 11 (23) 1853, permitting the Russian-American Company to "occupy Sakhalin Island on the same basis as it possessed other lands mentioned in its privileges, in order to prevent any foreign settlements in Sakhalin."[14]

WESTERN UNION TELEGRAPH PROPOSAL

One of the important concrete measures directed toward realizing both the American and the Russian geopolitical plans was the creation of a telegraph connection within the borders of both nations as well as between North America and Eurasia. The idea of creating this type of transcontinental telegraph line emerged almost simultaneously in Russia and the United States. Its realization took the form of the last large-scale scientific expedition in Russian America, launching Alaska studies by American scientists, and to some extent, prepared the ground for the purchase of the Russian transoceanic possessions by the United States.

Already "in 1854, Shafner, a citizen of the United States of North America, proposed to our Government to put through a telegraph line across Russia, Siberia, Kamchatka, and the Bering Strait, or alternatively from Irkutsk down the Amur along the Aleutian Islands to North America to link up with the American telegraph."[15] Mr. Shafner stood ready to execute the project at his own cost, provided that he would be supplied with a labor force and given the exclusive right of exploiting the telegraph for 100 years. The American's proposal was politely declined.

In 1856 captain of the First Mounted Pioneer Division, D. I. Romanov, drew up a "Russian-American Telegraph" project. He presented it to the governor-general of Eastern Siberia, Muraviev, at Kiakhta in 1857. The project was reported to the emperor and it was "ordered that it be admitted into print, with the exception of unsuitable opinions and expressions." One of the reasons for the decision to make the project public was a desire on the part of the Russian authorities to maintain national priority in advancing this idea,[16] especially since foreigners were continuously trying to interest the Russian leadership in their telegraph projects. In the end, they succeeded.

In 1858 the Russian government received three foreign requests for creating an intercontinental telegraph at the same time. Specifically, "the foreigner, Paris resident Liger de Libessart, who calls himself Chief Director of the Universal Telegraph Society" thought it possible to lay the cable on the bottom of the Bering Strait. "The London bankers DeVeau and Co. proposed the establishment of a telegraph across Russia, North America, and Oregon to California, without specifying more precisely the direction of the line."[17]

Most active was one Lieut. Col. Campbell-Sleigh, who came to Russia carrying the project of his brother, Colonel Campbell. In their opinion the telegraph link could be effected either along the bottom of the Bering Strait or along the Aleutian Island arc. Besides constructing the telegraph, Campbell also proposed building a railroad from Nizhny Novgorod to the Tartar Strait in the Pacific Ocean. Russia's official authorities received the British man's proposals rather cautiously and skeptically. This is evident in the letter of March 20, 1858, from "Chief Manager of Communications and Public Works" K. Chevkin to the Russian foreign minister, Alexander Gorchakov:

> My dear Sir, Alexander Mikhailovich!
>
> I am returning herewith to Your Excellency a dispatch by Actual State Councilor Baron Nikolai, [the Russian ambassador in London—AP], dated 4/16 March regarding the proposal by the foreigner Campbell to establish a telegraph line from Moscow to America via Okhotsk and the Bering Strait. I deem it my duty to report that I reported on this dispatch to the Lord Emperor that such a proposal is hardly feasible. Besides the 9 thousand versts to Okhotsk, there would remain several more thousand versts in the most severe and uninhabited lands not explored to this day. The statement that this huge enterprise can be accomplished in the course of a single year is so unrealistic that it raises doubts about the credibility of the author. In general, it is not possible to

contemplate without wonder the engagement of an English colonel in laying an *overland* telegraph line across Siberia to *America*, while the English army is in dire need of officers.

Your Excellency, please accept my assurance of my full respect and total loyalty,

Signed K. Chevkin.[18]

The governor-general of Eastern Siberia, Muraviev, however, gave decisive support to this project, especially for its Siberian component. He declared to Gorchakov that "the telegraph from S. Petersburg to the mouth of the Amur has become such a necessity that it is not permissible to begrudge for this the 5 million that it may cost." Muraviev stressed that "England, France and America have established themselves in Japan and China, and with all our policy art we shall not be able to outpace these gentlemen if our communications with the capital shall remain in its current primitive state."[19]

However, the governor-general of Eastern Siberia held that the telegraph construction on Russian territory must be accomplished by Russian companies and that the Siberian Russian Society of Amur stood ready to undertake the task. On December 11, 1858, the emperor ordered the telegraph proposal to be studied by the Siberian Committee.[20]

The members of the committee agreed, essentially unanimously, on the necessity of establishing a Siberian telegraph, especially so because on December 16, 1858, official agreement was received from China to put a telegraph line through from Kiakhta to Beijing.[21] It was decided to decline the service of the British colonel, however, as Russian diplomats learned from articles in the London newspapers *The Times* and *The Daily News* that Colonel Campbell was a swindler who engaged in monetary machinations.

On the basis of the decision made by the Siberian Committee, practical steps to investigate the proposed route were undertaken at once. For this purpose two engineers from the Office of Communication Routes were dispatched to Siberia.[22] The Campbell-Sleigh saga was not over, however. In 1862, having learned of the establishment of the telegraph line to Irkutsk and Kiakhta, Campbell applied for an award, through the emperor, for his idea and his expenses. Naturally, he failed in this, too.

Collins

The American author of the trans-Pacific telegraph line project, an energetic entrepreneur and political figure from San Francisco, Perry McDonough Collins, was more successful. His plans were based on the concept that following the annexation of the Amur region by Russia, excellent possibilities existed for American trade along the Amur, and that this opened the door to the unlimited markets of China and the other countries of southern Asia. In the winter of 1856 Collins traveled to Washington to secure support for his project from President Franklin Pierce and also from the Russian chargé d'affaires in the United States, Edward de Stoeckl.[23]

Collins proposed to the president that he be sent to the Amur region in order to "look at it in a commercial point of view and open it up to commerce by way of this river; present to the knowledge of our people the nature and extent of this country." He maintained that the Amur region was exceptionally important for the commerce of California, Oregon, and Washington. Collins's proposals found support from the secretary of state, William Marcy, and President Pierce. On March 24, 1856, Collins was appointed commercial agent (representative) of the United States on the Amur.[24] On April 12 of the same year he left New York for Liverpool, and on May 17 he arrived at Kronstadt with the first steamer.

Having secured support from the governor-general of Eastern Siberia, Muraviev-Amursky, Collins, together with the equally enterprising American Bernard Peyton (who also nurtured plans for the development of Amur trade but was in St. Petersburg independently from Collins),[25] set out in December 1856 by sled for Eastern Siberia. He reached its capital, Irkutsk, on January 7, 1857.[26] Somewhat later Peyton decided to return to St. Petersburg while Collins thoroughly investigated the situation in Siberia, got acquainted at Kiakhta with the state of commerce with China, and visited Chita and the Nerchinsk mines.

The high point of the U.S. commercial agent's sojourn in Eastern Siberia was his voyage from Chita to Nikolaevsk in the company of the region's military governor, Vice-Admiral Pyotr Kazakevich, and the head of Russian diplomatic mission to China, E. V. Putyatin. This voyage was along the entire course of the Amur to its mouth, where he encountered several American vessels from San Francisco and Boston engaged in trade.[27]

While in Chita, Collins advanced a bold proposal to construct a railroad link between Baikal and Amur, designed to link the Amur River with the center of Eastern Siberia, the city of Irkutsk. Collins's proposal, supported by Muraviev, was forwarded to the Siberian Committee, where it was rejected, as the committee deemed the first priority to be linking Eastern Siberia with the European part of the country. It also judged that the realization of this project "may in the future, prove very harmful in an additional respect by placing the internal interests of the eastern part of the Siberian region into dependence not on the metropolitan area, as was the case up to now, but on foreigners, specifically North Americans."[28]

Collins himself, however, did not give up on the idea of expanding U.S. commercial interests in the Russian Far East and wrote with enthusiasm to the old Secretary of State Marcy and also to the newly appointed Secretary of State, Lewis Cass, about the enormous possibilities for American commerce in that region. He stressed that the Amur was the Siberian Mississippi and that the Yankee trade developing at the time in Kamchatka would find an extension along the Amur and thus would plug into the wide commercial net in northern Asia. Collins returned to the United States via San Francisco in December of 1857, leaving behind as his replacement on the Amur the commercial vice-agent George Cushing.[29]

The most important result of Collin's voyage proved to be the grandiose Russian-American telegraph project he advanced that would have linked the telegraph systems of the two hemispheres via the Bering Strait. In order to realize this project, the enterprising American began to act energetically in several directions: in the USA, Russia, and Britain. He succeeded in interesting not only the governmental circles of these nations, but also the powerful president of the Western Union Telegraph Company, Hiram Sibley. In Russia this project was met initially with the same caution afforded all previous foreign initiatives of the same sort.

In 1859 Collins once again traveled to the Amur. In the following two years he attempted, without any visible success, to obtain the approval of the Russian government for the realization of the intercontinental telegraph. On the other hand, by 1861 he and Sibley had secured the support of President Abraham Lincoln's administration. Even the Civil War did not stop the realization of this project.[30] That year the Russian military engineer Romanov, who we encountered earlier, visited the United States to study the American experience of establishing and utilizing telegraph lines.

In the course of his business trip Romanov met the famous Samuel Morse, as well as Sibley and Collins, and discussed with them the question "of a telegraph between America and Siberia, especially in view of the fact that at that time a line across the entire continent had been completed and telegraph communication opened with California."[31] Moreover, Lt. Col. Romanov insisted on recommending that "a joint scientific expedition be organized (by Russia, America and England) for the study of the areas where the telegraph line would run." Baron de Stoeckl, the Russian ambassador in Washington, immediately brought Romanov's proposal to the attention of Grand Duke Konstantin.[32]

In 1862 the American government undertook official steps, through diplomatic channels, to secure the agreement of the Russian government for the construction of the trans-Pacific telegraph. On September 5, 1862, Simon Cameron, the American ambassador, approached the Russian minister of foreign affairs and vice-chancellor of the empire, Prince Alexander Gorchakov, with an official letter. The missive outlined the general idea of the telegraphic linkage of Asiatic Russia with the North American continent, with the cable being laid either along the bottom of the Bering Strait or along the Aleutian Island arc.

The first option was deemed preferable. It was stressed that "The enterprise has an importance beyond the intentions of its projectors—beyond national interests, even. It is another step in the grand forward movement of the world—another of those triumphs of a wise civilization, to the advancement of which neither Russia nor the United States have been the least contributors."[33]

Noting that Russia would receive substantial profit should the line be constructed from the telegram transmissions to Europe through its territory, the American representative did not conceal the United States' wide-ranging plans for Asia, the realization of which was linked to the construction of such a telegraph line: "The future construction of a Telegraph line from Irkutsk, via Kiachta to Peking, and thence to Shanghai and the other commercial ports of China, which may be regarded as certain, will render the proposed line the shortest communication between the United States and their important commercial interests in China and Japan."[34]

This last consideration did not arouse much enthusiasm on the part of the Russian authorities, who, quite rationally, saw in the United States a competitor for the Asian markets, especially as since time immemorial Kiakhta had been one of the largest commercial transit points in the trade between China and Russia. In future discussions of problems with regard to establishing a trans-Pacific telegraph line this was reflected in Russia's firm stand that all construction work in the Eastern Siberian territory be conducted by Russian specialists.[35] An exception was made only for the easternmost stretch (the Okhotsk Sea littoral and Chukotka).

The exceptional importance of this project for the United States was underscored by the fact that "in order to enable Mr. Collins to await the action of the Imperial Government, without incurring loss, the President has appointed him, temporarily, Vice-Consul at this port [St. Petersburg—AP]. The seals and archives of the Consulate, which have been hitherto in charge of the Undersigned [Bayard Taylor—AP] have been delivered to Mr. Collins, whose temporary official character, in the absence of a Consul, will, it is hoped, be recognized by the Imperial Government."[36]

William Seward was an enthusiastic propagandist within the governmental circles of the United States of the idea of constructing a trans-Pacific telegraph line. He viewed its realization in the light of his conviction about the special role of the United States in the world that was predestined by Providence. He stressed: "It is just as difficult to determine the limits of expansion of the national influence which will follow from these new and profitable conditions obtained by these means for the spread around the world of American ideas and the beginnings of political and private economy, politics, morals, philosophy, and religion."[37]

Without going into unnecessary detail about the prolonged negotiations between Russia and the United States, a matter far exceeding the framework of this study and examined in minute detail by Bolkhovitinov[38] on the basis of archival documents, we shall only note that the American argument was based, not in small part, on the above-mentioned partnership between Russia and the United States. At that time they were united against their common strategic adversary, Great Britain.

This fact is stressed specifically in a letter by Cassius Clay, the American ambassador in St. Petersburg, addressed to the Head of the Asiatic Department of Russia's

Ministry of Foreign Affairs, General Count N. P. Ignatiev. There it was noted that "the United States as well as Russia—are interested in not having the telegraphic intelligence of the world confined to the Atlantic line, in the sole possession of the British Nation."[39]

The negotiations were concluded on March 9/21, 1863, when the chief manager of the Postal Department, Count I. M. Tolstoy, Sibley, and Collins signed an agreement specifying the conditions of telegraph communication between Russia and America.[40] In 1865 these conditions were published in the official compendium of regulations and orders of the Russian government.

In the United States the transoceanic telegraph was widely discussed in the periodical press. A special map was published: *P. McD. Collins' Proposed Overland Telegraph via Behring's Strait and Asiatic Russia to Europe, under Russian & British Grants.*[41] In order to ensure the most benevolent atmosphere in carrying out the project, on September 6, 1865, Collins was issued a certificate, preserved in the original in the State Historical Archive of Russia in St. Petersburg:

> Certificate
>
> With His Imperial Majesty's consent, the construction of the telegraph line from America across the Russian possessions in the Primorye region, to the city of Nikolaevsk at the mouth of the Amur River, has been granted to the American Western Union Telegraph Company. Testifying to this fact, I humbly request the local authorities to offer all possible cooperation in the completion of this matter to the Company's representative, the American citizen Major Perry Collins.
>
> Minister of Posts and Telegraphs I. Tolstoy.[42]

THE TELEGRAPH EXPEDITION

Let us now examine the progress of the telegraph expedition itself and the geographical explorations conducted within its framework. Although the search for the route and construction of the telegraph line were to be effected simultaneously, scientific work had a major place. Within the governmental circles of the United States, where the question of the practicality of an Alaska purchase was being studied, there was great interest in firsthand data about the territory, specifically in data obtained by their own countrymen.

A young scientist from Illinois, Robert Kennicott (1835–1866) was appointed as head of the expedition's scientific sector.[43] Although he did not have a university education because of weak health, he had a solid theoretical scientific preparation thanks to a fine home upbringing and studies with important scholars—the naturalist Jared Kirtland and the ornithologist Philo Hoy. This allowed him to make a name for himself as a serious naturalist as early as 1855, having completed a detailed expeditionary study of Illinois. In 1857 he was appointed curator of the Natural History Museum of Northwestern University in Evanston, just outside Chicago.

In 1859 Kennicott decided to engage in natural history studies and collecting specimens in British Columbia and Russian America for his museum. He set out for Alaska at the expense of the Smithsonian Institution and the Academy of Sciences of Chicago (in actuality, the Museum of Natural History). He spent the winter of 1860–1861 at Fort Yukon. Kennicott planned to explore the Yukon River traveling downstream, within the Russian possessions, and for this purpose he took care to obtain the following document:

> Russian-American Company Head Office
> March 5/17, 1860
> No. 223
> St. Petersburg
>
> The person presenting this document, the young American scholar Robert Kennicott, has been sent from the United States to the possessions of the Hudson's Bay Company in order to conduct natural science research. Should he arrive in the north of the possessions of the Russian-American Company, specifically on the rivers *Kvikhpak*, *Kuskokwim*, *Nushagak* or elsewhere, it is ordered that the authorities of the Russian posts, redoubts and outposts in these areas render him all possible support, help and protection in order to facilitate the achievement of the goals of his scientific activities.
>
> Chairman of the Head Office of the Russian-American Company
> V. Politkovsky
> Office manager A. Timkovsky[44]

Kennicott had no occasion to use this document, as he did not reach the possessions of the Russian-American Company. He returned to the United States dreaming of going to Alaska again and studying the Yukon River valley in detail along its entire extent. Upon his return the scholar spent almost an entire year at the Smithsonian working on the rich natural science data he had collected, especially the ornithological data. He continued this work while occupying the post of curator of the Chicago Academy of Sciences (from 1864).[45]

In that period the Smithsonian began planning the scientific arm of the Telegraph Expedition under the guidance of Professor Spencer Baird. Besides its leader, Robert Kennicott, there were botanist Joseph Rothrock, entomologist Ferdinand Bischoff, zoologists Charles Pease and Henry W. Elliott, paleontologist Henry M. Bannister, and mollusk specialist William Healey Dall.[46] At that time active and confidential consultations regarding the purchase of the Russian-American possessions were being conducted and the United States' authorities needed objective information on the natural resources of Russian America.

As American scholar Morgan Sherwood remarked, Kennicott was "the right choice" as the leader of the scientific detachment as he already had some Alaska experience, but "the wrong choice temperamentally." He was a solitary scholar. The conflict between the need to simultaneously lead a scientific enterprise and administer construction of the telegraph line undermined his health and led to his untimely death. Signs of Kennicott's psychological instability became apparent even at the initial stages of the preparation for and implementation of the expedition.

Dall, who never experienced any antipathy toward Kennicott, noted the following in his diary on May 10, 1865:

> "He is certainly in a great state of excitement or else he was intoxicated. I should say of another man judging from the same data that he was either insane or incapable of doing the work set before him . . . He is absurdly suspicious of everyone."[47]

It should be noted that Kennicott was an example of self-sacrifice and zeal toward the collection and in-depth analysis of natural history data. Even at the early stage of the expedition's preparation he made the effort to organize the collection of flora, fauna, and mineral specimens in Russian America for the Telegraph Expedition as well as for the enrichment of collections at the Smithsonian Institution. For this reason Kennicott entered into an active correspondence with the Russian consul in San Francisco, Baron Carl-Robert von Osten-Sacken, specifically suggesting that he should recommend to the chief manager of the Russian colonies that the collecting of natural history specimens be organized in Russian America even prior to the arrival there of the Telegraph Expedition.

In this matter Kennicott wrote the following to Osten-Sacken:

> "I propose having some of the usual Smithsonian labels for birds etc. printed with the Governor's name. If the Governor seems at all interested in Natural History or if he has any liking for the reputation of doing scientific work I believe we could urge him to cause large collections to be made for the Smithsonian. . . . If he would even do so much as to authorize the officers at the outposts to make collections themselves or receive them from the Indians and forward them to Sitka by the fur boats I will guarantee that I can get the collections made for him. If he enters at all actively into the matter he can by his mere patronage cause to be sent to the Smithsonian very fine collections."[48]

Besides the above, Kennicott informed Osten-Sacken that he had met the secretary of the Russian embassy in Washington, Waldemar Bodisco, "about some Russian charts," adding, mysteriously, "but I prefer not to discuss this matter with anyone but Yourself."[49]

On March 21, 1865, Kennicott, now having the rank of major, and his subordinate members of the scientific detachment of the expedition, Rothrock, Bischoff, Pease, Elliott, Bannister, Dall, and G. W. Maynard included as a volunteer, left New York aboard the steamer *Golden Rule*. Ten days after the expedition's departure Kennicott sent a letter to Baird from aboard ship in which he outlined his preliminary plans for field work and stated categorically his wish to keep the expedition's activities a secret from the Russians, noting in particular that "I believe that I should not take along *any Russians* with me up river (along the Yukon)."[50] Kennicott's group went ashore at Graytown (Nicaragua), crossed the isthmus overland, and then sailed to San Francisco by ship, arriving on April 25, 1865.[51]

Because of unexpected heart trouble and the need to supplement the expedition's equipment, Kennicott was delayed in San Francisco for two-and-a-half months. While in San Francisco he established a close friendship with the family of the commissioner of the Russian-American Company in California and Russian consul, Captain Martin Klinkovstrem. The family was instrumental in correcting the negative opinion about Russian people Kennicott had already formed. No small role was played here by the presence of Klinkovstrem 's two charming and exceedingly well-educated daughters. The younger one, Olga, conquered Kennicott's heart. He freely speaks about this in his letter to Baird dated July 9, 1865:

> Were it not for Miss Olga, I should be glad to see San Francisco swallowed by an earthquake, but now I prefer that it remain safe as long as she is here. . . . I do not know, and do not even dare to presume, if I mean anything to her . . . but . . . however . . . Russian alliances right now are very desirable. Be it as it may, I shall while in the Arctic carry some pleasant memories and in my heart, in spite of its general hardening, there remains a tender spot.[52]

This socializing with the Klinkovstrems gave Kennicott the opportunity to obtain, besides his lyrical reminiscences, some additional information about the Russian explorations along the Yukon. It should be admitted, however, that this information was rather superficial and subjective, as it was presented by persons who themselves were not active investigators of Russian America: Captain Klinkovstrem and his elder daughter, the widow of Captain Koshkin, who from time to time visited the lower course of the Kvikhpak. Kennicott sent to Baird, appended to his letter of June 28, 1865, excerpts from the notes Koshkin kept while on a voyage to the Kvikhpak in July 1862.[53]

On the basis of the nonobjective evaluation by Klinkovstrem, Kennicott formed a rather skeptical attitude toward the person of Zagoskin, and this could not help but affect Kennicott's evaluation of Zagoskin's continental explorations in Alaska. Kennicott expressed to Baird his rather confused thoughts on the subject in the following sententious passage:

> Russians are unwieldy wagons, I expect little of them. Among others, Zagoskin, who explored the course of the Kvikhpak, proved to be to a great extent a humbug. Captain Klinkovstrem, the Russian consul in San Francisco who sailed with him, acknowledges that at least at sea he was a no good lad and showed the white feather. Klinkovstrem himself is a real rock, not to speak of the fact that he is the father of the charming Miss Olga. I am indebted to him and to his elder daughter, widow of Captain Koshkin (who served in the Russian navy and then entered the service of the Russian-American Company), for most valuable information about the lower course of the Yukon. If we take Klinkovstrem, Baron Osten-Sacken, and the young Abas—a member of our expedition—as a sample of Russians, I must admit that they are an excellent selection of human beings![54]

By request of the overall commander of the expedition, Colonel Charles L. Bulkley, the local leadership of the Russian-American Company put at the Americans' disposal "as a pilot the free hired navigator Kadin who has sailed numerous times in the Bering Sea and, in general, knows the colonies; also specifically for the Kvikhpak expedition one of the *promyshlenniki*, the Finn Grenberg, and an Aleut for the Anadyr detachment as an interpreter of the Chukchi."[55]

The Finn Grenberg was sent to the expedition as an interpreter from English into Russian. Somewhat later, to aid the explorers, "two boys from the north, Ivan Kozhevnikov, who speaks the Malemiut language, and Creole Ivan German from Paul Island (St. Paul)," were appointed.

These youngsters had studied the English language and were training as interpreters.[56] Moreover, "the officer for special assignments of the Head Office for Eastern Siberia, the collegiate assessor P. Anosov," was appointed as the official representative of the Russian authorities. His duties were to ensure complete support for Western Union Telegraph, oversight of the progress of the work, and of exact compliance with the contracts.

Anosov's instruction, signed by the governor-general of Eastern Siberia, Lieutenant-General Mikhail Korsakov, specified that it was necessary to collect "data on oppressive actions against the Natives" on the part of foreign vessels and to employees of the Russian-American Company, "Having constantly in mind the observance of the just interests of the Company, but especially the preservation of friendly and undemanding relationships with the local inhabitants."[57]

On September 8, 1865, on board the *George S. Wright*, which was at anchor at St. Michael, the leader of the Telegraph Expedition, Colonel Bulkley, issued the following instruction to the "Chief of Exploration" in Russian America, Major Kennicott:

Steamer Geo. S. Wright
At St. Michael. Sept. 8. 1865.
Dear Sir,

You are assigned to duty as Chief of Exploration in Russian America. Your party consists of Wm. H. Ennis, Thos. S. Denison, Geo. R. Adams, F. M. Smith, Frank E. Ketchum, J. Y. Dyer, Chas. Pease, Michael Lebarge, Lewis F. Green, Andrew Gronberg, Jay Chappell, O. de Bendeleben, Richard Cotter. Of this party you are chief in command and will assign to each his respective rank and duty. Major Pope and party or any part thereof, will report to you for duty while in your division. At this place (St. Michael) supplies will be left in charge of Storekeeper James M. Bean, from whom you will draw, giving him receipts for the same.

The object of your explorations is to explore the Kwichpak River [*sic*], prove whether it is the Yukon and determine how far we can use it for our purposes. The map accompanying this has the shortest line traced, for our work, and it is desirable to keep as near this, as topography of the country will admit, and to this end, your explorations will be directed. You will treat with the Indian tribes for labor and supplies, and pursue a course that will render them useful in a further prosecution of our work; and you are authorized to employ such persons as you may deem expedient for purposes of exploration, construction and repairs. I enclose with this, orders to Maj. Pope, failure on his part will not interfere with vigorous action on yours.

In case of his failure, a winter party be pushed forward from Yukon, through Dease House to Babine Lake if possible, and communications sent to any and all parts of the coast. Your Judgment and discretion must guide as facts unfold and the unknown regions you visit become familiar.

As already in our experience, the clouds, that were so threatening in the distance dissolved into thin mist on near approach, so I feel certain you will encounter no serious obstacle and the success will attend your patient persevering effort.

Res'y yours
Chas. L. Bulkley . . . (title)

To: Maj. Robert Kennicott.
Chief of Explorations in R. Am.[58]

This shows that the expedition's basic *geographic* assignment was to establish the identity of the Kvikhpak and the Yukon. Kennicott had formulated this question for himself already on his first trip to the Yukon in the winter of 1860–1861, when he quite rightly proposed that the Kvikhpak was the lower course of this great

Alaska river. We have seen that earlier Zagoskin had arrived at the same conclusion, although he pictured this on his map only in a hypothetical form.

When Kennicott went ashore at St. Michael, he quickly learned that his basic geographic task had been long ago solved by the old-timers of Russian America, foremost among them Ivan Semyonovich Lukin, who was destined to play an important role in the Telegraph Expedition. Lukin was dispatched by the local colonial authorities in 1863 to collect information about the activities of the British on the Upper Yukon. He traveled along with the regularly scheduled Russian trade party to Nukluklayet [at the mouth of the Tanana River—MF], then went on by himself to Fort Yukon. There, posing as a Russian deserter, Lukin collected detailed information about the extent of the British trade in Alaska. Thus, Lukin was the first to ascend the river from the sea to the Porcupine River. The information he obtained, like much other geographic data gathered by the Russians in Alaska, was not made public at the time.[59]

The Americans who had just arrived at St. Michael quickly understood that the Russian-American Company was firmly established in Alaska and that its employees were well acquainted with its geography. These first impressions of the situation in Russian America are vividly reflected in a letter by Perry Collins to Count I. M. Tolstoy, the Minister of Posts and Telegraphs, dated January 29, 1866, in which he specifically acknowledges that he was struck by the fact "that Russia could extend her nationality, her power, *and her visible dominion*, right on to Bering's Strait; plant her standards upon both shores Asia and America, thence along the Kvichpak [*sic*] up to the British frontier and thus open up a Country in many respects valuable to commerce and civilization."[60]

Collins goes on, following his temperament, to call for more active exploration of the northern shores of Asia and America, the development of fisheries there, and study of the Yukon (Kvikhpak) valley, foreseeing its future importance as a transportation artery and a source of mineral and biological resources. Collins ends his letter with the energetic wish "that Russian eagles should rise in the Bering Strait." Tolstoy's reply is evident from his ironic comments on the margins of the document: "Eagles are surely soaring above the Bering Strait. The Cossacks are not there only for the reason that it would be very expensive to feed them there and it would be very cold for them to live there while there will be no benefit whatsoever from their presence. Maybe now we shall have need for them there to see what the Americans are up to or to help the telegraph workers."[61] Besides the cited good wishes and advice, Collins offered his services in this letter in exploring the natural resources in both the Asian and Russian-American possessions. Tolstoy politely but firmly declined.[62]

Having left Kennicott's detachment at St. Michael, Bulkley sailed with the steamer *Wright* and the bark *Golden Gate* into the northern part of the Bering Sea, where Captain Scammon and Navigator Kadin began sounding the depths in the Bering Strait in the direction where it was proposed to lay the telegraph cable. The

Wright, having stopped at Plover Bay to replenish coal supplies, undertook depth soundings in the Bay of Anadyr.

After finishing this hydrographic task, on October 15 the *Wright* arrived at Petropavlovsk-Kamchatsky, where it was awaited by the Siberian detachment of the Telegraph Expedition led by Major Sergei Savich Abaza. Having transferred to Mr. Abaza all the necessary instructions for work to be conducted on the Asian shore, Bulkley returned to San Francisco with the *Wright* and the *Golden Gate*.[63]

In Petropavlovsk the American vessels were joined by the Russian steam-propelled corvette *Varangian*. Its officers, according to Whymper, astounded the Americans with their impeccable manners, hospitality, and their command of French and English. The members of the expedition began to refer to this hospitality in a proverbial manner. Whymper says: "Three months of Russian hospitality" accompanied by ample alcohol libations "would kill most men."[64]

Anosov and Knox set out aboard the *Varangian* for the Okhotsk Sea coast, arriving at Nikolaevsk-on-Amur. From there Knox went to St. Petersburg.[65] The American contingent of the Siberian detachment commanded by Macrae was sent to the mouth of the Anadyr directly from Novoarkhangelsk as soon as August 21 with the schooner *Milton Bodger*. The work of the "Canadian" detachment under Frank Pope and Edward Conway commenced at the same time, sailing aboard the bark *Clara Bell* for New Westminster in British Columbia.

The expedition's contingent working in Alaska was subdivided into the eastern and western sections. The eastern group was supposed to investigate the Yukon above Nulato and join with the party working in Canada; the western group was to explore the region between the Yukon and the Bering Strait.

The work of the "Canadian" detachment proceeded well, as reported by the official inspector dispatched from Russia, Anosov, the administrator for special assignments at the head office of Eastern Siberia:

> E. Conway, having determined the route for the line in the region of British Columbia which is already more or less known, proceeded in September with the construction of the line from New Westminster to Quesnel (the line from St. Francisco to New Westminster was already in operation, built by the Telegraph Company of the State of California; this company is expected to join, in a short time, the Western Union Telegraph Company). He finished the line by November 1. The Frank Pope party, consisting of 25 men, explored in the month of December the area between the Quesnel River and Lake Tapicha. The initial 175 miles of this expanse present almost totally level terrain with very sparse tree cover. In general, it is very convenient for laying the cable. Further along one encounters swamps which will force the cable to be laid in small curves and to deviate from the straight direction of the line. The timber is suited for posts. Indians are few and express friendliness. The

> direction of the line has been determined and according to the last dispatches from Conway construction has begun 40 miles from Quesnel. Pope's party has also explored and measured the Stikine River that flows into the Ocean, along which in the course of this navigational season all the materials for the line between Quesnel and the Kvichpak (Yukon) will be delivered. In the course of this summer Pope's party is to explore the area from Lake Tapicha to the upper reaches of the Kvichpak, where they are to join with Kennicott's party, which is employed in exploring the entire length of the Kvichpak and the area adjacent to the Bering Strait . . . Macrae's party, which landed at the mouth of the Anadyr, is to ascend to the Andayr ostrog and there meet the party dispatched by Abaza from Gizhiga Inlet. The results of the work of these parties are already partially known from the telegram sent by Mr. Abaza via Eastern Siberia. More reliable information will become known upon arrival in Petropavlovsk. Besides, by this time Mr. Abaza should have completed the exploration of the area between Okhotsk and the mouth of the Amur. To achieve this goal he dispatched two parties, one from the mouth of the Amur in the direction of Okhotsk, the second to the SW of Okhotsk, but at this time no news has been received from them.[66]

The *Map of the Northern Stretch of the Asian Line of the Russian-American Telegraph*, scale 100 versts to an inch, preserved in the Russian State Historical Archive in St. Petersburg, shows the results of the explorations conducted by the Russian parties in the winter of 1865/1866. On it is indicated that

> "Explorations were completed in the winter of 1865/1866 by three parties:
> 1. From Nikolaevsk to Okhotsk, Captain Miguda
> 2. From Okhotsk to Gizhiginsk, Abaza
> 3. From Gizhiginsk to the mouth of the Anadyr
>
> The line stretches from Nikolaevsk along the rivers Amgun and Tugir, then along the sea shore to Ayan, from Ayan along the River Maya to Okhotsk. They kept to the sea shore and large navigable rivers, avoiding mountainous and forested terrain."[67]

The expedition's western detachment in Alaska was headed, by Kennicott's order, by William H. Ennis, who transmitted his results to the head of the scientific section in a report signed on December 23, 1865, in the village of Unalakleet.[68] Ennis's group, which included de Bendeleben, Jay Chappell, Richard Cotter, and two Russians, Talmdzhik Nikifor and V. Khansen, departed from Unalakleet on December 2 with two Malemiut sleds across the ice on Norton Sound.

Two days later they arrived safely at the settlement of Shaktoolik-miut, where they were hospitably received by the Natives, who offered shelter for them to rest and dog food. One day later the group continued on their way. They conquered an extremely difficult trail in very cold and windy weather, but reached the settlement of Inglutalik-miut with frostbitten hands, hungry and tired. In this settlement, as was the case earlier, they found that the Indians were holding their festivities, and for this reason difficulties arose in supplying the detachment with the needed food provisions.

The explorers had to move westward, to the settlement of Kvik-miut, where, in spite of another seasonal festivity holding sway, they were able to buy a small quantity of fish. Then the detachment moved on toward the settlement of Koyuk-miut, located at the mouth of the river of the same name. Two days later they found that the settlement was abandoned, "finding it a deserted village, being so for the last 13 years on account of sickness and death."

Ennis halted in one of the abandoned dwellings and commenced to explore the surrounding locality and collect information about the course of the River Koyuk. On December 12 two Indians from Kvik-miut arrived, bringing along two dogs that the travelers had lost. In his report to Kennicott, Ennis wrote:

> From these men, who live on the Koyuk river, I learned, that there is a lake, a great distance from the mouth, but that the river originates in the mountains, from which a small stream takes its source, running to the Westward, flowing into Golovin Sound [that is, Golovnin, to judge by all available data the Natives meant Fish River—AP], and another stream, which empties as they suppose, into Kotzebue Sound. I determined however, to see something of the river myself, and with that intention I left with Mr. O. de Bendeleben, two Russians and one sled lightly loaded, and proceeded on the road.
>
> I had travelled but a short distance, when a South East gale, with snow, came upon us; which in addition to the water and slush on ice, rendered it impossible, with four men pushing, and 7 dogs pulling, to make any headway, so I deemed it prudent to return, as it would with such roads, take several days to go but a short distance, which would add nothing to my stock of information.
>
> In connection with the Kuyuk [*sic*] river I learned that there is no continuous water communication from Kuyuk-mute, to Grantley Harbor, as indicated on your chart. All seem to think the rivers from Golovin Sound to Grantley Harbor are not navigable for even skinboats, and this seems to me to be true, for were those streams navigable, the Indians would not take the route along the Coast, when trading to Port Clarence, when river travelling is so much shorter.[69]

When one takes a glance at a modern map of Alaska, it is not difficult to become convinced as to the accuracy of the geographical information given to Ennis by the Natives. In contrast, the chart used by Kennicott apparently had some very old data, since the Russian materials, from Zagoskin's map on, rejected with certainty any possibility of the existence of an interior water route linking the Norton and Kotzebue Sounds.

On his way from Inglutalik-miut to Shaktulik-miut Ennis "camped for the night on the banks of a river, about 125 yards wide, originating in the Mountains North East, and flowing into Norton Bay. It is called by the Indians Unwakhtulik, abounds in fish of large size, and has never been visited by white men. It is not laid down on any English or Russian charts."[70]

The travelers apparently spent the night on the river known today as Ungalik, and its discovery should be acknowledged as one of the very few achievements of this short journey, the results of which are summarized on a schematic chart of Norton Sound composed by de Bendeleben on the basis of questioning Natives.

Ennis's detachment was not able, because of their lack of a guide, severe weather conditions, and a shortage of food supplies, to undertake the crossing in the direction of the Koyukuk and Nulato. On December 23 Ennis returned to Unalakleet.

Kennicott himself initially planned to ascend the Yukon River by means of a small steam launch that had been brought to St. Michael aboard the *Golden Gate*. He had to abandon this plan as the vessel's engine proved to be out of order, and the mechanic brought along from San Francisco lacked the expertise to repair it. In the end the launch *Lizzie Horner* was beached, moved a safe distance from the water by the members of the expedition and the outpost's residents, and left on shore for the winter. Kennicott, with ten members of the expedition, established yet another base at Unalakleet, where they built a special house as their winter quarters. Bannister remained at St. Michael, being responsible for the equipment and also to make meteorological observations and to collect specimens for the Smithsonian Institution.

After suffering the mishap with the launch, Kennicott decided to reach the Yukon River overland. For this reason, on October 3 he dispatched the youngest member of the expedition, the pharmacist from San Francisco George Russell Adams (1845–1938)[71] to reconnoiter the route. Adams set out up the Unalakleet River in a three-hatch kayak in company with Ivan Lukin and an Eskimo. As the diary of the young American traveler attests, from the very first hours of his acquaintance with Lukin, Adams was astounded by the wide knowledge of the Creole, whom Adams called a Russian, about the regions that Kennicott proposed to explore as a pioneer.

Adams's admiration was aroused not only by Lukin's practical skills in living in the taiga, but also his impressive intellectual abilities that became apparent when the two taught each other Russian and English. Later Adams learned that Lukin had graduated from the seminary in Novoarkhangelsk, and besides complete fluency in Russian spoke Eskimo and two other local languages, including Ingalik, the language of the Natives they were about to visit.

During the trip up the Unalakleet Lukin paddled almost constantly, while Adams determined the magnetic azimuths of separate river sections in order to compose a schematic map. In the settlement of Ulukuk, *toion* Amilka agreed to accompany the expedition party to the Kvikhpak. The young American, largely due to his interaction with Lukin and the elementary knowledge of the Russian language he acquired in their mutual language instruction, concluded that the party's Finnish interpreter was deliberately exaggerating the difficulties in supplying the explorers with warm clothing and provisions for winter travel because he did not want to leave St. Michael in winter.

In the end Adams realized that if the Finn had correctly translated the information supplied by Lukin and other employees of the Russian-American Company for Kennicott, Adams's trip would have been totally unnecessary, as the expedition's leader could have acquired all the necessary information from the Russians at Unalakleet, for whom the way to the Kvikhpak was as well known as to the Natives themselves.[72]

Kennicott, supplied by the Russian manager Stepanov with the clothing he needed and seventy-five dogs, set out on his way to the Kvikhpak on October 31, 1865. On the journey along the overland trail used by Russian trading and hunting parties Kennicott was accompanied by Mike Lebarge, Adams, two employees of the Russian-American Company (Lukin and Khakerin), and two Indian bearers. The Russians and Natives hunted, cooked food, and in general provided for the expedition.

Kennicott did not always manage to exercise the much-needed self-control. Thus, when Khakerin began to object to a demand by Kennicott and offered the opinion that they were "pure fools" to attempt such a journey in wintertime, the American, without thinking twice, beat the undisciplined Russian with an axe handle. Later the imperturbable Lukin remarked to Adams that "it was all right to punish Hackerine [*sic*], but that the Major should have taken a stick and not an axe, because he might have used the wrong end."[73]

The Russian manager at Nulato met the travelers very hospitably. Kennicott spent five days there, making observations and collecting various information from the fort's residents and from the Natives. Then he departed to return to Unalakleet, leaving the following instruction for Adams:

> Russ. American Fur. Co. Post Nulato
> R.A. Nov 14th 1865
> Lieut Geo. R. Adams
>
> You will during my absence (of about forty days) with Mike Lebarge and Evan [i.e., Ivan—AP] Simeon Lukine make such explorations on or near the proposed route for telegraph from Koyukuk westward, as your limited means will permit. You will also do any work you may deem expedient, with the view of facilitating the rapid and safe progress of the expedition towards Fort Youkon.
>
> You will draw such supplies from the Bajdarjik [*sic*] of this post as you deem necessary. He will also furnish you with a sled and three dogs. I would

advise you to proceed first to Koyukuk and attempt to penetrate the country westward toward Norton Bay, striking southward to the Kivhpak [*sic*] in time to reach this post before your provisions are exhausted. All further detailes [*sic*] I leave entirely to your own sicretion [*sic*].

Robert Kennicott
Major, Comm. Russ. Am. Tel. Ex.[74]

The execution of this instruction was postponed, however, as Kennicott found it imperative to entrust Adams and Lukin with preparing the detachment's wintering at Nulato. For this reason, after obtaining five dogs and a sled from the Russian manager, they set out eastward on a tour of Indian settlements where it was possible to obtain fish and game. An Indian who was nicknamed Kurilka was their guide.[75] Returning from this trip, Adams and Lukin rejoined Kennicott at Unalakleet. On January 15, 1866, Kennicott, Adams, Lukin, Smith, and an Eskimo once again set out for Nulato with three dog teams. Following their arrival at the Russian factory (Fig. 69), Kennicott, accompanied by Adams, Lukin, and Kurilka, completed a ten-day journey up the Kvikhpak.

AURORAL LIGHT SEEN FROM NULATO, YUKON RIVER DEC. 27, 1866.

FIGURE 69 The aurora over Nulato in Whymper, *Travel and Adventure in the Territory of Alaska . . . 1868* Frontispiece. UAF Rare Book A0547

The major again dispatched Adams with Smith, Ketchum, and Pease to Unalakleet to fetch baidarkas for the future ascent of the Yukon. Only on May 1, 1866, did Adams and Dyer undertake to execute the above-cited instruction by Kennicott, traveling north to ascend the River Kaiekak and explore the watersheds and rivers flowing toward the Bering Strait in order to assess the possibility of laying the telegraph line in this direction. Adams was unable to complete this task because of the spring break-up.

While he was at Nulato Kennicott made several exploration trips in the mountains fringing the Yukon valley, notwithstanding the severe subarctic winter conditions. One of his companions, Lieutenant Charles Pease, later wrote the following about Kennicott's trips:

> Attended by an Indian or two he went out on the bleak, desolate mountains west of Nulato, looking for a pass to the sea-coast and gathering materials for a map of the vicinity . . . His sufferings, mental and physical, during the period when he was on those forbidding mountains can be but faintly realized from a few chance words he afterwards let fall. His was the Spartan courage which suffered and gave no sign . . . After his return he seemed to throw off his dejection. He began to enjoy the gradual approach of leaves, birds and salmon, and thought less of the cares and annoyances of the dreary winter season.[76]

In the morning of May 13, 1866, Major Kennicott's lifeless body was found by Pease and Lebarge not far from the Nulato post: "On the beach was placed the Major's pocket compass and lines indicating the bearings of the various mountains in sight, drawn in the soft alluvium, showing that he had been busy in adding to his material for a map of the country around Nulato when his death occurred."[77]

To this day the cause of Kennicott's death has not been determined. A friend and coworker of the scientific detachment's leader, William H. Dall, wrote: "His end was caused by heart failure, which grew more acute due to cold and privations from which he protected others but not himself."[78] Lieutenant Adams thought that Kennicott committed suicide, taking strychnine, which the scholar possessed in quantity but which disappeared after his death.

Whatever the case may be, Kennicott foresaw his near end. In a letter he left behind he instructed that in case of his death Ketchum should be appointed leader of the Yukon party and Ennis was instructed to lead the explorations in regions to the west and north of Nulato to determine the telegraph route in the direction of the Bering Strait.[79] Kennicott's body was brought by baidara down the Yukon to St. Michael, in a coffin made by Lukin and Pease from boards from the old Nulato Post. The remains of the American naturalist were interred at his estate, The Grove, near Northfield in Illinois.

Dall Takes Command

Dall was appointed general head of the scientific corps, which by this time was actually falling apart. Bannister, whose face was frostbitten during his single field trip, was sent home. Bischoff left for Washington after a year spent at Novoarkhangelsk. Rothrock and Elliott were out of reach in Canada. Although Bulkley promised Dall that he would send Elliott to assist in scientific explorations in Alaska, the latter preferred to engage in the construction of the Canadian section of the telegraph line, which brought in large monetary profits. Thus, in reality, Dall became the scientific leader of an expedition that lacked researchers.

Kennecott's plans for "conquering the Yukon" were made a reality by Frank Ketchum. On May 26, 1866, with Lukin and Lebarge he set out in a three-hatch baidarka from Nulato upstream along the Kvikhpak.[80] They were accompanied up to the Nukluklayet by a Russian *promyshlennik* on his regular annual trading trip for barter with the Natives along the Yukon. According to Whymper, Lukin proved to be the *de facto* leader of this expedition.[81] At Nukluklayet the exploring party arrived at the annual trade fair of the Tanana Natives: two headmen were present; one was acknowledged by the Russians, the other by the British.

Continuing on their route, the travelers overtook an English missionary thirty miles upstream along the Yukon. Together with him they came to Fort Yukon. Not finding the main trade agent in residence, the American travelers left after three days, arriving at Nulato eight days later. This expedition was strictly for reconnaissance. It is difficult to discern how many points were determined by geographical coordinates and whether any astronomical determinations of geographical coordinates were carried out.

The next year two journeys were undertaken, in the course of which a reconnaissance survey of the river was indeed carried out. In the winter of 1866–1867 Dall and the British artist Frederick Whymper joined the explorers at Nulato and busied themselves assembling natural history and ethnographic collections. The significant event that winter was a trip by Ketchum and Lebarge to Fort Yukon. Accompanied by four Natives, the travelers left Nulato on March 11. Their equipment was transported on four sleds with fourteen dogs.

The party arrived at Fort Yukon May 9, then Ketchum and Lebarge proceeded farther upstream, traveling by birch bark canoe. On June 25 they reached Fort Selkirk, located across from the mouth of the Pelly River. This was the expedition's farthest point up the Yukon. Returning to Fort Yukon (Fig. 70), the travelers met Dall and Whymper there, who had arrived a week earlier by boat out of Nulato. On July 8 the united detachment left Fort Yukon, and five days later (July 13) they were at Nulato, where orders to immediately cease work and return to St. Michael awaited them. At this main base of the expedition the travelers learned that all the work in exploration and construction of the line had been abandoned, as by July 27, 1866, the transatlantic cable had been successfully laid[82] and had begun to function.

FIGURE 70 Fort Yukon from Whymper *Travel and Adventure in the Territory of Alaska . . . 1868.* Facing p. 219. UAF Rare Book A0547

Dall preferred to continue his explorations in Alaska, however. On February 3, 1868, he learned from Lukin, who had just arrived from St. Michael, that Alaska had been sold to the United States of America. Dall solemnly raised the U.S. flag over the former Russian outpost in the depth of the American continent.[83]

The Telegraph Expedition, in spite of the fact that it was never completed, was of major scientific significance. It opened the new, American period of Alaska's study. Investigations and impressions by members of this expedition were published in the fields of geography, meteorology and hydrology, magnetism, history, bibliography, economics, geology, paleontology, the geography of flora and fauna, anthropology, ethnography, zoology, and botany. The most important were the studies by Dall, completed both as a member of the expedition and in the course of the years he spent in Alaska following the expedition's cancellation.

Specifically, Dall established the Canadian character of northern Alaskan flora. He also noted the absence of signs of glaciation activity and concluded that to the north of the Alaska Range (as was the case for part of Siberian territory) there had been no glaciation. At the same time he stressed the presence of permafrost a few feet below the surface, in which the remains of mammoths were embedded. Dall described coal deposits on the Yukon and in Cook Inlet. Whymper and Dall noted

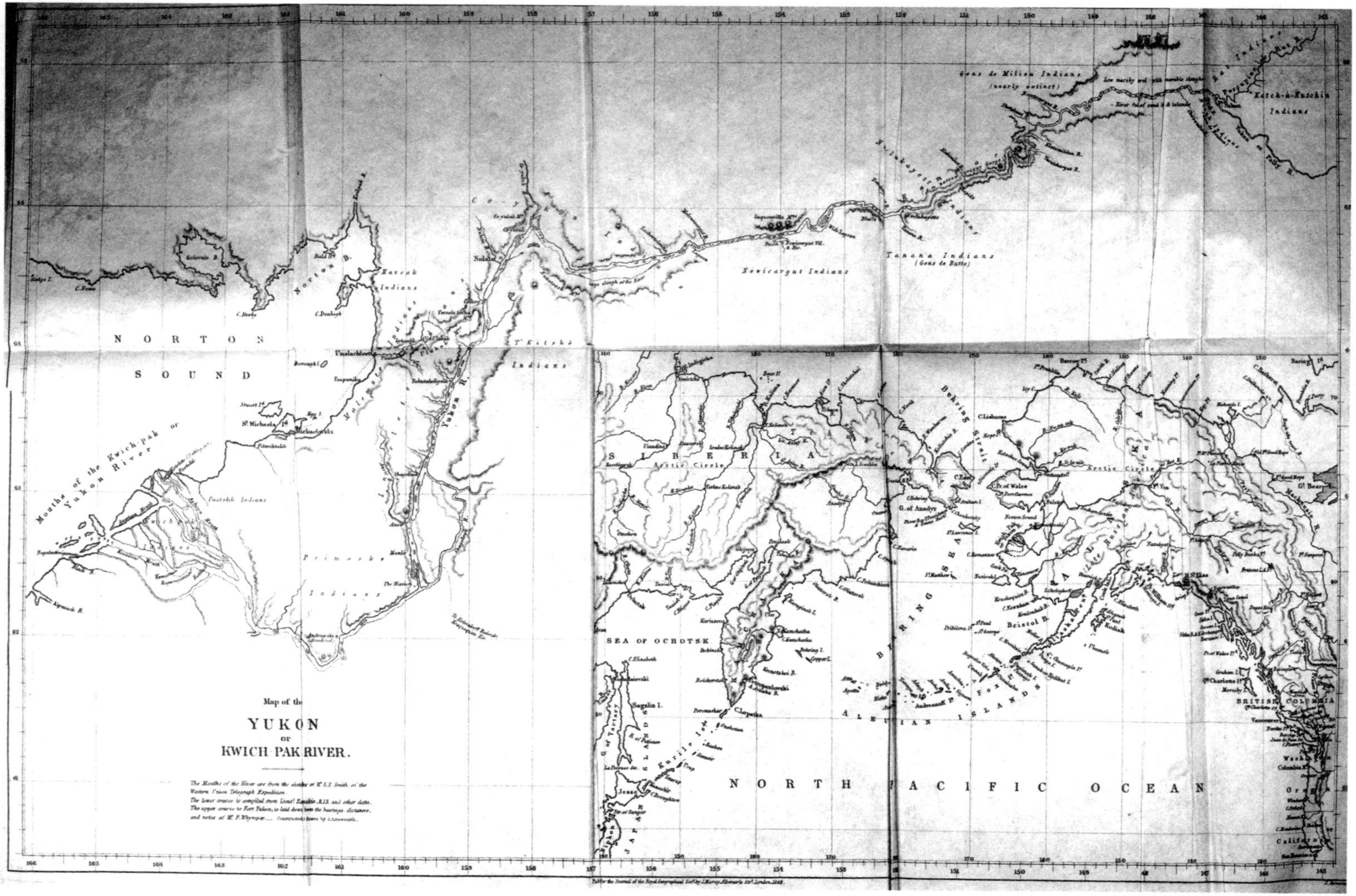

FIGURE 71 Map of Alaska from Whymper *Travel and Adventure in the Territory of Alaska . . . 1868* Endpapers. UAF Rare Book A0547

finds of gold on the Yukon River by Hudson's Bay Company employees. Members of the expedition also found gold nuggets on the Seward Peninsula.[84]

The summary results of the Yukon explorations of 1867–1868 are shown on the *Map of the Yukon or Kwich-pak River*, appended to Whymper's book (Fig. 71). The legend states that "the lower course was charted after Zagoskin and others [Russian sources—AP], the upper course on the basis of magnetic directions, distances and notes by F. Whymper."[85] The overall achievements by the Telegraph Expedition in Alaska were summarized in the *Map of Russian America or Alaska Territory, Compiled from Russian Charts and Surveys of W. U. Tel. Ex. Col. C. I. Bulkley, Eng.r in Chief. By J. F. Lewis, Chief Draughtsman. M. Cadin, Ass.t 1867, Lithog.t by Britton and Rey S. R.* and *Alaska and Adjoining Territory 1869. The Yukon River, Ranges of Mountains, Shores of Norton Sound and Many Features of the Interior from a Reconnaissance by W. H. Dall, Director of the Scientific Corps of the W. U. Tel. Expedition, 1865–1868. Drawn by H. Lindenkohl (Washington, U.S. Coast Survey, 1869).*

We must note that these three maps could not have been used in discussions by American lawmakers of the question of Alaska's acquisition, as they appeared after the decision was taken. However, the possibility cannot be ruled out that some early manuscript versions of these works were consulted. Various authors evaluate the role of the Telegraph Expedition in the Alaska Purchase differently. For obvious reasons, in their published works former participants tended to somewhat exaggerate the influence exerted upon the senators and congressmen by the geographical information about Alaska they obtained.

On the other hand, in preparing his massive campaign for the purchase of the Russian possessions in Alaska, Seward carefully studied the materials at the Smithsonian Institution that were just then being augmented by data from the Telegraph Expedition. Bannister and Bischoff testified to their personal impressions before the Senate Foreign Relations Committee, while Spencer Baird delivered important zoological data based on earlier collections that included the rich materials by Kennicott.

However, as the American scholar Morgan Sherwood reasonably suggests, the information from the Smithsonian Institution was clearly insufficient to ensure victory for the supporters of the purchase. Sherwood stresses: "There were few original materials that had been properly worked over, and the Telegraph explorers had after all barely examined the country. Bischoff could report with authority on Sitka, to be sure, and Bannister's meteorological data were vital, since climate was a major point of attack for opponents of the treaty. Whether, however, the Telegraph Expedition information was of more actual value than Russian and English materials is a moot point."[86]

It appears, however, to be beyond dispute that along with the basic geopolitical arguments advanced by the proponents of the Alaska Purchase, geographical data about the country also carried substantial weight, and in this sense the Telegraph Expedition "may indirectly have facilitated [Alaska's] purchase."[87] But this expedition was by no means the sole purveyor of geographical information to

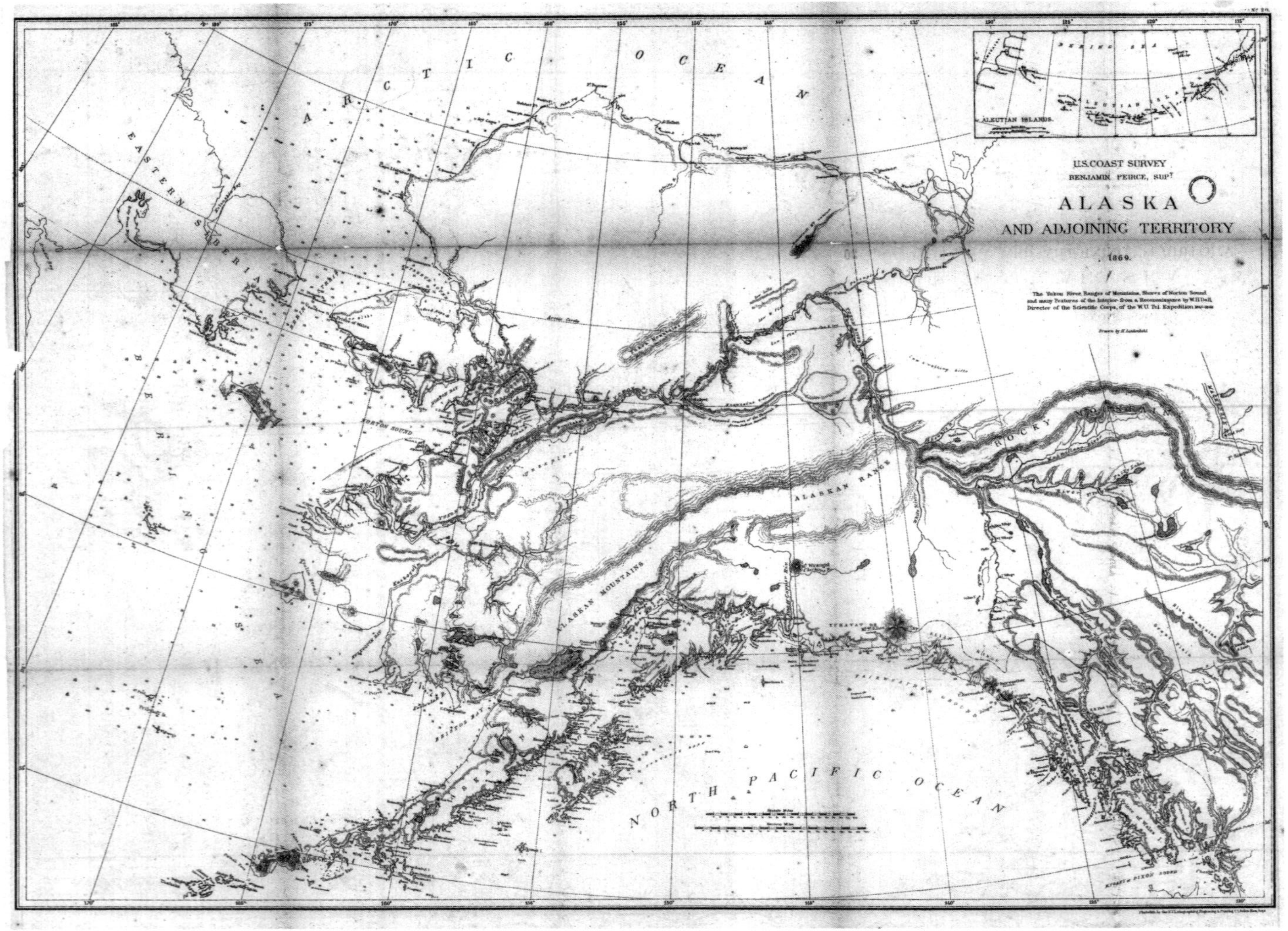

FIGURE 72 *Map of Russian America or Alaska Territory*. . . . U.S. Coast Survey, 1869. UAF Rare Map G4370 1869 U55a

the American authorities. Earlier we saw that back in 1863 William P. Blake had participated in the Russian expedition to the Stikine River. He presented a map of this river, along with his observations, to Seward.[88]

THE TRANSFER OF RUSSIAN INFORMATION TO THE UNITED STATES

Seward dispatched an expedition of the U.S. Coast Survey to Russian America in 1867. It was led by George Davidson aboard the steamer *Lincoln*. This expedition arrived in Sitka on August 11 and returned to San Francisco at the end of November.[89] In the short time allowed the expedition, Davidson himself could only complete some hydrographic surveys and compile a chart of St. Paul Harbor in Kodiak[90] and of Sitka Harbor on Baranov Island.[91] Astronomical, magnetic, and tide and ebb observations for these charts were made by Davidson, while he obtained soundings from Russian sources.

Judging by all available information, those sources were the main ones used in compiling the expedition's report, which was based more on literature than on field sources.[92] The full text of Davidson's *Report Relative to the Resources and the Coast Features of Alaska Territory* was included in the compendium of documents of the Second Session of the 40th Congress of the USA and an article about the expedition was published in 1868.[93]

We indicated above that Davidson had acquired a copy of Tebenkov's atlas in 1856.[94] It is quite possible that during the 1867 expedition the U.S. Coast Survey's collection of charts was enriched by yet another compilation of navigational charts published by the Hydrographic Department of the Ministry of the Navy of Russia and known in literature as *Atlas of the Eastern Ocean* by A. F. Kashevarov.[95] A copy of this compendium is in the cartographic department of the Library of Congress in the U.S.[96] The exact date of its creation is unclear.

The copy preserved in Russia contains thirty-six printed charts, engraved between 1844–1854, showing the coasts of the Arctic Ocean, of the Bering, Anadyr, Okhotsk, and Japanese seas, the Gulf of Alaska and San Francisco Bay, the Tlingit [Koloshenskie] Straits in Alexander Archipelago, and the entry into the Columbia River. The title page has been lost, while the table of contents was done by hand and does not contain the atlas compiler's name, nor the date and place of publication. All of these data are known only from V. V. Kolgushkin's catalogue, where it is noted that the atlas was kept in a cardboard case inscribed *Atlas of the Eastern Ocean, Compiled from Various Journals and Charts by A. Kashevarov.*

Alexsandr Filippovich Kashevarov, who we met earlier in command of an expedition along Alaska's Arctic coast, had subsequently commanded a series of Russian-American Company ships before being sent to the Hydrographic Department of the Naval Ministry in Saint Petersburg to make charts. He took a substantial amount of documentation from Alaska with him, and detailed charts of the North Pacific were published during his time at the Hydrographic Department, although his name does not appear on the maps themselves. Following a series of other assignments, he eventually retired with the rank of general in 1865. I believe the Kashevarov *Atlas* to be a compilation of the charts Kashevarov worked on that were issued by the Naval Ministry in the form of separate maps, but not intended to be published as a complete atlas for distribution [MF]. (Fig. 73 is one of these charts.)

The entry also indicates, without stating the source, that the *Atlas* was published by the Hydrographic Department of the Navy Ministry in October of 1862.[97] It can be stated with certainty only that this compendium was published not earlier than 1854.[98] Be that as it may, by 1867 Davidson had accumulated a considerable number of the contemporary Russian charts. Judging by everything, these materials, together with the data collected by members of the Telegraph Expedition, were freely used in preparing the famous three-hour speech of April 8, 1867, by the senator from Massachusetts[99] that ensured the ratification of the treaty by the United States Senate (thirty-seven votes for and only two against).

According to statements in the American press, which reported Senate debates in detail, Charles Sumner used "documents, charts, sketch drawings," and so on to illustrate his vivid presentation.[100] We may be certain that among these illustrations were some of the maps and charts we have discussed here. The materials collected by Davidson's expedition were used in the report by the head of the Coast Survey for 1867. It also formed the basis for the *Coast Pilot of Alaska*, published in 1869.[101]

Let us note that in the description of the resources of Russian America in Davidson's *Coast Pilot* there is a mention of the discovery of oil by a teacher from the Russian-American Company School—a natural resource that was destined to greatly enrich Alaska. The emergence of oil was noted at a point at 58°1′ N latitude and 154°54′ W longitude, and also adjacent to Kodiak Island.[102]

Naturally, not only cartography but also the opinions of well-known American scholars who had perused all the available information about Russian America, including data collected by the Telegraph Expedition, played an important role in preparing the arguments in favor of the acquisition of Alaska by the United States. Such a luminary as Professor G. L. R. Agassiz drew Sumner's attention to the vast natural resources of Russian America and benefits to the United States' commerce resulting from its acquisition. G. V. Fox, M. K. Migs, B. Pierce, Spencer Baird, and others also wrote to Sumner. Specialists from the Smithsonian Institution personally addressed meetings of the Foreign Relations Committee and provided convincing testimony in regard to the real value of the territory being acquired.[103]

The Russian side also devoted close attention to the territory that it planned to relinquish. But the imperial government was in essence interested in other aspects of the situation concerning Russian America than its American partners in the negotiations. The geographical specifics of the transoceanic colonial possessions were known to the Russians sufficiently well. Therefore their main focus of interest was the socioeconomic factors that determined the rationality of continuing

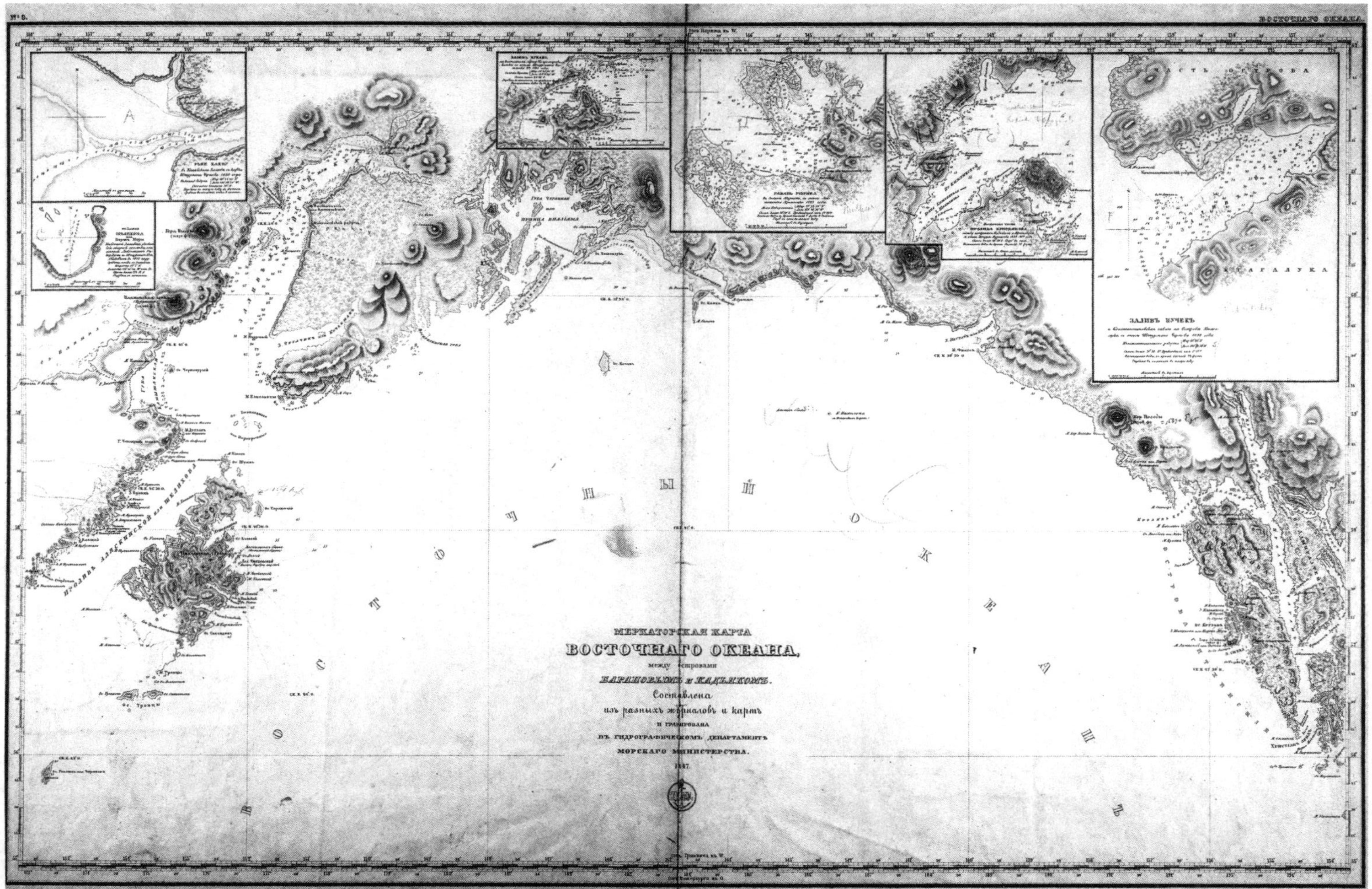

FIGURE 73 Mercator's Chart of the Eastern Ocean between the Islands of Baranov and Kodiak. Compiled from various journals and maps and engraved in the Hydrographic Department of the Naval Ministry. 1847 Ministry Map no. 1378 UAF Rare Map G4372 A4231 1847 G52

to maintain Russian America within the Russian Empire. The privileges of the Russian-American Company expired on January 1, 1862. This date coincided with a period of basic reforms taking place in Russia, the foremost being the abolition of serfdom. Progressive politicians had begun to criticize the colonial order established in Russian America with ever-greater frequency. They argued against the exploitation of the local population and the monopolistic privileges of the Russian-American Company.

In May 1860 a decision was taken to dispatch to Russian America two plenipotentiary inspectors, Active State Councilor S. A. Kostlivtsev, representing the Ministry of Finance, and Captain-Lieutenant (later Captain of the Second Rank) P. N. Golovin from the Navy Ministry, where the activities of the Russian-American Company were the subject of exceptionally harsh criticism.[104]

The main task of the inspectors, according to their secret instructions,[105] was to ascertain to what extent Russian America was ready for the changes taking place in Russia through the mechanism of reforms and for the condition of canceling the exclusive privileges of the Russian-American Company, and then, in general, if it was "rational to continue the dominion over Russian America?"[106] The core of the instruction to the inspectors was a questionnaire composed by Kostlivtsev in the form of a table in the tradition of the eighteenth- and nineteenth-century Russian questionnaire surveys: *Register about the Real Situation of the Russian-American Colonies, the Year 1860*. This *Register* demanded specifically that the following detailed information be given with respect to each place of habitation:

> 1. Name of the settlement, camp or village in sequence. 2. What tribes inhabit each. 3. Number of members of each tribe, separately (by gender male/female). 4. When and by whom was the fortification, camp or village founded. 5. At the time of founding of the post what was the number of souls (male/female). 6. Comparative decrease and if known causes of same. 7. What are the occupations engaged in. 8. What is the number of persons being sent out to work (males/females). 9. For how many days. 10. For what payment per day. 11. What is the average take of animals per year. 12. What is the total value according to the prices existing in the colonies or in the nearest markets. 13. Do the local inhabitants receive any kind of support beyond the monetary payments, and if yes, what does it consist of. 14. Who is in charge of assigning laborers to their tasks. 15. Who is the manager of the locality and who are his subordinates from among the employees. 16. What does each one of them receive in money and in other forms of support. 17. List the structures in the given locality and their valuation. 18. Besides structures, what kind of property is there in a given locality and what is its estimated valuation. 19. When and by what means are the local products obtained. 20. How are these products used. 21. Is any decrease in

catch noticed, and if so, what are the reasons. 22. What are the measures instituted to maintain the catch level.[107]

In addition, a general overview of the history of the Russian-American Company was to be produced, the explorations conducted by it in the colonies, a report in detail on the situation of the Natives and on the relations between them and company employees, and the activities of foreign citizens in the colonial territories and trade with the same.

The inspectors, Kostlivtsev and Golovin, returned from their inspection tours in the fall of 1861 and presented detailed reports about the situation in the Russian transoceanic possessions. In spite of the serious criticisms leveled against the Russian-American Company, both concluded, after an attentive study of the questions, that rationality called for the preservation of the company. The published materials by Kostlivtsev and Golovin gained the status of fundamental sources for the history of Russian America and are widely used by scholars to this day.[108]

As Bolkhovitinov demonstrated, Golovin's original report, preserved in the State Archive of the Russian Navy, is essentially different from the published version.[109] The concluding section of the report that is omitted in the printed version contains not only an overview of the well-known confidential negotiations between Senator Gwin and Edward de Stoeckl, but also Golovin's own thoughts about the fate of Russian America. Golovin's opinion tended decisively toward the preservation of Russian possessions on the American continent. In support of his position he cited the following convincing considerations:

> Public opinion in Russia to this day is up in arms about our having ceded the former factory in California, especially since the time that gold was discovered next to the Ross settlement. It can easily happen that after the reorganizations envisioned for our colonies, enterprising men, entering upon their tasks with reason and energy, will discover in our colonies riches not suspected to exist at this time. In respect to securing friendly relations between Russian and the United States, it is possible to state that their goodwill will last only as long as it does not bind them to anything or as long as it is to their benefit. Americans will never sacrifice their interests out of simple principles.
>
> In conclusion, in order to characterize the personalities of the American **politicians** [English in original—LB], I should add that Senator Gwin left the State of California after failing to attract the voters to his cause. Abandoning his convictions, he went to the southern states to try his luck among the separatists.
>
> Captain-Lieutenant Golovin, October 20 in the year 1861.[110]

The Committee for the Reorganization of the Russian American Possessions, created to study the reports by Kostlivtsev and Golovin, chaired by the director of the Department of Manufactures and Interior Trade, A. I. Butovsky, on the whole acknowledged the merits of the Russian-American Company and came out in favor of extending its charter for twelve (instead of twenty) years. However, the company was to be put under the Navy Ministry.

A military governor of the region was to be appointed and a permanent patrol by navy vessels established. Other improvements were to be introduced in the colonies too. In the final analysis, the Council of State, upon representations by the Department of the State Economy, supported the opinion that it was desirable to preserve the Russian-American Company. Tsar Alexander II approved this decision on June 14 (26), 1865, which was signed by Grand Duke Konstantin and published in the press.[111]

None of this, however, interfered with the decision taken barely a year-and-a-half later, on December 16 (28), 1866, to sell the Russian possessions in North America to the United States, made in a "special session" held in the main hall of the Ministry of Foreign Affairs and attended by Alexander II, Grand Duke Konstantin, Minister of Foreign Affairs Alexander Gorchakov, Finance Minister Mikhail Reitern, Minister of the Navy Nikolai Krabbe, and Ambassador to Washington de Stoeckl.[112] On March 18 (30), 1867, at four o'clock in the morning, the treaty was signed for the sale by Russia of her American possessions to the United States of America for $7,200,000.[113]

Later, Robert J. Walker,[114] an influential Washington lawyer, senator (1835–1845) and finance secretary of the USA (1843–1849) who was well known for his expansionist views, was to call the purchase of Alaska "the greatest act" of that period's American administration. Developing his expansionist plans, Walker stated that the theater for the "greatest triumphs" of the United States was the Pacific Ocean, where soon there would not be any "threatening European competitors." He did not hide the fact that he dreamt of final domination by the USA in politics and commerce throughout the entire world.[115]

On October 6 (18), 1867, at 11 o'clock, the American representative General Lovell Rousseau, the official representative of the Russian government, Captain Second Rank Alexei Peshchurov, commanding the armored frigate *Minin*, and Captain Second Rank F. F. Kosul, representing the interests of the Russian-American Company, arrived at Novoarkhangelsk.

The representatives of the two countries, in the presence of the last chief manager of the Russian colonies, Captain First Rank Prince Dmitri Maksutov, on the same day, at three-thirty in the afternoon, commenced the solemn ceremony of transferring Russian America to the United States. Military detachments were lined up on the plaza in front of the chief manager's residence, and while cannon shots were fired in salute, the ceremony of lowering the Russian flag and raising the American flag took place.[116] And thus the sole transoceanic colony of the Russian empire ended its existence.

FIGURE 74 Sitka from the Eastern Harbor *Coast Pilot* UAF Rare Book A1868 1869-120

FIGURE 75 Sitka from Western Harbor *Coast Pilot* UAF Rare Book A1868 1869-118

NOTES

1. N. N. Bolkhovitinov, *Doktrina Monro: Proiskhozhdenie i kharakter* (*The Monroe Doctrine: Its Origin and Character*). Moscow, 1959; Bolkhovitinov, *Rossiya i SShA: arkhivnye dokumenty i istoricheskie issledovaniya: Analitichesky obzor*, Moscow, 1984; an expanded edition in English: Bolkhovitinov, *Russia and the United States: An Analytical Survey of Archival Documents and Historical Studies.* Translated and edited by J. D. Hartgrove. Armonk, N.Y., 1989; Bolkhovitinov, "Obshchestvennost SShA i ratifikatsiya dogovora 1867 g." ("The American Public and the Ratification of the Treaty of 1867"), *Amerikansky ezhegodnik* (*American Yearly*), Moscow, 1987:157–174; Bolkhovitinov, "Kak prodali Alyasku" ("How Alaska was Sold"), *Mezhdunarodnaya zhizn* (*International Life*), 1988:7:120–131; Bolkhovitinov, "Alyaskinsky skandal" ("The Alaska Scandal"), *Voprosy istorii* (*Historical Affairs*), 1989, no. 4:37–54; Bolkhovitinov, *Russko-amerikanskie otnoshenie i prodazha Alyaski 1834–1867* (*Russian-American Relations and the Sale of Alaska*), Moscow: Nauka, 1990 in English as *Russian American Relations and the Sale of Alaska, 1834–1867.* Translated and edited by Richard A. Pierce. Kingston and Fairbanks, 1996.
2. For details, see Bolkhovitinov, 1959:68–69.
3. John Q. Adams, *Memoirs of John Quincy Adams*, vol. 1–12, edited by C. F. Adams. Philadelphia, 1874–1877, vol.4:438–439. Bolkhovitinov, 1990:50–51; On Adams's expansionist views, see also W. Lafeber, ed., *John Quincy Adams and American Continental Empire*, Chicago, 1965.
4. Cited after Howard I. Kushner, *Conflict on the Northwest Coast: American-Russian Rivalry in the Pacific Northwest*, 1790–1867, *Contributions in American History*, Number 4, London, Westport, 1975:3.
5. Kushner, ibid., 1975:59–60.
6. J. W. Pratt, "The origin of 'Manifest Destiny,'" *American Historical Review*, July 1927, vol. 32:4:793–798. See also J. W. Pratt, "The Ideology of American Expansion," *Essays in Honor of William E. Dodd.* Edited by A. Craven, Chicago, 1935 and Bolkhovitinov, 1990:49–50.
7. W. H. Seward, *The Works of William H. Seward.* Edited by G. E. Baker, vol. 3. New York, 1972:409.
8. Victor J. Farrar, *The Annexation of Russian America to the United States*, Washington, 1937:21–22.
9. Kushner, ibid., 1975:114–115.
10. H. M. McPherson, "The Interest of Wiliam McKendree Gwin in the Purchase of Alaska, 1854–1861." *Pacific Historical Review*, 3:1:31–32, March 1934. The life and activities of Gwin are examined in detail in the unpublished dissertation by the same author: H. M. McPherson, *William McKendree Gwin, Expansionist*, Ph.D. dissertation, University of California, Berkeley, 1931, VII. See also Bolkhovitinov, 1990:5–58.
11. A. B. Cole, ed., *Yankee Surveyors in the Shogun's Seas*, Princeton, 1947:55–56. Bolkhovitinov, 1990:143.
12. Russian-American relations during the years of the Crimean War have been examined in detail by V. N. Ponomarev. See V. N. Ponomarev, *Russko-amerikanskie otnosheniya v gody Krymskoi voiny, 1852–1856* (*Russian-American Relations during the Years of the*

Crimean War, 1852–1856). Dissertation. Candidate of Historical Sciences, Moscow, 1984; Ponomarev, V. N., "Russko-amerikanskie otnosheniya v gody Krymskoi voiny, 1852–1856." *Istoricheskie zapiski* (*Historical Notes*), vol. 110, 1984:232–281.

13. B. V. Struve, *Vospominaniya o Sibiri, 1848–1854 g.*, St. Petersburg, 1889. These same materials were used, with reference to Struve, by I. Barsukov and Bolkhovitinov. See I. Barsukov, *Graf Nikolai Nikolaevich Muraviev-Amursky* (*Count Nikolai Nikolaevich Muraviev-Amursky*), vol. 1, Moscow, 1891:322–323 and Bolkhovitinov, 1990:91–92.
14. Struve, 1889:157; Bolkhovitinov, 1990:92–93. On the activities of the Russian-American Company on Sakhalin Island, see P. A. Tikhmenev, *Istoricheskoe obozrenie obrazovaniya Rossiisko-Amerikanskoi kompanii i deistvii ee do nastoyashchego vremeni* (*Historical Review of the Creation of the Russian-American Company and its Activities to Date*), part 2. St. Petersburg: E. Veimar, 1862:99–128.
15. RGIA, fond 1289, opis 1, no. 1444: *Po predlozheniyu polkovnika Kembelya o provedenii telegrafnoi linii cherez Sibir dla soedineniya Evropy s Amerikoi* (*On the Proposal of Colonel Campbell to Establish a Telegraph Line across Siberia to Join Europe with America*). August 16, 1858–November 6, 1862 (196 folios), folio 48–48V.
16. Ibid., *Perepiska ob ustroistve telegrafa mezhdu Amerikoi i Sibiriyu* (*Correspondence on the Construction of the Telegraph between America and Siberia*), February 25–March 10, 1859 (in five folios), folios 1–1 verso. Romanov's materials were published without delay and he himself was sent to the USA to acquaint himself with the telegraph business. There he partook in discussion of the project, see D. I. Romanov, "Proekt rossko-amerikanskogo mezhdunarodnogo telegrafa" ("The Russian-American International Telegraph Project"), *Russkoe slovo* (*The Russian Word*), February 4, 1859; Romanov, "Perezabytyi vopros o telegrafe cherez Sibir mezhdu Starym i Novym Svetom" ("The Forgotten Issue of the Telegraph across Siberia between the Old and the New World"), *Sankt-Peterburgskie vedomosti* (*The St. Petersburg Register*), February 28, 1861; and Romanov, "Sibirsky telegraf" ("The Siberian Telegraph"), *Russky invalid* (*The Russian Invalid*), March 12, 1864.
17. RGIA, fond 1289, opis 1, no. 1444, folio 48 verso.
18. Ibid., folios 4–4 verso. The English text of Colonel Campbell's proposal is preserved in this file: *The Great Eastern and Western (via Behring's Straits) European and American Overland Telegraph Line*, folios 7–10 verso.
19. Ibid., folios 44–44 verso.
20. Ibid., folio 46.
21. Ibid., folio 54.
22. Ibid., folio 143.
23. Bolkhovitinov, 1990:143.
24. Kushner, 1975:122.
25. Peyton's letters have been published by Norman E. Soul, "American's Siberian Dream," *Russian Review*, 37:4:405–420, October 1987.
26. Bolkhovitinov, 1990:144–146. The United States Congress published Collins's correspondence in its entirety in 1858. Later Collins published his observations as a separate book: *35th Congress. 1st Session. House Ex. Documents 98*, Washington, 1858; P. M. Collins, *Voyage down the Amoor: with Land Journey through Siberia, and Incidental Notices of Manchooria, Kamchatka and Japan*, New York, 1860 (second edition 1864).

In 1962 Charles Vevier reissued this book with slight changes and scholarly annotations: P. M. Collins, *Siberian Journey down the Amoor to the Pacific, 1856–1857*, edited with introduction by C. Vevier. Madison, 1962.

27. Bolkhovitinov, 1990:146–147.
28. RGIA, fond 1265, opis 6, *Vypiska iz zhurnala Sibirskogo komiteta, 22 aprelya (4 maya) 1857 g.* (*Excerpt from the Journal of the Siberian Committee, April 22 [May 4], 1857*). Folios 17–18. Bolkhovitinov, 1990:147–148.
29. Kushner, 1975:123.
30. Ibid., 124.
31. Romanov, 1964:3–4.
32. AVPRI, fond Sankt-Peterburgsky Glavny Arkhiv I-9, opis 8, no. 10, 1862–1868: *Po predlozheniyu po provedeniyu telegrafa ot Zapadnago Pribrezhya Severnoi Ameriki v Vostochnuyu Sibir . . .* (*On the Proposal to Establish a Telegraph from the West Coast of North America to Eastern Siberia . . .*). Folios 1–1 verso. Letter from Baron de Stoeckl to Grand Duke Konstantin dated April 14 ,1861.
33. Ibid., folio 14 verso.
34. Ibid., folio 14.
35. Specifically, the head of Russia's department of Communication Routes firmly informed the American ambassador that all exploration of the proposed route in Russian territory would be carried out by Russian specialists. See AVPRI, fond Sankt-Peterburgsky Glavny Arkhiv I-9, opis 8, no. 10, folios 12–13.
36. Ibid., Letter. United States embassy, dated October 29–November 10, 1862, signed by Bayard Taylor.
37. *Communication of Hon. William H. Seward, Secretary of State, upon the Subject of an Intercontinental Telegraph Connecting the Eastern and Westerns Hemispheres by way of Bering's Strait, in reply to Hon. Z. Chandler, Chairman of Committee on Commerce of the United States, to which was Referred the Memorial of Perry McDonough Collins*. Washington, Government Printing Office, 1864. Cited after AVPRI, fond 339, RAK, opis 888, no. 408, *Perepiska i materialy otnositelno sdelannogo g. Kollinsom predlozheniya ob ustroistve elektricheskogo telegrafa ot ustya r. Amura cherez russkie i britanskie Severo-Amerikanskie vladeniya* (*Correspondence and Materials Relating to the Proposal Made by Mr. Collins on the Establishment of an Electrical Telegraph from the Mouth of the River Amur across the Russian and British North American Possessions*), folio 37. Such utterances were noticed not only by the official authorities in Russia, but also called forth lively interest and unease among the public of the land. This was reflected, for instance, in the publication of the above-cited materials of Seward, in Russian translation, in the newspaper *Sankt Peterburgskie vedomosti* from October 18, 1864, no. 235.
38. Bolkhovitinov, 1990:143–166.
39. AVPRI, fond Sankt-Peterburgsky Glavny Arkhiv I-9, opis 8, no. 10. Letter of the Embassy of the USA, St. Petersburg, dated May 13/1, 1863, folio 22.
40. Bolkhovitinov, 1990:161.
41. Furnished by J. H. Colton, New York: printed by H. P. Cooper, 1864. Besides the published maps, several manuscript maps connected with Collins's telegraph project are in the Elmer E. Rasmuson Library, University of Alaska Fairbanks. See Marvin

W. Falk, *Alaskan Maps: A Cartobibliography of Alaska to 1900*. New York, Garland Publishing, Inc., 1983:129.

42. RGIA, fond 1289, opis 1, no. 1957, folio 130.
43. On Kennicott, see James Alton, *The First Scientific Exploration of Russian America and the Purchase of Alaska*. Evanston and Chicago, Northwestern University, 1942:1–19.
44. The text of this certificate is reproduced from a copy put at the disposal of the author through the courtesy of Mr. Stephan Swanson, director of the Museum/Residence of Robert Kennicott, The Grove National Landmark, Illinois.
45. Morgan B. Sherwood, *Exploration of Alaska, 1865–1900*, New Haven and London: Yale University Press, 1965:17–18.
46. Ibid., 1965:18.
47. Cited after Sherwood, 1965:19.
48. Robert Kennicott to Osten-Sacken. TC/PC of ObMU 1626.10.3, Harvard, Museum of Comparative Zoology, folios 2742a-2742b, Smithsonian Institution, Washington, D.C., February 15, 1965. The materials cited and the following correspondence of the Telegraph Expedition, scientific detachment, were put at author's disposal through the courtesy of the Chicago Academy of Sciences.
49. Kennicott to Osten-Sacken. TC/PC of ObMu 1626.10.3 Harvard: Mus. Comp. Zoology, folio 2742b.
50. Ibid., folio 2759, Kennicott to Baird, March 31, 1865.
51. James A. James, *The First Scientific Exploration of Russian America and the Purchase of Alaska*. Chicago,1942:13–14.
52. TC/PC of ObMu 1626.10.3 Harvard: Mus. Comp. Zool., folio 2799.
53. Ibid., folio 2799.
54. Ibid., folio 2807, letter from Kennicott to Baird, July 23, 1865.
55. AVPRI, fond 339, opis 888, no. 408. Report of Captain of the Second Rank Gavrilov, folio 65.
56. Ibid., folio 78 verso (from the letter of Captain First Rank, D. P. Maksutov, October 26, 1866).
57. RGIA, fond 1289, opis 1, no. 1967, folios 185–189. *Instruktsiya Chinovniku osobykh poruchenii Glavnogo Upravelniya Vostochnoi Sibiri Kollezhskomu assessoru Anosovu* (*Instruction to Official on Special Assignment from the Head Office of Eastern Siberia Collegiate Assessor Anosov*), January 13, 1866.
58. Charles L. Bulkley to RK, St. Michaels. T/C Russian Telegraph Expedition, Notes and Correspondence, Box 82, Museum of Natural History Library, S1, folios 2831a-2831b. The copy of this document was kindly put at the author's disposal by the Chicago Academy of Sciences.
59. Sherwood, 1965:21.
60. RGIA, fond 1289, no. 1958, *O predlozhenii Telegrafnogo soobshcheniya Rossii s Amerikoi* (*On the Proposal for a Telegraph Link between Russia and America*). Part Three. February 11, 1863–February 20, 1867, folios 13 and 7 verso.
61. Ibid., folio 16.
62. Ibid., folio 24.
63. AVPRI, fond Sankt-Peterburgsky Glavny Arkhiv, I-9, no. 10. *Opisanie deistvii Kompanii Zapadnykh telegrafov soedinennykh Shtatov po sooruzheniiu Rossiisko-Amerikanskogo*

Telegrafa (ot Chinovnika Osobykh poruchenii pri Glavnom Upravlenii Vostochnoi Sibiri Anosova ministru Pocht i telegrafov) (Description of the Activities of the Western Telegraph Company of the United States for the Construction of the Russian-American Telegraph (from Official on Special Assignment from the Head Office of Eastern Siberia Anosov to the Minister of Post and Telegraph)). Folios 132–132 verso.

64. Frederick Whymper, *Travel and Adventure in the Territory of Alaska, Formerly Russian America—Now Ceded to the United States—and in Various Other Parts of the North Pacific*. London, 1868:108–109.
65. Whymper, 1868:110.
66. AVPRI, fond Sankt-Peterburgsky Glavny Arkhiv, I-9, no. 10, folios 133–134.
67. Ibid., folios 146–147. See also Abaza, S. S., "Russko-Amerikansky telegraf" ("The Russian-American Telegraph"), *Sankt Peterburgskie vedomosti* no. 54, 1867.
68. William H. Ennis to RK, TC/PC, William Healey Dall papers, 7073, box 6, folder 4, SIA, pp. 1–7.
69. William H. Ennis to RK. TC/PC of William Healey Dall Papers, 7073, box 6, folder 4, SIA, p. 3.
70. Ibid., p. 6.
71. G. R. Adams, *Life on the Yukon*. 1982:33. On the diaries of George Adams and Fred M. Smith preserved in the Manuscripts Section of the University of Washington Library.
72. Ibid., 1982:35–43.
73. Ibid., 1982:52.
74. A copy of this document was made available to the author by the Chicago Academy of Sciences, folio 2859, TC from George R. Adams, MS diary, November 14, 1865, University of Washington Library.
75. Adams, 1982:60.
76. Cited after James, 1942:17.
77. Ibid., 1942:17.
78. William H. Dall, *Biography of Spencer Fullerton Baird*, Philadelphia, J. B. Lippincott, 1915:268.
79. Adams, 1982:91; Sherwood, 1965:24–25.
80. Sherwood, 1965:25.
81. Frederick Whymper, 1868:195.
82. Phillip H. Ault, "The (almost) Russian-American Telegraph," *American Heritage*, XXVI:4 (June 1975):97.
83. Sherwood, 1965:26–31.
84. Ibid., 1965:34–35. For more details on the Telegraph Expedition, see publications by members: William H. Dall, *Alaska and its Resources*, Boston, 1870; *On the Trend of the Rocky Mountain Range North of Lat. 60° and its Influence on Faunal Distribution*, AAAS, Proceedings, 18:247 (1870); *Remarks upon the Natural History of Alaska*, Proceedings of the Boston Society of Natural History, 12:143–151 (1868) ; Frederick Whymper, *Russian America or 'Alaska': the Natives of the Yukon River and Adjacent Country*, Transactions of the Ethnological Society, London, 1:175–176 (1869), and others.
85. Whymper, 1868. Map by J. Arrowsmith.
86. Sherwood, 1965:32.

87. S. R. Tompkins, *Alaska: Promyshlennik and Sourdough*, Norman, Oklahoma, 1952:182.
88. *Sketch Map of the Stickeen River from the Mouth to the Little Canon. The Stickeen River from Survey in 1863, and Information Obtained by William P. Blake. To Accompany a Report to Hon. W. H. Seward. 1868. In: Message of the President in Answer to a Resolution of the House of the 19th Dec. last Transmitting Correspondence in Relation to Russian America*. 40th Congress, H. R. Ex. Doc. No 177, second session.
89. Ivan Petroff, *Report on the Population, Industries and Resources of Alaska*, Washington, 1884:119.
90. *St. Paul Harbor, Kadiak Island. Drawn by H. Lindenkohl; J. Bien, Photo Lith. Astronomical & Magnetic Observations and Topography by the Party in Charge of Assist. George Davidson in 1867. The Soundings are from Russian Authorities*, Washington: U.S. Coast Survey, 1869. Falk, 1983:134.
91. *Sitka Harbor, Alaska. Drawn by H. Lindenkohl; J. Bien, Photo Lith. Astronomical & Tidal Observations and Topography by the Party in Charge Assist. G. Davidson in 1867. The Soundings are from Russian Authorities*, Washington: U.S. Coast Survey, 1869. "No. 21," Falk, 1983:134.
92. M. B. Sherwood, "George Davidson and the Acquisition of Alaska," *Pacific Historical Review* 28:141–154, 1959.
93. Congressional Papers, House Executive Document 177, 40th Congress, Second Session. Serial 1339, Washington, 1867:219–314 and George Davidson, "Scientific Expedition to Alaska," *Lippincott's Magazine*, vol. II, November 1868. Davidson's expedition of 1867, his encounters with Russian geographers and navigators in Alaska, and his use of Tebenkov's charts are described in detail in the biography of Davidson by Professor William Forrest King, Ph.D. dissertation defended in 1973, Clermont Graduate School. I [AP] was kindly made aware of this work by James E. Ratcliff, Jr.
94. The Rasmuson Library has one of Davidson's annotated copies of Tebenkov as well as a substantial number of maps and atlases, also annotated by Davidson, that he acquired in Stika in 1867. [MF]
95. RGAVMF, fond 1331, opis 4, no. 719.
96. Described in the introduction to Donald J. Orth, *Dictionary of Alaska Place Names*, Geographical Survey Professional Paper 567, Washington: Government Printing Office, 1971.
97. V. V. Kolgushkin, comp., *Opisanie starinnykh atlasov, kart i planov 16, 17 18 vekov i poloviny 19 veka khranyashchikhsya v arkhive Tsentralnogo kartograficheskogo proizvodstva VMF. Upravlenie nachalnika Gidrograficheskoi sluzhby voenno-morskogo flota (Description of Old Atlases, Maps and Plans of the Sixteenth, Seventeenth and Eighteenth Centuries and First Half of the Nineteenth Century Preserved in the Archive of the Central Cartographic Manufacturer of the Navy. Office of the Head of the Hydrographic Service of the Navy)*, Leningrad, 1958, no. 893.
98. S. G. Fedorova, "Shturmany Ivany Vasilievy i ikh rol v izuchenii Alyaski (pervaya polovina 19 v.)" ("Ivan Vasiliev's Navigator and their Role in the Study of Alaska (First Half of the Nineteenth Century"), *Letopis Severa* 9:206, Moscow, Mysl, 1979.
99. Charles Sumner, *Speech of Honorable Charles Sumner on the Cession of Russian America to the United States*, Washington, 1867. Later, the speech was incorporated, albeit

with some changes, into Sumner's complete works. See C. Sumner, *Works of Charles Sumner*, vol. 11:186–349, Boston, 1883.

100. R. E. Neunherz, *The Purchase of Russian America: Reasons and Reactions*, Ph.D. dissertation, University of Washington, 1975:122. Bolkhovitinov, 1990:238.
101. George Davidson, *Report Relative to the Resources and the Coast Features of Alaska Territory. Coast Survey; Report of the Superintendent Showing the Progress of the Survey during the Year 1867.* Appendix 18, Washington, 1869:187–329; T. A. Blake, "General Topographic Features of the Northwestern Coast of America, from the Strait of Juan de Fuca to the Parallel of Sixty Degrees North Latitude." *Coast Survey; Report of the Superintendent showing the Progress of the Survey during the Year 1867.* Appendix 18:281–290, Washington, 1869; "Vocabularies of the Languages of the Natives of Kadiak, Unalaska, Kenai and Sitka." *Coast Survey; Report of the Superintendent Showing the Progress of the Survey during the Year 1867.* Appendix 18:293–298, Washington, 1869; A Kellog, "Report on the Botany of Alaska," *Coast Survey; Report of the Superintendent Showing the Progress of the Survey during the Year 1867.* Appendix 18:318–324; George Davidson, *United States Coast Survey, Benjamin Pierce Superintendent. Pacific Coast. Coast Pilot of Alaska (First Part) from Southern Boundary to Cook Inlet*, Washington, D.C.: Government Printing Office, 1869.
102. Davidson, *Coast Pilot* . . . 1869:36.
103. Bolkhovitinov, 1990:236. For more details, see James, 1942:35, 44–45; E. C. Herbert, "Spencer Fullerton Baird and the Purchase of Alaska," *Proceedings of American Philosophical Society*, 98:2:139–143, April 1954; *The Role of the Smithsonian Institution in the Purchase of Alaska*, manuscript, Smithsonian Institution Archives.
104. Bolkhovitinov, 1990:119.
105. AVPRI, fond Sankt-Peterburgsky Glavny Arkhiv, I–9, opis 8, 1860–1866, no. 10: *O komandirovanii desitvitelnogo statskogo sovetnika Kostlivtseva i kapitana-leitenanta Golovina dlya obozreniya Severo-Amerikanskikh Kolonii nashikh . . .* (*On the Trip of Actual State Councilor Kostlivtsev and Captain-Lieutenant Golovin to Survey our North American Colonies . . .*). Folios 3–89.
106. Ibid., folios 3–4 verso.
107. Ibid., folios 88–89.
108. RGIA, fond 18, opis 5, no. 1347a, folios 802–804, Letter from S. A. Kostlivtsev to A. M. Knyazhevich, of October 21 (November 2), 1861. For Kostlivtsev's report, see folios 804a–888 (printed). Golovin's original report is in RGAVMF, fond 410, opis 2, no. 2515, folios 1–255, *Obzor russkikh kolonii v Severnoi Amerike, 20 Oktyabrya (1 Noyabrya) 1861 g.* (*Survey of the Russian Colonies in North America, October 20 [November 1], 1861*). The reports were published as follows: P. N. Golovin, *Obzor russkikh kolonii v Severnoi Amerike*, St. Petersburg: Navy Ministry, 1862; *Doklad Komiteta ob ustroistve russkikh amerikanskikh kolonii* (*Report of the Committee on the Structure of the Russian American Colonies*), St. Petersburg, 1863, 1–267, 268–462. An English translation of the printed version of Golovin's report was published in the USA, see P. N. Golovin, *The End of Russian America: Captain P. N. Golovin's Last Report, 1862*. Translated with Introduction and Notes by B. Dmytryshyn and E. A. P. Crownhard-Vaughan. Portland, 1979. On Golovin, see also Pavel N. Golovin, *Civil and Savage Encounters:*

The Worldly Travel Letters of an Imperial Russian Navy Officer, 1860–1861. Translated and annotated by Basil Dmytryshyn and E. A. P. Crownhart-Vaughan, Introduction by Thomas Vaughan. Portland, 1983.

109. Bolkhovitinov, 1990:134.
110. RGAVMF, fond 410, opis 2, no. 2515, folios 252–255. Bolkhovitinov, 1990:135–136.
111. Bolkhovitinov, 1990:138.
112. Ibid., 1990:199.
113. PSZRI, Sobranie 2, vol. 42, section 1, 44518.
114. On Walker's expansionist views, see J. P. Shenton, *Robert John Walker: A Politician from Jackson to Lincoln*. New York, 1961.
115. Bolkhovitinov, 1990:246.
116. For more details on the transfer of Russian America to the United States, see Bolkhovitinov 1990:259–282.

CONCLUSION

Russian Heritage and the Influence of Geographic Explorations in Alaska

As has happened throughout history, Russian authorities undertaking reforms and executing sharp turns in the country's policy tend to look only at the global consequences of their decisions and not at those people who are broken in the process. The first years of Russian America's existence under the domain of the USA is a clear example.

In theory, according to the treaty of Alaska transfer, Russian colonists had the right to choose: either to depart for Russia or remain in Alaska, becoming American citizens.[1] In fact, not everyone could avail themselves of either choice. Only the Russian Orthodox Church's property and legal rights were protected. All other Russian organizations and persons (Russian-American Company foremost among them) at the stroke of an hour, lost all they possessed in the colonies.

In accordance with the protocol signed on October 26, 1867, by Captain Peshchurov and General Rousseau, the transfer of *papers and documents* from the colonial archive was fixed at the same time as the transfer of the territory.[2] Excluded were only cartographic materials that remained at Russia's disposal. All fortified posts and public buildings, including the house of the chief manager, wharves, storage structures, barracks, batteries, hospital, and school in Novoarkhangelsk, listed in a special register marked by letter "A," were transferred to the official American representative. In the register marked "B" were churches and clergy houses that, in accordance with the treaty, remained the property of the parishioners. Register "C" named persons (only twenty in all!) who received certificates ensuring their full property rights to houses and land they owned. Register "D" listed house owners who retained the right only to structures but not to land on which these structures stood. Russian-American Company property on Kodiak Island, transferred in full to the United States, was listed in register "E."[3]

Application of these clauses to real life was effected harshly and promptly, especially by the commander of the American garrison, Brigadier General Davis who, in the presence of the Russian representative, declared that *(I) do not intend to acknowledge the existence here of the Russian-American Company and its agents.* Not wishing to aggravate the relations with American authorities, Russian hurried to *clear all buildings designated for transfer to American government. This speedy vacation of the best dwellings caused considerable inconvenience to the Chief manager and to those persons who had to get out of their houses under the local rainy weather, majority of them moving to the ships.*[4] In spite of such harsh measures, in the beginning the Russian population of Baranov Island, which in December of 1867 numbered about 500 persons, appeared according to a report by General Rousseau *calm and law abiding.* But by this time the reaction of the Tlingit was rather different. One of the chiefs *remarked irritably* that *indeed, we permitted the Russians to own the island but we have no intention to give it to any and every fellow that comes along.*[5]

Regrettably, the calm and law-abiding mood of the Russians noted by General Rousseau was severely tried, and their relations with the Americans who had arrived developed badly. Many hospitable Russians who initially opened their doors to the new arrivals paid in bitter coin for their kindness, as along with the officers, traders, and *promyshlenniki*, crowds of scoundrels fell upon Alaska.[6] As Peshchurov testifies in his letter to Stoeckl: *Americans who rushed here behave rather disorderly, especially the soldiers, composed of all sorts of riff raff. Already there were many incidents of fighting and theft. From among the relatively small garrison, up to 30 men are constantly under arrest.*[7] The situation of the local population worsened with each passing day, and toward the end of 1867 a mass exodus of Russians from Novoarkhangelsk began. On November 1 a group of former employees of the Russian-American Company sailed aboard the steamer *John L. Stephens* for San Francisco. They hoped from there to travel to the home country via Panama. On the 14th of December, aboard the ship *Tsaritsa*, 168 Russians (sixty families, primarily employees of the Russian-American Company) sailed for Russia via London. On January 22, 1868, aboard the bark *Siana* (Nakhimov), the remaining sixty-nine soldiers of the military garrison sailed to Eastern Siberia. On November 30, 1868, more than thirty people left Novoarkhangelsk for Kronstadt aboard the ship *Winged Arrow*, purchased for the purpose in San Francisco.[8]

Not all Russians who wanted to leave Alaska could exercise their right within the stipulated three years. By 1869, Russian-American Company lost all its rights to act in the former colonies and the Russian government left its countrymen there (in Alaska) without any support.[9] The situation of Russians, former rank-and-file laborers for the Russian-American Company, by this time became desperate. As Emil Teichmann, who visited Alaska in 1868 testifies, the laborers who remained worked under military supervisors and, *as often happens under such patriarchal systems, the people lose any sense of responsibility for themselves that they used to have, their intellectual powers atrophy more and more and they sink into the state of animal apathy, knowing all too well that they will have their piece of the daily bread and will gain nothing more in the future, no matter how hard they try to work.* Teichmann asserted that practically all wives of the Russian laborers remaining in Alaska were prostitutes and their husbands drunkards. The morals of the Americans were no higher, but the Russians were distinguished from the former by their good nature.[10]

Prince D. P. Maksutov, the last chief manager of the colonies, returned to Russia in 1869. He was followed by Captain Koskul, who presented a detailed report in liquidation of the company's affairs.[11] From that time on, only the Russian Orthodox Church continued to pay attention to the Russian-speaking populations remaining resident in Alaska. Metropolitan Innokenty of Moscow followed closely the situation of the Orthodox flock in those now-foreign lands where he spent most of his self-sacrificing life as a missionary and as a scholar. He utilized every possible occasion to obtain information from these regions and to offer as much as he could both material and spiritual support to the clergy and the faithful. For example, in 1871, the retired Collegiate Assessor Stepan Nesterovich Buinitskii, who had visited the Aleutian Islands, compiled, on Innokenty's request *A memorandum of the situation (condition) of the Orthodox temples on the islands of Unalaska, St. Paul, and St. George in the territory of Alaska.* He stated that *Orthodoxy is a trace of Russian civilization in the Aleutian Archipelago of which the entire Russian nation can be justifiably proud. Seeds of Christian teaching sown by the Apostle to the Aleuts, His Reverence Ioann Veniaminov (now His Beatitude Metropolitan Innokenty) did not fail to bring fruit because he used, besides the word, most mighty tool in the field of people's education—the example of life lived in concordance with the teaching. Memory of his virtues to this day serve the Aleuts as an explanation of the truths set out in the Catechism.* Though he stressed the firmness of the Russian religious and cultural influence among the Aleuts, Buinitskii was compelled to note the sorrowful condition of the churches.[12] Henry W. Elliott, who in 1872 inspected the Pribilof Islands for the U.S. Treasury Department, recorded that all local residents, of rather mixed character (Aleuts, Russian, Kamchadal, and Creole) were Orthodox Christians. He found among them widespread knowledge of the Russian language and recorded the reluctance of the local inhabitants to abandon it and attend the English language school.[13]

Information that came to Metropolitan Innokenty from Baranov Island also demonstrated the strength of the Russian influence but stressed strongly the sorry state of morals and general deterioration of the economy and culture. According to the data from an inspection visit to Alaska undertaken in December 1870–January 1871 by the assistant to inspector general of the U.S. War Department Ludington, there were in Sitka 300 persons, among whom only 15 were Russian, the rest being Aleut and Creole. Ludington wrote that

> The women, almost all without exception are prostitutes, even girls aged 12 or 13. Almost every one of these people expresses the wish to resettle somewhere in Russian possessions and hope that next summer a ship will be sent to collect them. This hope has little basis . . . Their material situation is dreadful, the moral one even worse. They have no prospect of any kind of employment at the place of their residence and are thirsting only to be resettled somewhere in Russia.

The note also speaks of general drunkenness in Sitka, even among priests. The results of the inspection by the US War Department, with the recommendation to resettle Russian inhabitants of Sitka within the Russian Empire, were brought to the attention of the Russian Minister of Finance. The latter did not respond by any kind of direct action but forwarded all materials pertaining to the inspection trip to Metropolitan Innokenty.[14]

The Russian-speaking Orthodox population remaining in the former Russian colonial possessions for many years existed in relative isolation, which facilitated the preservation of many elements of Russian national culture. The steadfastness of these elements under the melting-pot conditions of the nation of immigrants that blends cultures and peoples in the USA is remarkable. There is, however, a heritage of Russian America that is not as obvious; modern scholars began to appreciate its importance only recently. Explorations and investigations conducted by pioneers during the period when Russian America existed made important contributions to the development of science. These found their most profound expression in the cartographic studies of Alaska and adjacent lands. The primacy of Russian hydrographers traveling in this area was acknowledged unconditionally by the Americans in the first years after acquisition of Alaska.

Cartographers have a special role to play in times of major changes in the fate of countries and peoples accompanied by territorial shifts; it is through maps that a country's leadership can visually and convincingly declare its achievements in expansion of the state's territory. We have seen earlier that in the incorporation of the Aleutian Islands and the territories on the American continent into the Russian Empire, the country's authorities more than once utilized this opportunity. The American government also did not tarry in publishing a whole series of maps—declarations announcing the acquisition of Alaska.

First among them is the map of *North Western America showing the territory ceded by Russia to the United States, compiled for the Department of State at the U.S. Coast Survey office. B. Pierce, Supt. 1867.* (Fig. 76). Two editions of this map were published in 1867, the first showing much less detail than the second. Both editions are based almost exclusively on Russian materials. On the basis of this official map, another series of maps in this same declaratory character were published in the years 1867–1869.[15]

Russian sources were used by William H. Dall in the early years of his activity regarding Alaska. Dall was responsible for bringing into foreign scholarly usage a number of Russian-language works. After his return to Washington from Alaska in 1868 and reworking the Smithsonian Institution collection of the Telegraph Expedition, his own discoveries, and the available literature, Dall published *Alaska and Its Resources*, a description of adventures and investigations by the Telegraph Expedition with a compendium of geographic, historical, and natural science information. This publication established the author's reputation as an Alaska expert, although a majority of the problems he presented as statements, not solutions.

In 1870, Dall undertook an expedition aboard the steamer *Humboldt* in accordance with the program for a hydrographic survey of Alaska that he had proposed to the U.S. Coast Survey. After concluding the field work, Dall and Marcus Baker compiled an appendix to the *Coast Pilot*. It consisted of extensive meteorological tables and an amazingly detailed bibliography on Alaska materials in many languages, including Russian.[16] This work was a major contribution to the store of knowledge about Alaska and has not lost its value to this day.[17]

Dall was in essence the first and last nineteenth-century scholar for whom the works of his Russian predecessors served as point of departure and source materials. After Dall, the Russian natural science materials, charts, and maps disappear rather rapidly from use by American investigators, with one exception that requires special explanation.

In the historiography of Alaska a special place is occupied by one Ivan Petroff, a personage in many respects mysterious and not reliable. Of the three most important books about Alaska written in the nineteenth century (the first one by Dall), Petroff is responsible for much of the material of one and is author of the second. These are the *History of Alaska* by Hubert H. Bancroft (1832–1918)[18] and *Report on the Population, Industries and Resources of Alaska* by Ivan Petroff from the results of the tenth census of 1880.[19] Both books, in the opinion of many modern scholars, are major contributions to the knowledge of history, geography, and natural resources of Alaska before the twentieth century.[20]

Little is known about Petroff, as he preferred not to talk about himself. He was born in 1824 in St. Petersburg and was in training as a military translator (interpreter?). Then he studied Sanscrit, Armenian antiquities, and literature. After scholarly research in Armenia, Petroff was sent to Paris and in 1861 he came to America. About his life in the United States before 1874 we have only fragmentary and contradictory information. We find in Bancroft information recorded from Petroff's accounts that he bravely served in the Union Army, was wounded twice, and earned the rank of lieutenant and that from 1865 to 1870 he served with the Russian-American Company in Cook Inlet. Documents, however, prove that in 1867 he was on Vancouver Island under arrest for desertion, but was freed and sent as a private to the Kenai Peninsula as an interpreter of Russian. He did not remain in Alaska long, as he enlisted in the army at Sacramento. After arrival in San Francisco he deserted once again and until 1874 worked as a journalist. After that, he was hired by Bancroft to collect materials for the history of Alaska. While studying archives in Washington, Petroff obtained an appointment to work for the census in Alaska. The tenth US census was not simply a population enumeration in the narrow sense of the word, but an attempt to put on record and describe the country's entire resources. For this purpose *special agents* were used who had a rather wide mandate. Petroff was appointed as one of these agents.[21]

In 1878–1880 Petroff visited Alaska several times. He interviewed various people and studied local archives in Sitka, Kodiak Island, and Unalaska. He traveled several times into the Interior. If one believes his own statements, Petroff also visited the Shumagin Islands, the Pribilof Islands, and St. Michael. From St. Michael he ascended the Yukon to the Novitna River, crossed to the Kuskokwim River, and ascended the latter to the old Lukin post. Traveling down the Kuskokwim, he then followed the coast to Naknek, crossed the Alaska Peninsula via Ketmia on Shelikof Strait, sailed across the latter, and arrived on Kodiak Island. The distance traveled according to Petroff's calculation was 8,700 miles, of which 2,500 were by water. If the special agent really covered this distance in two field seasons, as Petroff maintained, that was a considerable achievement, even though he did not visit any places not visited earlier by Russian and American traders and explorers. It is more likely that the scope of Petroff's journeys was much more modest, as according to testimonies by people who encountered him in that period in Alaska, his mode of travel through the country was rather unhurried and leisurely. The missionary stationed at Wrangell reported at the end of 1880 that Petroff spent several weeks consuming unbelievable quantities of rum and whisky without showing any apparent effects.

Petroff's career ended in an enormous scandal in 1892. While translating Russian documents in connection with the British-American dispute about the walrus hunt in the Bering Sea for the State Department, his translations interpolated fragments benefitting the American side and he falsified other documents. Petroff's falsifications were discovered, he confessed, and was dismissed from the State Department. After that he sank into obscurity. Another of Petroff's deeds that led the scholars to take a very skeptical approach when dealing with his data was his so-called "find" of a diary of the Russian missionary, Father Iuvenalii, who perished at Iliamna Lake [correction, at Kwhetluk—LB] at the end of the eighteenth century. The diary, "translated" by Petroff, is considered by modern scholars to be a falsification. There was no apparent reason for this. The tale paints the Russian missionary in very negative colors, showing Petroff's enmity toward the Russian Orthodox Church.[22]

In spite of the above-cited occurrences, which did not enhance Petroff's reputation as a truthful scholar, careful perusal of the published materials of the tenth census and especially *History of Alaska* by Bancroft convinces one that these works were an important contribution to the study of history, population, and natural resources of the territory newly acquired by the United States. It appears to us that a large part of their value lies in the fact that Petroff was the first to make Russian publications and sources available for English readers. His works contain, in translation, excerpts from Veniaminov, Doroshin, Grewingk, and other Russian authors. Sherwood, who checked at random Petroff's translation preserved in the Bancroft Library, concluded that they are correct, and that his translation of Zagoskin is better than the English text habitually used by modern ethnographers and historians.[23]

It is important to note that Petroff was one of the first who analyzed geographic and cartographic materials from the Russian period and rather convincingly demonstrated their role in the study of Alaska's environment and population. Looking even at the very early Russian materials, Petroff concluded that they are highly reliable, especially in transmitting local[24] geographic names. As an example, he cites the study by the Krause brothers of the local toponyms of Chukotka that proved to be identical with those recorded by the Russian cossacks at the end of the seventeenth century. Petroff's own experience, acquired in the process of compilation of statistical data and maps for the report on the tenth census in Alaska, concluded that Russian cartographic materials were highly accurate. Studying these materials, he conducted a random and even comparative map analysis, the results of which were summarized thus:

> A careful comparison of all the accessible authorities during the slow process of compilation naturally led to the discovery of errors in many of the maps and charts consulted, but throughout this work it has been the experience of the compiler that the Russian atlas of Tebenkof [*sic*] and to a certain extent the charts of Sarychef [*sic*], furnish the most reliable materials—in fact they are the basis of all maps of this vast territory. Wherever a point or coastline has been laid down as definitely known by Tebenkof, it may be relied upon as true in contour and latitudinal position. A curious instance confirming this assertion presents itself in the case of the southernmost outlet of the great Yukon river [*sic*]—the Kashunok—indicated as a broad arm on "chart 2" of Tebenkof's atlas. Mr. William H. Dall, in compiling his map for the United states [*sic*] coast survey in 1869, omitted this feature, but examination proved the Russian geographer to be correct. The outlet exists, but is less broad than indicated by Tebenkof.[25]

Discussing the history of cartography of the northern areas of the Bering Sea, Petroff quite rightly notes that the surveys conducted in the 1820s by Etholen and Khromchenko are still our best authorities for the coastline included in their labors. Briefly but objectively, Petroff characterizes other Russian expeditions and nineteenth century surveys, including Kotzebue's voyage; cartography of the northern coasts by Kashevarov, Tebenkov, and Rozenberg; as well as continental explorations by Glazunov, Makarov (Malakhov?), and especially by Zagoskin. He summarizes Tebenkov's cartographic work in Russian America noting that "*we owe (him) the best atlas of the coast of Alaska ever published. The hydrographic notes were very copious and correct, and nearly all subsequent charts and maps have been based upon his surveys.*"[26]

Petroff stressed the importance of the ethnographic studies by Veniaminov, Davydov, Zagoskin, Wrangell, and others, noting "*the first map showing in colors the distribution of native tribes in Russian America*" compiled by Captain Lieutenant of the Russian Navy Verman in 1863 (Petroff gives an erroneous date for this map of 1857). Petroff mentions "*the manuscript map, also in colors, compiled by Dr. George Gibbs from information obtained from the Russian authorities at Sitka.*"

The next step in the development of ethnographic cartography of Alaska was the *Map showing the distribution of the native tribes of Alaska and adjoining territory, compiled from the latest authorities by W. H. Dall* of 1875.[27] Minor differences between this and preceding maps are accounted for by the amassing of new field material by new researchers in ethnography.[28]

Petroff compiled charts and maps from the materials he collected. In total, his document represents the first complex Alaska atlas that summarizes the investigations from the Russian America period and initiates the beginnings of the American era of Alaska studies. This collection includes maps examining the topographic characteristics of the territory; general geographic and landscape as well as geobotanical regions (particularly distribution of timber, tundra, and glaciers); ethnographic maps; and areas where various economically important animals are distributed: bear (polar, brown, and black), beaver, fur seal, land otters, mink, muskrat, and fox.[29]

THE END RESULTS

In completing our work, let us try to note the most essential features and results of geographic investigations in the Russian America period.

America's "discovery" from Asia and subsequent exploration and cartography of the northern part of the Pacific Ocean and Alaska were processes involving many nations and countries, as was the case in other areas of the planet. The contributions of various nations to this process were uneven depending on the taxonomic level of the territory being explored.

At the local level the leading role, at least in the initial stages, belonged to the tribes that inhabited the islands and coasts of the northern part of the Pacific Ocean. They had mastered the local universe and territories necessary for economic enterprises and migrations, including responding effectively to change from natural causes such as climate, as well as changes brought about by human actions.

Native geographic knowledge, due to migrations and natural barter with neighboring tribes, extended to the *regional* level. However, it is difficult to establish with certainty, on the basis of data available in the scientific literature, what was the extent of information available to the tribal peoples about neighboring territories in the early stages of history. Obviously, their geographic horizon was directly dependent on the dominant ways of acquiring food. Representations of large regions in the form of descriptions, charts, and maps were created, as a rule, by nations chronicling policies of territorial enlargement. In our case, the actors were Russians, moving from the west to Siberia, to the coast of the Pacific Ocean and then, along the Aleutian Island chain to Alaska. Spanish conquerors moved from the south northward and claimed the entire North American continent on the basis of the "right of discovery."

Russian colonization of Siberia and the northern part of the Pacific Ocean lasted more than 150 years. It was brought about by ordinary mariners, cossacks, and hunters, who easily "adapted" to the surrounding natural and ethnic environments. For this reason, our pioneers, moving eastward, gradually absorbed the geographic knowledge accumulated by the local tribes in the areas of their settlement, adopted the latter's modes of orientation, and "incorporated into their arsenal" Native means of transportation, of which the most useful proved to be baidarkas, baidaras, and nartas [dog and reindeer sleds—LB].

Siberian sketch maps (*chertezhy*) during the 1660s–1690s were compiled on the basis of data supplied by pioneers and navigators and incorporated Native information about the terrain. The geographic scope of these sketch maps encompassed Eurasian territory from the Arctic Ocean to China and India. They are testimony to Russian inclusion of information obtained by questioning Native informants about regions far beyond the actual routes of Russian cossacks and hunters. These Siberian sketch maps include local names. The discovery of such names, and their preservation, was a characteristic feature of Russian colonization of Siberia and later of the northern part of the Pacific Ocean coastal regions. The overwhelming majority of these works show an unbroken expanse of water framing northeastern Asia from the Kolyma River to the Amur, solving in this way the ancient geographic problem regarding the presence of a strait between Asia and America.

The global level of generalization of geographic data, however, demands that these data be included in the scientific representations of the earth as a whole and in representations of the distribution of continental and sea masses over the planet's surface. Up to the beginning of the eighteenth century, there was practically no possibility in Russia to follow such procedures, as cartography was developing in isolation from the Ptolemaic paradigm by then dominant in Europe. Russian geographic sketch maps compiled without the use of the coordinate system of latitude and longitude and without permanent scale and cartographic projections, could not, strictly speaking, be incorporated into world maps compiled according to the established European cartographic tradition. Only in the early eighteenth century, with the advent of Peter the Great's transformation of economy, science, and culture, was Russian cartography restructured in concordance with European scientific and technical notions. The compilation of accurate maps through the use of geographic coordinates and maintenance of certain scale and mathematical rules in description allowed transition from the spherical sector of earth's surface to its representation on a flat map surface through geographic projection. From that time on, scientific hydrographic methods of western European expeditionary field work gradually assumed a growing role in the development of geographic exploration alongside the traditional methods of route descriptions and primitive compass surveys previously used by pioneers and mariners. In the course of the eighteenth century, the two methods approached each other, but the Russian long-term practices of reconnaissance exploration of new territories and compilation of schematic sketch maps which incorporated traditional knowledge gathered from local tribes enriched the strictly scientific method of geographic surveys and descriptions.

By the beginning of the eighteenth century, European cartography had developed through redefinition and supplementation of Ptolemy's map of the world. The Northern Hemisphere adequately reflected reality, with the exception of the northern part of America and adjacent parts of the Pacific Ocean. Empirical geographical material pertaining to these areas was lacking. As was often the case with "armchair" European cartography of the sixteenth and seventeenth centuries, lack of data led to the emergence of hypothetical and fantastical representation of the unexplored territory. In the case of the northern part of the Pacific Ocean, French, Spanish, English, and other charts and maps compiled in various European countries presented variant images. Many of these reflected ancient notions about the existence in the north of the Strait of Anian between the continental masses and about the "lands" of Esso, Da Gamma, Company, and others, that were in part fruits of travelers' fantasies, and partly incorrect placement of islands and continents framing the Pacific Ocean. Peter I, who strove to line up Russian with the European states not only in economic development but also in scientific achievements, understood the special possibilities offered by Russia to create an accurate chart of the northern part of the Pacific Ocean and to finally settle the question, *Is America joined to Asia*?

Russian marines and pioneers knew that the northeast extremity of Asia is washed by the waters of the ocean and showed these ideas on Siberian sketch charts, which were known to Peter I. But the first emperor of Russia wanted to obtain scientific confirmation of this knowledge by a competent naval exploratory expedition so that the results could be expressed in precise cartographic language and accepted in European scholarly circles. In order to achieve this goal, and also to reach America from Asia, two expeditions were organized: the First (1725–1730) and the Second (1732–1742) Kamchatka Expeditions, under the command of Vitus Bering and Aleksey Chirikov. In the course of these expeditions, and also through the expedition of Gvozdev and Fedorov (1732), Russian mariners gained priority in proving the existence of a strait between Asia and America. They also reached the coast of America at the northern and southern shores of Alaska.

Bering's expeditions facilitated the process of the elemental movement of cossacks and pioneers eastward so that they spilled across the waters of the Pacific Ocean. In the following decades, *promyshlenniki*, racing for the *soft gold* (furs), gradually expanded through the Aleutian Islands, reaching Kodiak and the Alaska Peninsula. The active process of mutual interaction between the traditional methods of Russian mariners and *promyshlenniki* began. Navigators of the fur-merchant's vessels now had the opportunity to receive elementary training in navigational sciences in the schools opened in Siberia by Bering. The Russian Admiralty began to pay serious attention to the schematic sketch charts of the *promyshlenniki*, treating them as sources for compilation of charts of the Aleutian Islands and the adjacent coasts of Alaska. Scientific expeditions organized in the 1760s–1790s, led by Krenitzin and Levashov as well as Billings and Sarychev, were tasked with determining the correct geography of the Pacific Ocean. In executing these expeditionary labors, Russian hydrographers began wide use of the Aleut vessels habitually favored by the *promyshlenniki*: baidarkas and baidaras. Sarychev set out scientific and technical methods in the first Russian textbook on sea surveys and the compilation of navigational charts, published in 1804.

The results of cartographic and geographic expeditions of the *promyshlenniki* and Russian government expeditions of the eighteenth century were kept secret by the authorities and in reality were not reflected even on the maps of the Pacific Ocean published by the Academy of Sciences in St. Petersburg. Most famous of these were the maps by Müller-Truscott and Staehlen. On these maps, hypothetical and mythological lands still appeared, while data obtained by the *promyshlenniki* and the secret government expeditions were not reflected at all, or if reflected, were very distorted. The Russian Admiralty had reasonably accurate charts of the Aleutian Islands, the southern part of the Alaska Peninsula, and the Bering Strait region on the eve of James Cook's voyage of 1778–1779. They were forced to publish this geographic information, presented in a generalized manner, in order to prove Russian priority in the exploration of the northern part of the Pacific Ocean immediately following Cook's expedition. The results of Cook and Clerke, and their charts, allowed the Russian hydrographers and cartographers to publish the concepts they already possessed about the distribution of land and sea in this part of the globe. Thus, summary charts published by the Russian Admiralty and St. Petersburg's Academy of Sciences beginning in the 1780s reflected rather accurately the basic geographic features of the northern part of the Pacific Ocean.

Subsequent improvement of cartographic imaging of coastal Alaska came with the increasing precision of astronomical and geodesical linkage of results of local surveys conducted by Russian merchant companies' navigators with existing charts. This was aided by the expedition of George Vancouver (1790–1795) and the later

Russian circumnavigating expeditions that began with the scientific voyage under Krusenstern and Lisyansky (1803–1806). The northern edges of Alaska, and the coasts of North America as a whole, were entered upon the global map as a result of voyages by Russians and Englishmen, such as that by Kotzebue (1815–1818), Beechey (1725–1727), and Dease and Simpson (1837–1838), as well as through voyages by the vessels of the Russian-American Company in the northern part of the Bering Sea and the seas of the Arctic Ocean, adjacent to Bering Strait. All of these voyagers utilized, at the local level, geographic information provided by Natives.

Beginning in the 1820s, the Russian-American Company and the Admiralty established a system of continuous collection of geographic information by company vessels as well as by means of expeditions into the interior of the continents for barter (*rastorzhki*) with the Natives. Special explorations of the continental interior by the Russian-American Company [and by the Lebedev-Lastochkin Company—LB] were already underway at the turn of the nineteenth century, but they found their full development in the1820–1840s with the expeditions by Korsakovsky (1818), I. Ia. Vasiliev (1829), Glazunov (1833), Ivan Malakhov (1832–1839), and Zagoskin's expedition (1842–1844). This was brought together in the *Atlas of the Northwest Coasts of America . . .* compiled by Tebenkov in Novoarkhangelsk and published, together with the volume of hydrographic descriptions, in St. Petersburg in 1852. This atlas became for many future years the basic navigational aid for mariners sailing Alaska waters as well as the main source for compilation of American charts in the first decades after the sale of Alaska to the United States.

The heritage of empirical sciences from the Russian America era became the basis for wide generalizations that influenced concepts of the structure of the earth's surface, global climatic processes, specifics of distribution of flora and fauna, and linkages between eastern Asian and northern American species. During the Second Bering Expedition (1841–1842), Georg Steller used the few hours he had on land on the coast of North America to obtain the first scientific information about the environment of the territory and the material culture of the Natives of southern Alaska. Later, during the winter spent on Bering Island, he made valuable observations about flora and fauna, most famous of which is discovery of Steller's sea cow, a mammal that was hunted to extinction by the end of the eighteenth century by the *promyshlenniki*. In the 1780s, geographic information collected by the *promyshlenniki* and members of the government expeditions, and the charts compiled as a result of these labors, became the basis of the wide scholarly generalizations by the Academician P. S. Pallas, of the St. Petersburg Academy of Sciences. On the basis of these materials Pallas was able to develop a general outline of the genesis and nature of morphological structure elements of northeastern Asia, the northern part of the Pacific Ocean, and Alaska, that is, mountain systems and the major mountain arcs. He also was the first to correctly propose the structural kinship between the mountain systems of Asia and America and the existence, in the past, of a land connection between the two continents in the area of the Bering

Strait and along the Aleutian Islands. These hypotheses were confirmed in the first half of the nineteenth century during investigations conducted by scientists participating in Russian circumnavigational voyages and by the ethnographic material of Veniaminov, who studied these materials in depth from this particular point of view.

The general scientific significance of the explorations in the northern part of the Pacific Ocean were clearly demonstrated for the whole world by Wrangell's book on the geography and ethnography of Russian America, which was published in German in 1839. In this work Wrangell compared data on the Alaska environment with materials from geographic explorations in other Arctic regions of the globe, basing his conclusions in part on his own experience as well as on those of his predecessors. This comparative analysis contributed much to the understanding of natural processes and facilitated the subsequent establishment and development of physical geography as a science. Wrangell's book is also commonly acknowledged as a fundamental contribution to ethnography. Baer, editor of this work and publisher of the series, added his own general account of the pioneering investigations by American scholars in the languages of and relationships among aboriginal inhabitants of North America. Baer also provided a fundamental description, the first of its kind, of Alaska's climate.

Following the publication of the work by Wrangell and Baer, the St. Petersburg Academy of Sciences made the decision to establish systematic collection of scientific data and natural science specimens. For this purpose the preparator of the Zoological Museum of the Academy of Sciences, I. G. Voznesensky, was sent to Russian America. He spent almost ten years conducting studies in zoology, botany, ethnography, and geography in California, at Kodiak Island, Baranov Island, Unga, Unalaska, the Pribilof Islands, St. Lawrence Island, in Cook Inlet, at St. Michael Redoubt, and in Kotzebue Sound. His drawings and ethnographic collections are preserved in the Museum of Anthropology and Ethnography in St. Petersburg. The 150 boxes of floral and faunal specimens he sent to Russia enriched the collections of the Zoological and Mineralogical museums of the Academy of Sciences. In the judgment of his contemporaries, Voznesenskii presented information on more than 400 new species of animals and plants. The results of geological and geomorphological investigations conducted in Russian America in the first half of the nineteenth century were summarized by Keeper Grewingk of the Cabinet of Mineralogy of the St. Petersburg Academy of Sciences in his works of 1848–1850.

Numerous natural science and ethnographic collections were created in the course of expeditions as well as through everyday activities of employees of the Russian-American Company. Such materials began to reach Russia and England in the eighteenth century. In the course of time, the ethnographic collection from the Aleutian Islands and Alaska in St. Petersburg (Museum of Anthropology and Ethnography) became the richest collection in the world. To this day, possibilities of the collection's use as a source for study of spiritual and material culture of aboriginal populations of these regions at the time of initial contact with Europeans have not been exhausted. Specimens of flora, fauna, minerals, and rocks, materials from occasional as well as regular meteorological and phonological observations preserved in various museums and institutes of the Russian Academy of Sciences and in the archives in St. Petersburg and Moscow deserve the closest attention on the part of modern scholars, especially those interested in the problems of global environmental changes and the role of human factors in those changes.

The history of geographic exploration and cartography in Russian America, as well as other parts of the globe, is but one example of the realization of efforts by many peoples and countries aimed toward gaining knowledge of their environment. Russia played a leading role in this process. The author cannot but agree with the opinion of Professor Richard A. Pierce, who stressed that in volume of information about its colonial possessions, Russian achievement exceeds any other in a territory comparable in size anywhere in North America and even in any other part of the globe undergoing colonization in that period. And all of this in spite of the fact that Russians were exploring an enormous territory with very limited personnel. The number of Russians in Alaska never numbered over 800 persons. At the time of the sale, they numbered only about 500. However, we must add to their number about 2,000 Creoles. Given that the main aim of the Russian-American Company was the procurement of furs, how were such outstanding achievements in scientific investigations in the territory of Russian America possible?

The answer to that question can be found in the exceeding interest on the part of Russia's imperial government in its transoceanic possessions. This factor conditioned the quasi-governmental character of the Russian-American Company. It was demanded that the company investigate natural resources to a scope and detail that a private enterprise could never afford. The nature of Russia's structure—the absolute monarchy—made it possible to demand, through various Russian ministries and the main office of the Russian-American Company, undertakings of many noncommercial enterprises.[30] As was traditional in the Russian Empire, the personnel who were to execute the orders of the central government in the colonial possessions were selected from among the most highly educated and gifted administrators, naval officers, and clergymen. Such talented scholars as Sarychev, Wrangell, Voznesenskii, and Veniaminov were closely tied to Russian America. Many of their activities contributed little, or nothing, to the company's commercial success, and at times even hindered growth of the company's well-being. But following generations benefitted and were enriched in knowledge thanks to their achievements—studies that brought no immediate material benefits.

NOTES

1. According to the testimony of Dall, in January 1867 there resided in Novoarkhangelsk in 120 dwellings, 968 persons, of whom 349 were Russians, the others Creoles and Aleuts. See Dall, 1870:254.

2. At present, this archive is preserved in Washington [in the National Archives of the USA—LB]. Microfilm copies of these documents are on hand in the Archive of the History of Foreign policy of the Russian Empire and in the Archive of Ancient Documents in Moscow.
3. Protocol of October 14 (26), 1867, signed by L. Ch. Rousseau and A. I. Peshchurov with appendices "A" to "E," in *Foreign Relations of the United States (Papers Relating to Foreign Affairs . . .)*, vol. 1, Washington, 1868:479–484.
4. From a letter from Peshchurov to Shtoeckl, October 17 (29), 1867, Library of Congress, Manuscript Division, Box 1, # 209. For English translation, see David H. Miller, *The Alaska Treaty*, Kingston, 1981:138–140.
5. From a letter by Russeau to Seward, from Portland, December 5, 1867, *Foreign Relations of the United States (Papers Relating to Foreign Affairs . . .)*, vol. 1, Washington, 1868:475–478.
6. Ivan Petroff, *Report on the Population, Industries and Resources of Alaska*, Washington, 1884:120.
7. Letter from Peshchurov to Shtoeckl, November 1 (13), 1867. Library of Congress, Manuscript Division, box 1, no. 210. English translation: *The Alaska Treaty*, 1981:140–141.
8. C. C. Hulley, *Alaska, 1741–1953*, Portland, 1953:203–205; Pierce, 1986:26–29; H. Chevigny, *Russian America: The Great Alaskan Venture, 1741–1867*, New York, 1965:259–261; R. A. Pierce, *Alaskan Shipping, 1867–1878: Arrivals and Departures at the Port of Sitka*, Kingston, Ontario: Limestone Press, 1972:7, 15. Bolkhovitinov, 1990:276–278.
9. S. G. Fedorova, "Russkoe nasledie v sud'bakh korennogo naseleniia Aliaski, *Traditsionnye kul'tury Severnoi sibiri i Severnoi Ameriki*, Moscow, 1981:248, 257–258.
10. Emil Teichmann, *A Journey to Alaska in the year 1868: being a diary of the late Emil Teichmann*. New York, 1962:186–187.
11. AVPRI, fond RAK, no. 181 [no opis given—LB], *otchet upolnomochennykh po likvidatsii del RAK obshchemu sobraniiu aktsionerov*, folios 303–317.
12. AVPRI, fond Innokentii (Veniaminov, Ivan Evseevich), opis 874, no. 1, Washington, February 12/24, 1871, folios 2–5.
13. Henry W. Elliott (Assistant Agent, Treasury Department), *Report on the Prybilov Group, or Seal Islands, of Alaska*, Washington: Government Printing Office, 1873.
14. RGIA, fond 797, opis 41, II-III, no. 64. *Po povodu preprovozhdeniia Ministerstvom Finansov vypiski iz bumah Voennago ministerstva Severo-Amrikanskikh S. Sh. Kasatel'no natsoiashchago sostoianiia byvshikh russkikh kolonii i o dukhovenstve nashem v Sitkhe.* March 19–27, 1871, folios 2–5.
15. Drawn by Lindenkohl. Washington, 1867. Second edition May 1867.
16. U.S. Coast Survey . . . *Pacific coast pilot. Coast and islands of Alaska. Second Series.* Compiled by William H. Dall, assisted by Marcus Baker. Washington: Government Printing Office, 1879.
17. M. B. Sherwood, *Exploration of Alaska, 1865–1900*. 1965:40–53.
18. Huber H. Bancroft, *History of Alaska. 1730–1885.*, San Francisco: A. L. Bancroft and Co., 1886.
19. Ivan Petroff, special agent. United States Department of the Interior, 10th Census of U.S., vol. 8. Washington: Government Printing Office, 1884.

20. Sherwood, 1965:57–58.
21. For another assessment of Petroff's accuracy, see Richard Pierce, "New Light on Ivan Petroff, Historian of Alaska," *Pacific Northwest Quarterly* 59, no. 1 (1968):1–10. [MF]
22. Sherwood, 1965:59–68.
23. Ibid., 1965:64.
24. Petroff, 1884:83.
25. Ibid., 1884:83.
26. Ibid., 1884:112–115.
27. *Map showing the distribution of the native tribes of Alaska and adjoining territory, compiled from the latest authorities by W. H. Dall.* Washington, 1875, includes an insert map of the Aleutian Islands.
28. Petroff, 1884:123–124.
29. *Map of Alaska and adjoining regions. Compiled by Ivan Petroff, 1880 . . . Russian charts . . . French charts . . . English charts . . . Charts and maps of U.S. . . . and personal notes of the compiles.* Drawn by Harry King. Petroff, 1884, Map 1; *Map of Alaska and adjoining regions. Compiled by Ivan Petroff. Showing the distribution of native tribes . . .* Drawn by Harry King, Petroff, 1884, Map 2; *Map of Alaska . . . Showing distribution of Timber, Tundra and Glacier.* Petroff, 1884:75, map; *Map of Alaska . . . Showing Distribution of Mink and Marten*, Petroff, 1884:59, map; *Map of Alaska . . . Showing distribution of Foxes.* Petroff, 1884:58, map; *Map of Alaska . . . Showing the Distribution of Polar Bears, Brown and Black Bears*, Petroff, 1884; *Map of Alaska . . . Showing Distribution of Beaver, Land Otter and Sea Otter*, Petroff, 1884:55, map.
30. Richard A. Pierce, "Russian Exploration in North America." *In Exploration in Alaska: Captain Cook Commemorative Lectures, June -November 1978.* Edited by Antoinette Shalkop. Anchorage, 1980:125.

Bibliography

RUSSIAN ARCHIVE ABBREVIATIONS

ARGO	**Archiv Russkogo geograficheskogo obshchestva** (Archive of the Russian Geographical Society) St. Petersburg
AVPRI	**Arkhiv vneshnei politiki Rossiiskoi Imperii** (Archive of Russian Foreign Policy) Moscow
BAN	**Biblioteka Rossiiskoi Akademii nauk** (Library of the Russian Academy of Sciences) St. Petersburg
GIM	**Gosudarstvennyi Istoricheskii muzei** (State Historical Museum) Moscow
IRGO	**Imperial Russian Geographical Society**
LOII	**Leningradskoe otdelenie Instituta istorii SSSR AN SSSR** (Leningrad Division of the Institute of History of the USSR). Since 2000, **Sankt-Peterburgskii Institut istorii RAN (SPbII).**
MAE	**Muzei antropologii i etnografii im. Petra Velikogo RAN (Kunstkamera/MAE)** St. Petersburg
RAK	**Rossiskaya Amerikanskaya Kompaniya** (Russian-American Company)
RAN	**Sankt-Peterburgskii Institut istorii RAN (SPbII)** (St. Petersburg Institute of History) St. Petersburg
RGADA	**Rossiiskii gosudarstvennyi arkhiv drevnikh aktov** (Russian State Archives of Ancient Documents) Moscow
RGAVMF	**Rossiiskii gosudarstvennyi arkhiv Voenno-Morskogo Flota** (Russian State Naval Archives) St. Petersburg
RGB	**Rossiiskaia gosudarstvennaia biblioteka** (Russian State Library) Moscow
RGIA	**Rossiiskii gosudarstvennyi istoricheskii Arkhiv** (Russian State Historical Archive) St. Petersburg
RGVIA	**Rossiiskii gosudarstvennyi voenno-istoricheskii arkhiv** (Russian State Military History Archive) Moscow
VUA	**Voenno-ucheny arkhiva** (Military Educational Archive)

REFERENCE BIBLIOGRAPHY

Adams, G. R. *Life on the Yukon, 1865–1867.* Edited by Richard A. Pierce. Kingston, Ontario: Limestone Press, 1982.

Adams, John Q. *Memoirs of John Quincy Adams,* vol. 1–12. Edited by C. F. Adams. Philadelphia: J. B. Lippincott & Co., 1874–1877.

Afonsky (Gregory). *A History of the Orthodox Church in Alaska (1794–1917).* Kodiak: St. Herman's Theological Seminary, 1977.

Alekseev, Aleksandr I. *Beregoviia Cherta (Shore Charts).* Magadan: Magadanskoe knizhnoe izd-vo, 1987. A study of the mapping of the shores of the North Pacific and Bering Sea, especially on the Asian side.

———. "Brat'ia Shmalevy i ikh deiatel'nost'." *letopis' severa,* vol. 3, 1962.

———. "Chertezhy i karty Rossii XVII veka, naidennye v poslevoennye gody" ["Charts and Maps of Russia of the Seventeenth Century, Discovered in the Post-War Years"]. In *Voprosy ekonomiki i klassovykh otnoshenii v Russkom gosudarstve XII–XVII vekov (Issues of Economics and Class Relationships in the Russian State of the Twelfth to Seventeenth Centuries).* Number 2 (Trudy Leningradskogo otdeleniya Instituta istorii Akademii Nauk SSSR). Moscow/Leningrad, 1960:80–90.

———. *Fedor Petrovich Litke.* Translated from the Russian edition of 1970 by Serge LeComte and edited by Katherine Arndt. Rasmuson Historical Translation Series 10. Fairbanks: University of Alaska Press, 1996.

———. *Gavriil Andreevich Sarychev.* Moscow: Nauka, 1966.

———. *Ilya Gavrilovich Voznesensky.* Moscow: Nauka, 1977.

———. *The Odyssey of a Russian Scientist: I. G. Voznesenskii in Alaska, California and Siberia, 1839–1849.* Translated by Wilma C. Follette and edited by Richard A. Pierce. Alaska History Series 30. Kingston, Ontario: Limestone Press, 1987. Twenty-five of Voznesenskii's drawings are reproduced.

———. *Okhotsk—kolybel russkogo Tikhookeanskogo flota (Okhotsk—the Cradle of the Russian Pacific Fleet).* Khabarovsk: Khabarovskoe knizhnoe izd-vo, 1958.

———. *Russian Discoveries in the Pacific and North America in the Eighteenth and Nineteenth Centuries.* Translated by Carl Ginsberg. Ann Arbor: Published for the American Council of Learned Societies by J. W. Edwards, 1952. Originally published as *Russkie Otkrytiia v Tikhom Okeane i Severnoi Amerike v XVIII–XIX Vekakh.* Moscow and Leningrad: AN SSSR, 1944.

Alekseev, P. *Sibir v izvestiyakh zapadno-evropeiskikh puteshestvennikov i pisatelei, XIII–XVII vv. (Siberia in the Reports of Western European Travelers and Writers, Thirteenth–Seventeenth Centuries).* Irkutsk: Krajgiz, 1936.

Alton, James. *The First Scientific Exploration of Russian America and the Purchase of Alaska.* Evanston and Chicago: Northwestern University, 1942.

Andreev, A. I. "Ekspeditsiya Beringa." In *Izvestiya VGO (News of the All-Union Geographical Society),* vol. 75, no. 2, 1943.

———. "Pervye karty Sibiri XVII veka" ("The First Maps of Siberia of the Seventeenth Century)." In *Globus. Geograficheskу ezhegodnik dlya detei (The Globe. Geographical Annual for Children).* Moscow-Leningrad, 1939:309–313.

———. *Russkie otkrytiia v Tikhom okeane i v severnoi Amerike v XVIII–XIX vekakh.* Moscow: Izd-vo Akademii nauk SSSR, 1944.

———. *Russkie otkrytiia v Tikhom okeane i v severnoi Amerike v XVIII veke.* Moscow: OGIZ, 1948.

Andreev, V. A. "Dokumenty po ekspeditsii kapitana-komandora Beringa v Ameriku v 1741 g." (Documents of the Expedition of Captain-Commander Bering to America in 1741). *Morskoi Sbornik*, vol. CCLV, no. 5 (1893):1–16.

Andrews, Clarence Leroy. *Sitka, the Chief Factory of the Russian American Company*. 3rd edition. Caldwell, Idaho: Caxton, 1945.

Archer, Christon A. "Russians, Indians, and Passages. Spanish Voyages to Alaska in the Eighteenth Century." *Exploration in Alaska*. Anchorage: Cook Inlet Historical Society, 1980:129–143.

Arndt, Katherine L. "Dynamics of the Fur Trade on the Middle Yukon River, Alaska, 1837–1867." Ph.D. dissertation, University of Alaska Fairbanks, 1996.

Arndt, Katherine L., and Richard A. Pierce. *A Construction History of Sitka, Alaska, as Documented in the Records of the Russian-American Company*. 2d ed. Anchorage: U.S. Department of the Interior, National Park Service, 2003.

Ault, Phillip. "The (Almost) Russian-American Telegraph," *American Heritage*, XXVI:4 (June 1975).

Baddeley, J. F. *Russia, Mongolia, China. Being Some Record of the Relations Between Them from the Beginning of XVIIth Century to the Death of the Tsar Alexei Mikhailovich, A.D. 1602–1676*. . . . 2 vols. with maps. London and New York: Macmillan, 1919.

Baer, Karl E. von. "Ueber das Klima von Sitcha und den russischen Bestizungen an der Nordwestkuste von America uberhaupt" (Concerning the Climate of Sitka and the Russian Possessions on the Northwest Coast). *Beiträge Zur Kenntniss Des Russischen Reiches* (1839):290–320.

Bagrow, Leo. A Few Remarks on Maps of the Amur, the Tartar Strait and Sakhalin." In *Imago Mundi*, vol. XII (1955):127–136.

———. "The First Russian Maps of Siberia and Their Influence on the West European Cartography of N. E. Asia." *Imago Mundi (Leiden)* 9 (1952):83–93.

———. *History of Cartography*. Revised and enlarged by R. A. Skelton, 2nd edition. Chicago: Precedent, 1985.

———. *A History of Russian Cartography up to 1800*. 2 vols. Edited by Henry W. Castner. Wolfe Island, Ontario: Walker Press, 1975.

———. *Karty Aziatskoi Rossii. Istoricheskie zametki* (*Maps of Asiatic Russia. Historical Notes*). Petrograd: Artisticheskoe zavedenie T-va A.F. Marks, 1914.

———. "Semyon Remezov—a Siberian Cartographer." *Imago Mundi (Leiden)* 12 (1954):111–126.

———. "Sparwenfeld's Map of Siberia." *Imago Mundi* 4 (1947):65–70.

Baker, Marcus. *Geographic Dictionary of Alaska*. U.S.G.S. Bulletin 229. Washington: Government Printing Office, 1902. 2nd edition prepared by James McCormick.

———. "Sarichef's Atlas, 1826." *National Geographic Magazine* 13 (1902):86–92.

Bancroft, Hubert Howe. *History of Alaska. 1730–1885*. The Works of Hubert Howe Bancroft, XXXIII. San Francisco: A. L. Bancroft & Company, 1886. Published as a separate volume by Bancroft in 1890. Reprinted separately by various publishers including the Antiquarian Press in 1959 and Hafner Pub. Co. in 1970.

———. *History of the Northwest Coast*. 2 vols. San Francisco: A. L. Bancroft & Company, 1886.

Barratt, Glynn. *Russia in Pacific Waters, 1715–1825. A Survey of the Origins of Russia's Naval Presence in the North and South Pacific*. Pacific Maritime Studies 1. Vancouver and London: University of British Columbia Press, 1981.

Barrow, Sir John. *A Chronological History of Voyages into the Arctic Regions; Undertaken Chiefly for the Purpose of Discovering a Northeast, Northwest, or Polar Passage Between the Atlantic and Pacific*. London: J. Murray, 1818.

Barsukov, Ivan Plantonovich. *Graf Nikolai Nikolaevich Muraviev-Amursky* (*Count Nikolai Nikolaevich Muraviev-Amursky*), vol. 1. Moscow: V' Sinodal'naia tip., 1891.

———. *Tvoreniia Innokentiia Mitropolita Moskovskago (Works of Innokenti, Metropolitan of Moscow)*. Moscow: V' Synodal'noi tip., 1886–1888.

Basanoff, V. "Archives of the Russian Orthodox Church in Alaska in the Library of Congress." *Pacific Historical Review*, II, no. 1, 1933:72–84.

Beaglehole, John Cawte, comp. *The Journals of Captain James Cook*. Hakluyt Society. Extra Series, 34–37. Cambridge: Cambridge University Press, 1955–1974.

Beechey, Frederick William. *Narrative of a Voyage to the Pacific and Beering's Strait, to Co-Operate With the Polar Expeditions: Performed in His Majesty's Ship* Blossom, *Under the Command of Captain F. W. Beechey . . . in the Years 1825, 26, 27, 28 . . .* Reprint of the 1831 London edition. Amsterdam: Nico Israel, 1969.

Belcher, Edward. *Narrative of a Voyage Round the World, Performed in Her Majesty's Ship Sulpher, During the Years 1836–42 . . . .* London: H. Colburn, 1843. Belcher visited Sitka and other points in Alaska in 1837.

Belov, M. I. "Daniya i Vitus Bering" ("Denmark and Vitus Bering"). In *Puteshestviya i geograficheskie otkrytiya v XV–XIX vv.* (*Journeys and Geographical Discoveries in the Fifteenth–Nineteenth Centuries*). Moscow-Leningrad, 1956.

———. *Istoriya otkrytiya i osvoeniya severnogo morskogo puti* (*The History of the Discovery and Mastery of the Northern Sea Route*). Moscow: Morskoj transport, 1956.

———. "Novye dannye o sluzhbakh V. Atlasova i pervykh pokhodakh russkikh na Kamchatku" ("New Data About the Services of Vladimir Atlasov and the First Russian Journeys to Kamchatka"). In *Letopis Severa* (*Chronicle of the North*), vol. 2:79–107, 1957.

———. "O sostavlenii Generalnoi karty Vtoroi Kamchatskoi ekspeditsii 1746 g." ("On the Creation of the General Map of the Second Kamchatka Expedition of 1746"). In *Geografichesky sbornik* (*Geographical Collection*), vol. 3, 1954.

———. *Otkrytie Kamchatki I expeditsii Beringa 1725–1742 gg.* Moscow and Leningrad: Akademii nauk SSR, 1946.

———. *Russians in the Bering Strait, 1648–1791*. Translated by Katerina Solovjova and edited by J. L. Smith. Anchorage: White Stone Press, 2000.

———. *Semen Dezhnev*. Moscow: Morskoi Transport, 1955.

Benyowski, Mauritius Augustus. *Memoirs and Travels*. First English ed. London: G.G.J. and J. Robinson, 1790. The first French edition was put out by the same publisher in London in 1790. It became very popular with another English edition in Dublin in 1790; German editions in 1790, 1791 (three in different cities); 1796, 1797, 1806, 1807 and 1816; Dutch in 1792; Swedish in 1791; Slovakian in 1808. The first Polish edition, translated from the 1791 French edition, appeared in 1797.

Berg, L. S. *Otkrytie Kamchatki i ekspeditsii Beringa, 1725–41* (*The Discovery of Kamchatka and Bering's Expeditions*). Moscow-Leningrad: AN SSSR, 1946.

Bering, Vitus. "The Logbook of Bering's Vessel, the 'St. Peter' and of Her Successor, the Hooker 'St. Peter,'" In *Bering's Voyages*, edited by Frank A. Golder, 36–269. New York: American Geographical Society, 1922–1925. Republished by Octagon Books in 1968. This is a combination of logs kept by assistant navigator Kharlam Yushin and mate Sofron Khitrov.

———. *Until Death Do Us Part: The Letters and Travels of Anna and Vitus Bering.* Edited by Peter Ulf Møller and Natasha Okhotina Lind and translated by Anna Halager. Rasmuson Historical Translation Series, 14. Fairbanks: University of Alaska Press, 2008.

Berkh, Vasilii Nikolaevich. "Captain-Lieutenants Vasil'ev and Shishmarev, 1819." In *Ethnohistory in the Arctic: The Bering Strait Eskimo.* Edited by Dorothy Jean Ray and translated by Rhea Josephson. Alaska History 23. Kingston, Ontario: Limestone Press, 1983.

———. *Khronologicheskaia Istoriia Otkrytiia Aleutskikh Ostrovov Ili Podvigi Rossiiskago Kupechestva, s Prisovokupleniem Istoricheskago Izvestiia i Mekhovoi Torgovle* (*A Chronological History of the Discovery of the Aleutian Islands or the Exploits of the Russian Merchantry, With the Addition of Historical Information About the Fur Trade*). St. Petersburg: N. Grech., 1823. Published in English as *A Chronological History of the Discovery of the Aleutian Islands.* Translated by Dmitri Krenov. Materials for the Study of Alaska History 5. Kingston, Ontario: Limestone Press, 1974. An earlier translation was published by the Works Progress Administration in 1938.

———. *Khronologicheskaya istoriya vsekh puteshestvii v Severnuyu polyarnuyu stranu* (*A Chronological History of All the Journeys to the Northern Polar Country*), part I. St. Petersburg: Imperatorskogo velichestva, 1823.

———. *Pervoe Morskoe Puteshestvie Rossiian, Predpriniatoe Dlia Resheniia Geograficheskoi Zadachi: Soediniaetsia Li Aziia s Amerikoiu? i Sovershennoe v 1727, 28 i 29 Godakh Po Nachal'Stvom Flota Kapitana I-Go Ranga Vitus Beringa. S Prisovokupleniem Kratkago Biograficheskago Svedeniia o Kapitana Beringa i Byvshikh s Nim Ofitserov* (*The First Russian Maritime Voyage, Undertaken to Decide the Geographical Question: Is Asia Joined to America? and Completed in 1727, 1728, and 1729 Under the Command of Fleet Captain of the First Rank Vitus Bering. With the Addition of Brief Biographical Information About Captain Bering and Officers Who Were With Him*). St. Petersburg: Imperatorskaia akademiia nauk, 1823.

——— ed. *Puteshestvie po Severnoi Amerike k Ledovitomu moryu i Tikhomu okeanu, sovershennoe gospodinom Khernom i Makenziem* (*Journey across North America to the Icy Sea and the Pacific Ocean, Completed by Mssrs. Hearne and Mackenzie*). Translated and with an introduction by V. N. Berkh, parts 1 and 2. St. Petersburg, 1808.

———. *Zhizneopisanie admirala Alekseia Ivanovicha Nagaeva* (*Biography of Admiral Alexei Ivanovich Nagaev*). St. Petersburg: Morska͡ia tip, 1831.

Besprozvannykh, E. L. *Priamur'ie v sisteme russko-kitaiskikh otnoshenii.* Moscow: Izd-vo "Nauka," Glav. red. vostochnoĭlit-ry, 1983.

Black, Dawn Lea, and Alexander Yu. Petrov, editors and translators. *Natalia Shelikhova: Russian Oligarch of Alaska Commerce.* Rasmuson Library Historical Translation Series, 15. Fairbanks: University of Alaska Press, 2010.

Black, Lydia T. "Ivan Pan'kov, an Architect of Aleut Literacy." *Arctic Anthropology*, XIV, no. 1, 1977:94–107.

———. *The Journals of Iakov Netsvetov. The Atkha Years, 1828–1844.* Kingston, Ontario: Limestone Press, 1980.

———, trans. *The Lovtsov Atlas of the North Pacific Ocean, Compiled at Bol'sheretsk, Kamchatka, in 1782.* Alaska History Series 38. Fairbanks: Limestone Press, 1991.

———. *Russians in Alaska, 1732–1867.* Fairbanks: University of Alaska Press, 2004.

Blake, William P. "General Topographic Features of the Northwestern Coast of America, from the Strait of Juan de Fuca to the Parallel of Sixty Degrees North Latitude." *Coast*

Survey; Report of the Superintendent showing the Progress of the Survey during the Year 1867. Appendix 18:281–290. Washington, 1869.

———. *Geographical Notes upon Russian America and the Stickeen River: being a Report Addressed to the Hon. W. H. Seward, Secretary of State.* Washington: Government Printing Office, 1868.

Blomkvist, E. E. *Risunki I. G. Voznesenskogo Eksedititsiia 1839–1849* (*Voznevski Expedition 1839–1849*). Reprinted from *Sbornik Muzeia antropologii I etnografii.*

Bockstoce, John R. *Furs and Frontiers in the Far North: The Contest Among Native and Foreign Nations for Control of the Intercontinental Bering Strait Fur Trade.* New Haven: Yale University Press, 2009.

———, ed. *The Journal of Rochfort Maquire, 1852–1854: Two Years at Point Barrow, Alaska, Aboard HMS Plover in the Search for Sir John Franklin.* Works Issued by the Hakluyt Society, Second Ser., 169–170. London: Hakluyt Society, 1989. Rich in ethnographic detail, this is one of the first extended accounts of life at Pt. Barrow.

———. "The Search for Sir John Franklin in Alaska." In *The Franklin Era in Canadian Arctic History, 1845–1859*, edited by Patricia D. Sutherland, 93–113. Mercury Series. Ottawa: National Museums of Canada, 1985.

———. *Whales, Ice and Men: The History of Whaling in the Western Arctic.* Seattle: University of Washington Press in association with the New Bedford Whaling Museum, Massachusetts, 1986.

Bolkhovitinov, Nikolai N. "Alyaskinsky skandal" ("The Alaska Scandal"). *Voprosy istorii* (*Historical Affairs*), 1989, no. 4:37–54.

———. *Doktrina Monro: Proiskhozhdenie i kharakter* (*The Monroe Doctrine: Its Origin and Character*). Moscow, 1959; *Rossiya i SShA: arkhivnye dokumenty i istoricheskie issledovaniya: Analitichesky obzor*, Moscow, 1984. An expanded edition in English: *Russia and the United States: An Analytical Survey of Archival Documents and Historical Studies*, translated and edited by J. D. Hartgrove. Armonk, N.Y., 1989.

———, ed. *Istoriia Russkoi Ameriki, 1732–1867.* 3 vols. Moscow: Mezhdunarodnye otnoshniia, 1997–1999.

———. "Kak prodali Alyasku" ("How Alaska was Sold"). *Mezhdunarodnaya zhizn* (*International Life*), 1988:7:120–131.

———. "Obshchestvennost SShA i ratifikatsiya dogovora 1867 g." ("The American Public and the Ratification of the Treaty of 1867"). *Amerikansky ezhegodnik* (*American Yearly*), Moscow, 1987: 157–174.

———. *Rossiia Otkryvaet Ameriku, 1732–1799* (Russian discovery of America, 1732–99). Moscow: Mezhdunarodnyie Otnosheniia, 1991.

———. *Russko-Amerikanskie otnoshoniia I prodazha Aliaski, 1834–1867.* Moscow: Nauka, 1990. Translated by Richard A. Pierce as *Russian-American Relations and the Sale of Alaska, 1834–1867.* Kingston, Ontario and Fairbanks: Limestone Press, 1996.

Bolotov, Ivanov. *Kratkoe opisanie ob Amerikanskom ostrove Kadiake, sobrannoe iz dostovernykh zapisok i raspolozhennoe na topograficheskoe, klimaticheskoe, statisticheskoe i esteticheskoe otdeleniya* (*Short Description of the American Island of Kodiak, Collected from Reliable Notes and Arranged in Topographical, Climatic, Statistical and Aesthetic Sections*). Moscow, 1805.

Borodin, A. V. "Moskovskaya grazhdanskaya tipografiya i bibliotekari Kipriyanovy" ("The Moscow Civil Printing House and the Kipriyanov Librarians"). In *Stati i materialy po istorii knigi v Rossii* (*Articles and Materials on the History of the Book in Russia*). Moscow/Leningrad: AN SSSR, 1936:59–87.

Bouguer, Pierre. *Traité complet de la navigation*. Paris: Heuqueville, 1706.

Bragin, Dmitri. "Report of a Four-year Voyage Undertaken in 1772 to the Islands Situated between Kamchatka and America, under the Command of the Peredovshik Dmitri Bragin." In *Bering's Successors 1745–1780*, translated by James Masterson and Helen Brower, 67–76. Seattle: University of Washington Press, 1948 (c.f. Pallas below).

Bronshtein, Yuri, and N. Shnakenburg. "Zapiski doktora K. Merka uchastnika ekspeditsii Billings-sarycheva v 1785–1792 gg." *Sovetskaia Arktika*, no. 4, 1941:76–88.

Buache, Philippe. *Exposé des découvertes au Nord de la Grande Mer, soit dans Nord-Est de l'Asie, soit dans le Nord-Ouest de l'Amerique*. Paris: Imprimerie de Ballard, 1753.

Bucareli, Antonio Maria. *La administración de d. frey Antonio María de Bucareli y Ursúa, cuadragésimo sexto virrey de México*. Mexico: Talleres graficos de la nacion, 1936–.

Buffon, Georges-Louis Leclerc de. *Histoire naturelle générale et particulière, servant de suite à l' histoirie naturelle de l' homme [. . .]/ Supplément. t. 4*. Paris: de l' Imprimerie royal, 1777.

Burney, James. *A Chronological History of the North-Eastern Voyages of Discovery; and of the Early Navigations of the Russians*. 5 vols. London: Payne and Foss, 1819. Reprinted, Amsterdam: Nico Israel, 1969. Burney sailed with Captain Cook's second and third voyages.

Burykin, A. F. "Four Voyages of Captain Khromchenko." *Priroda* (*Nature*), no. 3, 1957:77–80.

Büsching, A. F., ed. *Woechentliche Nachrichten von neuen Landcharten, geographischen, statischen und historischen Buechern und Sachen*. 14 vols. Berlin, 1773–1786.

Campbell, J. "A distinct account of part of the north-east frontier of the Russian Empire, commonly called the country of Kamchatka or Kamschatska." In J. Harris, *Navigantium at que itinerarium bibliotheca of voyages and travels*. 2nd edition, vol. II, London 1748; third ed., London, 1764:1016–1041.

Chamisso, Adelbert von. *A Voyage Around the World With the Romanzov Exploring Expedition in the Years 1815–1818 in the Brig Rurik, Captain Otto Von Kotzebue*. Translated and edited by Henry Kratz. Honolulu: University of Hawaii Press, 1986. Chamisso published his German language account in 1836, long after the voyage. It includes accounts of life aboard ship.

Chaplin, Piotr. *The Journal of Midshipman Chaplin: A Record of Bering's First Kamchatka Expedition*. Edited and translated by Tatiana S. Fedorova, Peter Ulf Møller, Victor G. Sedov, and Carol L. Urness. Copenhagen: Arhus University Press, 2010.

Cherikov, Alexi. "Journal Kept on the Ship 'St. Paul' in Command of Captain Alexei Chirikov at the Time It Made the Voyage East From Kamchatka Harbor of the Holy Apostles Peter and Paul to the Western Coast of America in the Year 1741." In *Bering's Voyages*, edited by F. A. Golder, 283–311. New York: American Geographical Society, 1922.

Chernenko, M. B. "Puteshastviia po Chukotskoi zemle I plavanie na Aliasku kazak'ego sotnika Ivana Kobeleva v 1779 I 1789–1791 gg." *Letopis' Severa* 2 (1957): 121–141.

Chernenko, M. B., and G. A. Agranat. *Lavrenty Alekseevich Zagoskin. Puteshestviya Zagoskina v Russkoi Amerike v 1842–1844 gg.* (*Lavrenty Alekseevich Zagoskin. Zagoskin's Journey in Russian America in 1842–1844*). Moscow, 1956.

Chevigny, Hector. *Russian America: The Great Alaskan Venture, 1741–1867*. New York: Viking, 1965.

Choris, Ludovik. *Voyage Pittoresque Autour Du Monde, Avec Des Portraits de Sauvages D'Amerique, D'Asie, D'Afrique, et des Iles Du Grad Ocean; des Paysaques, des Vues*

Maritimes, et Plusieurs Objets D'Histoire Naturelle; Accompagne des Descriptions par M. Le Baron Cuvier, et M. A. de Chamisso, et D'Observations sur les Cranes Humains, par M. le Docteur Gall. Contributors include Georges Cuvier, Adelbert von Chamisso, Franz Josef Gall, and Jean Baptiste Eyries. Paris: Didot, 1822. The work was originally issued in parts (livraisons) between 1821 and 1823. The total number of plates varies in different bound copies.

Cole, A. B., ed. *Yankee Surveyors in the Shogun's Seas.* Princeton, N.J.: Princeton University Press, 1947.

Colección de documentos inéditos para la historia de España. Madrid: Academia de la Historia, 1842–1895.

Collins, P. M. *Voyage down the Amoor: with Land Journey through Siberia, and Incidental Notices of Manchooria, Kamchatka and Japan.* New York, 1860 (2nd edition 1864). In 1962 Charles Vevier reissued this book with slight changes and scholarly annotations: P. M. Collins, *Siberian Journey Down the Amoor to the Pacific, 1856–1857,* edited with introduction by C. Vevier. Madison: University of Wisconsin Press, 1962.

Collinson, Richard. *Journal of H.M.S. Enterprise, on the Expedition in Search of Sir John Franklin's Ships by Bering Strait, 1850–55.* Edited by T. B. Collinson. London: S. Low, Marston, Searle & Rivington, Ltd., 1889. Collinson's account was published nearly 40 years after his voyage. The *Enterprise* spent three years in the Arctic and sailed along the entire northern coast of Alaska and into the Central Canadian Arctic from Bering Strait.

Cook, James, and James King. *A Voyage to the Pacific Ocean. Undertaken by Command of His Majesty, for Making Discoveries in the Northern Hemisphere, to Determine the Position and Extent of the West Side of North America; Its Distance From Asia; and the Practicability of a Northern Passage to Europe. Performed under the Direction of Captains Cook, Clerke, and Gore, in His Majesty's Ships the Resolution and Discovery, in the Years 1776, 1777, 1778, 1779 and 1780. Vols I and II written by Captain James Cook, F.R.S., vol III by Captain James King, L.L.D., F.R.S. . . . Published by order of the Lords Commissioners of the Admiralty.* 3 vols. London: G. Nicoll & T. Cadell, 1784. The official edition. There are numerous abridgements, translations and unauthorized editions. The first to reach print was the unauthorized Dublin (396 pp.) and London (388 pp.) editions of 1781. The most complete modern version, based upon original journals is the Beaglehole edition in three volumes.

Cook, Warren L. *Flood Tide of Empire: Spain and the Pacific Northwest, 1543–1819.* New Haven and London: Yale University Press, 1973. A study of why Spain did not successfully exercise its claims to the Oregon country and beyond to Alaska.

Coxe, William. *An Account of the Russian Discoveries Between Asia and America, to Which Are Added, the Conquest of Siberia and the History of the Transactions and Commerce Between Russia and China.* One of the most important sources for knowledge of Russian discoveries available outside of Russia at the time. Published in 4 major editions, each with newly available information, through 1803 (1780, 1780 [quarto], 1787, 1803). French edition in 1781 and German in 1783.

———. *Travels in the Northern Countries of Europe.* Five vols. London: Printed for T. Cadell, Jun. and W. Davies, 1802 (reprint Philadelphia, 1812).

Crowell, Aron L. *Archaeology and the Capitalist World System: A Study From Russian America.* New York: Plenum Press, 1997. The book is based upon historical archeology conducted at Three Saints Harbor on Kodiak Island, the site of the second continuously occupied Russian center in Alaska, founded by fur merchant Grigorii Shelikhov in 1784.

Dall, William H. *Alaska and its Resources*. Boston: Lee and Shepard, 1870.

———. *Biography of Spencer Fullerton Baird*. Philadelphia: J. B. Lippincott, 1915.

———. "Notes on an original manuscript chart of Bering's Expedition of 1725–1730, and on an original manuscript chart of his second expedition, kept by Peter Chaplin, and now first rendered into English from Berkh's Russian version." U.S. Coast and Geodetic Survey. *Annual Report*, 1890. Washington, 1891. Appendix No. 19.

———. "On the Trend of the Rocky Mountain Range North of Lat. 60° and Its Influence on Faunal Distribution." *AAAS Proceedings*. Vol. 18, 1870.

———. "Remarks upon the Natural History of Alaska." *Proceedings of the Boston Society of Natural History* 12 (1868).

Dauenhauer, Nora, et al., eds. *Anoooshi lingit ka = Russians in Tlingit America: the battles of Sitka, 1802 and 1804*. Seattle: University of Washington Press, 2008.

David, Andrew, ed. "Russian Charts and Captain Cook's Third Voyage." *The Map Collector*, Autumn 1990, no. 52:2–6.

———. *The Voyage of the Resolution & Discovery, 1776–1780, Vol. 3 of The Chartes & Coastal Views of Captain Cook's Voyages With a Descriptive Catalog of All the Known Original Surveys and Coastal Views and the Original Engravings Associated With Them Together With the Running Journal of James King 1779–80*. The Hakluyt Society Extra Series, 46. London: The Hakluyt Society in association with the Australian Academy of the Humanities, 1997. The front matter includes an essay on Russian charts of the North Pacific by Alexei V. Postnikov.

Davidson, George. *Pacific Coast. Coast Pilot of Alaska, From the Southern Boundary to Cook's Inlet*. Washington: U.S. Coast Survey, 1869.

———. *Report Relative to the Resources and the Coast Features of Alaska Territory. Coast Survey; Report of the Superintendent Showing the Progress of the Survey during the Year 1867*. Appendix 18. Washington 1869:187–329.

———. "Scientific Expedition to Alaska." *Lippincott's Magazine*, vol. II, November 1868:467–485.

———. *United States Coast Survey, Benjamin Pierce Superintendent. Pacific Coast. Coast Pilot of Alaska (First Part) from Southern Boundary to Cook Inlet*. Washington, D.C.: Government Printing Office, 1869.

Davydov, G. I. *Two Voyages to Russian America, 1802–1807*. Translated by Colin Bearne. Materials for the Study of Alaska History, 10. Kingston, Ontario: Limestone Press, 1977. Translated from the Russian edition of 1810–1812.

Davydov, Yuri. *Ferdinand Vrangel' (Ferdinand Wrangell)*. Moscow: Gos. Izd-vo geogr. Litry, 1959.

Dease, Peter Warren. *From Barrow to Boothia: The Arctic Journal of Chief Factor Peter Warren Dease, 1836–1839*. Edited and annotated by William Barr. Montreal and Kingston: McGill-Queen's University Press, 2002.

De Laguna, Frederica. *The Prehistory of Northern North America as Seen From the Yukon*. Memoirs of the Society for American Archaeology, 3. Menasha, Wisconsin, 1947. Supplement to *American Antiquity*, vol. XII, no. 3, part 2, 1947. Reprinted by Kraus Reprint in 1974 and by AMS Press in 1980.

Delisle, Joseph Nicolas. *Explication de la Carte des Nouvelles Decouvertes au Nord de la Mer du Sud*. Paris: Desaint et Saillant, 1752. Accompanying map (37 x 63 cm.): "Carte de nouvelles decouvertes au nord de la mer du Sud, tant a l'est de la Siberie et du Kamchatka, qu'a l'ouest de in Nouvelle France."

———. *Nouvelles Cartes Des Decouvertes De L'Admiral DeFonte, Et Autres Navigateurs Espagnols, Portugais, Anglois, Hollandois, Francois & Russes, Dans Les Mers Septentrionales*. Paris, 1753.

Divin, Vasilli A. *To the American Coast: The Voyages and Explorations of M. S. Gvozdev, the Discoverer of Northwestern America*. Translated by Anatoli Perminov and edited by J. L. Smith. Anchorage: White Stone Press, 1997. Originally published as *K Beregam Ameriki; Plavanniia i Issledovaniia M. S. Gvozdeva, Pervoot-Kryvatelia Severo-Zapadnoi Ameriki*. Moscow: Gos. izdvo geog. lit-ry, 1956.

———. *The Great Russian Navigator, A. I. Chirikov*. Translated and annotated by Raymond H. Fisher. Rasmuson Library Historical Translation Series, 6. Fairbanks: University of Alaska Press, 1993.

———. *Russkaia tikhookeanskaia epopeia*. Khabarovsk: khizhnoe izd-vo, 1979.

Dixon, George. *A voyage round the World but more particularly to the north west coast of America; performed in 1785, 1786, 1787 and 1788, in "King George" and "Queen Charlotte captaines Portlock and Dixon*. London: G. Goulding, 1789.

Documents on the History of the Russian-American Company. Translated by Marina Ramsey and edited by Richard Pierce. Kingston, Ontario: Limestone Press, 1976.

Dorosh, John T. "The Alaskan Russian Church Archives." *Quarterly Journal of the Library of Congress*, XVIII, no. 4, 1861:193–203.

Doroshin, P. "Iz zapisok vedennykh v Russkoi Amerike" ("From Notes Kept in Russian America"). *Gorny Zhurnal* no. 3, 1866:365–400.

———. "Neskolko podrobnostei o rasprostranenii zolota v russkikh Severo-Amerikanskikh vladeniyakh" ("Some Details about the Extent of Gold in the Russian Possessions in North America"). *Gorny Zhurnal*, 2, 1866:277–282.

———. "O nekotorykh vulkanakh, ikh izvezheniyakh i zemletryaseniyakh v byvshikh amerikanskikh vladeniyakh" ("On Volcanoes, their Eruptions and Earthquakes in the Former American Possessions"), *Zapiski Sankt Peterburgskogo mineralogicheskogo obshchestva* (*Notes of the St. Petersburg Mineralogical Society*). Second series, part 6, 1870:25–44.

———. "Zoloto v Verkhnei Kalifornii" ("Gold in Upper California"). *Gorny Zhurnal* 2, 1850:133–162.

Du Halde, Jean-Baptiste. "Relation Succincte Du Voyage Du Capitaine Beering Dans La Siberie." In *Description Geographique, Historique, Chronologique, Politique Et Physique De L'Empire De La Chine Et De La Tartarie Chinoise*, vol. IV, 452–458, map. First folio ed. 4 vols. Paris: P. G. Lemercier, 1735. This is the first printed report of Vitus Bering's 1725–28 expedition. Du Halde was given a copy of a report provided to the King of Poland by Russia. Du Halde decided to include it in his work on China. The 1736 English translation (London: J. Watts) contains the same report and map.

Duflot de Mofras, Eugene. *Duflot De Mofras' Travels on the Pacific Coast*. Translated and edited by Marguerite Eyer Wilber. Calafia Series, 2. Santa Anna: The Fine Arts Press, 1937. Based on the French edition below and "collated with a transcription of his ten original memoires, on which the French publication was based . . . and further compared with a supplementary holographic manuscript of certain portions dealing with Fort Ross, found in his Melanges, volume II."

———. *Exploration Du Territoire De L'Oregon, Des Californies Et De La Mer Vermeille, Executee Pendant Les Annees 1840, 1841 Et 1842* . . . . Paris: A. Bertrand, 1844. It is accompanied by an atlas.

Dumitrashko, N. V. "Y. F. Lisyansky and Russian Circumnavigating Voyages." In *Puteshestvie vokrug sveta na korable "Neva" v 1803–1806 godakh* (*A Voyage Around the World on the Ship "Neva" in 1803–1806*). Moscow, 1947.

Edwards, Clinton R. "Wandering Toponyms: El Puerto de la Bodega and Bodega Bay." *Pacific Historical Review*, vol. 33, no. 3 (Aug. 1964):253–272.

Efimov, Aleksei V. *Atlas Geograficheskikh Otkrytii v Sibiri i v Severo-Zapadnoi Amerike, XVII–XVIII Veka. Atlas of Geographical Discoveries in Siberia and North-Western America, XVII–XVIII Centuries.* Moscow: Nauka, 1964. This atlas reproduces the most significant manuscript maps depicting the emerging knowledge of Siberia and the North Pacific, including the coast of Alaska. A summary of the foreword and a list of maps are printed in English.

———. *Is Istorii Velikikh Russkikh Geograficheskikh Otkrytii v Severnom Ledovitom i Tikhom Okeanakh XVIII-Pervaia Polovina XVIII V.* (*History of the Great Russian Geographical Discoveries in the Arctic and Pacific Oceans in the Seventeenth and First Half of the Eighteenth Centuries*). Moscow: Gos. izd-vo geogr. lit-ry, 1949.

———. *Iz Istorii Russkikh Ekspeditsii na Tikhom Okeane (Pervaia Polovina XVIII Veka) (History of Russian Expeditions on the Pacific Ocean [First Half of the Eighteenth Century])*. Moscow: Voennoe izadel'stvo Ministerstva voruzhennykh sil SSSR, 1948.

———. *Iz Istorii Velikikh Russkikh Geograficheskikh Otkrytii v Severnom Ledovitsom i Tikhom Okeane* (*History of Russian Geographical exploration of the Northern Icy and Pacific Oceans*). Moscow: Nauka, 1950. Revised edition issued in 1971.

Elliott, Henry W. *Report on the Prybilov Group, or Seal Islands, of Alaska*. Washington: Government Printing Office, 1873.

Engel, Samuel. *Memoires et Observations. Anecdotes russes*. Lausanne: A. Chapuis, 1765.

Esakov, V. A., A. F. Plakhotnik, and A. I. Alekseev. *Russkie okeanicheskie i morskie issledovaniya. XIX—nachalo XX veka* (*Russian Ocean and Sea Explorations. Nineteenth—early Twentieth Centuries*). Moscow, 1964.

Evteev, O. A. *Pervye russkie geodesisty na Tikhom okeane* (*The First Russian Geodesists on the Pacific Ocean*). Moscow, 1950.

Falk, Marvin W. *Alaskan Maps: A Cartobibliography of Alaska to 1900*. New York: Garland Publishing Co, 1983.

Farrar, Victor J. *The Annexation of Russian America to the United States*. Washington, D.C.: 1937.

Fedorova, Svetlana G. "K voprosu rannikh russkikh poselenii na Aliaske" ("On the Issue of the Early Russian Settlement in Alaska"). In *Letopis Severa* (*Chronicle of the North*), vol. 4. Moscow, 1964.

———. *New Data on Russian Geographic and Ethnographic Investigations in Alaska (First Half of the Nineteenth Century)*. Paper read at the 9th International Congress of Anthropological and Ethnological Sciences, 1973.

———. *Russkoe naselenie Aliaski i Kalifornii. Konets XVIII v.—1867*. Moscow: Nauka, 1971.

———. *The Russian Population in Alaska and California (Late Eighteenth Century to 1867)*. Translated from the Russian edition of 1972 by Richard A. Pierce and A. S. Donnelly. Materials for the Study of Alaska History 4. Kingston, Ontario: Limestone Press, 1973.

———. "Shturmany Ivan Vasil'ev i ikh rol' v izuchanii Aliaski (pervia polovina XIX v." (Navigator Ivan Vasil'ev and his role in the exploration of Alaska"). *Letopis Severa* 9 (1979):67–210; 10(1982):141–160.

Fel', S. E. "Kartografiia Rossii XVIII v. (kak sintez astronomo-geodezicheskikh i graficheskikh protsessov)." (Russian Cartography in the Eighteenth Century) In *Trudy Instituta istorii estestvoznaniia i tekhniki AN SSSR*, vol. 37, no. 2, Moscow, 1961.

Fischer, Johann Eberhard. *Sibirskaia istoriia s samago otkrytia Sibiri do zavoevaniašei zemli rossiskim oruzhīem*. St. Petersburg: Akademia Nauk, 1774.

———. *Sibirische Geschichte*. St. Petersburg: Akademie der Wissenschaften, 1768.

Fisher, Raymond H. *Bering's Voyages: Wither and Why*. Seattle: University of Washington Press, 1977.

———. *The Voyage of Semen Dezhnev in 1648: Bering's Precurser, With Selected Documents*. Works Issued, second series, no. 159. London: Hakluyt Society, 1981.

Fleurieu, Comte Charles Pierre de Claret. *Voyage atour du Monde, pendant les anee's 1790, 1791 et 1792 par E'tienne Marchand*. 5 vols. Paris, An VI-VIII [1798–1800].

Funakoshi, Akio. "O karte Sibiri perioda Kansi, khranyashcheisya v lichnom arkhive Lo Chzhen'yuya" ("On the Map of Siberia of the Kansi Period, Preserved in the Personal Archive of Lo Chzhenyuya"). In *Takho takukho*, no. 33 (Tokyo 1963):199–218.

Garrett, Paul D. "St. Innocent and the Mission on the Nushagak River." *Orthodox Alaska* VII, no. 2, 1979:13–21.

Glazunov, Andrei. "Extract of a Diary of Ship's Mate Andrei Glazunov, Written During a Journey into the Northwest American Interior." In *Russian America Statistical and Ethnography Information By Rear Admiral Ferdinand Petrovich Wrangell With Additional Material by Karl-Ernst Baer*, translated by Mary Sadowski and Richard A. Pierce, 69–89. Materials for the Study of Alaska History 15. Kingston, Ontario: Limestone Press, 1980.

———. "Russian Exploration in Interior Alaska. An Extract From the Journal of Andrei Glazunov." Edited by James W. VanStone. *Pacific Northwest Quarterly* 50, no. 2 (1959):37–47. First published in the *Zhurnal Manafaktur i Torgovli* in 1836.

Glushankov, L. V. *Sekretnaia ekspeditsiia: ekseditsiia Krenitsyna-Levashova*. Magadan: Magadanskoe knizhnoe izd-vo, 1972.

Gnucheva, V. F. *Geograficheskii departament Akademii Nauk XVIII veka* (*The Geographical Department of the Academy of Sciences in the Eighteenth Century*). Trudy Arkhiva AN SSSR, issue 6. Moscow/Leningrad, 1946.

Goldenberg, L. A. "The Atlases of Siberia by S. U. Remezov as a Source for Old Russian Urban History." In *Imago Mundi*, vol. 25 (1971):39–46.

———. *Goldenberg's Gvozdev: The Russian Discovery of Alaska in 1732*. Translated by James L. Smith. Anchorage: White Stone Press, 1990.

———. *Izograf Zemli Sibirskoi. Zhizn i trudy Semyona Remezova Isograph of the Siberian Land* (*The Life and Work of Semyon Remezov*). Magadan: Magadanskoe knizhnoe izd-vo, 1990.

———. "Izuchenie kart Sibiri XVII–XVIII vv. v Iaponii" ("The Study of Maps of Siberia of the Seventeenth–Eighteenth Centuries in Japan"). In *Voprosy istorii* (*Historical Issues*) 8 (1981):163–168.

———. *Katorzhanin-sibirskii gubernator*. Magadan, 1979.

———. *Mezhdu dvumya ekspeditsiyami Beringa* (*Between Bering's Two Expeditions*). Magadan: Magadanskoe Knizhnoe izd-vo, 1984.

———. "Podlinnaya rospis chertezha Sibiri 1667 g" ("The Original Description of the Chart of Siberia of 1667"). *Trudy instituta istorii estestvozniya i tekhniki AN SSSR* (*Works of the Institute of the History of Natural Science and Technology of the Academy of Sciences of the USSR*). Vol. 42, no. 3. Moscow, 1962:252–271.

———. "Russian Maps and Atlases as Historical Sources." *Cartografica*, 3. Toronto, 1971.

———. *Semen, Ul'i͡anovich Remezov, sibirskiĭ kartograf I geograf, 1942-posle 1720 gg.* (*Semyon Ulyanovich Remezov: Siberian Cartographer and Geographer, 1642–after 1720*). Moscow: Nauka, 1965.

———. "Zdes byl geodesist Gvozdev 1732 goda" ("Geodesist Gvozdev was here in 1732"). ("Kartograficheskie traditsii XVIII v. v izobrazhenii pervootkrytii v Beringovom prolive"); ("The Cartographic Traditions of the Eighteenth Century in the Depiction of the First Discoveries in the Bering Strait"). In *Izvestiya VGO*, vol. 114, issue 6, 1982:526–532.

Golder, Frank A. *Bering's Voyages: An Account of the Efforts of the Russians to Learn the Relation of Asia and America*. New York: American Geographical Society, 1922. A compilation of original accounts including translations of logbooks from both Bering expeditions. Bering, Chirikov, Waxell, and Georg Steller are featured.

———. *Russian Expansion on the Pacific 1641–1850: An Account of the Earliest and Later Expeditions Made by the Russians Along the Pacific Coast of Asia and North America, Including Some Related Expeditions to the Arctic Regions*. Cleveland: Arthur H. Clark, 1914. Reprinted, Gloucester, Mass.: Peter Smith, 1960.

Golovin, Pavel Nikolaevich. *Civil and Savage Encounters: the Worldly Travel Letters of an Imperial Russian Navy Officer, 1860–1861*. Translated by Basil Dmytryshyn, and E. A. P. Crownhart-Vaughan from the 1863 Russian edition. North Pacific Studies Series, 5. Portland: Western Imprints, Oregon Historical Society, 1983.

———. *Doklad Komiteta ob ustroistve russkikh amerikanskikh kolonii* (*Report of the Committee on the Structure of the Russian American Colonies*), St. Petersburg, 1863.

———. *The End of Russian America: Captain P. N. Golovin's Last Report, 1862*. Translated by Basil Dmytryshyn, and E. A. P. Crownhart-Vaughan. North Pacific Studies, 4. Portland: Oregon Historical Society, 1979. Translated, with introduction and notes, from the Russian originally published in *Morskoi Sbornik*, vol. 57 (1862), no. 1, part 3, 19–192. A German translation was published in the *Archiv für wissenschaftliche Kunde von Russland*, vol. 22, 1863, 47–70. An American abridgement was published as "Survey of the Russian Colonies in North America." Congressional Papers, 40th Congress Second session, Executive Document 177:109–112.

———. *Obzor russkikh kolonii v Severnoi Amerike* (*Survey of the Russian colonies in N. America*). In Morskoi Subornik lcii, 1 (iii) 19–192. 1862. English translation in Congressional Papers, H. P. 40th Cong., 2nd Session, Ex. Doc. 177.

Golovnin, Vasilii Mikhailovich. *Sochineniia i Perovody* (works). 5 vols. Saint Petersburg: V tip. Morskago Ministerstva, 1864–1865. Vol. 1 includes a biography of Golovnin by N. Grech; vol. 2 has his account of his captivity in Japan; vol. 3 contains his voyage to Alaska and California; vol. 4 contains accounts of shipwrecks; and vol. 5 contains remarks on the Russian-American Company.

Gough, Berry M. *The Northwest Coast: British Navigation, Trade and Discoveries to 1812*. Vancouver: University of British Columbia Press, 1992.

———, ed. *To the Pacific and Arctic with Beechey: The Journal of Lieutenant George Peard of HMS* "Blossom" *1825–1828*. Cambridge: Hakluyt Society, 1973.

Green, John. *Remarks in the Support of the New Chart of North and South America*. London: J. Jeffreys, 1753.

Grekov, V. I. *Ocherki iz istorii russkikh geograficheskikh issledovanii v 1725–1765 gg.* (*Essays from the History of Russian Geographical Research in 1725–1765*). Moscow: Akademii nauk, 1960.

Grewingk, Constantin Caspar Andreas. *Grewingk's Geology of Alaska and the Northwest Coast of America: Contributions Toward Knowledge of the Orographic and Geognostic Condition of the Northwest Coast of America, with the Adjacent Islands.* Edited by Marvin W. Falk and translated by Fritz Jaensch. Rasmuson Library Historical Translation Series 11. Fairbanks: University of Alaska Press, 2003. Originally published as *Beitrag zur Kenntniss der Orographischen und Geognostischen Beschaffenheit der Nord-West-Kuste Amerikas, mit den anliegenden Inseln . . .* Verhandlung Der Mineralogischen Gesellschaft Zu St. Petersburg, 1848–1849. St. Petersburg: K. Kray, 1850.

Grigoriev, A. V. "Podlinnaya karta Sibiri XVII veka (raboty Semyona Remezova)" ("The Original Map of Siberia of the Seventeenth Century [the Work of Semyon Remezov"]). In *Zhurnal Ministerstva narodnogo prosveshcheniya*. Part II (1907):374–381.

Grinev, A. V. "Zabytaia ekspeditsiia Dmitriia Tarkhanova n Mednuiu reku." *Sovetskaia etnografiia*, 1987, no. 4: 88–110. Translated by Richard Bland under the title "The Forgotten Expedition of Dmitrii Tarkhanov on the Copper River. *Alaska History* 12, no. 1 (1997):1–17.

———. *The Tlingit Indians in Russian America 1741–1867*. Translated by Richard L. Bland and Katerina G. Solovjova. Lincoln and London: University of Nebraska Press, 2005.

Haswell, Robert. "Journal of Robert Gray's voyage in 1788–9 and 1790–2." Appendix to vol. 1, Bancroft's *History of the Northwest Coast*, 1886.

Hearne, Samuel. *A Journey from Prince of Wales' Fort in Hudson Bay to the Northern Ocean Undertaken by Order of the Hudson Bay Company, for the Discovery of Copper Mines, North West Passage, in the Years 1769, 1770, 1771 & 1772*. London: A. Strahan and T. Cadell, 1795.

Herbert E. C. "Spencer Fullerton Baird and the Purchase of Alaska." *Proceedings of American Philosophical Society*, 98:2:139–143, April 1954.

Holmberg, H. J. *Ethnographische Skizzen ueber die Voelker des Russischen Amerika*, Erste Abteilung (Part One), Acta Societatus Scuentarum Fennicae, Vol. 4, Helsingfors (Helsinki) 1856:410–417. Published in English as *Holmberg's Ethnographic Sketches*. Edited by Marvin Falk and translated by Fritz Jaensch. Fairbanks: University of Alaska Press, 1985.

Hooker, Sir William Jackson, and George A. Walker Arnott. *The Botany of Captain Beechey's Voyage*. London: H. G. Bohn, 1841. Reprint, London, 1965.

Hopkins, David M., ed. *The Bering Land Bridge*. Stanford, Calif.: Stanford University Press, 1967.

Howay, Frederick William, ed. *Voyages of the Columbia to the Northwest Coast 1787–1790 and 1790–1793*. Boston: Massachusetts Historical Society, 1941.

Hulley, Clarence C. *Alaska, 1741–1953*. Portland: Binfords & Mort, 1953.

Ivanov, B. P. "K istorii kartografirovaniya russkogo Severa" ("On the History of the Cartography of the Russian North"). In *Vestnik Kharkovskogo universiteta. Seriya geologo-geograficheskaya Kharkov University Bulletin (Geological and Geographical Series)*. Issue 2, no. 25, Kharkov 1967.

J. L. S. *Neue Nachrichten von denen neuentdekten Insuln in der See zwischen Asien und Amerika; aus mitgetheilten Urkunden und Auszügen verfasset von J. L. S.* Hamburg and Lepzig: F. L. Gleditsch, 1776.

James, James Alton. *The First Scientific Exploration of Russian America and the Purchase of Alaska.* Evanston and Chicago: Northwestern, 1942.

Kabanov, P. I. *Amurskie voprosy.* Blagoveshchensk, 1959.

Kajdanski, E. "The authenticity of Maurice Benyowski's account of His Voyage through the Bering Sea: The earliest Description and the Earliest Drawings of St. Lawrence Island." *Terrae Incognitae. The Journal for the History of Discoveries*, vol. XXIII (1991):51–79.

Kari, James, and James A. Fall. "The Russian Presence in Upper Cook Inlet" in *Shem Pete's Alaska. The Territory of Upper Cook Inlet Den'ina.* Compiled and edited by James Kari and James A. Fall. Fairbanks: Alaska Native Language Center, University of Alaska Fairbanks, 1987.

Kashevarov, Aleksandr Filippovich. *A. F. Kashevarov's Coastal Explorations in Northwest Alaska, 1838.* Edited by James W. VanStone. Translation of the 1879 Russian edition. Fieldiana: Anthropology 69. Chicago: Field Museum of Natural History, 1977.

———. "Obozrenie beregov Severnoi Ameriki i mysa barrova sovershennoe Rossiiskoi ekspeditsiei v 1838 godu" ("Overview of the Coasts of North America and Cape Barrow Completed by a Russian Expedition in 1838"). *Syn otechestva* no. 1, 1840:127–144.

———."Otryvki iz dnevnika, vedennago pri obozrenii poliarnogo berega Rossiiskoi Ameriki po porucheniyu Rossiisko-Amerikanskoi kompanii v 1838 godu" ("Excerpts from a Diary Kept during the Overview of the Polar Coast of Russian America by Order of the Russian-American Company in 1838"). *Sankt Peterburgskie vedomosti. 1846*:190–192, 195.

———. "Zametki ob eskimosakh v Russkoi Amerike" ("Notes on the Eskimos in Russian America"). *Severnaya pchela* (*The Northern Bee*), no. 227, 1846:907–908; no. 228, 1846:911–912.

———. Zhurnal, vedennoi pri baidarochnoi ekspeditsii, naznachennoi dlya opisi severnogo berega Ameriki, 1838 goda . . ." ("Journal Kept during a Baidar Expedition Sent to Describe the Northern Coast of America, 1838 . . ."). Imperatorskoe geograficheskoe obshchestvo (IRGO) (Imperial Geographical Society), *Zapiski po obshchei geografii* (*Notes on General Geography*), 8. St. Petersburg, 1879:275–361.

Kellog, A. "Report on the Botany of Alaska." *Coast Survey; Report of the Superintendent Showing the Progress of the Survey during the Year 1867.* Appendix 18:318–324.

Keuning, Johannes. "Nicolaas Witsen as a Cartographer." *Imago Mundi*, vol. 11, 1954.

Khlebnikov, Kyrill T. *Khlebnikov's Notes. O Amerike (On America). Materialy dlya istorii russkikh zaselenii po beregam Vostochnogo okeana*, III, St. Petersbur g, 1861:137. In English as *Colonial Russian America: Kyrill T. Khlebnikov's Reports, 1817–1832.* Translated and edited by Basil Dmytryshyn and E. A. P. Crownhart-Vaughan. Portland: Oregon Historical Society, 1976.

———. *Russkaya Amerika v neopublikovannykh zametkakh K. T. Khlebnikova* (*Russian America in the Unpublished Notes of K. T. Khlebnikov*). Compiled with an introduction and commentaries by R. G. Lyapunov and S. G. Fedorova. Leningrad, 1979.

———. *Russkaya Amerika v Zapiskakh Kirila Khlebnikova* (*Khlebnikov's notes on Russian America*). Compiled, with an introduction and commentaries by S. G. Fedorova and V. A. Aleksandrov. Moscow: Nauka, 1985.

Khmyrov, M. D. "Svedeniya o Vasily Kipriyanove, bibliotekare Moskovskoi grazhdansoi tipografii pri Pyotre I" ("Information about Vasily Kipriyanov, Librarian of the

Moscow Civil Printing House under Peter I"). In *Russkii arkhiv* (*Russian Archive*), no. 8/9 (1866).

King, William Forrest. "George Davidson: Pacific Coast scientist for the U.S. Coast and Geodedetic survey, 1845–1895." Ph.D. dissertation, Claremont Graduate School, 1973.

Kirillov, E. A. "Geologicheskie issledovaniya v Russkoi Amerike, (konets 18 v.–1867)" ("Geological Investigations in Russian America, end of the Eighteenth Century–1867"). *Tikhookeanskaya geologiya* (*Pacific Ocean Geology*) no. 2, 1993.

Kirwan, Laurence P. *A History of Polar Exploration*. New York: Norton, 1960.

Kitchin, Thomas. *A chart of the discoveries made by the late Capt. Cook & other European navigators in the great Pacific Ocean between Asia & America* by. Printed for R. Baldwin in *London Magazine* (July 1780).

Kittlitz, F. H. von. *Denkwuerdigkeiten einer Reise nach dem russischen Amerika*. Gotha, 1858.

———. *Twenty-Four Views of the Vegetation of the Coast and Islands of the Pacific, with Explanatory Descriptions taken during the Exploring Voyage of the Russian Corvette "Senjawin" under the Command of Captain Luetke, 1827–1829*. London, 1861.

———. *Vierundzwanzig Vegetations-Ansichten* . . . , 2 vols. Siegen und Wiesbaden: Friedrichsche Verlagsbuchhandlung, 1844.

Kolgushkin, V. V., comp. *Opisanie starinnykh atlasov, kart i planov 16, 17 18 vekov i poloviny 19 veka khranyashchikhsya v arkhive Tsentralnogo kartograficheskogo proizvodstva VMF. Upravlenie nachalnika Gidrograficheskoi sluzhby voenno-morskogo flota* (*Description of Old Atlases, Maps and Plans of the Sixteenth, Seventeenth and Eighteenth Centuries and First Half of the Nineteenth Century Preserved in the Archive of the Central Cartographic Manufacturer of the Navy. Office of the Head of the Hydrographic Service of the Navy*). Leningrad: Upr. Nachal′nika gidrograficheskoĭ sluzhby Voenno-Morskogo Flota, 1958.

Konstantinov, V. M. *Orosiyakoku Suimudan (Sny o Rossiim)*. Pamyatniki literatury narodov Vostoka (*Literary Documents of the People of the East*). Teksty. Malaya Seriya. XI, Moscow: 1961.

———. "Svedeniya ob ekipazhe korablya 'Sinse-Maru'" ("Information about the Crew of the 'Sinse-Maru'"). *Problemy vostokovedeniya* (*Problems of Oriental Studies*), no. 3, 1959.

Korsakovsky, Petr. *Russian Exploration in Southwest Alaska: The Travel Journals of Petr Korsakovskiy (1818) and Ivan Ya. Vasilev (1829)*. Edited by James VanStone and translated by David Kraus. Rasmuson Library Historical Translation Series, 4. Fairbanks: University of Alaska Press, 1988.

Kotzebue, Otto von. *Entdeckung-Reise in die Sued-See und nach Bering-Strasse zur Erforschung einer nordoestlichen Durchfahrt. Untenommen in den jahren 1815, 1816, 1817 und 1818 auf Kosten Sr. Erlaucht des herrn Reichs-Kanzlers Grafen Rumanzoff auf dem Schiffe Rurich*. 3 vols., Weimar 1821. English translation Kotzebue, ibid., 1–3 vols. London, 1821.

———. *A New Voyage Round the World, in the Years 1823, 24, 25 & 26*. 2 vols. London: H. Colburn & R. Bentley, 1830.

———. *Puteshestvie v Yuzhny okean i v Beringov proliv dlya otyskaniya severo-vostochnogo morskogo prokhoda, predprinyatoe v 1815, 1816, 1817 i 1818 godakh izhdiveniem ego siyatelstva grafa Nikolaya Petrovicha Rumiantseva* . . . , 3 vols. St. Petersburg, 1821–1823

———. *A Voyage of Discovery into the South Sea and Bering's Straits, Undertaken in the Years 1815–1818*. London: Longman, Hurst, Rees, Orme and Brown, 1821.

Krasheninnikov, Stepan P. *Explorations of Kamchatka: North Pacific Scimitar*. Edited and translated by E. A. P. Crownhart-Vaughan. North Pacific Studies No. 1. Portland: Oregon Historical Society, 1972. Originally published as *Opisanie Zemli Kamchatki, s Prilozheniem Raportov, Donesenii i Drugikh Neopublikovannykh Materialov*. St. Petersburg, 1755.

Kristofovich, A. N. *Geologia Tikhii okean*. Leningrad, 1929.

Kromchenko, V. S. *V. S. Khromchenko's Coastal Explorations in Southwestern Alaska, 1822*. Edited by James VanStone and translated from the 1824 Russian edition by David Kraus. *Fieldiana*: Anthropology 64. Chicago: Field Museum of Natural History, 1973.

Krusenstern, Ivan Fedorovich. *Atlas iuzhnago moria, sochinennyĭ flota kapitan" komandorom" Kruzenshternom*. 2 vols. Saint Petersburg: Izdannyĭ po Vysochaĭshemu povel͡eni͡iu, 1824–1826.

———. Atlas de l'ocean Pacifique dressé par M. de Krusenstern. 2 vols. in 1. St. Petersbourg: publié par ordre de Sa Majesté impériale, 1827 [i.e., 1835?].

Kulikovsky, P. G. *M. V. Lomonosov—astronom i astrofizik M.V.* (*Lomonosov, Astronomist and Astrophysicist*). Moscow/Leningrad (n.d.).

Kushner, Howard I. *Conflict on the Northwest Coast: American-Russian Rivalry in the Pacific Northwest*, 1790–1867, Contributions in American History, Number 4. London, Westport, Conn.: Greenwood Press, 1975.

La Perouse, Jean Francois Galop de. *A Voyage Round the World 1715–1788* . . . reprinted. Amsterdam, New York: Nico Israel; Da Capo Press, 1969. This is a reprint of the first complete English edition of 1799; the French was published in 1797.

Lafeber, W., ed. *John Quincy Adams and American Continental Empire*, Chicago: Quadrangle Books 1965.

Langsdorff, Georg Heinrich von. *Bemerkungen einer Reise um die Welt inden Jahren 1803 bis 1807*. Frankfurt: F. Wilmans, 1812

———. *Voyages and travels in various parts of the world during the years 1803, 1804, 1805, 1806 amd 1807*. London: Henry Colburn, 1814. A modern and more accurate translation of the original German has been translated by Victoria Joan Moessner and edited by Richard A. Pierce. Fairbanks: Limestone Press, 1993.

Lavrentsova, T. D. *Russkie arkticheskie ekspeditsii XVII–XX vv.* (*Russian Arctic Expeditions, Seventeenth–Twentieth Centuries*). Leningrad: Gidrometizdat, 1964.

Lazarev, A. P. *Zapiski o plavanii voennogo shlyupa Blagonamerennyi* . . . (*Notes on the Voyage of the Naval Sloop Loyal*). Moscow, 1950.

Lebedev, D. M. *Geografiya v Rossii Petrovskogo vremeni* (*Geography in Russia in the Era of Peter I*). Moscow/Leningrad: zd-vo Akademii nauk SSSR, 1950.

———. *Ocherki po istorii geografii v Rossii XV i XVI vv.* (*Essays on the History of Geography in Russia in the Fifteenth and Sixteenth Centuries*). Moscow: Izd-vo Akademii nauk SSSR, 1956.

———. *Plavanie A. I. Chirikova na paketbote Sv. Pavel k poberezhiyam Ameriki* (*A. I. Chirikov's Voyage on the Small Boat St. Paul to the Shores of America*). Moscow: Изд-во Академии наук СССР, 1951.

Liapunova, Roza G. "Ekspeditsiya I. G. Voznesenskogo i ego znachenie dlya etnografii Russkoi Ameriki" ("I. G. Voznesensky's Expedition and its Significance for the Ethnography of Russian America"). *Sbornik Muzeia Antropologii i Etnografii*, vol. 24. Leningrad, 1967.

———. *Essays on the ethnography of the Aleuts : (at the end of the eighteenth and the first half of the nineteenth century)*. Translated by Jerry Shelest, with the editorial assistance of William B. Workman and Lydia T. Black. Rasmuson Library Historical Translation Series No. 9. Fairbanks : University of Alaska Press, 1996.

Library of Congress. *The Alaskan Orthodox Church Archives: Records of the Russian Ortodox Greek Catholic Church of North America—Diocese of Alaska*. Washington: Manuscript Division, Library of Congress, 1984. Prepared by Antionette Shalkop.

Limonov, Y. A. "Rospis pervogo obshchego chertezha Sibiri. (Opyt datirovki)" (Description of the First Complete Chart of Siberia). In *Problemy istochnikovedeniya* (*Problems of Sourcing*), vol. 8, 1959.

Lipshits, B. A. "Kashevarov as an Alaskan Explorer" ("Kashevarov kak issledovatel Alyaski"). *Sovetskaya etnografiya*, no. 1, 1952:175–178.

———. "On collections in the Museum of Anthropology and Ethnography Assembled by Russian Travelers and Explorers in Alaska and California" ("O kollektsiyakh Muzeia antropologii i etnografii, sobrannykh russkimi puteshestvennikami i issledovatelyami Alyaski i Kalifornii). *Sbornik Muzeia antropologii i etnografii* (*Anthology of the Museum of Anthropology and Ethnography*) 16, 1953:358–369.

———. "Zagoskin kak issledovatel etnografii Alyaski" ("Zagoskin as an Ethnographic Explorer of Alaska"). In *Puteshestvie i issledovaniya leitenanta Lavrentiya Zagoskina v Russkoi Amerike v 1842–1844 gg.* Moscow, 1956.

Lisyanski, Y. F. *A Voyage Round the World in the Years 1803, 4, 5, & 6, Performed in the Ship "Neva."* London: Weybridge, 1814. The English edition was translated by Lisyanski himself, who had earlier served with the British Navy. The Russian edition (*Puteshestvie vokrug Svieta . . . .* St. Petersburg: Drekhsler, 1812) was accompanied by an atlas (*Sobranie kart i risunkov . . .*) St. Petersburg: Morskoi Tipografiia, 1812. A 1947 edition was issued without the accompanying atlas.

Litke, Frederic. *Puteshestvie vokrug sveta, sovershennoe po poveleniyu Imperatora Nikolaya I, na voennom shlyupe "Senyavin" v 1826–1829 gg. Flota kapitanom Fyodorom Litke. Otdelenie Istoricheskoe* (*A Voyage around the World, Completed by Command of Emperor Nicholas I, on the Naval Sloop Senyavin in 1826–1829. By Fleet Captain Fyodor Litke. Historical Department*). 3 vols. with an atlas. St. Petersburg: Gintse, 1834, 1835, 1835.

———. *Puteshestvie vokrug sveta sovershennoe po poveleniyu Imperatora Nikolaya I, na voennom shlyupe "Senyavin" v 1826, 1827, 1828 i 1829 godakh . . . Otdelenie morekhodnoe* (*A Voyage Around the World, Completed by Command of Emperor Nicholas I, on the Naval Sloop Senyavin in 1826, 1827, 1828 and 1829. . . Nautical Department*). With an atlas. St. Petersburg: Gintse, 1836.

———. *Voyage autour du monde, exécuté par order de S. M. l'Majesté Nicolas Ier, sur la corvette "le Séniavine" dans les années 1826, 1827, 1828 et 1829 par Frédéric Lüke, capitaine de vaisseau, aide-de-camp de s. m. l'Empereur, commandant de 'lExpédition*. 3 vols. with an atlas. Paris: Engelmann,1835–1836. This is the historical part of the voyage report. The nautical part was published as *Voyage autour du monde . . . partie nautique*. St. Petersburg: C. Hintze, 1836. It was also accompanied by an atlas of maps. The historical atlas contains fifty-one plates made from drawings by Baron Kittlitz and Alexander Postels. An English translation of vol. 1: *A Voyage Around the World in the Years 1826–1829: To Russian America and Siberia*. Translated by Renée Marshall, supplemented with a parallel account by E. H. Baron von Kittlitz, translated from the German with an Introduction by Joan Moessner. Edited by Richard A. Pierce. Kingston, Ontario: Limestone Press, 1987.

Lyon, G. F. *The Private Journal of Captain G. F. Lyon of H.M.S. Hecla . . .* London: J. Murray, 1824.

Mackenzie, Alexander. *Voyages from Montreal, on the River St. Lawrence, through the Continent of North America, to the Frozen and Pacific Oceans, in the Years 1789 and 1793. With a Preliminary Account of the Rise, Progress, and Present State of the Fur Trade of that Country.* London: T. Cadell, 1801.

Makarova, Raisa V. *Russians on the Pacific, 1743–1799.* Translated from the 1968 Russian edition (*Russkie na Tikhom okeane vo vtoroi polovine XVIII v*) by Richard A. Pierce and A. S. Donnelly. Materials for the Study of Alaska History 6. Kingston, Ontario: Limestone Press, 1974.

Malaspina, Alessandro. *The Malaspina Expedition 1789–1794: The Journal of the Voyage*, vol. 2. *Panama to the Philippines.* Edited by Andrew David, Felipe Fernandez-Armesto, Carlos Novi, and Glyndwr Williams. Hakluyt Society, Third Series, No. 11. London and Madrid, 2003.

Markova-Dvoichenko, E. "Shturman Gerasim Izmailov" ("Navigator Gerasim Izmailov"). *Morskie zapiski* vol. 13, no. 4, 14–27.

Masterson, James, and Helen Brower. *Bering's Successors, 1745–1780.* Seattle: University of Washington Press, 1948.

McPherson, H. M. "The Interest of Wiliam McKendree Gwin in the Purchase of Alaska, 1854–1861." *Pacific Historical Review*, 3:1:31–32, March 1934.

———. *William McKendree Gwin, Expansionist.* Ph.D. dissertation, University of California, Berkeley, 1931.

Meares, John. *Voyages Made in the Years 1788 and 1789, From China to the North West Coast of America. To Which Are Prefixed, an Introductory Narrative of a Voyage Performed in 1786, From Bengal, in the Ship Nootka: Observations on the Probable Existence of a North West Passage; and Some Account of the Trade Between the North West Coast of America and China; and the Latter Country and Great Britain.* London: Printed at the Logographic Press, 1790. Reprinted in 1967 by Nico Israel in Amsterdam and Da Capo Press in New York as Bibliotheca Australiana 22.

Medushevskaia, O. M. *Kartograficheskie istochniki XVII–XVIII vv. Uchebnoe posobie po istochnikovedeniyu SSSR* (*Cartographic Sources of the Seventeenth–Eighteenth Centuries, Instructional Textbook on Sourcing for the USSR*). Moscow, 1957.

———. "Russkia geograficheskie otkrytiia na Tikhm okeane I v Severnoi Amerike (50-e nacalo 80-kh godov XIII v.)." Russian geographical exploration of the Pacific Ocean and North America (1750s to 1780s) Ph.D. dissertation, Moscow State University, 1952.

Merck, Carl Heinrich. *Siberia and Northwestern America, 1788–1792. The Journal of Carl Heinrich Merck, Naturalist With the Russian Scientific Expedition Led by Captain Joseph Billings and Gavriil Sarachev.* Translated by Fritz Jaensch and edited by Richard A. Pierce. Materials for the Study of Alaska History 17. Kingston, Ontario: Limestone Press, 1980.

Mikami, M. "Russkaya karta Sibiri 1687 g." ("1673 Rissian Map of Siberia"). In *Dzimbun tirigaku-no semondai.* Tokyo, 1968:425–436.

Miller, David H. *The Alaska Treaty.* Kingston, Ontario: Limestone Press, 1981.

Mostakhov, S. E. *Uchastie mestnogo naseleniia v geograficheskom izuchenii severo-vostochnoi Sibiri (XVII—nachalo XX v.).* Dissertation in geographic sciences. Moscow, 1965.

Müller, Gerhard Frederich. *Bering's Voyages: The Reports from Russia.* Translated with commentary by Carol Urness. The Rasmuson Library Historical Translation Series, 3. Fairbanks: University of Alaska Press, 1986. It includes his "Lettre d'un officier de la marine Russienne . . . Nouvelle bibliotheque germanique, ou histoire litteraire de l'Allemagne, de la Suisse et de Pays Nord, par Mr. Samuel Formay." Published in Berlin in 1752 and in Amsterdam in 1753.

———. "Nachrichten von Land- und See-Carten, die das Russische Reich und die zunächst angränzende Länder betreffen" (Reports of Land and Sea Charts of Russia and neighboring lands). *Sammlung Russicher Geschichte* 6, no. 1, 1761:1–108.

———. *Sochineniya i perevody* (*Works and Translations*). Part 1. St. Petersburg: Akademia Nauk, 1758.

———. *Voyages from Asia to America for Completing the Discoveries of the North West Coast of America. To Which is Prefixed a Summary of the Voyages Made by the Russians on the Frozen Sea, in Search of a North East Passage, Serving as an Explanation of a Map of the Russian Discoveries, Published by the Academy of Sciences at Petersburg.* London, 1761.

Nakamura, S. *Yapontsy i russkie. Iz istorii kontaktov* (*The Japanese and the Russians. From the History of Contacts*). Moscow, 1983.

Neunherz, R. E. *The Purchase of Russian America: Reasons and Reactions*, Ph.D. dissertation, University of Washington, 1975.

Nielsen, N. "Vitus Berings Stordoad." *Danmarkposten* 6 (1942).

Nordenskiöld, A. E. "Pervaya karta Severnoi Asii, osnovannaya na deistvitelnykh nablyudeniyakh" ("The First Map of Northern Asia Based on Actual Observations"). In *Zapiski Voenno-topograficheskago otdela Glavnogo shtaba* (*Notes of the Military Topographical Department of the Chief Headquarters*), vol. XILIV, Part 7, 1889.

Ogloblin, N. N. "Dve 'skazki' V. Atlasova ob otkrytii Kamchatki" ("Vladimir Atlasov's Two 'Fairytales' About the Discovery of Kamchatka"). *Chteniia v Imperatorskom obshchestva istorii i drevnostei rossiskikh pri Moskovskom universitete* (*Lectures at the Imperial Society of the History and Ancient Times of the Russians at Moscow University*). Book 3. Moscow, 1891.

Olson, Wallace M. *Through Spanish Eyes: the Spanish Voyages to Alaska, 1774–1792.* Auke Bay, Alaska: Heritage Research, 2002. A compilation of Spanish accounts translated, edited and interpreted by Olson.

Orth, Donald J., comp. *Dictionary of Alaska Place Names.* Geological Survey Professional Paper 567. Washington: U.S. Geological Survey, 1967.

Osgood, C. *Ingalik Material Culture.* Yale University Publications in Anthropology no. 22, 1940.

Oswalt, Wendell H. *Eskimos and Explorers.* Novato, Calif.: Chandler & Sharp, 1979.

———. *Mission of Change in Alaska. Eskimos and Moravians on the Kuskokwim.* San Marino, Calif.: Huntington Library, 1963.

———. "Population in Western Alaska and Migration Theory." *Anthropological Papers of the University of Alaska*, vol. 11, no. 1, 1962:1–14.

Pallas, Peter Simon. *Neue Nordische Beytraege* (New Northern Contributions). 7 vols. St. Petersburg and Leipzig: Johann Zaharias Logan, 1781–1796.

———. O rossiiskikh otkrytiiakh na moriakh mezhdu Aziei i Amerikoi. *Mesiatseslov istoricheskii i geogrficheskii na 1781 g.* St. Petersburg, 1781.

———. *Zoographia rosso-asiatica* (*Zoology of Russian Asia*). 3 vols. 2nd ed. St. Petersburg: Academy of Science, 1811.

Pamiatniki Sibirskoi istorii XVIII v. (*Relics of Eighteenth-Century Siberian History*), Book 1, 1700–1713, St. Petersburg, 1882.

Parry, William E. *Journal of a Second Voyage for the Discovery of the North-West Passage.* London: J. Murray, 1824.

———. *Journal of a Voyage for the Discovery of the North-West Passage*. London: J. Murray, 1821.

———. *Journals of the First, Second and Third Voyages for the Discovery of a Northwest Passage*. 5 vols. London: J. Murray, 1828.

Pelley, J. H. "An Account of the Recent Arctic Discoveries by Messrs Dease and Simpson." *Journal of the Royal Geographical Sociey*. Vol. 8, 1837.

Perry, J. *Sostoyanie Rossii pri nyneshnem tsare (The State of Russia Under the Present Czar)*. Moscow, 1871.

Petersen, Edward S. *The Russian-Indian Wars, 1741–1855*. The Western Brand book, 1994. Vol. 1, no. 1.

Petroff, Ivan. *Report on the Population, Industries and Resources of Alaska*. Washington Government Printing Office, 1884.

Pierce, Richard A. *Alaskan Shipping, 1867–1878: Arrivals and Departures at the Port of Sitka*. Kingston, Ontario: Limestone Press, 1972.

———. *Documents on the History of the Russian-American Company*. Translated by Marina Ramsey. Kingston, Ontario: Limestone Press, 1976.

———. "New Light on Ivan Petroff, Historian of Alaska." *Pacific Northwest Quarterly* 59, no. 1 (1968):1–10.

———. *Russian America: A Biographical Dictionary*. Alaska History Series 33. Kingston, Ontario: Limestone Press, 1990. An invaluable tool for the study of Russian America, Pierce has assembled brief biographies of 675 individuals who were active in Alaska, or who had significant influence over events in Alaska.

———. "Russian Exploration in North America." In *Exploration in Alaska: Captain Cook Commemorative Lectures, June–November 1978*, edited by Antoinette Shalkop, 115–127. Anchorage: Cook Inlet Historical Society, 1980.

———. *The Russian Governors: Builders of Alaska, 1818–1867*. Kingston, Ontario: Limestone Press, 1986.

———. *The Russian Orthodox Religious Mission in America, 1794–1837 with Materials Concerning the Life and Works of the Monk German and Ethnographic Notes by Hieromonk Gedeon*. Translated by Colin Bearne. Kingston, Ontario: Limestone Press, 1978.

Pokrovsky, A. A. *250-letie otkrytiya Alyaski, Izvestiya Vsesoyuznogo Geograficheskogo obshchestva (250th anniversary of the discovery of Alaska, News of the All-Union Geographical Society)* (cited as Izv. VGO), vol. 114, no. 5. Leningrad, 1982.

———. "Commemorating the 300th Anniversary of the Godunov Map of Siberia." Translated by James R. Gibson. *The Canadian Cartographer*, vol. 8, no.1 (June 1971):19–26.

———. "Ekspeditskya Beringa" ("Bering's Expedition"). Moscow, 1941.

———. "Gipoteza o *godunovskom* atlase Sibiri 1667 g" ("Hypothesis on the Godunov Map of Siberia of 1667"). In *Izvestiya AN SSSR. Seriya geograficheskaya (News of the Academy of Sciences of the USSR Geographical Series)*, 4 (1966):123–132.

———. "Pyotr I, Nikolai Vitsen i problema soshlas li Amerika s Asiei" ("Peter I, Nicolaas Witsen and the Problem of whether America was Joined with Asia"). In *Strany i narody Vostoka (Countries and Peoples of the East)*, No. 17, book 3, 1975.

———. "Sibirskaya kartografiya XVII v. i problema Bolshogo chertezha" ("Siberian Cartography of the Seventeenth Century and the Problem of the Great Chart"). In *Strany i narody Vostoka (Countries and Peoples of the East)*, No. 18 (1976):213–227.

Polevoi, B. P. "Iz istorii otkrytiia severo-zapadnoi chasti Ameriki" ("From the History of the Discovery of the Northwest Part of America"). In *Ot Aliaski do Ognennoi zemli (From Alaska to the Land of Fire)*. Moscow, 1967.

———. "Pyotr Pervy, Nikolai Vitsen i problema soshlas li Amerika s Azici." In *Strany i narody Vostoka*, no. 17, book 3. Moscow, 1975:29.

Polnyi svod zakonov Rossiiskoi imperii. 1 vypusk (*Complete Code of Laws of the Russian Empire*). S.t Peterburg : Zakonoviēdiēnie, 1911.

———. *Razvitie krupnomasshtabnoi kartografii v Rossii* (*The Development of large scale Cartography in Russia*). Moscow: Nauka, 1989.

Ponomarev, V. N. *Russko-amerikanskie otnosheniya v gody Krymskoi voiny, 1852–1856* (*Russian-American Relations during the Years of the Crimean War, 1852–1856*). Dissertation. Candidate of Historical Sciences, Moscow, 1984.

———. "Russko-amerikanskie otnosheniya v gody Krymskoi voiny, 1852–1856." *Istoricheskie zapiski* (*Historical Notes*), vol. 110, 1984:232–281.

Pratt, J. W. "The Ideology of American Expansion." In *Essays in Honor of William E. Dodd*, edited by A. Craven. Chicago: Chicago University Press, 1935.

———. "The Origin of 'Manifest Destiny,'" *American Historical Review*, July 1927, vol. 32:4:793–798.

Prisadsky, A. *K istorii pravoslavnoi tserkvi v Amerike. 200-letie otkrytiya Alyaski* (*Towards a History of the Orthodox Church in America. The 200th Anniversary of the Discovery of Alaska*). Izdanie Russkogo istoricheskogo obshchestva v Amerike pod predsedatelstvom M. D. Sedykh (Publication of the Russian Historical Society in America under the Chairmanship of M.D. Sedykh). San Francisco, 1943.

Puteshestvie na shlyupe Diana iz Kronshtadta na Kamchatku, sovershennoe pod nachalstvom flota leitenanta Golovnina v 1807–1811 gg. (*Voyage on the Sloop Diana from Kronstadt to Kamchatka, Completed under the Leadership of Fleet Lieutenant Golovnin in 1807–1811*). Moscow ,1960.

Rathburn, Robert R. "Indian Education and Acculturation in Russian America." *Orthodox Alaska*, VII, nos. 3 and 4, 1979:79–101.

———. *The Russian Orthodox religious mission in America, 1794–1837 : with materials concerning the life and works of the Monk German, and ethnographic notes by the Hieromonk Gedeon*. Translated by Colin Bearne and edited by Richard A. Pierce. Kingston, Ontario: Limestone Press, 1978.

Ray, Dorothy Jean. "Eskimo Place-Names in Bering Strait and Vicinity." *Names* 19(1) 1974:4–7.

———. *The Eskimos of Bering Strait, 1650–1898*. Seattle: University of Washington Press, 1975.

———. *Ethnohistory in the Arctic: the Bering Strait Eskimo*. Edited by Richard A. Pierce. Alaska History, 23. Kingston, Ontario: Limestone Press, 1983. This is a compilation of previously published articles and new translations that complement Ray's earlier books on ethnography and Eskimo art. The diverse subject matter includes accounts of the little-known Vasil'ev-Shishmarev expedition of 1819–1822.

Rezanov, N. P. Pervoe puteshestvie rossiyan okolo sveta (The First Russian Voyage Around the World). *Otechestvennye zapiski* (*Notes from the Fatherland*), X, XI, XII, 1823; XX, 1824.

Richardson, J., et al. *The Zoology of Captain Beechey's Voyage; compiled from the collections and notes made by Captain Beechey, the officers and naturalist of the expedition . . .* London: Lords Commissioners of the Admiralty, 1839.

Richardson, Sir John. *Fauna Boreali-Americana, or Zoology of the Northern Parts of British America*. 4 vols. London: J. Murray, 1829–1837.

Richeson, A. W. *English Land Measuring to 1800: Instruments and Practices*. Cambridge, Mass.: MIT Press, 1966.

Rochcan, Vsevolod. "The Origins of the Orthodox Church in Alaska, 1820–1840." *Orthodox Alaska III*, no. 1, 1971:1–23 and III, no. 2, 1–15.

Romanov, D. I. "Perezabytyi vopros o telegrafe cherez Sibir mezhdu Starym i Novym Svetom" ("The Forgotten Issue of the Telegraph across Siberia between the Old and the New World"). *Sankt-Peterburgskie vedomosti* (*The St. Petersburg Register*), February 28, 1861.

———. "Proekt rossko-amerikanskogo mezhdunarodnogo telegrafa" ("The Russian-American International Telegraph Project"). *Russkoe slovo* (*The Russian Word*), February 4, 1859.

———. "Sibirsky telegraf" ("The Siberian Telegraph"). *Russky invalid* (*The Russian Invalid*), March 12, 1864.

Russkaia istoricheskaya biblioteka, izdavaemaya Arkheograficheskoi komissieyu (*Russian Historical Bibliography Published by the Archeographic Commission*), vol. 2. St. Petersburg, 1875.

Russkie ekspeditsii po izucheniyu severnoy chasti Tikhogo okeana v pervoi polovine XVIII. Sbornik dokumentov (*Russian Expeditions for the Study of the Northern Part of the Pacific Ocean in the First Half of the Eighteenth century. Collected Documents*). 2 vols. Moscow: Nauka, 1984, 1989.

Rybakov, B. A. "Geograficheskie znaniya" ("Geographical Knowledge"). In *Ocherki russkoi kultury XVI v.* (*Essays on Russian Culture of the Sixteenth Century*), Part 2. Moscow: Izd-vo, 1977.

Sarychev, Gavriil Andreevich. *Account of a Voyage of Discovery to the North-East of Siberia, The Frozen Ocean and the North-East Sea*. London: Printed for R. Phillips by J. G. Barnard, 1806.

———. *Puteshestvie flota kapitana Sarycheva po severovostohnoi chasti Sibiri, Ledovitommu moriu, i vostochnomu Okeanu, v prodolzhenia Osmi liet, pri geografcheskoi astronomicheskoi expeditsii Pod Nachal'stvom flota Kapitana Billinga s 1785 po 1793 God* (*Sarychev's Voyage to NE Siberia, the Arctic and Pacific Oceans, on an eight year geographical and astronomical expedition under Captain Billings*). Saint Petersburg: Shora, 1802.

———. *Puteshestvie Kapitana Billingsa cherez Chukotskuiu zemliu Ot Beringovo proliva do Nizhnekolymskogo ostroga, i plavanie Kapitana Galla na sudne "Chornom Orle" po severo-vostochnomu okeanu v 1791 godu* (*Billings's journey through the Chukotsk country from Bering Strait to the Nizhne-Kolymsk post, and Captain Hall's cruise to the north Pacific Ocean*). Saint Petersburg: Morskoi tipografii, 1811.

———. *Puteshestvie po severo-vostochnoi chasti Sibiri, Ledovitomu moriu i Vostochnomu okeany*. Moscow: Gos. Izd-vo geograficheskoi lit-ry, 1952.

Sauer, Martin. *An Account of a Geographical and Astronomical Expedition to the Northern Parts of Russia Performed by Commodore Joseph Billings in the Years 1785–1794*. London: Cadell, 1802.

Semyonov, Yuri. *Siberia: Its Conquest and Development*. London, 1963.

Sergeev, V. K. "Moskovskaia matematiko-navigatskaia shkola (1701–1752)" ("Moscow mathematical-navigational school (1701–1752)"). *Voprosy geografii*, No. 34, 1954:150–151.

Seward, W. H. *The Works of William H. Seward*, edited by G. E. Baker, vol. 3. New York: AMS Press, 1972.

Shelikhov, Grigory Ivanovich. *A Voyage to America, 1783–1787*. Translated by Marina Ramsay and edited by Richard A. Pierce. Alaska History 19. Kingston, Ontario:

Limestone Press, 1981. The first Russian and German editions were published in St. Petersburg in 1793. There was a second Russian edition in 1812 containing some additional information. A reprint edition with an essay by B. P. Polovoi was published in Khabarovsk in 1971.

Shenton, J. P. *Robert John Walker: A Politician from Jackson to Lincoln*. New York: Columbia University Press, 1961.

Sherwood, Morgan B. *Exploration of Alaska, 1865–1900*. New Haven: Yale University Press, 1965.

———. "George Davidson and the Acquisition of Alaska." *Pacific Historical Review*, 1959. Vol. 28.

Shibanov, F. A. "Russkaya polevaya astronomiya v 18 veke" ("Russian Field Astronomy in the Eighteenth Century"). In *Uchenye zapiski Leningradskogo gosudarstvennogo universiteta. Seriya geograficheskikh nauk Scholarly* (*Notes from Leningrad State University. Geographical Sciences Series*). No. 226, issue 12, kartografiya, 1958.

Shirina, D. A. *Letopis' ekspeditsii Akademii nauk na severo-vostok Azii v dorevoliutsionnyi period* (*Chronicle of the Academy of Science expdition to northeast Azia in pre-Revolutionary times*). Novosibirsk: Nauka Sibirskoe otd-nie, 1983:36–48.

Simpson, Thomas. *Narrative of the discoveries on the north coast of America; effected by the officers of the Hudson's Bay company during the years 1836–39*. London, R. Bentley, 1843.

Skelton, R. A., ed. *The Journals of Captain Cook on his Voyages of Discovery. Charts and Views by Cook and his Officers reproduced from the original manuscripts*. Cambridge: Cambridge University Press, 1955).

Sokolov, A. "Exspeditsiia k Aleutskim ostrovam kapitanov Krenitsyna; Levashova, 1764–1769 (The expeditions to the Aleutian Islands by captians Krenitsyn and Levashov"). *Zapiski Gidrograficheskago departamenta Morskago ministerstva*, 10 (1852):70–103.

———. "Pervyi pokhod russkikh k Amerike 1732 goda ("The first Russians to America in 1732"). *Zapiski Gidrograficheskago departamenta Morskago ministerstva* 9 (1851): 78–107.

———. "Pokhod geodezista Mikhaila Gvozdeva v Beringov proliv 1732 goda" ("Geodesist Mikhail Gvozdev's Voyage to the Bering Strait in 1732"). In *Morskoi sbornik*, vol. 4, no. 11, 1850.

———. "Severnaia expeditsiia 1733–1741 godov" ("Northern expediions 1733–1741"). *Zapiski Gidrograficheskago departamenta Morskago ministerstvo*, 9 (1851):190–469 with charts.

Soul, Norman E. "American's Siberian Dream," *Russian Review*, 37:4:405–420, October 1987.

Staehlin, Jakob von. *An account of the new northern archipelago: lately discovered by the Russians in the seas of Kamtschatka and Anadir*. London: Printed for C. Heydinger, 1774.

Starr, Jerome Lincoln. *Education in Russian Alaska*. Juneau: Alaska Historical Library, 1972.

Steller, Georg Wilhelm. *Journal of a Voyage With Bering, 1741–1742*. Edited by O. W. Frost and translated by Margritt A. Engel and O. W. Frost. Stanford: Stanford University Press, 1988. Based upon an alternative version of Steller's journal, a copy of which is housed at the Library of Congress. Extensive comments and annotations.

———. "Reise von Kamtschatka nach Amerika mit dem Commandeur-Capitain Bering." In *Neue Nordische Beytraege*, edited by Peter Simon Pallas, 129–236; 6:1–26. Saint Petersburg, Leipzig: 1793.

———. *Steller's History of Kamchatka: Collected Information Concerning the History of Kamchatka, Its Peoples, Their Manners, Names, Lifestyle, and Various Customary Practices.* Translated by Margritt Engel and Karen Willmore, and edited by Marvin Falk. Rasmuson Historical Translation Series 12. Fairbanks: University of Alaska Press, 2003.

———. "Steller's Journal of the Sea Voyage From Kamchatka to America and Return on the Second Expedition 1741–1742." Translated by Leonhard Stejneger. In *Bering's Voyages*, edited by F. A. Golder, vol. 2. New York: American Geographical Society, 1925.

Struve, B. V. *Vospominaniya o Sibiri, 1848–1854 g* (*The revival of Siberia, 1848–1854*). St. Petersburg: Obshchestvennaia͡pol'za, 1889.

Sumner, Charles. *Speech of Honorable Charles Sumner on the Cession of Russian America to the United States.* Washington, 1867. Later, the speech was incorporated, albeit with some changes, into Sumner's complete works. See C. Sumner, *Works of Charles Sumner*, vol. 11:186–349, Boston, 1883.

Svet, Y. M. *Tretie plavanie kapitana Dzheimsa Kuka. Plavanie v Tikhom okeane v 1776–1780 gg.* (*The Third Voyage of Captain James Cook. The Voyage in the Pacific Ocean in 1776–1780*). Moscow, 1971.

Svet, Y., and S. Fedorova. "Captain Cook and the Russians." In *Pacific Studies*, vol. 11, no. 1, 1978.

Tatishchev, V. N. *Izbrannye trudy po geografii Rossii* (*Collected Works on the Geography of Russia*). Moscow, 1950.

Tebenkov, Mikhail Dmitrievich. *Atlas of the Northwest Coasts of America. With Hydrographic Notes.* Alaska History Series 21. Kingston, Ontario: Limestone Press, 1981. A facsimile of the original 1852 Russian edition: *Gidrograficheskie zamechaniya k Atlasu Severovostochnykh beregov Ameriki, ostrovov Aleutskikh i nekotorykh drugikh mest Severnogo Tikhogo Okeana Kapitana I-go Ranga Tebenkova* (*Hydrographic Notes to the Atlas of the Northwestern Coasts of America, the Aleutian Islands and some other Places in the North Pacific by Captain of the 1st Rank Tebenkov*). St. Petersburg: v Tipografii Morskogo kadetskogo Korpusa, 1852.

Teichmann, Oskar. *A Journey to Alaska in the Year 1868: Being the Diary of the Late Emil Teichmann.* New York: Argosy-Antiquarian Ltd, 1963.

Thompson, Robert A. *The Russian settlement in California, Fort Ross, Sonoma County.* Santa Rosa: Sonoma Democrat, 1896.

Tikhmenev, Petr Alexandrovich. *The History of the Russian-American Company.* Translated by Richard A. Pierce and Alton S. Donnelly. Seattle: University of Washington Press, 1978. Translated from the Russian edition of 1861–1863. The second volume (documents) published separately (q.v.).

———. *A History of the Russian-American Company. Vol. 2: Documents.* Translated by Demitri Krenov and edited by Richard A. Pierce and Alton S. Donnelly. Materials for the Study of Alaska History 13. Kingston, Ontario: Limestone Press, 1979.

———. *Istoricheskoe obozrenie obrazovaniya Rossiisko-Amerikanskoi kompanii i deistvii ee do nastoyashchego vremeni* (*Historical Review of the Creation of the Russian-American Company and its Activities to Date*), part 2. St. Petersburg: E. Veimar, 1862.

Titov, A. *Sibir v XVII veke. Sbornik strannykh russkikh statei o Sibiri i prilezhashchikh k nei zemliakh* (*Siberia in the Seventeenth Century. Anthology of Odd Russian Articles on Siberia and the Lands Adjacent to it*). Moscow: Tip. L. i A. Snegirevykh, 1890.

Titova, Z. D. *Etnograficheskie materialy Severo-Vostochnoi geograficheskoi ekspeditsii, 1785–1795 gg.* (*Ethnographic materials from the Northeastern geographical expedition of 1785–1795*). Magadan: Kn. izd-vo, 1978.

Tomkins, Stuart R. *Alaska: Promyshlennik and Sourdough.* Norman: University of Oklahoma Press, 1952.

Tompkins, Stuart R., and Max L. Moorhead. "Russia's Approach to America. Part I: From Russian Sources, 1741–61." *British Columbia Historical Quarterly* 13, no. 2 (April 1949):55–66.

———. "Russia's Approach to America. Part II: From Spanish Sources, 1761–75." *British Columbia Historical Quarterly* 13, no. 3–4 (July–October 1949):231–255.

Torrubia, José. *I Moscoviti Nella California.* Rome: G. Salomoni, 1759.

U.S. Coast Survey. *Pacific coast pilot. Coast and islands of Alaska. Second Series.* Compiled by William H. Dall, assisted by Marcus Baker. Washington, D. C.: Government Printing Office, 1879.

Vancouver, George. *A Voyage of Discovery to the North Pacific Ocean and Round the World, 1791–1795.* Edited by W. Kaye Lamb. Works Issued by the Hakluyt Society, second series, no. 163–166. London: Hakluyt Society, 1984.

———. *A voyage of discovery to the North Pacific Ocean, and round the world: in which the coast of north-west America has been carefully examined and accurately surveyed . . . .* London: printed for G. G. and J. Robinson, 1798.

VanStone, James W. *A. F. Kashevarov's Coastal Explorations in Northwest Alaska, 1838.* Edited with Introduction by James W. VanStone and translated by David H. Kraus. *Fieldiana Anthropology.* Field Museum of Natural History, vol. 69. Chicago, 1977.

———. *Russian Exploration in Southwest Alaska: the Travel Journals of Petr Korsakovskiy (1818) and Ivan Ya. Vasiliev (1829).* Translated by David H. Kraus. Rasmuson Historical Translation Series 4. Fairbanks: University of Alaska Press, 1988.

———. *V. S. Khromchenko's Coastal Explorations in Southwest Alaska, 1822.* Translated by David H. Kraus. *Fieldiana Anthropology. A Continuation of the Anthropological Series of the Field Museum of Natural History*, vol. 64. Chicago, 1973.

Varjola, Pirjo (with contributions by Julia P. Averkieva and Roza G. Liapunova). *The Etholen Collection . . .* Helsinki: National Board of Antiquities, 1990.

Vaughan, Thomas. *Voyages of Enlightenment: Malaspina on the Northwest Coast, 1791–1792.* Portland: Oregon Historical Society, 1977.

Venegas, M. *Noticia de la Califernia, y de su conquista temporal* (*Notes on California and its conquest*). 3 vols. Madrid: En la imprenta de la viuda de m. Fernandez, y del Supremo consejo de la Inquisicion, 1757.

Veniaminov, Ioan. *Journals of the Priest Ioann Veniaminov in Alaska, 1823 to 1836.* Translated by Jerome Kisslinger, introduction and commentary by S. A. Mousalimas. Rasmuson Library Historical Translation Series 7. Fairbanks: University of Alaska Press, 1993.

———. *Notes on the Islands of the Unalaska District.* Translated by Lydia T. Black and R. H. Geoghegan and edited by Richard A. Pierce. Alaska History 27. Kingston, Ontario: Published jointly by The Elmer Rasmuson Library Translation Program and Limestone Press. 1984. Translation of *Zapiski Ob Ostravakh Unalashkinskago Otdela*, St. Petersburg, 1840. Included is a short biography by Richard Pierce and explanatory footnotes by Lydia Black.

———. *Zapiski ob ostrovakh Unalashkinskogo otdela* (*Notes on the Islands of the Unalashka Section*). 2 vols. St. Petersburg: Tip Imp. Akad., 1840.

Verner, Coolie. *Explorers' Maps of the Canadian Arctic 1818–1860*. Toronto: Cartographica, 1972.

Vila Vilar, E. *Los Rusos en America* (*Russians in America*). Seville: Publicaciones de la Escuela de Estudios Hispano-americanos de Sevilla. no. 169, 1966.

Vishnevsky, B. P. "Iz istorii izucheniya rek i ozer Severo-Zapadnoi Ameriki" ("From the History of the Study of the Rivers and Lakes of Northwest America"). *Izvestiya Vsesoyuznogo Geograficheskogo obshchestva*, vol. 94, no. 6, 1962:507–510.

Volkl, E. *Russland und Lateinamerika, 1741–1841*. Wiesbaden: O. Harrassowitz, 1968.

Wagner, Henry R. *The Cartography of the Northwest Coast of America to the Year 1800*. Berkeley: University of California Press, 1937.

Waldseemueller, Martin. *Universalis Cosmographia*. Preserved at Schloss Wolfegg, Wuertemberg (circa 1507).

Watrous, Stephen D., ed. *John Ledyard's journey through Russian and Siberia. 1787–1788. The journal and selected letters*. Madison: University of Wisconsin Press, 1966.

Waxell, Sven. *The Russian Expedition to America*. Translated by M. A. Michael. New York: Collier Books, 1962. Originally published in 1952 from Johan Skalberg's Danish version, with an introduction and note by M. A. Michael. Also published in New York by Macmillian. His German account was finished in 1756 and remained relatively unknown until a Danish translation was published in 1948.

———. *Vtoraya Kamchatskaya ekspeditsiya Vitusa Beringa* (*The Second Kamchatka Expedition of Vitus Bering*). Moscow/Leningrad: Glavsevmorput, 1940.

Weber, F. *Das verandertern Russlands dritter Theil*. Vol. 1, Hannover: Foerster, 1738, vol. 2, 1740.

Wheat, Carl. *Mapping the Transmississippi West 1540–1861*, vol. 1, *The Spanish Entranda to the Louisiana Purchase, 1540–1804*. San Francisco: Institute of Historical Cartography, 1957–1963.

Wheeler, Mary E. "The Russian American Company and the Imperial Government: Early Phase." In *Russia's American Colony*, edited by S. Frederick Starr. Durham, N.C.: Duke University Press, 1987:43–62, 374–378.

Whymper, Frederick. *Russian America or 'Alaska': the Natives of the Yukon River and Adjacent Country*, Transactions of the Ethnological Society, London, 1(1869).

———. *Travel and Adventure in the Territory of Alaska, Formerly Russian America—Now Ceeded to the United States—and in Various Other Parts of the North Pacific*. London: J. Murray, 1868. An American and a French edition were published in 1871.

Williams, Glynwr. "Alaska Revealed: Cook's Exploration in 1778." In *Exploration in Alaska. Captain Cook commemorative lectures*, Anchorage, Alaska, 1980:69–87.

———. *The British Search for the Northwest Passage in the 18th Century*, London: Published for the Royal Commonwealth Society by Longmans, 1962.

Wrangell, Ferdinand Petrovich von. "The Inhabitants of the Northwest Coast of America." Translated and edited by J. W. VanStone. *Arctic Anthropology* 6(2):13–19, 1970.

———. *Russian America. Statistical and Ethnographic Information*. Translated by Mary Sadouski and edited by Richard A. Pierce. Materials for the Study of Alaska History 15. Kingston, Ontario: Limestone Press, 1980. Wrangell, a chief manager of the Russian-American Company, was also an active explorer. This is translated from the German edition of 1839 and includes additional material contributed by Petr Kostromitinov, a manager at the Ross colony in California; an extract of Andrei Glazunov's expedition of 1834 in interior Alaska, material on the Aleuts by Ioann Veniaminov, and Native vocabularies.

Zagoskin, Lavrentii A. "Expedition auf dem festlande von Nord-Amerika" ("Expedition to the continent of North America"). *Denkschriften der russischen geographischen gesellschaft zu St. Petersburg*, vol. 1. Weimar, 1849 [See also Petermann, A.,] *Geographischen Mittheilungen*, 1857, heft IV und V, 211–312.

———. *Lieutenant Zagoskin's Travels in Russian America, 1842–1844*. Edited by Henry M. Michael. Translated from the 1956 Russian edition. Arctic Institute of North America Anthropology of the North: Translations From Russian Sources 7. Toronto: University of Toronto Press, 1967. This has become one of the major sources for the study of the exploration and ethnography of the lower Yukon and Kuskokwim rivers. There are some differences between various previous Russian and German language sources in the nineteenth century and the 1956 Soviet edition that remain unresolved.

———. "Redut Sv. Mikhaila na yuzhnom beregu zaliva Norton" ("St. Michael's Redoubt on the South Coast of Norton Sound"). *Zapiski Gidrograficheskago departamenta*, vol. 4. St. Petersburg, 1846:86–101.

Zamyslovsky, E. E. "Chertezhy Sibirskikh Zemel' XVI–XVII vv ("Charts of the Siberian Lands Sixteenth–Seventeenth Centuries"). In *Zhurnal Ministerstva narodnago prosveshcheniia* (*Journal of the Ministry of People's Enlightenment*), vol. 257 (June 1891): 334–347.

Index

Page numbers with an *n* denote an endnote; *f* refers to a figure.